Lecture Notes in Computer Science 10317

Commenced Publication in 1973
Founding and Former Series Editors:
Gerhard Goos, Juris Hartmanis, and Jan van Leeuwen

More information about this series at http://www.springer.com/series/7412

Fakhri Karray · Aurélio Campilho
Farida Cheriet (Eds.)

Image Analysis and Recognition

14th International Conference, ICIAR 2017
Montreal, QC, Canada, July 5–7, 2017
Proceedings

 Springer

Editors
Fakhri Karray
University of Waterloo
Waterloo, ON
Canada

Farida Cheriet
Politechnique Montreal
Montreal, QC
Canada

Aurélio Campilho
University of Porto
Porto
Portugal

ISSN 0302-9743   ISSN 1611-3349   (electronic)
Lecture Notes in Computer Science
ISBN 978-3-319-59875-8   ISBN 978-3-319-59876-5   (eBook)
DOI 10.1007/978-3-319-59876-5

Library of Congress Control Number: 2017943053

LNCS Sublibrary: SL6 – Image Processing, Computer Vision, Pattern Recognition, and Graphics

Printed on acid-free paper

This Springer imprint is published by Springer Nature
The registered company is Springer International Publishing AG
The registered company address is: Gewerbestrasse 11, 6330 Cham, Switzerland

Preface

ICIAR 2017 was the 14th edition in the series of annual conferences on image analysis and recognition, offering a forum for participants to interact and present their latest research contributions in theory, methodology, and applications of image analysis and recognition. ICIAR 2017, the International Conference on Image Analysis and Recognition was held in Montreal, Canada, July 5–7, 2017. ICIAR is organized by AIMI - Association for Image and Machine Intelligence, a not-for-profit organization registered in Ontario, Canada.

We received a total of 133 papers from 33 countries. Before the review process, all the papers were checked for similarity using a comparison database of scholarly work. The review process was carried out by members of the Program Committee and other reviewers. Each paper was reviewed by at least two reviewers, and checked by the conference chairs. A total of 73 papers were finally accepted and appear in these proceedings. We would like to express our gratitude to the authors for their contribution, and we thank the reviewers for the careful evaluation and feedback provided to the authors. It is this collective effort that resulted in the strong conference program and high-quality proceedings.

We were very pleased to include three outstanding keynote talks: "The Role of Ultrasound and Augmented Reality for Intra-cardiac Beating Heart Interventions" by Terry Peters, Western University, Canada; "Matrix-Tensor Representation and Deep Learning in Large-Scale Image Data Analysis" by Tien Bui, Concordia University, Canada; and "Discovering Deep Knowledge from Biosequence and Temporal Sequence Data" by Andrew K. C. Wong, University of Waterloo, Canada. We have also included in this year's program, a special session on Machine Learning for Medical Image Computing, organized by Alex Wong, of the University of Waterloo, Canada, and Farzad Khalvati of Sunnybrook Research Institute in Toronto, Canada. We would like to express our gratitude to the keynote speakers and the special session organizers for accepting our invitation to share their vision and recent advances in their areas of expertise.

We would like to thank Khaled Hammouda, the webmaster of the conference, for maintaining the website, managing the registrations, interacting with the authors, and preparing the proceedings. We are also grateful to Springer's editorial staff, for supporting this publication in the LNCS series. We also would like to acknowledge the members of the local Organizing Committee for their assistance and support during the organization of the conference, and the École Polytechnique de Montreal for making the venue available.

Finally, we were very pleased to welcome all the participants to ICIAR 2017 in Montreal. For those who were not able to attend, we hope this publication provides a good view into the research work presented at the conference, and we look forward to meeting you at the next ICIAR conference.

July 2017

Fakhri Karray
Aurélio Campilho
Farida Cheriet

Organization

General Chairs

Fakhri Karray
University of Waterloo, Canada
karray@uwaterloo.ca

Aurélio Campilho
University of Porto, Portugal
campilho@fe.up.pt

Farida Cheriet
École Polytechnique de Montréal, Canada
farida.cheriet@polymtl.ca

Webmaster

Khaled Hammouda
Waterloo, Ontario, Canada
khaledh@aimiconf.org

Supported by

AIMI – Association for Image and Machine Intelligence

Department of Electrical and Computer Engineering
Faculty of Engineering
University of Porto
Portugal

CPAMI – Centre for Pattern Analysis and Machine Intelligence
University of Waterloo
Canada

Center for Biomedical Engineering Research
INESC TEC - INESC Technology and Science
Portugal

Program Committee

A. Abate	University of Salerno, Italy
J. Alba-Castro	University of Vigo, Spain
E. Alegre	University of Leon, Spain
L. Alexandre	University of Beira Interior, Portugal
H. Araujo	University of Coimbra, Portugal
A. Barkah	University of Waterloo, Canada
J. Barron	University of Western Ontario, Canada
J. Batista	University of Coimbra, Portugal
S. Bedawi	University of Waterloo, Canada
J. Boisvert	CNRC, Ottawa, Canada
G. Bonnet-Loosli	UCA, Université Clermont Auvergne, France
F. Camastra	University of Naples Parthenope, Italy
M. Camplani	University of Bristol, UK
J. Cardoso	INESC TEC and University of Porto, Portugal
M. Coimbra	University of Porto, Portugal
A. Dawoud	University of Southern Mississippi, USA
J. Debayle	Ecole Nationale Supérieure des Mines de Saint-Etienne (ENSM-SE), France
G. Doretto	West Virginia University, USA
L. Duong	École de technologie supérieure, Canada
M. El-Sakka	University of Western Ontario, Canada
P. Fallavollita	University of Ottawa, Canada
A. Farahat	Hitachi America, Ltd. R&D, USA
J. Fernandez	CNB-CSIC, Spain
R. Fisher	University of Edinburgh, UK
D. Frejlichowski	West Pomeranian University of Technology, Szczecin, Poland

G. Giacinto	University of Cagliari, Italy
V. Gonzalez-Castro	University of Leon, Spain
G. Grossi	University of Milan, Italy
L. Heutte	University de Rouen Normandie, France
T. Hurtut	Polytechnique Montreal, Canada
L. Igual	University of Barcelona, Spain
D. Jin	National Institutes of Health, USA
F. Khalvati	University of Toronto, Canada
A. Khamis	CPAMI, University of Waterloo, Canada
Y. Kita	National Institute AIST, Japan
A. Kong	Nanyang Technological University, Singapore
M. Koskela	CSC, IT Center for Science Ltd., Finland
A. Kuijper	TU Darmstadt and Fraunhofer IGD, Germany
P. Langlois	Polytechnique Montreal, Canada
H. Lombaert	École de technologie supérieure, Canada
J. Lorenzo-Ginori	Universidad Central Marta Abreu de Las Villas, Cuba
A. Marçal	University of Porto, Portugal
J. Marques	Instituto Superior Tecnico, Portugal
M. Melkemi	University of Haute Alsace, France
A. Mendonça	University of Porto, Portugal
M. Mignotte	University of Montreal, Canada
P. Morrow	University of Ulster, UK
M. Nappi	University of Salerno, Italy
H. Ogul	Baskent University, Turkey
C. Ou	University of Waterloo, Canada
C. Ouali	University of Waterloo, Canada
M. Penedo	CITIC, Universidade da Coruña, Spain
P. Pina	Instituto Superior Técnico, Universidade de Lisboa, Portugal
A. Pinho	University of Aveiro, Portugal
J. Pinto	Instituto Superior Técnico, Portugal
L. Prevost	ESIEA (Ecole d'ingénieurs du monde numérique), France
P. Radeva	Universitat de Barcelona, CVC, Spain
B. Remeseiro	Universitat de Barcelona, Spain
H. Ren	Samsung Research USA, USA
J. Rouco	INESC TEC, Portugal
K. Roy	North Carolina A&T State University, USA
A. Sappa	Compute Vision Center, Spain
G. Schaefer	Loughborough University, UK
P. Scheunders	University of Antwerp, Belgium
L. Seoud	CNRC, Ottawa, Canada
J. Sequeira	Ecole Supérieure d'Ingénieurs de Luminy, France
J. Silva	University of Porto, Portugal
J. Sousa	Instituto Superior Técnico, Portugal
S. Sural	Indian Institute of Technology, India
A. Taboada-Crispi	Universidad Central Marta Abreu de Las Villas, Cuba

X. Tan	Nanjing University of Aeronautics and Astronautics, China
J. Tavares	University of Porto, Portugal
A. Torsello	Università Ca' Foscari Venezia, Italy
A. Uhl	University of Salzburg, Austria
M. Vento	Università di Salerno, Italy
E. Vrscay	University of Waterloo, Canada
Z. Wang	University of Waterloo, Canada
M. Wirth	University of Guelph, Canada
A. Wong	University of Waterloo, Canada
X. Xie	Swansea University, UK
J. Xue	University College London, UK
P. Yan	Philips Research, USA
P. Zemcik	Brno University of Technology, Czech Republic
H. Zhou	Queen's University Belfast, UK
R. Zwiggelaar	Aberystwyth University, UK

Additional Reviewers

H. Bendaoudi	Polytechnique Montreal, Canada
C. Chevrefils	Polytechnique Montreal, Canada
X. Clady	UPMC, France
A. Cunha	University of Trás-os-Montes-e-Alto-Douro, Portugal
A. Elmogy	University of Tanta, Egypt
R. Farah	Polytechnique Montreal, Canada
A. Galdran	INESC TEC Porto, Portugal
F. Girard	Polytechnique Montreal, Canada
M. Hortas	University of A Coruña, Spain
C. Kandaswamy	University of Porto, Portugal
Y. Miao	University of Waterloo, Canada
F. Monteiro	Polytechnic Institute of Bragança, Portugal
P. Negri	UADE, Argentina
J. Novo	University of A Coruña, Spain
H. Oliveira	INESC TEC, Portugal
N. Rodriguez	Universidade da Coruña, Spain
L. Teixeira	University of Porto, Portugal
L. Zhang	Hong Kong Polytechnic University, Hong Kong, SAR China
F. Zhou	Tsinghua University, China

Contents

3D Computer Vision

Feature Extraction

Detection and Classification

Remote Sensing

Applications

Machine Learning in Image Recognition

A Weight-Selection Strategy on Training Deep Neural Networks for Imbalanced Classification

Antonio Sze-To[1,2] and Andrew K.C. Wong[1,2(✉)]

[1] Centre for Pattern Analysis and Machine Intelligence, University of Waterloo,
Waterloo, ON N2L 3G1, Canada
{hy2szeto,akcwong}@uwaterloo.ca
[2] Department of Systems Design Engineering, University of Waterloo,
Waterloo, ON N2L 3G1, Canada

Abstract. Deep Neural Networks (DNN) have recently received great attention due to their superior performance in many machining-learning problems. However, the use of DNN is still impeded, if the input data is imbalanced. Imbalanced classification refers to the problem that one class contains a much smaller number of samples than the others in classification. It poses a great challenge to existing classifiers including DNN, due to the difficulty in recognizing the minority class. So far, there are still limited studies on how to train DNN for imbalanced classification. In this study, we propose a new strategy to reduce over-fitting in training DNN for imbalanced classification based on weight selection. In training DNN, by splitting the original training set into two subsets, one used for training to update weights, and the other for validation to select weights, the weights that render the best performance in the validation set would be selected. To our knowledge, it is the first systematic study to examine a weight-selection strategy on training DNN for imbalanced classification. Demonstrated by experiments on 10 imbalanced datasets obtained from MNIST, the DNN trained by our new strategy outperformed the DNN trained by a standard strategy and the DNN trained by cost-sensitive learning with statistical significance ($p = 0.00512$). Surprisingly, the DNN trained by our new strategy was trained on 20% less training images, corresponding to 12,000 less training images, but still achieved an outperforming performance in all 10 imbalanced datasets. The source code is available in https://github.com/antoniosehk/WSDeepNN.

Keywords: Weight selection · Training strategy · Deep neural networks · Imbalanced data · Classification

1 Introduction

Deep Learning [1] has recently achieved great success by achieving superhuman performance in many places such as beating the world champion in GO [2]. It has been shown to outperform other machine-learning techniques in a variety of tasks [1]. In essence, Deep Learning refers to the use of Deep Neural Networks

© Springer International Publishing AG 2017
F. Karray et al. (Eds.): ICIAR 2017, LNCS 10317, pp. 3–10, 2017.
DOI: 10.1007/978-3-319-59876-5_1

(DNN), defined as artificial neuron networks with at least 3 hidden layers [3]). Its success is built upon the advent of fast graphics processing units (GPU), and the availability of large amount of data [1]. Nevertheless, if the data is highly skewed, or imbalanced [4], even DNN could not avoid such negative impact.

Imbalanced classification [5–7] refers to the problem that one class contains a much smaller number of samples than the other classes in classification. Such phenomenon exists in many real-world machine learning applications [6], e.g. fraud detection, medical diagnostics and equipment failure prediction. Due to the difficulty in recognizing patterns of the minority class, imbalanced classification poses a great challenge to existing classifiers including DNN [8].

Throughout the years, many methods [5–7] have been developed for tacking imbalanced classification. They can be categorized into two types: (1) re-sampling [9], which changes the balance between classes by considering the local characteristics of the samples, and (2) cost-sensitive learning [10], which assigns higher misclassification costs to the minority class.

However, studies on training DNN on imbalanced classification are still scanty. Most of them applied traditional methods such as re-sampling [8,11] or cost-sensitive learning [12] or both [13]. Recent trend includes regularizing the loss function [14,15] during training or representation learning [4,16].

In this study, we propose a new strategy to reduce over-fitting in training DNN for imbalanced classification based on weight selection. In training DNN, by splitting the original training set into two subsets, one used for training to update weights, and the other for validation to select weights, the weights that render the best performance in the validation set would be selected. To our knowledge, it is the first systematic study to examine a weight-selection strategy on training DNN for imbalanced classification. Our objective is to investigate if such strategy can improve the predictive performance of DNN on imbalanced classification, even when the DNN are trained on a reduced training set.

2   Methodology

2.1   Problem Definition

Let $S_{train} = \{(x_i, y_i)\}_{i=1}^{N}$ be a set of N training samples, where $x_i \in X$ is an n-dimensional vector and $y_i \in Y = \{1, \ldots, C\}$ is a class label associated with the instance x_i. Given S_{train}, the problem is to learn a function $f : X \rightarrow Y$, such that

$$E_{(x,y) \sim D}[L(f(x); y)] \tag{1}$$

is minimized, where D is the data distribution over $X \times Y$, $x \in X$, $y \in Y$ and $L(z; y)$ is a loss function that measures the loss if we predict y as z. Equation 1 represents the generalization error of a learnt function. In practice, it would be approximated by a set of testing samples $S_{test} = \{(x_i, y_i)\}_{i=1}^{M}$ as follows:

$$Test_f(S_{test}) = \frac{1}{M} \sum_{i=1}^{M} [L(f(x_i); y_i)] \tag{2}$$

2.2 Proposed Method

Deep Neural Networks (DNN) [1] can be abstractly represented as a function $g_\theta : X \to Y$, where g is the function that represents the neuron network configuration (e.g. the number of hidden layers...) while θ is the model parameter that represents the weights. By applying back-propagation algorithm [1] on S_{train}, a set of weights can be obtained. To achieve a smaller loss value, a standard strategy is to run the back-propagation algorithm iteratively on S_{train}, and in each iteration (epoch), an updated set of weights would be obtained. According to [17], the procedure can be simplified as shown in Algorithm 1. We denote this procedure as a standard strategy to train DNN in this study.

Algorithm 1. DNN

Input: a set of training samples S_{train}, a function g, initialized model parameter θ_1

Output: a model parameter θ corresponds to function g.
initialize $\theta = \theta_1$
for t = 1 to epochs **do**
    $\theta_{t+1} = \text{backprop}(S_{train}, \theta_t)$
    $\theta = \theta_{t+1}$
end for
return θ

The problem of the standard strategy is that it is prone to over-fitting, particularly for imbalanced data. In addition to traditional approaches such as re-sampling and cost-sensitive learning, here we propose a new strategy known as Validation-Loss (VL) strategy, as shown in Algorithm 2. It is to split the input training set into two sets, one for training the DNN to update weights, and the other for performance validation to select weights. The weights that render the the minimum loss value on the validation set would be selected.

Algorithm 2. DNN+VL

Input: a set of training samples S_{train}, a function g, initialized model parameter θ_1

Output: a model parameter θ corresponds to function g.
split S_{train} into T_{train} and $T_{validate}$
initialize $\theta = \theta_1$
initialize $v_{loss} = Test_{g_\theta}(T_{validate})$
for t = 1 to epochs **do**
    $\theta_{t+1} = \text{backprop}(S_{train}, \theta_t)$
    if $v_{loss} > Test_{g_\theta}(T_{validate})$ **then**
        $v_{loss} = Test_{g_\theta}(T_{validate})$
        $\theta = \theta_{t+1}$
    end if
end for
return θ

---

3 Experiments and Results

3.1 Datasets

Ten imbalanced image datasets were obtained from the widely used MNIST dataset [18] following the procedure described in [19]. It is a handwritten digit dataset, containing 60,000 training images and 10,000 testing images. Each image is associated with a distinct digit as its class. There are 10 digit classes (0, 1, 2,..., 9) in total. All images are in black and white with the dimension of 28×28. Following [19], we obtained an imbalanced image dataset (Dataset-i) by referring the digit class i as positive (+ve) class, and the other digit classes as negative (−ve) class, with the imbalance ratio (ImR) provided (Table 1). The ImR was calculated via dividing the number of negative samples (majority) by the number of positive samples (minority). The higher the ImR is, the more imbalance the dataset is. All image gray levels were divided by 255 to make each pixel value to be within 0 and 1.

Table 1. A summary of the 10 imbalanced image dataset obtained from MINST [18]. Each dataset has two classes: −ve and +ve classes, where the +ve class represents a digit (minority class) and the −ve class represents the remaining digits (majority class). The imbalance ratios (ImR) were calculated via dividing the no. of −ve samples by the no. of +ve samples. The higher the ImR is, the more imbalance the dataset is.

	Training −ve	Training +ve	Training ImR	Testing −ve	Testing +ve	Testing ImR
Dataset-0	54077	5923	9.13	9020	980	9.20
Dataset-1	53258	6742	7.90	8865	1135	7.81
Dataset-2	54042	5958	9.07	8968	1032	8.69
Dataset-3	53869	6131	8.79	8990	1010	8.90
Dataset-4	54158	5842	9.27	9018	982	9.18
Dataset-5	54759	5421	10.10	9108	982	10.21
Dataset-6	54082	5918	9.14	9042	958	9.44
Dataset-7	53735	6265	8.58	8972	1028	8.73
Dataset-8	54149	5851	9.25	9026	974	9.27
Dataset-9	50000	5949	9.09	8991	1009	8.91

3.2 Experiments

Five classifiers were adopted in experiments under a binary classification setting: (1) support vector machine (SVM) with a radial basis function (RBF) kernel; (2) decision tree constructed by CART (Classification and Regression Trees) algorithm; both of them were adopted as baseline algorithms and were trained on the entire training set of each dataset. (3) DNN: a DNN trained by a standard strategy, i.e. Algorithm 1, training on the entire training set of each dataset (which contains 60,000 images). (4) DNN-CL: a DNN trained by a standard strategy with cost-sensitive learning, i.e. setting the misclassification cost of the minority class to the majority class as ImR to 1; (5) DNN-VL: a DNN trained

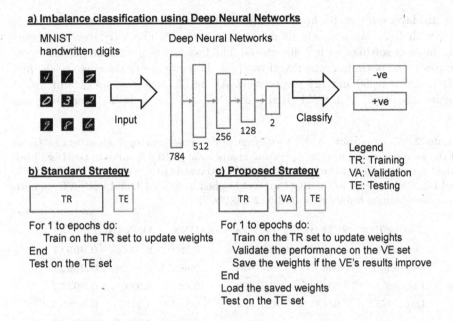

Fig. 1. (a) DNN with 3 hidden layers were trained to classify if a digit image pertains to the −ve class (minority) or +ve class (majority) on 10 imbalanced datasets obtained from MNIST (Table 1). (b) An illustration of the standard strategy, where the DNN is trained on the entire training set to update weights. (c) An illustration of the proposed strategy, where the original training set is split into two subsets, one for training to update weights and the other for weight selection. The weights which obtain the best performance on the validation set would be selected.

by the proposed strategy, i.e. Algorithm 2, training only on a randomly selected 48,000 (80%) images from the training set of each dataset for weight updates, and the remaining 12,000 (20%) images for weight selection. These classifiers were tested on the testing 10,000 images of each dataset.

3.3 Evaluation

Following [20], the prediction performance of the classifiers was evaluated by the area under the receiver operating characteristic (ROC) curve to enable the comparison over a range of prediction thresholds, and also by the area under the precision-recall curve (PRC) to provide a more informative representation of performance assessment for highly skewed datasets.

3.4 Implementation and Parameter Setting

To build the DNN, the deep learning library Keras (http://keras.io/) with Theano backend [21] was adopted. The DNN has 5 layers, as illustrated in Fig. 1. The 1st layer is an input layer with 784 (=28 × 28) dimension. The 2nd layer, the 3rd layer and the 4th layers are hidden layers of rectified linear units (ReLu) [22] with 512, 256 and 128 dimension respectively. The 5th layer is an output layer of softmax with 2 dimension. The loss function was categorical cross-entropy. The optimizer was RMSProp [20]. The number of training epochs and batch size were 50 and 128 respectively. All parameters were set default unless further specified. Also, SVM (RBF) and Decision Tree were implemented using

Table 2. A comparison of the prediction performance among 5 classifiers in terms of the area under the receiver operating characteristic (ROC) curve in **testing**. Two-tailed Wilcoxon Signed-Rank Test was conducted on the ROC values obtained by DNN and DNN-VL, as well as the ROC values obtained by DNN-CL and DNN-VL. In both tests, we obtained a p-value of **0.00512 < 0.05**.

Testing ROC	SVM (RBF)	Decision tree	DNN	DNN-CL	DNN-VL
Dataset-0	0.99864	0.96504	0.99549	0.99555	**0.99917**
Dataset-1	0.99857	0.97613	0.99671	0.99436	**0.99921**
Dataset-2	0.98468	0.91820	0.99297	0.99088	**0.99874**
Dataset-3	0.98835	0.91336	0.99299	0.99558	**0.99865**
Dataset-4	0.99323	0.92614	0.99383	0.99363	**0.99817**
Dataset-5	0.98109	0.92016	0.98830	0.99558	**0.99921**
Dataset-6	0.99577	0.95732	0.99120	0.99261	**0.99880**
Dataset-7	0.98999	0.94502	0.99244	0.98820	**0.99602**
Dataset-8	0.98351	0.89631	0.98670	0.97505	**0.99835**
Dataset-9	0.97748	0.90388	0.98679	0.98825	**0.99575**

Table 3. A comparison of the prediction performance among 5 classifiers in terms of the area the under precision-recall curve (PRC) in **testing**. Two-tailed Wilcoxon Signed-Rank Test was conducted on the PRC values obtained by DNN and DNN-VL, as well as the PRC values obtained by DNN-CL and DNN-VL. In both tests, we obtained a p-value of **0.00512 < 0.05**.

Testing PRC	SVM (RBF)	Decision tree	DNN	DNN-CL	DNN-VL
Dataset-0	0.99630	0.96852	0.99709	0.99526	**0.99847**
Dataset-1	0.99715	0.97564	0.99749	0.99591	**0.99851**
Dataset-2	0.97197	0.92987	0.99457	0.99034	**0.99696**
Dataset-3	0.98099	0.91459	0.99400	0.99251	**0.99708**
Dataset-4	0.98123	0.93479	0.99355	0.99020	**0.99675**
Dataset-5	0.97172	0.92722	0.99148	0.99414	**0.99741**
Dataset-6	0.99136	0.96317	0.99349	0.99367	**0.99740**
Dataset-7	0.98191	0.94941	0.99232	0.98696	**0.99429**
Dataset-8	0.96266	0.90656	0.99249	0.97733	**0.99520**
Dataset-9	0.95822	0.91355	0.99095	0.98462	**0.99363**

sci-kit learn [23] with default parameter setting. All experiments were run on a computer with 8.0 GB RAM, a i5-2410M-2.30 GHz CPU (4 Cores) and a GT520M Graphics card. These settings were used in all experiments unless further specified.

3.5 Results

Table 2 shows the comparative results on the area under the ROC curve in the testing images. We observe that in all datasets DNN-VL obtained a higher ROC than DNN and DNN-CL, and the baseline algorithms, SVM (RBF) and Decision Tree. We conducted a Two-tailed Wilcoxon Signed-Rank Test on the ROC values obtained by DNN and DNN-VL, as well as the ROC values obtained by DNN-CL and DNN-VL. Both tests are statistically significant ($\mathbf{p = 0.00512 < 0.05}$). Similar results evaluated by the area under the PRC in the testing images are observed in Table 3.

4 Conclusion

In this study, we proposed a new strategy, known as weight-selection strategy, for training Deep Neural Networks (DNN) on imbalanced classification. To our knowledge, it is the first systematic study to examine a weight-selection strategy on training DNN for imbalanced classification. Illustrated by experiments on 10 imbalanced datasets obtained from MNIST, the DNN trained by the new strategy outperformed the DNN trained by a standard strategy and the DNN trained by cost-sensitive learning with statistical significance ($p = 0.00512$). It should be noted that the DNN trained by our new strategy was trained on 20% less training images, corresponding to 12,000 training images less, but still achieved an outperforming performance. Future extension of this work includes incorporating the proposed strategy with existing techniques to investigate if the performance can be further improved.

References

1. LeCun, Y., Bengio, Y., Hinton, G.: Deep learning. Nature **521**(7553), 436–444 (2015)
2. Silver, D., Huang, A., Maddison, C.J., Guez, A., Sifre, L., Van Den Driessche, G., Schrittwieser, J., Antonoglou, I., Panneershelvam, V., Lanctot, M., et al.: Mastering the game of go with deep neural networks and tree search. Nature **529**(7587), 484–489 (2016)
3. Bengio, Y., et al.: Learning deep architectures for AI. Foundations and trends®. Mach. Learn. **2**(1), 1–127 (2009)
4. Huang, C., Li, Y., Change Loy, C., Tang, X.: Learning deep representation for imbalanced classification. In: Proceedings of the IEEE Conference on Computer Vision and Pattern Recognition, pp. 5375–5384 (2016)
5. Sun, Y., Wong, A.K., Kamel, M.S.: Classification of imbalanced data: a review. Int. J. Pattern Recogn. Artif. Intell. **23**(04), 687–719 (2009)

6. He, H., Garcia, E.A.: Learning from imbalanced data. IEEE Trans. Knowl. Data Eng. **21**(9), 1263–1284 (2009)
7. Haixiang, G., Yijing, L., Shang, J., Mingyun, G., Yuanyue, H., Bing, G.: Learning from class-imbalanced data: review of methods and applications. Expert Syst. Appl. **73**, 220–239 (2017)
8. Masko, D., Hensman, P.: The impact of imbalanced training data for convolutional neural networks. In: Degree Project in Computer Science, pp. 1–28. KTH Royal Institute of Technology (2015)
9. Chawla, N.V., Bowyer, K.W., Hall, L.O., Kegelmeyer, W.P.: SMOTE: synthetic minority over-sampling technique. J. Artif. Intell. Res. **16**, 321–357 (2002)
10. Ting, K.M.: A comparative study of cost-sensitive boosting algorithms. In: In Proceedings of the 17th International Conference on Machine Learning. Citeseer (2000)
11. Oquab, M., Bottou, L., Laptev, I., Sivic, J.: Learning and transferring mid-level image representations using convolutional neural networks. In: Proceedings of the IEEE Conference on Computer Vision and Pattern Recognition, pp. 1717–1724 (2014)
12. Khan, S.H., Bennamoun, M., Sohel, F., Togneri, R.: Cost sensitive learning of deep feature representations from imbalanced data. arXiv preprint arXiv:1508.03422 (2015)
13. Zhang, C., Gao, W., Song, J., Jiang, J.: An imbalanced data classification algorithm of improved autoencoder neural network. In: 2016 Eighth International Conference on Advanced Computational Intelligence (ICACI), pp. 95–99. IEEE (2016)
14. Shen, W., Wang, X., Wang, Y., Bai, X., Zhang, Z.: Deepcontour: a deep convolutional feature learned by positive-sharing loss for contour detection. In: Proceedings of the IEEE Conference on Computer Vision and Pattern Recognition, pp. 3982–3991 (2015)
15. Wang, S., Liu, W., Wu, J., Cao, L., Meng, Q., Kennedy, P.J.: Training deep neural networks on imbalanced data sets. In: 2016 International Joint Conference on Neural Networks (IJCNN), pp. 4368–4374. IEEE (2016)
16. Ng, W.W., Zeng, G., Zhang, J., Yeung, D.S., Pedrycz, W.: Dual autoencoders features for imbalance classification problem. Pattern Recogn. **60**, 875–889 (2016)
17. Sutskever, I.: Training recurrent neural networks. Ph.D. thesis, University of Toronto (2013)
18. LeCun, Y., Bottou, L., Bengio, Y., Haffner, P.: Gradient-based learning applied to document recognition. Proc. IEEE **86**(11), 2278–2324 (1998)
19. An, J., Cho, S.: Variational autoencoder based anomaly detection using reconstruction probability. In: Special Lecture on IE, vol. 2, pp. 1–18. SNU Data Mining Center (2015)
20. Tieleman, T., Hinton, G.: Lecture 6.5-rmsprop: divide the gradient by a running average of its recent magnitude. COURSERA: Neural Netw. Mach. Learn. **4**(2) (2012)
21. Bastien, F., Lamblin, P., Pascanu, R., Bergstra, J., Goodfellow, I., Bergeron, A., Bouchard, N., Warde-Farley, D., Bengio, Y.: Theano: new features and speed improvements. arXiv preprint arXiv:1211.5590 (2012)
22. Nair, V., Hinton, G.E.: Rectified linear units improve restricted Boltzmann machines. In: Proceedings of the 27th International Conference on Machine Learning (ICML-10), pp. 807–814 (2010)
23. Pedregosa, F., Varoquaux, G., Gramfort, A., Michel, V., Thirion, B., Grisel, O., Blondel, M., Prettenhofer, P., Weiss, R., Dubourg, V., et al.: Scikit-learn: machine learning in python. J. Mach. Learn. Res. **12**(Oct), 2825–2830 (2011)

End-to-End Deep Learning for Driver Distraction Recognition

Arief Koesdwiady[✉], Safaa M. Bedawi, Chaojie Ou, and Fakhri Karray

Center for Pattern Analysis and Machine Intelligence, University of Waterloo,
Waterloo, ON, Canada
{abkoesdw,sbedawi,c9ou,karray}@uwaterloo.ca

Abstract. In this paper, an end-to-end deep learning solution for driver distraction recognition is presented. In the proposed framework, the features from pre-trained convolutional neural networks VGG-19 are extracted. Despite the variation in illumination conditions, camera position, driver's ethnicity, and genders in our dataset, our best fine-tuned model, VGG-19 has achieved the highest test accuracy of 95% and an average accuracy of 80% per class. The model is tested with leave-one-driver-out cross validation method to ensure generalization. The results show that our framework avoided the overfitting problem which typically occurs in low-variance datasets. A comparison between our framework with the state-of-the-art XGboost shows that the proposed approach outperforms XGBoost in accuracy by approximately 7%.

Keywords: Driver distraction · Deep learning · Intelligent transportation system

1 Introduction

Distracted driving is much more dangerous than most people realize. According to the most recent published World Health Organization (WHO) report, it was estimated that, in 2015, over 1.25 million people were killed on the roads worldwide, making road trac injuries a leading cause of death globally [6]. Driver errors still remain the main cause of accidents in the roads. Using the cellphone for texting, talking and navigation as well as drowsiness are dierent types of activities that drastically decrease drivers attention to the road. In this background, research has focused on help to improve these alarming statistics. Research includes, but is not limited to: research on identifying driver behaviour to help the industry creating solutions to reduce the effect of driver distraction [3] and research on self-driving cars [1], where lane marking detection, path planning, and control are the area of enhancements. In [3], a review is provided on algorithms used in driver distraction detection. The main methods are focused on face detection, face/hand tracking and detection of facial landmarks [9]. Among the algorithms used are: SVM, Histogram of Oriented Gradients, Artficial Neural Networks (ANN, and Deep Neural Networks (DNN). The reported results for

© Springer International Publishing AG 2017
F. Karray et al. (Eds.): ICIAR 2017, LNCS 10317, pp. 11–18, 2017.
DOI: 10.1007/978-3-319-59876-5_2

these methods show that they are affected to various degrees, by illumination, skin colour and camera position inside the car.

In this paper, an end-to-end deep learning based classifier is investigated for driver distraction detection. This is motivated by its performance in self-driving car [1] and to have the system operated without direct human interaction. The purpose of this paper is to investigate the robustness of the suggested frame-work under different illuminations, different drivers type and different camera positions. To our best knowledge, an end-to-end deep learning framework has not been used for driver distraction detection.

The rest of this paper is organized as follows. An overview on deep learning framework, off-the-shelf feature extraction, dimensionality reduction and how it is used in classification are introduced in Sect. 2. In Sect. 3, a description of our dataset, experimental setup and fine-tuning approaches are described. Analysis of our results is in Sect. 4 and finally the conclusion and future work are presented in Sect. 5.

2   End-to-End Deep Learning Framework

Driver distraction recognition problem can be treated as a multi-class classification process to map the input observations to a driver state. The developed system includes three main components, as shown in Fig. 1. The first component is a variant of convolutional deep neural network for high-abstracted feature extraction. Followed by a max pooling layer which reduces the dimension of features. The last component includes 6 fully-connected layers and a softmax layer.

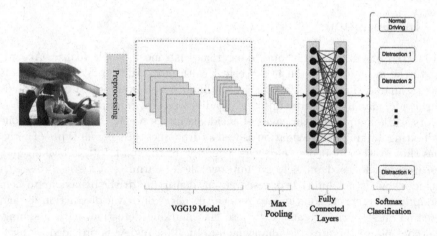

Fig. 1. End-to-end deep learning framework.

2.1 Feature Extraction

A common ConvNet is a stack of convolutional layers, pooling layers, most often followed by several fully-connected layers. The convolutional layer and pooling layer operate on small local input patches, and the combination of these two layers makes the network more robust to location variation of objects in the given images.

As a variant of ConvNet, the VGG 19 network is firstly proposed for images classification, object detection, and objects localization in competition ImageNet [7]. It is quickly accepted in many other computer vision and image processing researches, due to its simple structure and moderate number of parameters. There are two common methods to adopt this network: by fine-tuning all parameters in VGG and extracting highly-abstracted features for pre-trained VGG models. The research work in this paper follows the second method and extract represented features from the VGG19 model.

The architecture and configuration of VGG19 can be found in [7] is also roughly summarized here. The input should be a 224×224 RGB picture. The size of kernel is 3×3, which makes the following layers contains small local patch information, the stride for convolution is 1. The max-pooling is performed on a 2×2 patch with stride of 2. The last three fully-connected layers in VGG19 are dropped, and the remaining structure works as a feature extractor.

2.2 Classification

The classifier in the original VGG19 is a three-layers fully-connected network, which is designed and trained for classification of images that contains different objects. The dimension of features maps after the last max-pooling layer in VGG19 is $7 \times 7 \times 512$. To reduce the dimension of features and speed up the learning process, another max-pooling layer is connected to the last pooling layer in the VGG19 model and before the DNN classifier. The max-pooling is also performed over a 2×2 pixel window, with stride 2. In this work, the XGBoost and a six-layers fully-connected networks are exploited as classifiers to classify distracted drivers. The classification results of this two classifiers is presented and compared in Sect. 4.

The fully-connected network classifier contains 6 layers with 1000 nodes in each layer, and it is trained by back-propagation with stochastic gradient descent optimization. The XGBoost is a fast-implementation of gradient boosting trees [2,4]. Many successful solutions in Kaggle competitions are developed with this additional tree boosting method. However, the learning process for gradient boosting tree is time-consuming, thus it is not suitable for working directly on pixel level of large images.

3 Experiments

This section presents descriptions of the dataset gathered using a video camera, data pre-processing, and experimental setups.

3.1 Dataset

Taking into account the comfort of the drivers, a video camera is used in this work to capture driver situations. The camera is located such that the upper-body, hand positions, and rear-part of the car are captured and available to analyze. From the camera, sets of 640 × 480 - RGB video images with a frame rate 15 frames per second are obtained. In this experiment, two different cars are used in different lighting conditions. This way, the proposed system is forced to learn real-world driving situations. These situations are illustrated in Fig. 2.

As can be seen in Fig. 2, the experiment is carried out by participants from different ages, genders, and ethnicity. Ten drivers were involved in the experiments. Each driver was asked to perform or mimic the following driving activities: Normal/safe driving, Text messaging using left and right hand, Phone calling using left and right hand, Operating radio/navigation systems, Reaching objects at the rear-part of the car and Drinking using left or and right hand.

Fig. 2. Distraction types (from top-left to right-bottom): normal driving, text messaging (right-hand), drinking, reaching object at the back, calling on the phone, operating radio.

3.2 Experimental Settings

In this work, the driving activities are grouped into three types: distraction involving left hand, right hand, and distraction while reaching object at the back of the car. Together with normal driving, 4 classes of driving states are selected. For simplicity, these 4 classes will be called normal driving, distracted left, distracted right, and distracted back throughout the paper. As grouping the classes has caused imbalance in our data set, we increased the number of images for normal driving cars by flipping the images vertically and horizontally and sharpening the images to avoid imbalance classification problems. The images generated by the camera are reshaped to 224 × 224 pixels so that it can be fed in the feature extraction module VGG19. Subsequently, a pre-processing stage involving mean subtraction of RGB values was implemented.

After the features are extracted using VGG19 model, a max-pooling layer is implemented for dimensionality reduction. Furthermore, fully-connected layers are stacked for the classifications using a soft-max layer. The fully-connected layers are trained using back-propagation with stochastic gradient descent optimization. To obtain the best performance, the parameter tuning was done by varying the number of hidden layers and neurons. However, the number of neurons was kept fixed throughout the hidden layers. For example, six hidden layers with each layer consists of 1000 neurons.

The most popular non-linear activation function, namely rectified linear unit (ReLU), is implemented so that the deep neural network can learn much faster without unsupervised pre-training and, in the same time, avoid the vanishing gradient problem [5]. Moreover, to avoid over-fitting during the training, the drop-out method is introduced [8]. The training of the deep neural networks in this work uses a computing system that is powered by Intel Core i5 Quad Core 3.5 Ghz, 16 GB of RAM, and a GTX1070 8 GB GPU card.

To analyze the performance of the proposed model, XGBoost [2] method is used for comparison. XGBoost is a machine learning algorithm that has recently been winning major machine learning competitions such as Kaggle. This method is selected so that the state-of-the-art algorithm, i.e., deep learning, is fairly compared with another state-of-the-art non-connectionist machine learning algorithm. This method is implemented for classification after the features are extracted using the VGG19 model. The parameters of the model such as number of rounds, maximum depth of the tree, minimum child weight, and β (minimum loss reduction required to make a further partition on a leaf node of the tree) are tuned using similar cross-validation procedure as the proposed deep neural network.

The performance of both models are measured based on classification accuracy and multi-class log-loss. These metrics were applied to 1 test driver after the model is trained using the other 9 drivers. This scheme was applied for all drivers and is also known as leave-one-subject-out cross validation to ensure the model to generalize under different types of drivers.

4   Results and Analysis

Table 1 shows the performance of two different classifiers on each class. The results shows that the DNN classifier is dominating XGBoost on three classes (Normal Driving, Distraction Right, and Distraction Back); while on Distraction Left class, the difference is minimal. The performance of these two classifiers on the average precision is more close than that on recall and F1-measure.

In addition, the class of Distraction Left is well discriminated for both classifiers, the F1-Measure of both classifiers are almost 1. The table also shows that the precision for Distraction Right and Normal Driving classes are larger than their recall values, while the opposite can be found for Distraction Back. This difference shows that the system classifies more samples into Distraction Back, not the opposite.

Table 1. Per class performance comparison between DNN and XGBoost

Class	DNN			XGBoost		
	Precision	Recall	F1-Measure	Precision	Recall	F1-Measure
Normal driving	**0.96**	0.65	0.75	0.94	**0.67**	**0.76**
Distraction left	0.99	0.99	0.99	**1.00**	**1.00**	**1.00**
Distraction right	**0.80**	**0.71**	**0.70**	0.64	0.59	0.54
Distraction back	**0.68**	**0.85**	**0.73**	0.59	0.75	0.61
Average	**0.86**	**0.80**	**0.80**	0.79 ·	0.75	0.73

The performance of both classifiers on different drivers is presented in Table 2. As can be seen, the accuracy on different drivers by DNN classifier changes from 68.88% to 98.58%; while in XGBoost, it changes in a range of 64.25% to 85.38%. This significant disparity on different drivers shows that the learning process is limited by the magnitude of the dataset, and pictures of the distracted driver are related.

The variance of accuracy of DNN is 0.101, and for XGBoost it is 0.0635. The smaller variance shows that XGBoost is more stable but the average accuracy is 75.04%, also the standard deviation of Log-Loss by XGBoost is larger than that of DNN. The receiver operating characteristic curve (ROC) by changing the minimum probability of accepting the result is illustrated in Fig. 3. The Fig. 3(a) is based on the testing result of driver 3, while Fig. 3(b) is based on driver 8. The ROC for the class of Distraction Left in Fig. 3(b) reveals that by changing acceptance probability it is possible to achieve a good accuracy on Distraction Left class.

Table 2. Driver performance comparison between DNN and XGBoost

Driver	DNN		XGBoost	
	Accuracy	Log-loss	Accuracy	Log-loss
1	**94.63**	**0.740**	82.08	0.395
2	**80.75**	**0.619**	76.89	0.547
3	**98.58**	0.602 ·	73.80	**0.816**
4	76.92	**0.609**	**85.38**	0.459
5	69.79	0.674	**73.71**	**0.791**
6	**85.38**	0.633	76.63	**0.647**
7	79.21	**0.758**	**78.92**	0.606
8	**68.88**	**0.911**	64.25	0.784
9	**70.63**	0.588	70.50	**0.595**
10	**77.83**	0.683	68.29	**0.911**
Average	**80.258**	**0.682**	75.04	0.655
Standard deviation	10.10	**0.10**	**6.35**	0.17

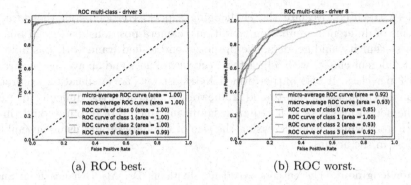

(a) ROC best.   (b) ROC worst.

Fig. 3. ROC of results

    The reason for the difference in performance of different classes can be found by checking the position between the camera and the driver. The camera is mounted on the dash board before the assistant's seat, and it takes pictures of the driver from the side view, so the activity of the hand close to the camera is represented more clearly than the activity of the other hand. By changing the position of the camera, the occlusion between two hands in images can be mitigated and classification accuracy of other classes can be improved.

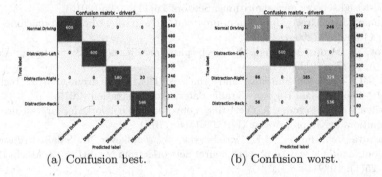

(a) Confusion best.   (b) Confusion worst.

Fig. 4. Confusion matrices for different drivers

    Figure 4 are confusion matrices of the best testing result (Driver 3) and the worst result (Driver 8). Figure 4(b) shows clearly that many samples of Normal Driving and Distraction Left are classified as Distraction Back. This result means the Normal Driving and Distraction Back share many similar states or images. This is also verified by checking the data.

5   Conclusion

In this paper, an end-to-end deep learning solution for driver distraction recognition is suggested in which the pre-trained convolutional neural networks

VGG-19 are used. Despite the challenging aspects considered in the dataset in terms of different illumination conditions, camera positions and variations in driver's ethnicity, and genders, the proposed end-to-end framework was able to detect different classes with a best test accuracy of 95% and an average accuracy of 80% per class. It also outperformed XGBoost by 7% classification accuracy. The main challenge of end-to-end framework comes from the difficulty of tuning the neural networks as it requires significant amount of resources and time. Future work will include increasing the size of the dataset and using ensembles to boost in the accuracy.

Acknowledgment. The authors would like to thank Dr. Alaa Khamis from Suez University, Egypt for his generous assistance with the data collection process.

References

1. Bojarski, M., Del Testa, D., Dworakowski, D., Firner, B., Flepp, B., Goyal, P., Jackel, L.D., Monfort, M., Muller, U., Zhang, J., et al.: End to end learning for self-driving cars. arXiv preprint arXiv:1604.07316 (2016)
2. Chen, T., Guestrin, C.: XGBoost: a scalable tree boosting system. In: Proceedings of the 22nd ACM SIGKDD International Conference on Knowledge Discovery and Data Mining, pp. 785–794. ACM (2016)
3. Fernández, A., Usamentiaga, R., Carús, J.L., Casado, R.: Driver distraction using visual-based sensors and algorithms. Sensors **16**(11), 1805 (2016)
4. Friedman, J.H.: Greedy function approximation: a gradient boosting machine. Ann. Stat. 1189–1232 (2001)
5. Glorot, X., Bordes, A., Bengio, Y.: Deep sparse rectifier neural networks. In: Aistats, vol. 15, p. 275 (2011)
6. National Center for Statistics and Analysis. Distracted driving 2013. Technical report, The National Highway Traffic Safety Administration (2015)
7. Simonyan, K., Zisserman, A.: Very deep convolutional networks for large-scale image recognition. arXiv preprint arXiv:1409.1556 (2014)
8. Srivastava, N., Hinton, G.E., Krizhevsky, A., Sutskever, I., Salakhutdinov, R.: Dropout: a simple way to prevent neural networks from overfitting. J. Mach. Learn. Res. **15**(1), 1929–1958 (2014)
9. Viola, P., Jones, M.: Rapid object detection using a boosted cascade of simple features. In: Proceedings of the 2001 IEEE Computer Society Conference on Computer Vision and Pattern Recognition, CVPR 2001, vol. 1, p. I. IEEE (2001)

Deep CNN with Graph Laplacian Regularization for Multi-label Image Annotation

Jonathan Mojoo[1]([⊠]), Keiichi Kurosawa[2], and Takio Kurita[1]

[1] Department of Information Engineering, Hiroshima University, Hiroshima, Japan
jonathanmojoo@yahoo.com, tkurita@hiroshima-u.ac.jp
[2] Faculty of Engineering, Hiroshima University, Hiroshima, Japan
lab.keiichikurosawa@gmail.com

Abstract. To compensate for incomplete or imprecise tags in training samples, this paper proposes a learning algorithm for the convolutional neural network (CNN) for multi-label image annotation by introducing co-occurrence dependency between tags as a graph Laplacian regularization term. To exploit the co-occurrence dependency, we apply Hayashi's quantification method-type III to the tags in the training samples and use the distances between the acquired representative vectors to define the weights for graph Laplacian regularization. By introducing this regularization term, the possibility of co-occurrence between tags with high co-occurrence frequency can be increased. To confirm the effectiveness of the proposed algorithm, we have done experiments using Corel5k's dataset for multi-label image annotation.

Keywords: Convolutional neural network · Multi-label classification · Laplacian regularization · Hayashi's quantification method · Co-occurrence between tags

1 Introduction

With the prevalence of social networks and digital photography in recent years, countless pictures are uploaded to various photo sharing services, e.g. Flickr [1] and Instagram [2]. Typically, users can provide contextual tags for describing their picture's semantic content and these tags are utilized for managing the pictures. However users sometimes provide incomplete or imprecise tags because of the time-consuming tagging process and the arbitrariness of users [3].

Many methods for multi-label image annotation have been proposed [4] such as methods using generative models [5], methods using discriminative models [6,7] and nearest neighbor based methods [8].

Recently, deep convolutional neural networks (CNN) have become a very popular tool in image recognition, movie recognition and so on, because A. Krizhevsky et al. applied the extended version of LeNet [9] to object recognition and won the ILSVRC 2012 with a much higher score than the traditional methods [10]. Inspired by the great success of deep CNNs in single-label image

© Springer International Publishing AG 2017
F. Karray et al. (Eds.): ICIAR 2017, LNCS 10317, pp. 19–26, 2017.
DOI: 10.1007/978-3-319-59876-5_3

classification, deep CNNs have been applied to multi-label image annotation problems. For example, nearest neighbor based methods were extended by using deep CNNs and achieved good results. Examples include deep convolutional ranking [11] and Hypothesis-CNN-Pooling [12].

To further improve the performance of image annotation, many researchers proposed various methods to exploit co-occurrence dependency between tags [13–15]. This is because it was noted that multi-label classification problems often exhibit strong co-occurrence dependency between tags [13]. For instance, sky and cloud usually co-occur, while plane and tiger almost never co-occur.

In this paper, to compensate for incomplete or imprecise tags in the training samples, we propose a multi-label image annotation method using a deep CNN by introducing the co-occurrence dependency between tags as the regularization term. To exploit the co-occurrence dependency between tags, we apply Hayashi's quantification method-type III to tags in the training samples and use the distances between the acquired representation vectors to define the weights for graph Laplacian regularization. By introducing this regularization term, the possibility of co-occurrence between the tags with high co-occurrence frequency can be increased. To confirm the effectiveness of the proposed algorithm, we have performed experiments using Corel5k's dataset for multi-label image annotation.

The rest of this paper is organized as follows. Section 2 explains the related works such as Hayashi's quantification method-type III, and Laplacian eigenmaps. The proposed method for image annotation using label co-occurrence dependency is explained in Sect. 3. Experiments and results are presented in Sect. 4.

2   Related Works

2.1   Hayashi's Quantification Method-Type III

Hayashi's quantification method-type III (HQ-III) was proposed by C. Hayashi in 1954 [16]. This method was developed to analyze categorical data, including cross tabulation or contingency tables, and give the vector representations of each row and column by using the information of co-occurrences. It is known that HQ-III is mathematically equivalent to Correspondence Analysis [17] and Dual Scaling [18].

Suppose the cross tabulation that records the frequency of co-occurrence is given by $F = \left[f_{ij} \right], (i = 1, \ldots, M; j = 1, \ldots, N)$. From this table, we can obtain the vector representations p_i and q_j for the i-th row and the j-th column by applying HQ-III. It is expected that the distances between these vectors become close if the pattern of responses in the cross tabulation are similar. In this paper, the distances between the vectors obtained by HQ-III for each label are used to define the similarity between each pair of labels.

2.2   Laplacian Eigenmaps

Belkin et al. [19] proposed the Laplacian eigenmaps. This is one of the manifold learning algorithms and it constructs the vector representation of each sample from the similarities between the samples.

This algorithm is based on spectral graph theory. Assume that we have N samples and the similarity between sample i and j is given as s_{ij}. Then we construct a graph (G, E) where each node in G corresponds to a sample and the edge $(i, j) \in E$ connecting two nodes i and j represents the similarity s_{ij} between these two samples. Then the 1-dimensional representation y_i for each sample is obtained by minimizing the objective function

$$\frac{1}{2} \sum_{i,j}^{N} s_{ij}(y_i - y_j)^2 = \boldsymbol{y}^T L \boldsymbol{y} \tag{1}$$

where $\boldsymbol{y} = \begin{bmatrix} y_1 \cdots y_N \end{bmatrix}^T$. The matrix L is Laplacian matrix defined as $L = D - S$, where $S = \begin{bmatrix} s_{ij} \end{bmatrix}$ and D is a diagonal matrix with diagonal elements $d_{ii} = \sum_{j=1}^{N} s_{ij}$.

To preserve the locality of the input space in the outputs of the Multi-nominal Logistic Regression, K. Watanabe et al. introduced a regularization term based on this objective function of Laplacian eigenmaps [20]. In this paper we also use this objective function of Laplacian eigenmaps as the regularization term to introduce the similarities between the labels obtained from the co-occurrences between labels in the training data.

3   Deep CNN with Graph Laplacian Regularization

3.1   Estimation of the Similarity Between Each Pair of Labels

Let $\{< \boldsymbol{x}_i, \boldsymbol{t}_i > | i = 1, \ldots, M\}$ be the set of training samples, where $\boldsymbol{x}_i$ is the input image and $\boldsymbol{t}_i = \begin{bmatrix} t_{i1} \cdots t_{iN} \end{bmatrix}^T$ is the binary vector of the labels. The number of training samples and the number of labels are denoted by M and N respectively.

We assume that $t_{ij} = 1$ if the j-th label is assigned and $t_{ij} = 0$ otherwise. Let $T = \begin{bmatrix} t_{ij} \end{bmatrix}, (i = 1, \ldots, M; j = 1, \ldots, N)$ be the matrix of the labels in the training samples. Then the frequency table of the labels in the training samples can be obtained by $F = T^{\mathrm{T}} T$.

By applying HQ-III to this cross tabulation, we can obtain the vector representation of each label. Let $\boldsymbol{q}_j, (j = 1, \ldots, N)$ be the vector representations of the labels, obtained by HQ-III.

From the vector representations of the labels in the training sample, we define the similarity between the i-th tag and the j-th tag as

$$s_{ij} = \exp(-\beta \|\boldsymbol{q}_i - \boldsymbol{q}_j\|^2) \tag{2}$$

where β is a parameter to control the range of influence.

3.2   The Deep CNN with the Graph Laplacian Regularization

The deep CNN is trained to estimate the label vector as y for a given input image x. The objective function for training the deep CNN is defined by combining the sum of the sigmoid cross entropy of each label and the graph Laplacian regularization term.

We use the sum of the sigmoid cross entropy of each label because the posterior probability of each label should be estimated for multi-label image annotation. It is defined as

$$E = -\sum_{i=1}^{M}\sum_{j=1}^{N}\{t_{ij}\log(y_{ij}) - (1 - t_{ij})\log(1 - y_{ij})\}. \tag{3}$$

To compensate for incomplete or imprecise labels in the training samples, we also introduce the graph Laplacian regularization term as

$$G = \sum_{i=1}^{N}\left(\frac{1}{2}\sum_{j=1}^{M}\sum_{k=1}^{M} s_{jk}(y_{ij} - y_{ik})^2\right) = \sum_{i=1}^{M} y_i^T L y_i \tag{4}$$

where L is called the Laplacian matrix of the graph and is defined as

$$L = D - S \tag{5}$$

where $S = [s_{ij}]$ and D is a diagonal matrix with diagonal elements $d_{ii} = \sum_{j=1}^{N} s_{ij}$.

Then the objective function to train the deep CNN is given as

$$Q = E + \alpha G \tag{6}$$

where α is the tuning parameter to control the balance between the accuracy of the labels of the training samples and the graph Laplacian regularization calculated from the co-occurrence between labels in the training samples.

For the learning algorithm, we used the gradient based optimization algorithm.

4   Experiments

4.1   Dataset and the Structure of the Deep CNN

To investigate the effectiveness of the proposed method, we performed experiments with the Corel5k data set which is one of the standard benchmark data sets for image annotation. Corel5k contains 4500 training images and 499 test images. All images are normalized to 127×127 pixels. The number of possible labels is 260 in this data set. Each image is annotated with 1–5 labels and the average number of keywords per image is 3.4.

In the following experiments, we used the Alex network with 5 convolutional layers and 1 fully connected layer and a classifier layer. For the first 5 convolutional layers, we used parameters from a pre-trained model. The number of neurons in the fully connected layer was set to 1024. We trained this network with Corel5k's training data set starting from the weights of the pre-trained model.

4.2 Results of HQ-III Applied to Labels in the Training Samples

To demonstrate the effectiveness of HQ-III, we applied it to label data of the training samples. A sample of the results are shown in Table 1. In this table, top 5 labels with the highest co-occurrence frequency with the label 'bear' are shown in the first row. Also the top 5 labels with the minimum squared Euclidean distances between their representative vectors obtained by HQ-III are shown in 2nd to 5th rows. The dimension of the representative vectors is changed from 40 to 160. From this table, we notice that the co-occurrence between tags in the training samples is roughly reflected in the Euclidean distances between the extracted representative vectors.

Table 1. The results of HQ-III applied to labels in the training samples. This table shows top 5 labels with the highest co-occurrence frequency with label 'bear' (1st row), and top 5 labels whose representative vectors have the minimum square Euclidean distance with the representative vector of the label 'bear'.

	1st	2nd	3rd	4th	5th
Co-occurrence frequency	Polar	Snow	Water	Black	Grizzly
	122	92	42	34	30
Euclidean distance					
dim = 40	Polar	Black	Face	Snow	Nets
	2.83	6.21	7.29	13.01	13.79
dim = 110	Polar	Snow	Water	Grass	Sky
	19.25	30.42	31.65	34.74	34.99
dim = 160	Polar	Water	Snow	Sky	Tree
	24.27	33.53	34.41	36.87	37.49

4.3 Effectiveness of the Laplacian Regularization

To show the effectiveness of the proposed multi-label annotation algorithm, we have performed some experiments using Corel5k data set. We trained the deep CNN by changing the tuning parameter α, which controls the balance between the accuracy of the labels of the training samples and the graph Laplacian regularization. The results are compared with the standard deep CNN. In this experiment, the parameter β, which controls the range of influence, is set to 1.0 and the number of dimensions of the representative vectors is fixed at 40.

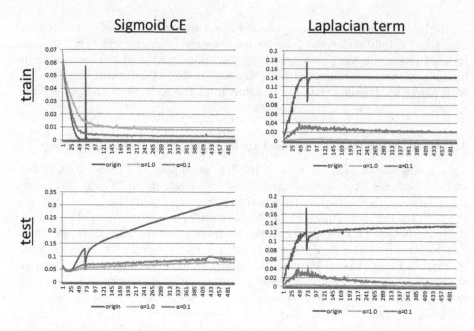

Fig. 1. The learning curves for the sigmoid cross entropy loss and the graph Laplacian regularization term. The tuning parameter α is changed. Upper: graphs for training data. Lower: graphs for test data.

The learning curves for the sigmoid cross entropy and the graph Laplacian term, G, are shown in Fig. 1. The results of the learning curves for the standard deep CNN are also shown in these graphs, in blue. From this figure we notice that the test errors are improved by introducing the graph Laplacian regularization term. Also, the graph Laplacian term is much lower when the graph Laplacian regularization term is introduced. These results show that the proposed algorithm is effective for the learning of a deep CNN for multi-label image annotation.

We also investigated the estimated tags for training and test images. We trained the CNN with and without the graph Laplacian regularization term until 500 epochs on the Corel5k training data set. The parameters are set to $\alpha = 1.0$, $\beta = 1.0$ and the dimension of the representative vectors is set to 40.

Figure 2 shows examples of tag estimations for images in corel5k's training data set. The left most column shows the tags given in the data set, which may be imprecise or incomplete, and the right and middle columns show the top 5 tags estimated by the deep CNN with and without the Laplacian regularization term respectively. From this figure, we notice that although the standard deep CNN without Laplacian regularization can perfectly estimate the tags given in the training data, the proposed method with the Laplacian regularization adds other relevant tags such as plants for the upper image, and anemone and fish for the lower image.

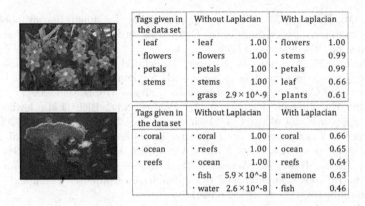

Tags given in the data set	Without Laplacian		With Laplacian	
· leaf	· leaf	1.00	· flowers	1.00
· flowers	· flowers	1.00	· stems	0.99
· petals	· petals	1.00	· petals	0.99
· stems	· stems	1.00	· leaf	0.66
	· grass	2.9×10^{-9}	· plants	0.61

Tags given in the data set	Without Laplacian		With Laplacian	
· coral	· coral	1.00	· coral	0.66
· ocean	· reefs	1.00	· ocean	0.65
· reefs	· ocean	1.00	· reefs	0.64
	· fish	5.9×10^{-8}	· anemone	0.63
	· water	2.6×10^{-8}	· fish	0.46

Fig. 2. Examples of tag estimation for images in the training data set. The tags given in the data set, and the tags estimated by the deep CNN with and without Laplacian regularization are shown in the tables on the right of each image.

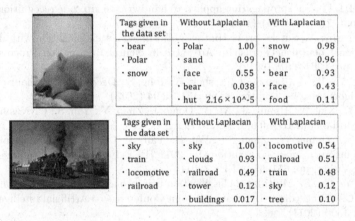

Tags given in the data set	Without Laplacian		With Laplacian	
· bear	· Polar	1.00	· snow	0.98
· Polar	· sand	0.99	· Polar	0.96
· snow	· face	0.55	· bear	0.93
	· bear	0.038	· face	0.43
	· hut	2.16×10^{-5}	· food	0.11

Tags given in the data set	Without Laplacian		With Laplacian	
· sky	· sky	1.00	· locomotive	0.54
· train	· clouds	0.93	· railroad	0.51
· locomotive	· railroad	0.49	· train	0.48
· railroad	· tower	0.12	· sky	0.12
	· buildings	0.017	· tree	0.10

Fig. 3. Examples of tag estimation for the images in the test data set. The tags given in the data set, tags estimated by the deep CNN with and without the Laplacian regularization are shown in the tables on the right of each image.

Figure 3 shows examples of tag estimation for images in the test data set. Similarly the estimated tags seem to be improved by the proposed method.

Acknowledgement. This work was partly supported by JSPS KAKENHI Grant Number 16K00239.

References

1. Flicker. https://www.flickr.com
2. Instagram. https://www.instagram.com

3. Ames, M., Naaman, M.: Why we tag: motivations for annotation in mobile and online media. In: Proceedings of the Conference on Human Factors in Computing Systems, pp. 971–980. ACM (2007)
4. Guillaumin, M., Mensink, T., Verbeek, J., Schmid, C.: TagProp: discriminative metric learning in nearest neighbor models for image auto-annotation. In: 2009 International Conference Computer Vision, pp. 309–316. IEEE (2009)
5. Barnard, K., Duygulu, P., Forsyth, D., Freitas, N.D., Blei, D.M., Jordan, M.I.: Matching words and pictures. J. Mach. Learn. Res. **3**(Feb), 1107–1135 (2003)
6. Grangier, D., Bengio, S.: A discriminative kernel-based approach to rank images from text queries. IEEE Trans. Pattern Anal. Mach. Intell. **30**(8), 1371–1384 (2008)
7. Hertz, T., Bar-Hillel, A., Weinshall, D.: Learning distance functions for image retrieval. In: 2004 Computer Society Conference on Computer Vision and Pattern Recognition, vol. 2, p. II. IEEE (2004)
8. Makadia, A., Pavlovic, V., Kumar, S.: A new baseline for image annotation. In: Forsyth, D., Torr, P., Zisserman, A. (eds.) ECCV 2008. LNCS, vol. 5304, pp. 316–329. Springer, Heidelberg (2008). doi:10.1007/978-3-540-88690-7_24
9. LeCun, Y., Boser, B., Denker, J.S., Henderson, D., Howard, R.E., Hubbard, W., Jackel, L.D.: Backpropagation applied to handwritten zip code recognition. Neural Comput. **1**(4), 541–551 (1989)
10. Krizhevsky, A., Sutskever, I., Hinton, G.E.: Imagenet classification with deep convolutional neural networks. In: Advances in Neural Information Processing Systems, pp. 1097–1105 (2012)
11. Gong, Y., Jia, Y., Leung, T., Toshev, A., Ioffe, S.: Deep convolutional ranking for multilabel image annotation arXiv:1312.4894 (2013)
12. Wei, Y., Xia, W., Huang, J., Ni, B., Dong, J., Zhao, Y., Yan, S.: CNN: single-label to multi-label arXiv:1406.5726 (2014)
13. Xue, X., Zhang, W., Zhang, J., Wu, B., Fan, J., Lu, Y.: Correlative multi-label multi-instance image annotation. In: 2011 IEEE International Conference Computer Vision, pp. 651–658. IEEE (2011)
14. Guo, Y., Gu, S.: Multi-label classification using conditional dependency networks. In: IJCAI Proceedings-International Joint Conference on Artificial Intelligence, vol. 22, p. 1300 (2011)
15. Wang, J., Yang, Y., Mao, J., Huang, Z., Huang, C., Xu, W.: CNN-RNN: a unified framework for multi-label image classification. In: IEEE Conference on Computer Vision and Pattern Recognition, pp. 2285–2294 (2016)
16. Hayashi, C.: Multidimensional quantification. I. Proc. Jpn. Acad. **30**(2), 61–65 (1954)
17. Benzecri, J.P.: Lanalyse des donnees, tome II. Lanalyse des correspondances (1973)
18. Nishisato, S.: Analysis of Categorical Data: Dual Scaling and Its Applications. JSTOR (1980)
19. Belkin, M., Niyogi, P.: Laplacian eigenmaps and spectral techniques for embedding and clustering. In: NIPS, vol. 14, pp. 585–591 (2001)
20. Watanabe, K., Kurita, T.: Locality preserving multi-nominal logistic regression. In: ICPR, pp. 1–4. IEEE (2008)

Transfer Learning Using Convolutional Neural Networks for Face Anti-spoofing

Oeslle Lucena[✉], Amadeu Junior, Vitor Moia, Roberto Souza,
Eduardo Valle, and Roberto Lotufo

University of Campinas, Campinas, Brazil
{oeslle,amadeu,vghmoia,rmsouza,dovalle,lotufo}@dca.fee.unicamp.br

Abstract. Face recognition systems are gaining momentum with current developments in computer vision. At the same time, tactics to mislead these systems are getting more complex, and counter-measure approaches are necessary. Following the current progress with convolutional neural networks (CNN) in classification tasks, we present an approach based on transfer learning using a pre-trained CNN model using only static features to recognize photo, video or mask attacks. We tested our approach on the REPLAY-ATTACK and 3DMAD public databases. On the REPLAY-ATTACK database our accuracy was 99.04% and the half total error rate (HTER) of 1.20%. For the 3DMAD, our accuracy was of 100.00% and HTER 0.00%. Our results are comparable to the state-of-the-art.

Keywords: Face anti-spoofing · Transfer learning · Deep learning · Face recognition

1 Introduction

In the last few years, the usage of face recognition systems for surveillance and authentication increased significantly due to the advances in computer vision technologies. As usage grows, the complexity of spoofing attacks also arises, and more complex counter-measure approaches are built. For instance, some of them consist of presenting to the vision sensor a fake image, video or even a 3D mask.

Deep Learning methods are representation-learning methods with multiple levels of representation along the neurons in a deep neural network (DNN) [1]. In a DNN, each level utilizes a non-linear module that transforms the representation at one level into a higher and more abstract representation at the next level [2]. Also, in contrast to conventional machine learning algorithms, DNNs are fed with raw data, and they discover the representations needed for detection or classification [1].

Inspired by this ability, this work proposes a method to identify spoofing attacks based on transfer learning using a pre-trained convolutional neural network (CNN). The main contributions of this paper are (1) a CNN approach based on transfer learning, using only static features, in other words no time relation

© Springer International Publishing AG 2017
F. Karray et al. (Eds.): ICIAR 2017, LNCS 10317, pp. 27–34, 2017.
DOI: 10.1007/978-3-319-59876-5_4

between frames was used (2) evaluation of half total error rate (HTER) on two public databases that outperformed or at least matched the state-of-the-art.

This work is organized as follow: Sect. 2 presents the related work with research that motivated our study of face anti-spoofing. The description of the proposed CNN method is detailed in Sect. 3. The experiments and results to validate our architecture, including pre-processing methods and evaluation metrics are reported in Sect. 4. Finally, the conclusions and future works are presented in Sect. 5.

2   Related Work

Due to many spoofing attack forms, elaborated approaches have been developed to identify and block these attacks. Some of them rely on extra sensors [3,4]. Others can be categorized into two main groups: feature level static and dynamic [5].

Feature level static methods focus on the analysis of images without considering the time relation between them. Those comprise techniques that use Fourier analysis, Lambertian models, Difference-of-Gaussian (DoG) and Local Binary Patterns (LBP). Li *et al.* [6] analyze the Fourier spectrum and introduced a high-frequency descriptor to identify spoofing attacks through images sequences. Using Fourier analysis combined with Lambertian model features, Tan et al. [7] were capable of extracting reflectance to recognize attacks. Peixoto et al. [8] used DoG filters and Sparse Logistic Regression Model to improve previous results in extreme light environments. Erdogmus et al. [9] proposed a solution based on LBP to recognize attacks on the 3D-MAD database, and their results showed an HTER of 0.95%.

Feature level dynamic methods explore the time relation between sequential frames of a video. In many works, the authors have used motion on detected faces as cues to recognize the attack (e.g.: eye blink movement [10,11], tracking face natural movements [12] or lips movement [13]). Anjos *et al.* [14] used correlation between background and foreground optical flows. Pereira et al. [15] used LBP-TOP operator combining space and time into a single texture descriptor. In this work, the authors reported an HTER of 7.6% on the REPLAY-ATTACK database. Komulainen et al. [16] presented a method that combined through a linear logistic regression the LBP operator and the correlation between background and face movements. The authors reported an HTER of 5.11% on the REPLAY-ATTACK database. Feng *et al.* [17] presented a neural network that fuses features such as shearlet-based image quality, face motion, and scene motion clues. The neural network is a pre-trained layer-wise sparse autoencoder. In this approach, the neural network is fine-tuned with a softmax layer classifier and labeled data using backpropagation. The HTER for REPLAY-ATTACK and 3DMAD database were of 0.00%.

Except for Feng *et al.*'s [17] work, aforementioned methods rely on hand-crafted features to determine attacks. Other approaches take advantage of CNNs abilities to identify features from images, thus recognizing attacks. Yang *et al.* [18] proposed an approach that uses a CNN architecture of AlexNet [19] for feature extraction, and Support Vector Machines (SVM) for classification. Also, the authors deployed a different method for pre-processing images varying faces' bounding boxes sizes, and number of successive frames used on the CNN. The HTER in their best case scenario on REPLAY-ATTACK database was of 2.81%.

Menotti *et al.* [20] proposed architecture optimization (AO) and filter optimization (FO). The second approach uses fine-tuning on CifarNet CNN [20]. However, their best case scenario was using AO that seeks for an optimal architecture of CNNs with filters weights set randomly. This approach achieve an HTER of 0.76% on REPLAY-ATTACK and 0.00% on the 3DMAD database. Alotaibi *et al.* [21] used a non-linear diffusion operator as a pre-processing step, then the processed image was applied to a custom six layers CNN. Using the REPLAY-ATTACK database the HTER was of 10.00%.

3   Proposed Method

Our approach is based on transfer learning a pre-trained CNN [22–24]. Transfer learning passes learned "knowledge" from a Machine Learning model trained on one task to another one [25]. For CNNs, there are two approaches to do transfer learning. The simpler uses the source model as a "off-the-shelf" feature extractor [26], using the output of a chosen layer as input for the target model, which is the only one trained for the new task. A more sophisticated approach "fine-tunes" the source model, in whole or in part, retrain its weights via backpropagation.

Transfer learning may be used to avoid overfitting a large network if there is not enough data to train it from scratch [25]. Transfer learning also saves computational resources, since training from scratch may take from days to weeks. In our case, we chose the CNN architecture VGG-16 that was pre-trained on ImageNet database [27], and its team secured the first and the second places in the localization and classification tasks of the ImageNet ILSVRC-2014 [28]. As the transfer learning approach, we used the fine-tuning.

3.1   CNN Architecture

The proposed face anti-spoofing network (FASNet) follows the VGG-16 architecture except for the top layers. The FASNet architecture is shown in Fig. 1 with its top layers highlighted in yellow. Our model was built based on Keras tutorials [29]. The FASNet code is available at https://github.com/OeslleLucena/FASNet.

The VGG-16 architecture is a 2D CNN for a 224×224 image size as input. In total, it has 16 convolutional layers, 64 for filters in the first block, 128 filters in the second, 256 in the third, 512 in the fourth and fifth blocks. Each convolution has a kernel of size 3×3. All max-pooling layers are performed in a 2×2 window, with stride 2. The activation functions are rectifier linear unit (ReLU). There are three fully connected (FC) layers, the sizes of each one are 4096, 1000, and 1000. The FASNet changes compared to the VGG-16 are only at the top layers.

One FC layer was removed, and the size of the other two was modified to 256 and 1. Moreover, we adopted the Adam optimizer [30] with the configurations provided by the paper, changing learning rate to 10^{-4} and weight decay to 10^{-6}. Also, the decision function has been modified from a softmax to a sigmoid, which often performs better for binary classification [31].

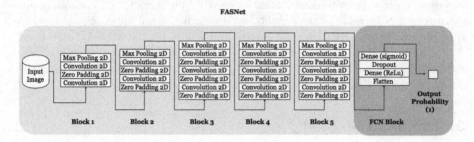

Fig. 1. FASNet architecture. As previously mentioned, the top layers of VGG-16 were changed to perform the binary classification for the anti-spoofing task. The layers highlighted in yellow represent the modified top layers (Color figure online)

3.2   Databases

3DMAD. This database is comprised of 76,500 frames of 17 persons in a total of 255 videos (300 frames per video) recorded using a Kinect for both genuine subjects and mask attacks [32]. Each frame consists of a depth image, corresponding RGB image, and manually annotated eye positions. For our purposes, only color images were used. All data is split into 3 sessions, which the first two sessions only have genuine subjects, and the last session has the mask attacks. The two first sessions have both 7 videos while the third session has only 5 videos. Some samples of this benchmark are shown in Fig. 2.

(a) Session 1   (b) Session 2   (c) Session 3

Fig. 2. Example of images from 3DMAD database from each session [32]. (a) and (b) are genuine images and (c) is a mask attack.

REPLAY-ATTACK. This database consists of 1,300 video clips of photo and video attack attempts of 50 subjects [33]. The subjects of this database were collected under two different illumination conditions, controlled and adverse.

The first condition was with office light turned on, blinds were down and homo-geneous background. The second condition had the following characteristics: blinds up, complex backgrounds, and office lights were out. Also, the attack pro-tocols are classified according to the type of device used to generate the attack, for example: print, mobile (phone), and high-definition (tablet). Some samples of this benchmark are shown in Fig. 3.

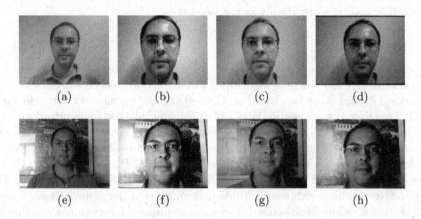

Fig. 3. Example of images from REPLAY-ATTACK database [33]. The first row repre-sent controlled images and the second row represents adverse images. In the first column is shown examples of genuine images. From the second column to the fourth are shown example of the following protocols of attacks: print, mobile and high definition.

4 Experiments and Results

Our experiments were conducted on the train and test folders provided by each face anti-spoofing database. Since the public databases are composed of video clips and our architecture is 2D, some pre-processing was needed. All the details of the adopted procedures and evaluation metrics used are found in Sect. 4.1.

To train the FASNet we froze the weights from the bottom layers up to the third block of our face anti-spoofing network (FASNet), and we fine-tuned weights from the fourth block up to the top layers via backpropagation. Then, we evaluate our method on REPLAY-ATTACK and 3DMAD test folders. The comparison with the state-of-the-art results are discussed in Sect. 4.2.

The implementation of the algorithm was built using Keras library [29] with Theano [34] as backend. Furthermore, all analyses were conducted using a com-puter equipped with Intel(R) Xeon(R) E5506 2.13 GHz (6 GB RAM), using a NVIDIA Tesla K40c GPU (12 GB).

4.1 Pre-processing and Metrics

The pre-processing steps adopted were two: (1) subsampling and (2) face detec-tion. Step one consisted of extracting half of the frames per second in each video.

In step two, we first used the OpenFace face detector [35] algorithm to find the region-of-interest (ROI) corresponding to a face. Next, using OpenFace algorithms we cropped to a window sized 96 pixels and aligned the faces to center based on the nose and eyes position.

Our evaluation was based on accuracy (ACC) and half total error rate (HTER), which are often used for assessing biometrics systems. Since the CNN output scores are probabilities [20], the computation of HTER was done assuming $\tau = 0.5$, and its estimation was not needed.

4.2 Results

We compare our method with six other approaches, three of them based on conventional machine learning algorithms [9,15,16], one based on shallow neural networks [17], and other two based on CNNs [20,21]. Our method outperformed HTER for all of the conventional machine learning methods submitted on the 3DMAD and REPLAY-ATTACK databases, which are reported in the first three lines of Table 1. FASNet has beaten almost all the state-of-art methods, losing only for the Multi-cues Integration approach [17] on REPLAY-ATTACK benchmark, and our method reached an HTER of 0.00% on 3DMAD database. All comparative results are shown in Table 1. Remark that our approach only uses static features, while Multi-cues Integration approach, for example, combined both static and dynamic features.

Table 1. Comparative test results for different databases among FASNet and state-of-the-art methods.

Databases	3DMAD		REPLAY-ATTACK	
	ACC (%)	HTER (%)	ACC (%)	HTER (%)
LBP-TOP + SVM [15]	-	-	-	7.60
Texture-based countermeasures [9]	-	0.95	-	-
Context Based [16]	-	-	-	5.11
Spoofnet [20]	100.00	0.00	98.75	0.70
Multi-cues Integration [17]	-	0.00	-	**0.00**
Non-linear Diffusion [21]	-	-	-	10.00
FASNet	100.00	0.00	**99.04**	1.20

5 Conclusions

In this paper, we introduced a new approach that uses transfer learning in convolutional neural networks to address face anti-spoofing methods. Our experimental results showed a HTER on REPLAY-ATTACK and 3DMAD databases equal to 1.20% and 0.00%, respectively, outperforming almost all state-of-the-art methods. For future work, we intend to investigate deeper CNN architectures, for example ResNet [36], incorporate dynamic features, and also explore other complex benchmarks.

References

1. LeCun, Y., Bengio, Y., Hinton, G.: Deep learning. Nature **521**(7553), 436–444 (2015)
2. Schmidhuber, J.: Deep learning in neural networks: an overview. Neural Netw. **61**, 85–117 (2015)
3. Seal, A., Ganguly, S., Bhattacharjee, D., Nasipuri, M., Basu, D.K.: Automated thermal face recognition based on minutiae extraction. CoRR abs/1309.1000 (2013)
4. Zhang, Z., Yi, D., Lei, Z., Li, S.Z.: Face liveness detection by learning multispectral reflectance distributions. In: Face and Gesture 2011, pp. 436–441, March 2011
5. Galbally, J., Marcel, S., Fierrez, J.: Biometric antispoofing methods: a survey in face recognition. IEEE Access **2**, 1530–1552 (2014)
6. Li, J., Wang, Y., Tan, T., Jain, A.K.: Live face detection based on the analysis of fourier spectra, vol. 5404, pp. 296–303 (2004)
7. Tan, X., Li, Y., Liu, J., Jiang, L.: Face liveness detection from a single image with sparse low rank bilinear discriminative model. In: Daniilidis, K., Maragos, P., Paragios, N. (eds.) ECCV 2010. LNCS, vol. 6316, pp. 504–517. Springer, Heidelberg (2010). doi:10.1007/978-3-642-15567-3_37
8. Peixoto, B., Michelassi, C., Rocha, A.: Face liveness detection under bad illumination conditions. In: 2011 18th IEEE International Conference on Image Processing, pp. 3557–3560, September 2011
9. Erdogmus, N., Marcel, S.: Spoofing 2D face recognition with 3D masks. IEEE Trans. Inf. Forensics Secur. **9**(7), 1084–1097 (2014)
10. Pan, G., Sun, L., Wu, Z., Lao, S.: Eyeblink-based anti-spoofing in face recognition from a generic webcamera. In: 2007 IEEE 11th International Conference on Computer Vision, pp. 1–8, October 2007
11. Li, J.W.: Eye blink detection based on multiple gabor response waves. In: 2008 International Conference on Machine Learning and Cybernetics, vol. 5, pp. 2852–2856, July 2008
12. Liting, W., Xiaoqing, D., Chi, F.: Face live detection method based on physiological motion analysis. Tsinghua Sci. Technol. **14**(6), 685–690 (2009)
13. Kollreider, K., Fronthaler, H., Bigun, J.: Verifying liveness by multiple experts in face biometrics. In: 2008 IEEE Computer Society Conference on Computer Vision and Pattern Recognition Workshops, pp. 1–6, June 2008
14. Anjos, A., Chakka, M.M., Marcel, S.: Motion-based counter-measures to photo attacks in face recognition. IET Biom. **3**(3), 147–158 (2014)
15. Freitas Pereira, T., Anjos, A., Martino, J.M., Marcel, S.: *LBP – TOP* based countermeasure against face spoofing attacks. In: Park, J.-I., Kim, J. (eds.) ACCV 2012. LNCS, vol. 7728, pp. 121–132. Springer, Heidelberg (2013). doi:10.1007/978-3-642-37410-4_11
16. Komulainen, J., Hadid, A., Pietikäinen, M.: Context based face anti-spoofing. In: 2013 IEEE Sixth International Conference on Biometrics: Theory, Applications and Systems (BTAS), pp. 1–8, September 2013
17. Feng, L., Po, L.M., Li, Y., Xu, X., Yuan, F., Cheung, T.C.H., Cheung, K.W.: Integration of image quality and motion cues for face anti-spoofing: a neural network approach. J. Vis. Commun. Image Represent. **38**, 451–460 (2016)
18. Yang, J., Lei, Z., Li, S.Z.: Learn convolutional neural network for face anti-spoofing. CoRR abs/1408.5601 (2014)
19. Krizhevsky, A., Sutskever, I., Hinton, G.E.: Imagenet classification with deep convolutional neural networks. In: Advances in Neural Information Processing Systems, pp. 1097–1105 (2012)

20. Menotti, D., Chiachia, G., Pinto, A., Schwartz, W.R., Pedrini, H., Falcão, A.X., Rocha, A.: Deep representations for iris, face, and fingerprint spoofing detection. IEEE Trans. Inf. Forensics Secur. **10**(4), 864–879 (2015)

21. Alotaibi, A., Mahmood, A.: Deep face liveness detection based on nonlinear diffusion using convolution neural network. Sig. Image Video Process. **11**, 1–8 (2016)

22. Girshick, R., Donahue, J., Darrell, T., Malik, J.: Region-based convolutional networks for accurate object detection and segmentation. IEEE Trans. Pattern Anal. Mach. Intell. **38**(1), 142–158 (2016)

23. Noh, H., Hong, S., Han, B.: Learning deconvolution network for semantic segmentation. In: The IEEE International Conference on Computer Vision (ICCV), December 2015

24. Shin, H.C., Roth, H.R., Gao, M., Lu, L., Xu, Z., Nogues, I., Yao, J., Mollura, D., Summers, R.M.: Deep convolutional neural networks for computer-aided detection: CNN architectures, dataset characteristics and transfer learning. IEEE Trans. Med. Imaging **35**(5), 1285–1298 (2016)

25. Yosinski, J., Clune, J., Bengio, Y., Lipson, H.: How transferable are features in deep neural networks? In: Ghahramani, Z., Welling, M., Cortes, C., Lawrence, N.D., Weinberger, K.Q. (eds.) Advances in Neural Information Processing Systems 27, pp. 3320–3328. Curran Associates, Inc. (2014)

26. Sharif Razavian, A., Azizpour, H., Sullivan, J., Carlsson, S.: CNN features off-the-shelf: an astounding baseline for recognition. In: The IEEE Conference on Computer Vision and Pattern Recognition (CVPR) Workshops, June 2014

27. Russakovsky, O., Deng, J., Su, H., Krause, J., Satheesh, S., Ma, S., Huang, Z., Karpathy, A., Khosla, A., Bernstein, M., Berg, A.C., Fei-Fei, L.: ImageNet large scale visual recognition challenge. Int. J. Comput. Vis. (IJCV) **115**(3), 211–252 (2015)

28. Simonyan, K., Zisserman, A.: Very deep convolutional networks for large-scale image recognition. CoRR abs/1409.1556 (2014)

29. Chollet, F.: Keras (2015). https://github.com/fchollet/keras

30. Kingma, D.P., Ba, J.: Adam: a method for stochastic optimization. CoRR abs/1412.6980 (2014)

31. Duch, W., Jankowski, N.: Survey of neural transfer functions. Neural Comput. Surv. **2**, 163–213 (1999)

32. Chingovska, I., Anjos, A., Marcel, S.: On the effectiveness of local binary patterns in face anti-spoofing. In: 2012 BIOSIG - Proceedings of the International Conference of Biometrics Special Interest Group (BIOSIG), pp. 1–7, September 2012

33. Erdogmus, N., Marcel, S.: Spoofing in 2D face recognition with 3D masks and anti-spoofing with kinect. In: 2013 IEEE Sixth International Conference on Biometrics: Theory, Applications and Systems (BTAS), pp. 1–6, September 2013

34. Bastien, F., Lamblin, P., Pascanu, R., Bergstra, J., Goodfellow, I.J., Bergeron, A., Bouchard, N., Bengio, Y.: Theano: new features and speed improvements. In: Deep Learning and Unsupervised Feature Learning NIPS 2012 Workshop (2012)

35. Amos, B., Ludwiczuk, B., Satyanarayanan, M.: Openface: a general-purpose face recognition library with mobile applications. Technical report, CMU-CS-16-118, CMU School of Computer Science (2016)

36. He, K., Zhang, X., Ren, S., Sun, J.: Deep residual learning for image recognition. In: The IEEE Conference on Computer Vision and Pattern Recognition (CVPR), June 2016

Depth from Defocus via Active Quasi-random Point Projections: A Deep Learning Approach

Avery Ma[(✉)], Alexander Wong, and David Clausi

Vision and Image Processing Lab, Systems Design Engineering,
University of Waterloo, East Campus 4, 295 Phillip Street,
Waterloo, ON N2L 3G1, Canada
b24ma@uwaterloo.ca

Abstract. Depth estimation plays an important role in many computer vision and computer graphics applications. Existing depth measurement techniques are still complex and restrictive. In this paper, we present a novel technique for inferring depth measurements via depth from defocus using active quasi-random point projection patterns. A quasi-random point projection pattern is projected onto the scene of interest, and each projection point in the image captured by a cellphone camera is analyzed using a deep learning model to estimate the depth at that point. The proposed method has a relatively simple setup, consisting of a camera and a projector, and enables depth inference from a single capture. We evaluate the proposed method both quantitatively and qualitatively and demonstrate strong potential for simple and efficient depth sensing.

Keywords: Depth from defocus · Convolutional neural network · 3D reconstruction · Depth sensing

1 Introduction

Depth information allows us to understand the three-dimensional relationship of objects in world space. It has many applications in fields such as robotics, remote sensing, action recognition and product inspection. As such, depth cameras have received much attention both academically and in industry with constant advancements to depth camera technology.

There are various depth sensing techniques including stereo-vision, depth-from-motion, depth-from-shape, depth-from-focus and projector-camera structured light systems such as the Microsoft Kinect [5]. While recent efforts have lowered the cost of depth camera sensors, depth sensing still requires specialized complex hardware that are not compact. Motivated by these challenges, in this paper we present a system for inferring sparse depth measurements with the acquisition of a single image by leveraging active quasi-random point projections and camera defocus. The proposed method of sparse depth inference has a relatively simple setup, thus can potentially lead to very portable and low cost active depth sensing systems. The proposed method involves projecting a

© Springer International Publishing AG 2017
F. Karray et al. (Eds.): ICIAR 2017, LNCS 10317, pp. 35–42, 2017.
DOI: 10.1007/978-3-319-59876-5_5

quasi-random point pattern onto the scene of interest. The quasi-random point pattern can for instance be generated by using a light source with a point pattern mask, resulting in much simpler and cheaper hardware configuration. The detected projected points on the imaged scene are used to train a convolutional neural network, which is then used to estimate depth.

2   Related Work

Depth from defocus (DFD) methods generally estimate depth by analyzing the difference in blurriness of two images captured at different focal lengths [8,11,12], with different methods using different filters for determining the measure of blur. A major drawback to such DFD methods is the unreliable detection of blur, especially in untextured areas of the image. This problem is mitigated in active depth sensing, where an optical projection is used to find correspondences for triangulation. A review of structured light patterns for depth measurement is provided by Salvi *et al.* [10]. Therefore, using active projection patterns does not depend on the objects in the scene and is also effective in untextured regions. On the other hand, active depth sensing systems suffer from occlusion and require complex hardware. As such, we are motivated in the proposed system to takes advantage of the benefits from both DFD and active depth sensing methods to design a system that has a simple setup yet reliable in the depth measurements.

The concept of using DFD using active projections have also been explored in literature. Pentland *et al.* [9] used evenly spaced line projections to determine depth from line spread. This simple method is able to create low resolution depth maps. Nayar *et al.* [7] used a dual sensor plane with optimized projection and camera setup to produce a dense depth map and reduce front/back focal ambiguity. Ghita *et al.* [2] used a dense projected pattern with a tuned local operator for finding the relationship between blur and depth. Moreno *et al.* [6] used an evenly spaced point pattern with defocus to obtain an approximate depth map used for automatic image refocusing. These methods use a high density projection pattern which require either a projector or more specialized calibrated hardware, neither of which are required by the method presented here.

Promising results were shown in the preliminary approach by Ma *et al.* [4]. However, it occurs that the point pattern experiences a radial distortion through projector-lens system, which results in deterioration in depth estimation accuracy as it gets further away from the center of the point pattern. In the previous work, it made use of active quasi-random point projection pattern and blurry projected pattern as captured by the camera was modelled using the standard deviation of a circularly-symmetric 2D Gaussian which did not account for the distortion of the pattern after projection.

3   Method

The proposed system can be described as follows. A quasi-random point pattern is projected onto the scene, which is then captured by a cellphone camera.

The camera's focus is fixed such that the degree of focus of each point in the quasi-random point pattern as it appears in the captured image is dependent on the depth of the surface. A deep learning convolutional neural network (CNN) architecture is constructed as a calibration model, and trained using captured images of point pattern. The trained network can be used for depth inference at each point to produce sparse depth measurements. A one-time calibration step is required to train the CNN model.

3.1 Calibration

When out of focus, a projected point will appear blurred, with the degree of blurriness correlated with the depth of the scene at that point. The purpose of the calibration procedure is to learn features that can effectively characterize the blurriness of dot pattern at different depths. Those features coupled with trained weights and biases can be used to construct a CNN model to predict depth given an input image of dot pattern. The blur effect of a projected point is visualized in Fig. 1.

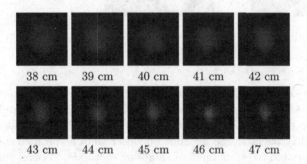

Fig. 1. Projected points on an object at various distances (38 cm to 47 cm) away from the setup, as captured by the camera.

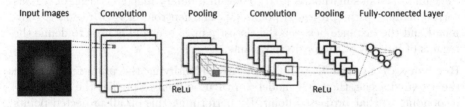

Fig. 2. Visualization of the construction of the CNN model

The network is trained using camera captures of point pattern projection with 10 discrete depth labels. Specifically, a point pattern is projected onto surfaces at known distances away from the projector-camera setup. Captured cellphone images are cropped to 30×30 pixel patches, with dot pattern placed at the center of the patch. The network consists of two 5×5 convolutional layers with 10 kernels, each followed by a rectified linear unit (ReLu) layer and a 2×2 max-pooling layer. The architecture has one fully-connected layer at the end of the network. Figure 2 illustrates the training process of the CNN model.

3.2 Sparse Depth Estimation Pipeline

With the CNN model, the proposed system can then be used to estimate sparse depth of the scene. To this end, the proposed depth recovery method can be divided into 3 main stages outlined in Fig. 3 and described as follows.

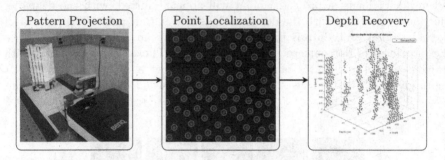

| Pattern Projection | Point Localization | Depth Recovery |

Fig. 3. Flowchart of the depth recovery pipeline

Active Quasi-random Point Projection: A quasi-random point pattern is projected onto the scene. In the current system, a Poisson Disk sampling method was utilized to generate the quasi-random point pattern such that the random points are tightly packed together, but no closer than a specified minimum distance [1]. Compared to other random sampling methods, Poisson Disk sampling method significantly reduces the chances of having overlaps between blurred projected points, which would result in erroneous depth recovery.

Point Localization: After the projected point pattern has been captured by the camera, Otsu's method is used to obtain a binary map consisting of regions corresponding to the projected points [3]. The centroid of each region is computed and the distance between the closest pair of centroid is used to define the regions of interest of the projected points.

Recovery of Sparse Depth Map: After identifying the projected point in the acquired scene, the CNN model can then be used to predict the depth corresponding to that projected point. By performing this on all projected points in the quasi-random point projection pattern, the sparse depth map can be obtained.

4 Experiment

The main goal of this current realization of the proposed technique is to build a compact and portable system to obtain depth information of the scene. For this purpose, the scene is imaged using an iPhone 5S camera (resolution: 3264×2448) and the quasi-random point pattern is projected using a BENQ MH630 Digital Projector (resolution: 1920 × 1080).

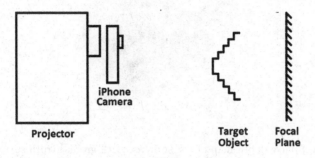

Fig. 4. Experimental setup

The experimental setup is described in Fig. 4. The iPhone camera is mounted 3 cm before the projector. The camera is in-line with the projection lens. The focal plane of both projector and the camera are on a plane 50 cm away from the projector.

5 Results

The test scene was a two-way staircase with 1 cm step-size away from the setup, and the quasi-random point density is approximately 0.20% of the projector resolution.

Ground truth surfaces were constructed from known points. Figure 5 shows the depth recovery of the object compared to the ground truth surfaces. To evaluate the performance of the proposed method, the root mean square error (RMSE) and correlation coefficient were computed between the estimated depth and the ground truth depth at corresponding locations.

To evaluate the reconstruction errors, the RMSE value is calculated as:

$$RMSE = \sqrt{\frac{1}{n}\sum_{i=1}^{n}(y_i - \hat{y}_i)^2} \tag{1}$$

over the n captured dot pattern. The RMSE value of the system is ±**0.45 cm**, which demonstrates that the proposed depth recovery pipeline is capable of achieving 3D reconstruction with high accuracy. Figure 6 shows the top view of the sparse depth recovery of the test scene.

The correlation coefficient was computed to be **0.9578**, which demonstrates that a strong positive correlation between the ground truth depth and the estimated depth. The mathematical formulation for correlation coefficient is:

$$r = \frac{1}{n-1}\sum_{i=1}^{n}\frac{(y_i - \mu)}{\sigma}\frac{(\hat{y}_i - \hat{\mu})}{\hat{\sigma}} \tag{2}$$

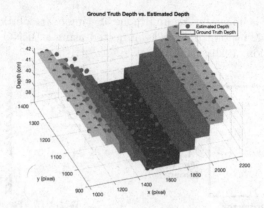

Fig. 5. Sparse depth map of the staircase with ground truth surfaces

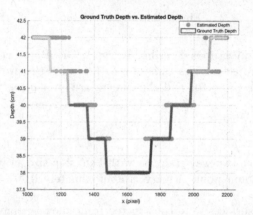

Fig. 6. Top view of the sparse depth map of the staircase

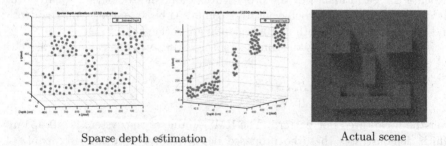

Sparse depth estimation   Actual scene

Fig. 7. Depth estimation of LEGO smiley face

Figure 7 illustrates the depth estimation of a smiley face made of LEGO. It can be observed that the proposed method can also achieve great results when imaging complex geometric shapes and objects. Comparison of the depth estimation accuracy using the proposed Deep Learning approach and previous Gaussian

Modelling approach [4] can be found in Table 1. With the proposed CNN architecture, there is a drastic improvement in the depth estimation accuracy which is reflected in the significant decline in RMSE value.

Table 1. Comparison of system accuracy between using Deep Learning approach and Gaussian Modelling approach

	Deep learning approach	Gaussian modelling approach
RMSE (cm)	0.45	1.28
Correlation coefficient	0.9578	0.9533

6 Conclusion

In this paper, we present a sparse depth measurement system based on active quasi-random point projection and depth from defocus using a deep learning approach. The proposed system can be used to reconstruct full sparse depth map with the acquisition of a single image. The main limitation of the method arise from the vulnerability to ambient light condition and the reflectance of different materials. Despite these limitations, the advantage of the proposed method is its simplicity in hardware and computation, requiring merely a camera and projected dot pattern. We are currently exploring ways to increase the density of sparse measurements. This requires the knowledge of maximum number of dot projections that can be captured and correctly identified in the acquired image. The use of such high density quasi-random projection pattern can allow us to leverage compressive sensing theory to produce full depth maps in future applications.

Acknowledgments. This work was supported by the Natural Sciences and Engineering Research Council of Canada, Canada Research Chairs Program, and the Ontario Ministry of Research and Innovation.

References

1. Bridson, R.: Fast poisson disk sampling in arbitrary dimensions. In: ACM SIGGRAPH 2007 Sketches, p. 22. ACM (2007)
2. Ghita, O., Whelan, P.F., Mallon, J.: Computational approach for depth from defocus. J. Electron. Imaging **14**(2), 023021 (2005)
3. Jianzhuang, L., Wenqing, L., Yupeng, T.: Automatic thresholding of gray-level pictures using two-dimension Otsu method. In: 1991 International Conference on Circuits and Systems, Conference Proceedings, China, pp. 325–327. IEEE (1991)
4. Ma, A., Li, F., Wong, A.: Depth from defocus via active Quasi-random point projections. J. Comput. Vis. Imaging Syst. **2**(1) (2016)
5. Microsoft. Kinect hardware (2016). https://developer.microsoft.com/en-us/windows/kinect/hardware

6. Moreno-Noguer, F., Belhumeur, P.N., Nayar, S.K.: Active refocusing of images and videos. ACM Trans. Graph. (TOG) **26**(3), 67 (2007)
7. Nayar, S.K., Watanabe, M., Noguchi, M.: Real-time focus range sensor. IEEE Trans. Pattern Anal. Mach. Intell. **18**(12), 1186–1198 (1996)
8. Pentland, A., Darrell, T., Turk, M., Huang, W.: A simple, real-time range camera. In: IEEE Computer Society Conference on Computer Vision and Pattern Recognition, Proceedings CVPR 1989, pp. 256–261. IEEE (1989)
9. Pentland, A., Scherock, S., Darrell, T., Girod, B.: Simple range cameras based on focal error. JOSA A **11**(11), 2925–2934 (1994)
10. Salvi, J., Pages, J., Batlle, J.: Pattern codification strategies in structured light systems. Pattern Recogn. **37**(4), 827–849 (2004)
11. Watanabe, M., Nayar, S.K.: Rational filters for passive depth from defocus. Int. J. Comput. Vis. **27**(3), 203–225 (1998)
12. Xiong, Y., Shafer, S.A.: Depth from focusing and defocusing. In: 1993 IEEE Computer Society Conference on Computer Vision and Pattern Recognition, Proceedings CVPR 1993, pp. 68–73. IEEE (1993)

Machine Learning for Medical Image Computing

Discovery Radiomics via a Mixture of Deep ConvNet Sequencers for Multi-parametric MRI Prostate Cancer Classification

Amir-Hossein Karimi[1]([✉]), Audrey G. Chung[2], Mohammad Javad Shafiee[2], Farzad Khalvati[4], Masoom A. Haider[4], Ali Ghodsi[3], and Alexander Wong[2]

[1] Department of Computer Science, University of Waterloo, Waterloo, Canada
a6karimi@uwaterloo.ca
[2] Department of Systems Design Engineering, University of Waterloo, Waterloo, Canada
[3] Department of Statistics and Acturial Sciences, University of Waterloo, Waterloo, Canada
[4] Department of Medical Imaging, University of Toronto, Toronto, Canada

Abstract. Prostate cancer is the most diagnosed form of cancer in men, but prognosis is relatively good with a sufficiently early diagnosis. *Radiomics* has been shown to be a powerful prognostic tool for cancer detection; however, these radiomics-driven methods currently rely on hand-crafted sets of quantitative imaging-based features, which can limit their ability to fully characterize unique prostate cancer tumour traits. We present a novel *discovery radiomics* framework via a mixture of deep convolutional neural network (ConvNet) sequencers for generating custom radiomic sequences tailored for prostate cancer detection. We evaluate the performance of the mixture of ConvNet sequencers against state-of-the-art hand-crafted radiomic sequencers for binary computer-aided prostate cancer classification using real clinical prostate multi-parametric MRI data. Results for the mixture of ConvNet sequencers demonstrate good performance in prostate cancer classification relative to the hand-crafted radiomic sequencers, and show potential for more efficient and reliable automatic prostate cancer classification.

Keywords: Discovery radiomics · Computer-aided prostate cancer classification · Multi-parametric magnetic resonance imaging (mpMRI) · Mixture ConvNet

1 Introduction

Prostate cancer is the most diagnosed form of cancer (excluding non-melanoma skin cancers) in North American men, accounting for an estimated 21,600 new cases and 4,000 deaths in Canada [4] and an estimated 180,890 new cases and 26,120 deaths in the U.S. [2] in 2016. However, prognosis is relatively good if prostate cancer is detected early, making fast and reliable prostate cancer screening methods crucial. Magnetic resonance imaging (MRI) has recently grown in

© Springer International Publishing AG 2017
F. Karray et al. (Eds.): ICIAR 2017, LNCS 10317, pp. 45–53, 2017.
DOI: 10.1007/978-3-319-59876-5_6

popularity as a non-invasive imaging-based prostate cancer screening method; however, a diagnosis through MRI requires an experienced medical professional to extensively review the data.

Computer-aided prostate cancer classification or *radiomics*-driven methods for prostate cancer classification have been developed to help streamline the diagnostic process and to increase diagnostic consistency. *Radiomics* [13] refers to the high-throughput extraction and analysis of large amounts of quantitative features from medical imaging data to characterize tumour phenotypes, allowing for a high-dimensional mineable feature space that can be utilized for cancer classification and prognosis, and has previously been applied to lung, and head-and-neck cancer patients, indicating the potential of radiomic features for personalized medicine and predicting patient outcomes [1,9].

A number of radiomic-driven methods for prostate cancer classification have been proposed in literature [3,15–18,22]. Recently, Khalvati *et al.* [10] proposed a multi-parametric MRI texture feature model for radiomics-driven prostate cancer analysis. The texture feature model, based on the one proposed by Peng *et al.* [19], comprises of 19 low-level texture features extracted from each MRI modality, including features extracted from the gray-level co-occurrence matrix. Khalvati *et al.* [11] extended the previous texture feature model to radiomics-driven models. An attempt at designing comprehensive quantitative feature sequences, the radiomics-driven models include additional MRI modalities, additional low-level features, and feature selection.

While current radiomic-driven methods for prostate cancer classification have been shown to be highly effective, the reliance on hand-crafted quantitative features can limit their ability to fully characterize unique prostate cancer tumour phenotype. Motivated by this, the concept of *discovery radiomics* [5,6,12,21,23] was introduced, where the authors forgo the notion of predefined feature models by discovering customized, tailored radiomic feature models directly from the wealth of medical imaging data already available. One approach to discovery radiomics used a convolutional neural network (ConvNet) sequencer for automatic feature discovery, coupled with a classifier for predicting whether a tumour candidate tissue is cancerous [5]. The authors demonstrated the potential of ConvNets for discovering abstract imaging-based features that characterize the highly unique tumour phenotype beyond what can be captured using predefined feature models; however, the unbalanced nature of the image dataset limits the capacity of ConvNets from accurately diagnosing complex prostate images. This is a common issue in computer-aided clinical decision support systems where datasets often contain more healthy tissues than cancer tissues [20], biasing the discovered sequencer and reducing its effectiveness.

In this study, we set out to build a discovery radiomics framework that can potentially overcome the challenge of unbalanced datasets and biased radiomics discovery. Motivated to explore a mixture of deep ConvNet sequencers (mConvNet) by leveraging boosting [8], our approach individually trains a number of ConvNet sequencer-classifier pairs (SC) to uncover and classify unique and abstract imaging-based tumour bio-markers. In an attempt to overcome dataset bias, the SCs are trained sequentially, and successive SCs are trained via data weighted by the misclassified candidates of previous SCs. Finally, the weighted vote (based on feature expertise) of the individual SCs is used as the prediction of the mixture for a given test tumour candidate.

2   Methods

The proposed discovery radiomics framework for prostate cancer classification is comprised of a mixture of sequencer-classifier pairs (SC) that are trained on the available prostate cancer multi-parametric MRI (mpMRI) data, and collectively classify a new tumour candidate (see Fig. 1). In this section, we motivate this learning strategy and describe each component in detail.

2.1   Patient Data Collection and Data Pre-processing

A requirement of the proposed discovery radiomics framework is the presence of standardized mpMRI data from past patients. In this work, mpMRI data of 20 patients was acquired using a Philips Achieva 3.0T machine at Sunnybrook Health Sciences Centre, Toronto, Ontario, Canada. Institutional research ethics

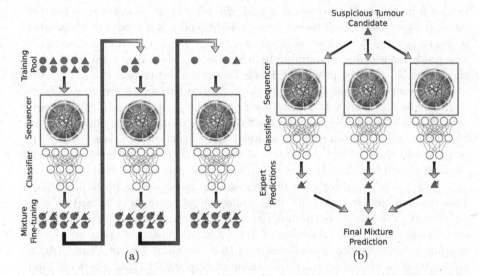

Fig. 1. General setup for training of mixture model (a) and classifying a test sample using a trained mixture of deep ConvNet sequencers (b). In part (a), the original training samples are passed into a sequencer-classifier pair (SC) to be learnt. Based on the performance of earlier SCs, future SCs may be trained on different samples of initial training pool, while adjusting sample weights to guide later SCs to focus on hard-to-classify samples. Meanwhile a weight is assigned to every SC based on its performance on the initial training pool, and these weights are used in part (b) when voting on the classification of a test sample.

board approval and written informed consent was waived by the Research Ethics
Board of Sunnybrook Health Sciences Centre. The patients' ages ranged from
53 to 83. A total of 80 cancerous regions and 714 healthy regions were identified
as tumour candidates from the patients and used in this work. As the number
of available labelled data was limited, data augmentation was performed where
each tumor candidate was rotated 8 times at 45° angles and flipped on the
vertical axis, resulting in a total of 1280 cancerous regions and 1280 healthy
regions[1]. In this work, the mpMRI dataset consists of the following modalities:
apparent diffusion coefficient (ADC) maps, computed high-b diffusion weighted
imaging (CHB-DWI), and correlated diffusion imaging (CDI).

2.2   Base Sequencer and Classifier

Prior to the introduction of our proposed base sequencer and classifier, we first
formally introduce the concept of supervised classification and define *sequencer*.
In machine learning, *classification* aims to learn a system capable of predicting a
class label for an input that the system has not encountered. Generally, a classification system or *classifier*, takes as input a set of instances $x_1, x_2, \ldots, x_N \in \mathcal{X}$
each characterized by a number of features $f_{i1}, f_{i2}, \ldots, f_{iM} \in \mathcal{F} \ \forall \ i \in [1, N]$
which have numerical or nominal values. The relation between an input and
its set of features, $\mathcal{X} \rightarrow \mathcal{F}$, can either be extracted manually via *hand-
crafted feature extractors*, or learned using *discovered feature sequencers*. Furthermore, a *supervised* classification system additionally has access to class labels
$y_1, y_2, \ldots, y_N \in \mathcal{Y}$ for the set of inputs $x_1, x_2, \ldots, x_N \in \mathcal{X}$, and uses these labels
as a guide for learning a mapping between the input's features and the output
class label, $\mathcal{F} \rightarrow \mathcal{Y}$, for predicting labels for unseen instances.

One prominent class of sequencers or feature extractors are convolutional
neural networks (ConvNets), which are a particular realization of artificial neural
networks (ANNs). ANNs are comprised of a series of layers of artificial neurons,
non-linear activation functions, and weighted connections between each layer's
neurons and those of the previous and next layers. Equipped with a powerful
learning algorithm, namely *gradient descent*, the network's classification error
$|y_{i,prediction} - y_{i,actual}|$ for each training sample x_i is incrementally propogated
back through the network while fine-tuning model parameters toward an overall network performance optimium. ConvNets additionally exploit spatially local
correlations in the input image by enforcing a local connectivity pattern between
neurons of neighbouring layers, ensuring that the filters (learnt through back-
propagation as above) produce the strongest responses to spatially local input
patterns [14]. Because of these architectural characteristics, ConvNets allow for
both the automatic discovery of deep sequencers (i.e., $\mathcal{X} \rightarrow \mathcal{F}$) and a classifier
(i.e., $\mathcal{F} \rightarrow \mathcal{Y}$) to be trained together for an end-to-end sequencing and classification (SC) system (i.e., $\mathcal{X} \rightarrow \mathcal{Y}$).

---

[1] A total of 11,424 (i.e., $714 \times 8 \times 2$) augmented healthy samples were generated, of
which 1280 candidates were drawn at random to balance the training set.

2.3   Sequentially Trained Mixture Model

The general intuition behind a mixture model is that a number of base SCs, where each SC discovers certain features and is an expert on a certain part of the input feature space, are collectively more accurate than any individual SC. Using a strategy known as AdaBoost [7], a number of SCs are trained sequentially, where each is trained on a slightly different dataset from that of its predecessors. More specifically, successive SCs are trained via data weighted by the misclassified candidates of previous SCs. An interesting side-effect of mixture models of this form is that each SC becomes an expert in a particular subspace of the input feature space and discovers unique features for the input image [8]. This diversity allows for the collective vote of the mixture to perform better across a wider variety of unseen inputs.

The setup above is achieved through assigning a weight $w_{i,t} \in \mathbb{R}$ to each training sample x_i prior to training SC_t, the tth SC in a mixture of T SCs. As illustrated in Algorithm 1, sample weights are adjusted in accordance with performance of each SC_t so that later SCs focus to a greater extent (through either weighted cost functions or weighted sampling) on those samples that have higher residual weight.

Algorithm 1. Sample Weights Updater

Inputs:

- SC_t, the tth sequencer-classifier pair in the mixture
- sequence of N examples $(x_1, y_{1,a}, w_{1,t}), ..., (x_N, y_{N,a}, w_{N,t})$ with *actual* labels $y_{i,a} \in \{0,1\}$ and weights $w_{i,t} \in \mathbb{R}$ prior to training the tth SC

forall $x_i \in [1, N]$ **do**
  let $y_{i,p} = SC_t(x_i)$ % the predicted label of example x_i
  if $y_{i,p} == y_{i,a}$ **then**
    $w_{i,t+1} = w_{i,t} \cdot \frac{\epsilon_t}{1-\epsilon_t}$ % according to (1)
  else
    $w_{i,t+1} = w_{i,t}$

2.4   Weighted Vote on Test Data

In addition to using training sample weights as a guide for future SCs to focus on hard-to-classify tumour candidates, training sample weights are used to assign a voting weight $\alpha_t \in \mathbb{R}$ for each SC. This weight follows from the intuition that the prediction of better performing SCs should be more influential relative to their worse performing counterparts. Thus, model weights α_t are calculated as follows:

$$\alpha_t = \ln\left(\frac{1 - \epsilon_t}{\epsilon_t}\right) \quad \text{where,} \quad \epsilon_t = \sum_{i, y_i \neq SC_t(x_i)} w_{i,t} \tag{1}$$

where ϵ_t is the pseudoloss for SC_t. It is worth noting that a pseudoloss greater than 0.5 suggests that the SC should be outright rejected as it was not able to sequence and learn at least half of the samples (weighted by their weights).

Finally, given a mixture of trained SCs for prostate cancer classification, each with its own weight, the mixture prediction for a tumour candidate tissue x^* is decided as follows:

$$\text{mixture prediction} = \sum_{t=1}^{T} \alpha_t \cdot SC_t(x^*) \tag{2}$$

3   Results

3.1   Experimental Setup

To assess the usefulness of the proposed mixture of deep ConvNet sequencers (mConvNet) for prostate cancer classification, mConvNet was compared against Peng et al.'s [19] and Khalvati et al.'s [11] hand-crafted radiomic sequencers, and Chung et al.'s [5] discovery radiomic sequencers, using a feedforward neural network classifier with a single hidden layer of 100 nodes. While [11,19] evaluated the hand-crafted radiomic sequencers for voxel-based classification via a linear discriminant analysis (LDA) classifier and support vector machine (SVM), respectively, only the hand-crafted radiomic sequencers themselves were included in this study to assess the use of hand-crafted and discovered radiomic sequencers. The performance of the classifier learned based on each radiomic sequencer was evaluated using leave-one-patient-out cross-validation using the collected data set, and quantitatively assessed via sensitivity, specificity, and accuracy metrics.

3.2   Results and Comparisons

As depicted in Table 1, Khalvati et al. [11] performed better than Peng et al. [19] in terms of all performance metrics. While [11] has a high specificity at 92.31%, it is worth noting that the hand-crafted radiomic sequencer also has a relatively low sensitivity (i.e., proportion of correctly identified cancerous tumour candidates)

Table 1. Comparison of hand-crafted radiomic sequencers [11,19], earlier discovery radiomic sequencers with a single sequencer [5], and mixture of ConvNet sequencers (mConvNet) for tumour candidate classification.

	Sensitivity	Specificity	Accuracy
Peng et al. [19]	0.1824	0.9005	0.5821
Khalvati et al. [11]	0.3568	0.9231	0.6730
Chung et al. [5]	0.6400	0.8203	0.7365
mConvNet	**0.6385**	**0.8947**	**0.8497**

at 35.68%, missing approximately two thirds of the cancerous candidates. This indicates that the hand-crafted radiomic sequencer produce radiomic sequences that better represent tumour candidates consisting of healthy tissue than tumour candidates consisting of cancerous tissue.

Table 1 also shows that the proposed mConvNet framework maintained state-of-the-art sensitivity at 63.85% (compared to 64.00% in Chung *et al.* [5]), while improving on specificity at 89.47%, and accuracy at 84.97%. In addition, mConvNet has noticeably more consistent performance across the metrics relative to both hand-crafted radiomic features [11,19]. This suggests that the custom radiomic sequences generated using discovery radiomics via mConvNet are better able to represent both healthy and cancerous prostate tissue in a more balanced fashion, as opposed to favouring healthy tissue.

4   Conclusion

In this paper, we proposed a new method for discovery radiomics through the use of a mixture of deep ConvNet sequencers. The use of discovery radiomics for prostate cancer and lung cancer classification was first demonstrated in [5,6,12,21,23] where the proposed systems discovered more accurate abstract imaging-based tumour features compared with state-of-the-art hand-crafted feature extractors [11,19] and earlier discovery radiomic sequencers [5]. In this work, we attempt to address the shortcoming of earlier works when discovering radiomic sequences from limited unbalanced data. The results we present demonstrate that by sequentially training and consulting a mixture of deep ConvNet sequencers, we are able to both discover highly unique image features for prostate cancer classification and deliver comparatively high classification performance in terms of sensitivity, specificity, and accuracy. The results further demonstrate the potential of discovery radiomics for computer-aided clinical decision support systems used in personalized medicine and predicting patient outcomes.

References

1. Aerts, H.J., Velazquez, E.R., Leijenaar, R.T., Parmar, C., Grossmann, P., Cavalho, S., Bussink, J., Monshouwer, R., Haibe-Kains, B., Rietveld, D., et al.: Decoding tumour phenotype by noninvasive imaging using a quantitative radiomics approach. Nat. Commun. **5** (2014)
2. American Cancer Society: Cancer Facts and Figures 2016 (2016)
3. Artan, Y., Haider, M., Langer, D.L., van der Kwast, T.H., Evans, A.J., Yang, Y., Wernick, M.N., Trachtenberg, J., Yetik, I.S., et al.: Prostate cancer localization with multispectral MRI using cost-sensitive support vector machines and conditional random fields. IEEE Trans. Image Process. **19**(9), 2444–2455 (2010)
4. Canadian Cancer Society: Canadian Cancer Statistics (2016)
5. Chung, A.G., Shafiee, M.J., Kumar, D., Khalvati, F., Haider, M.A., Wong, A.: Discovery radiomics for multi-parametric MRI prostate cancer detection. arXiv preprint arXiv:1509.00111 (2014)

6. Chung, A.G., Shafiee, M.J., Kumar, D., Khalvati, F., Haider, M.A., Wong, A.: Discovery radiomics via layered random projection (LaRP) sequencers for prostate cancer classification. In: Annual Meeting of the Imaging Network of Ontario (2016)
7. Freund, Y., Schapire, R.E.: A decision-theoretic generalization of on-line learning and an application to boosting. J. Comput. Syst. Sci. **55**, 119–139 (1997)
8. Galar, M., Fernández, A., Barrenechea, E., Bustince, H., Herrera, F.: A review on ensembles for class imbalance problem: bagging, boosting and hybrid based approaches. IEEE Trans. Syst. Man Cybern. **42**(4), 463–484 (2012)
9. Gevaert, O., Xu, J., Hoang, C.D., Leung, A.N., Xu, Y., Quon, A., Rubin, D.L., Napel, S., Plevritis, S.K.: Non-small cell lung cancer: identifying prognostic imaging biomarkers by leveraging public gene expression microarray datamethods and preliminary results. Radiology **264**(2), 387–396 (2012)
10. Khalvati, F., Modhafar, A., Cameron, A., Wong, A., Haider, M.A.: A multiparametric diffusion magnetic resonance imaging texture feature model for prostate cancer analysis. In: O'Donnell, L., et al. (eds.) Computational Diffusion MRI, pp. 79–88. Springer, Heidelberg (2014)
11. Khalvati, F., Wong, A., Haider, M.A.: Automated prostate cancer detection via comprehensive multi-parametric magnetic resonance imaging texture feature models. BMC Med. Imaging **15**(1), 27 (2015)
12. Kumar, D., Shafiee, M.J., Chung, A.G., Khalvati, F., Haider, M.A., Wong, A.: Discovery radiomics for computed tomography cancer detection. arXiv preprint arXiv:1509.00117 (2014)
13. Lambin, P., Rios-Velazquez, E., Leijenaar, R., Carvalho, S., van Stiphout, R.G., Granton, P., Zegers, C.M., Gillies, R., Boellard, R., Dekker, A., et al.: Radiomics: extracting more information from medical images using advanced feature analysis. Eur. J. Cancer **48**(4), 441–446 (2012)
14. LeCun, Y., Bottou, L., Bengio, Y., Haffner, P.: Gradient-based learning applied to document recognition. Proc. IEEE **86**(11), 2278–2324 (1998)
15. Liu, X., Langer, D.L., Haider, M., Yang, Y., Wernick, M.N., Yetik, İ.Ş., et al.: Prostate cancer segmentation with simultaneous estimation of Markov random field parameters and class. IEEE Trans. Med. Imaging **28**(6), 906–915 (2009)
16. Madabhushi, A., Feldman, M.D., Metaxas, D.N., Tomaszeweski, J., Chute, D.: Automated detection of prostatic adenocarcinoma from high-resolution ex vivo MRI. IEEE Trans. Med. Imaging **24**(12), 1611–1625 (2005)
17. Ozer, S., Haider, M.A., Langer, D.L., van der Kwast, T.H., Evans, A.J., Wernick, M.N., Trachtenberg, J., Yetik, I.S.: Prostate cancer localization with multispectral MRI based on relevance vector machines. In: IEEE International Symposium on Biomedical Imaging: From Nano to Macro, ISBI 2009, pp. 73–76. IEEE (2009)
18. Ozer, S., Langer, D.L., Liu, X., Haider, M.A., van der Kwast, T.H., Evans, A.J., Yang, Y., Wernick, M.N., Yetik, I.S.: Supervised and unsupervised methods for prostate cancer segmentation with multispectral MRI. Med. Phys. **37**(4), 1873–1883 (2010)
19. Peng, Y., Jiang, Y., Antic, T., Giger, M.L., Eggener, S., Oto, A.: A study of t2-weighted MR image texture features and diffusion-weighted MR image features for computer-aided diagnosis of prostate cancer. In: SPIE Medical Imaging, p. 86701H. International Society for Optics and Photonics (2013)
20. Rahman, M.M.L., Davis, D.N.: Addressing the class imbalance problems in medical datasets. Int. J. Mach. Learn. Comput. **3**(2), 224–228 (2013)
21. Shafiee, M.J., Chung, A.G., Kumar, D., Khalvati, F., Haider, M.A., Wong, A.: Discovery radiomics via stochasticnet sequencers for cancer detection. In: NIPS Workshop on Machine Learning for Healthcare (2015)

22. Vos, P., Barentsz, J., Karssemeijer, N., Huisman, H.: Automatic computer-aided detection of prostate cancer based on multiparametric magnetic resonance image analysis. Phys. Med. Biol. **57**(6), 1527 (2012)
23. Wong, A., Chung, A.G., Kumar, D., Shafiee, M.J., Khalvati, F., Haider, M.A.: Discovery radiomics for imaging-driven quantitative personalized cancer decision support. Vis. Lett. **1**(1) (2015)

Discovery Radiomics for Pathologically-Proven Computed Tomography Lung Cancer Prediction

Devinder Kumar[1]([✉]), Audrey G. Chung[1], Mohammad J. Shaifee[1],
Farzad Khalvati[2,3], Masoom A. Haider[2,3], and Alexander Wong[1]

[1] Department of Systems Design Engineering, University of Waterloo,
Waterloo, ON, Canada
{devinder.kumar,agchung,mjshafie,alexander.wong}@uwaterloo.ca
[2] Department of Medical Imaging, University of Toronto, Toronto, ON, Canada
farzad.khalvati@sri.utoronto.ca
[3] Sunnybrook Health Sciences Centre, Toronto, ON M4N 3M5, Canada
masoom.Haider@sunnybrook.ca

Abstract. Lung cancer is the leading cause for cancer related deaths. As such, there is an urgent need for a streamlined process that can allow radiologists to provide diagnosis with greater efficiency and accuracy. A powerful tool to do this is radiomics: a high-dimension imaging feature set. In this study, we take the idea of radiomics one step further by introducing the concept of *discovery radiomics* for lung cancer prediction using CT imaging data. In this study, we realize these custom radiomic sequencers as deep convolutional sequencers using a deep convolutional neural network learning architecture. To illustrate the prognostic power and effectiveness of the radiomic sequences produced by the discovered sequencer, we perform cancer prediction between malignant and benign lesions from 97 patients using the pathologically-proven diagnostic data from the LIDC-IDRI dataset. Using the clinically provided pathologically-proven data as ground truth, the proposed framework provided an average accuracy of 77.52% via 10-fold cross-validation with a sensitivity of 79.06% and specificity of 76.11%, surpassing the state-of-the art method.

Keywords: Radiomics · Discovery radiomics · Deep convolutional neural network · Lung nodules

1 Introduction

According to the American Cancer Society [1], lung cancer is the second most diagnosed form of cancer in the United States, second only to prostate cancer in males and breast cancer in females. Furthermore, lung cancer remains the leading cause of cancer-related deaths in the United States, accounting for approximately 27% of all cancer-related deaths. A similar report by the Canadian Cancer Society [2] estimates the number of new cases of lung cancer to be the highest of all cancers in Canada, and it is also the leading cause of cancer-related deaths in

© Springer International Publishing AG 2017
F. Karray et al. (Eds.): ICIAR 2017, LNCS 10317, pp. 54–62, 2017.
DOI: 10.1007/978-3-319-59876-5_7

Canada, accounting for approximately 26% of all cancer-related deaths. Early screening and diagnosis of lung cancer at a more treatable stage of the disease can play a pivotal role in improving the survival rates for such patients.

One of the biggest emerging areas in recent years related to quantitative cancer screening and diagnosis is radiomics [3,4], which involves the high-throughput extraction and analysis of a large number of imaging-based features for quantitative characterization and analysis of tumour phenotype. The use of radiomics-driven approaches allow for a more objective and quantitative evaluation and diagnosis of cancer, which can significantly reduce inter-observer and intra-observer variability and improve diagnostic accuracy and efficiency compared to current qualitative cancer assessment strategies. In a comprehensive study by Aerts et al. [5] involving more than 1000 patients across seven datasets, it was shown that radiomics can be used to obtain phenotype differences between tumours that can have clinical significance and prognostic value. The idea behind the success of radiomics as revealed by radio-genomics is the radiomic sequences being generated were able to capture quantitative features that define intra-tumour heterogeneity, which is directly related to the underlying gene-expression patterns [5]. Other radiomics-driven approaches have been investigated for lung cancer prediction [6–8]. For example, Anirudh et al. [6] used automated weakly labeled data with 3D CNNs to classify lung nodules for SPIE-LUNGx challenge. In [9], Orozco et al., used a radiomic sequence consisting of wavelet features and illustrated its effectiveness on a smaller sub-set of CT images from the LIDC-IDRI dataset. Shen et al. [7] introduced a convolutional radiomic sequencer, based on multi-crop windows around the lung nodules effectively and later extends the framework [8] to rate the suspiciousness of nodules. One important thing to note here is that, all of the above mentioned literature explored the radiologist driven nodule ratings for predicting the malignant of nodules.

In this study, we use the concept of *discovery radiomics* for lung cancer prediction, where we discover custom radiomic sequencers tailored for lung cancer characterization and prediction using pathologically-proven diagnostic dataset. To realize the concept of discovery radiomics for lung cancer prediction in this study, we introduce a deep convolutional radiomic sequencer that is discovered using a deep convolutional neural network architecture based on CT imaging data and pathologically-proven diagnostic data from past patients in a medical imaging archive. This methodology of using a more challenging pathologically-proven diagnostic data as ground truth distinguishes this study from the above mentioned literature that are based only on radiologist provide malignancy scores. In recent past, only two studies were proposed [10,11], which also explores the pathologically-proven dataset from LIDC-IDRI dataset for predicting lung nodule malignancy. In [11], the authors proposed an autoencoder based unsupervised approach for predicting malignancy of the lung nodules. More recently, Shen et. al [10] proposed a method using CNN incorporated with multiple instance learning for classifying the pathologically-proven diagnostic data. In [10], the authors performed knowledge (weight) transfer from CNN trained on large non-diagnostic data to a different CNN for the task for diagnostic malignancy prediction. The authors compared different methodologies and prove that their method is the current state-of-the art in this domain. In this study, we surpass the state-of-the art results stated in [10] which proves the effectiveness for the proposed discovery radiomics approach.

The main contributions of this study can be summarized as follows: (i) An effective malignancy prediction framework can be developed with limited pathologically proven diagnostic data without any external knowledge or large radiologist based annotated dataset. (ii) We provide experimental evidence that the proposed discovery radiomics framework is more effective in predicting pathologically-proven malignancy than the state-of-the art method [10].

2   Methodology

This section presents the methodology used in this study in detail. The various steps involved in the proposed discovery radiomics framework for lung cancer prediction is shown in the Fig. 1. First, using the wealth of CT imaging data, radiologist annotations, and pathologically-proven diagnostic results for past patients available in a medical imaging archive, a custom radiomic sequencer is discovered for generating radiomic sequences composed of abstract, quantitative imaging-based features that characterize tumour phenotype. Once we obtain the discovered radiomic sequencer, we can apply this sequencer to any new patient case to extract radiomic sequences tailored for lung cancer characterization and prediction.

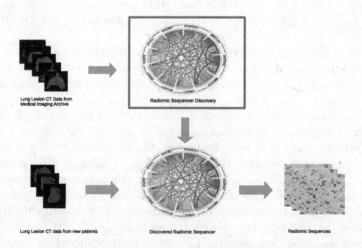

Fig. 1. Overview of the proposed discovery radiomics framework for CT lung cancer prediction. A custom radiomic sequencer is discovered directly using the wealth of lung CT data, hand annotations by the radiologist, and diagnostic results available from the imaging archive. The discovered radiomic sequencer can then be used to generate custom radiomic sequences for new patients using their lung CT data.

2.1 Patient Data Collection

For learning the custom radiomic sequencer in the proposed discovery radiomics framework, a set of lung CT images, radiologist annotations, and pathologically-proven diagnostic results from past patients is needed. In this study, we use a subset of LIDC-IDRI [12] dataset. Seven different academic and eight medical imaging companies collaborated to build the LIDC-IDRI database, which consists of data for 1,010 patients, each of which contains clinical helical thoracic CT scans and associated data of two-phase annotation process performed by four experienced thoracic radiologists. For annotating the nodules in the lung CT scans, a two phase annotation process was finalized for obtaining the interpretation from up to four different radiologists for a single CT scan. The radiologists marked each identified lesion as either a nodule of size ≥ 3 mm, <3 mm or non-nodule ≥ 3 mm. The database contains a total of 7,371 lesions marked as nodule by at least one radiologists with 2,699 of them marked ≥ 3 mm. We selected a subset of the LIDC-IDRI of 97 patient cases, for which definite pathologically-proven diagnostic results were available. The diagnostic results were obtained at two levels: (i) patient level and (ii) nodule level. At each level, the lesions in lung were marked as either: 0 - Unknown (no data),1 - Benign, 2 - Primary malignancy or 3 - Metastatic lesion with extra-thoracic primary malignancy.

These ratings were obtained by using either biopsy, surgical resection, progression or review of radiological images to show two years of stable nodule size. In this study, we take the lesions marked ≥ 3 mm for only the cases with ratings of 1, 2 or 3 and classifying the ratings 2 and 3 as malignant and rating 1 as benign. This resulted in 69 patient cases for malignant and 28 patient cases for benign lesions, with a total of 608 benign lesions and 3,698 malignant lesions. It is important to note here that lesions ≥ 3 mm were considered to facilitate for a fair comparison with other state-of-the art methods [10,11].

2.2 Data Augmentation

For building an effective radiomics sequencer that can well represent and characterize different types of the lung lesions and have a prognosis efficacy similar to the pathologically-proven diagnosis level results, there is a need of larger datasets for the radiomic sequencer discovery process. Therefore, to further enrich our dataset, we take each lesion candidate associated with pathologically-proven diagnostic dataset and perform spatial deformation using rotations. Using this, we ended with a total of 42,340 lesion candidates to be used for radiomic sequencer discovery. More specifically, each lesion for the malignant cases is rotated by $45°$ in total eight different variations for the particular lesion to get 29,956 malignant lesions. As the initial number of cases marked with benign were very less and it is ideal to have a balanced dataset for the radiomic sequencer discovery process, we rotated each benign lesion by $10°$ from 0 to $360°$ resulting in 22,384 benign lesions for creating near equal lesion candidate ratio for both the categories.

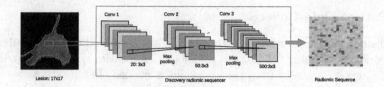

Fig. 2. Architecture of the deep convolutional radiomic sequencer consists of 3 convolutional sequencing layers and 2 max pooling sequencing layers, which outputs a radiomic sequence with 500 abstract imaging-based features.

Table 1. Summary of number of receptive fields and receptive field size at each convolutional sequencing layer.

Sequencing layer	Number of receptive fields	Receptive field size
1	20	3×3
2	50	3×3
3	500	3×3

2.3   Radiomic Sequencer

Given the CT imaging data, radiologist annotations, and pathologically-proven diagnostic results for past patients, the next step is to discover a custom radiomic sequencer for generating radiomic sequences tailored for lung cancer prediction. To realize the concept of discovery radiomics for lung cancer prediction in this study, we take inspiration from deep network architectures, which has been shown to be effective for medical image analysis [7,13] and introduce a deep convolutional radiomic sequencer that is discovered using a deep convolutional neural network architecture. The architecture of the deep convolutional radiomic sequencer is shown in Fig. 2. The proposed radiomic sequencer has 3 convolutional sequencing layers; the number of receptive fields and receptive field size used in each convolutional sequencing layer is specified in Table 1.

In between each pair of convolutional sequencing layers there is a max-pooling sequencing layer for improving translational tolerance. The final output of the deep convolutional radiomic sequencer is a radiomic sequence with 500 abstract imaging-based features.

Radiomic Sequencer Discovery. To discover the aforementioned deep convolutional radiomic sequencer, we construct a deep convolutional neural network architecture for the radiomic sequencer discovery process, where the radiomic sequencer is directly embedded in the sequencer discovery architecture and learned based on the available CT imaging data, radiologist annotations, and diagnostic results from past patients. In the radiomic sequencer discovery architecture, a fully-connected layer, a rectified linear unit layer, and a loss layer are augmented at the end of the radiomic sequencer to be discovered for the learning process. The receptive fields in the convolutional sequencing layers of the radiomic sequencer along with the fully-connected layer in the sequencer discovery architecture are learned in a supervised manner based on the input CT imaging data. This process allows us to learn specialized receptive fields in the custom radiomic sequencer that better characterize the unique tumour traits captured in the CT imaging data in an abstract fashion beyond that can be captured using pre-defined, hand-engineered features in current radiomics-driven approaches. A softmax loss function is used in the loss layer of the sequencer discovery process, and stochastic gradient descent (SGD) is used for optimization in learning the radiomic sequencer. Furthemore, a rectified linear unit (ReLU) layer is used to introduce non-saturating nonlinearity into the sequencer. We initialize the learning process with a learning rate of 0.001 and complete the learning process in 60 epochs using a batch size of 100. The momentum is fixed to 0.9 with weight decay parameters set to 0.0005 through-out the learning process.

Radiomic Sequence Generation. After the custom deep convolutional radiomic sequencer has been discovered by the radiomic sequencer discovery process, it can then be used for generating radiomic sequences. As shown in Fig. 2, for a new patient case, we take the CT imaging data pertaining to the patient and feed it into radiomic sequencer to obtain a final custom radiomic sequence with 500 abstract imaging-based features.

3 Experimental Results

This section provides details about the experimental setup and obtained results that are used to evaluate the efficacy of the proposed discovery radiomics framework for the purpose of classifying lung lesions as either malignant or benign based on CT imaging data and pathologically-proven diagnostic data. The first subsection explains the experimental setup including the performance metrics, and the later subsections present the results obtained using the experimental setup for the task of lesion classification.

To create the input data for discovering the radiomics sequencer in the proposed discovery radiomics framework, we first extract the lung CT scans from the LIDC-IDRI dataset [12] for particular patients which have diagnostic data associated with their cases. Based on the diagnostic result, we were able to get lung CT scan images for 97 patients and divided them into two groups: (i) malignant, and (ii) benign. From each individual scan, we extracted the lesion based on the provided radiologist annotations of the lesion in the particular CT scan image by up to four radiologists for each image which was later verified by pathological data. To mitigate the inter-observability differences, we include all of the annotations provided by each radiologist. As stated earlier, after the data augmentation process, the enriched dataset contains 42,340 lung lesions composed of 29,956 of malignant lesions and 22,384 benign lesions. To be consistent with the past approaches in this domain, for evaluation purposes, we divided the dataset into two parts: 90% of the dataset is used for discovering a custom radiomic sequencer, and 10% of the dataset is used for testing classification performance using the discovered radiomic sequencer. We further divide the 90% into two parts: 80% for training and 10% as a validation set for validating

Table 2. Comparison of Patient-Level Classification Results Obtained using Discovered Radiomics Sequencers (DRS), CNN Multiple Instance Learning (CNN-MIL) [10] and Deep Autoencoding Radiomic Sequencers (DARS) [11]. The best results are highlighted in bold.

	DRS	CNN-MIL [10]	DARS [11]
Sensitivity (%)	79.06	–	**83.14**
Specificity (%)	**76.11**	–	20.18
Accuracy (%)	**77.52**	70.69	75.01

the training process. A binary decision tree classifier is used for evaluating the efficacy of utilizing the discovered radiomic sequencers for classifying between benign and malignant tumours. Performance is quantitatively determined via sensitivity, specificity and accuracy metrics.

3.1 Results and Analysis

The detailed results for the experiment are presented in Table 2. Through the 10 fold cross validation using the same setup we obtain an average accuracy of 77.52% with 79.06% sensitivity and 76.11% specificity using the discovered radiomic sequencer. Furthermore, we also present a comparison of the results obtained using the discovered radiomic sequencer (DRS) along with the results obtained using Deep Autoencoding Radiomic Sequencers (DARS) [11] and the current state-of-the art CNN Multiple Instance Learning approach (CNN-MIL) [10] (see Table 2). It is important to note here that while other past works in the area of lung nodule classification have reported higher accuracy for a similar experimental setup and dataset [7,8,14], these methods are based on ratings provided by radiologists as ground truth for malignancy, which can affect the reliability of the experimental results as it introduces additional inter-observer variability from the radiologists. To have a just comparison with the presented approach, we perform comparisons with CNN-MIL and DARS as these methods are based on pathologically-proven lung cancer prediction, which is considered to be the gold standard for ground truth. This results in a much more difficult but more reliable evaluation configuration similar to [10,11]. From Table 2, it is evident that proposed discovery radiomic sequencers framework attains much better overall accuracy compared to CNN-MIL [10] and outperforms the DARS [11] method on both accuracy and sensitivity while producing similar results for sensitivity. These results show that the proposed radiomics discovery framework has strong potential for improving diagnostic accuracy for lung cancer prediction.

4   Conclusion

In this study, we presented a discovery radiomics framework designed for pathologically proven lung cancer prediction using CT imaging data and diagnostic data. Since the custom radiomic sequencers are dependent on the data on which they are learned, the discovered radiomic sequencers produce highly tailor-made radiomic sequences for the tumour type, in this case lung nodules. Experimental results show that the discovered radiomic sequencers, when used for pathologically proven lung nodule classification, outperform state-of-the art approach when evaluated using the LIDC-IDRI diagnostic dataset. As such, the presented discovery radiomics framework can be a low cost, fast and repeatable way of producing quantitative characterizations of tumour phenotype that has the potential to speed up the screening and diagnosis process while improving consistency and accuracy.

In terms of future work, an area that is worth investigating is on the use of discovery radiomics for risk stratification. Furthermore, the discovery radiomics framework has the potential to be of high clinical impact in the field of tumour grading and staging analysis.

Acknowledgment. This research has been supported by the OICR, Canada Research Chairs programs, NSERC, and the Ministry of R&I, ON.

References

1. Cancer facts and figures 2015, pp. 1–55 (2015)
2. Canadian cancer statistics 2015, Committee on Cancer Statistics, pp. 1–151 (2015)
3. Lambin, P., Rios-Velazquez, E., Leijenaar, R., Carvalho, S., van Stiphout, R.G., Granton, P., Zegers, C.M., Gillies, R., Boellard, R., Dekker, A., et al.: Radiomics: extracting more information from medical images using advanced feature analysis. Eur. J. Cancer **48**(4), 441–446 (2012)
4. Balagurunathan, Y., Antic, S., Chen, H., Schabath, M., Gu, Y., Wang, H., Walker, R., Gillies, R., Massion, P., Atwater, T.: Radiomic analysis for improved lung cancer prediction of indeterminate pulmonary nodules. Am. J. Respir. Crit. Care Med. **191**, A6119 (2015)
5. Aerts, H.J., Velazquez, E.R., Leijenaar, R.T., Parmar, C., Grossmann, P., Carvalho, S., Bussink, J., Monshouwer, R., Haibe-Kains, B., Rietveld, D., et al.: Decoding tumour phenotype by noninvasive imaging using a quantitative radiomics approach. Nat. Commun. **5** (2014). Article no. 4006
6. Anirudh, R., Thiagarajan, J.J., Bremer, T., Kim, H.: Lung nodule detection using 3D convolutional neural networks trained on weakly labeled data. In: SPIE Medical Imaging. International Society for Optics and Photonics, p. 978532 (2016)
7. Shen, W., Zhou, M., Yang, F., Yang, C., Tian, J.: Multi-scale convolutional neural networks for lung nodule classification, pp. 588–599 (2015)
8. Shen, W., Zhou, M., Yang, F., Yu, D., Dong, D., Yang, C., Zang, Y., Tian, J.: Multi-crop convolutional neural networks for lung nodule malignancy suspiciousness classification. Pattern Recogn. **61**, 663–673 (2017)
9. Orozco, H.M., Villegas, O.O.V., Sánchez, V.G.C., Domínguez, H.D.J.O., Alfaro, M.D.J.N.: Automated system for lung nodules classification based on wavelet feature descriptor and support vector machine. Biomed. Eng. Online **14**(1), 9 (2015)
10. Shen, W., Zhou, M., Yang, F., Dong, D., Yang, C., Zang, Y., Tian, J.: Learning from experts: developing transferable deep features for patient-level lung cancer prediction. In: Ourselin, S., Joskowicz, L., Sabuncu, M.R., Unal, G., Wells, W. (eds.) MICCAI 2016. LNCS, vol. 9901, pp. 124–131. Springer, Cham (2016). doi:10.1007/978-3-319-46723-8_15

11. Kumar, D., Wong, A., Clausi, D.A.: Lung nodule classification using deep features in CT images. In: 2015 12th Conference on Computer and Robot Vision (CRV), pp. 133–138 (2015)
12. Armato, S.G., McLennan, G., Bidaut, L., McNitt-Gray, M.F., Meyer, C.R., Reeves, A.P., Zhao, B., Aberle, D.R., Henschke, C.I., Hoffman, E.A., et al.: The lung image database consortium (LIDC) and image database resource initiative (IDRI): a completed reference database of lung nodules on CT scans. Med. Phys. **38**(2), 915–931 (2011)
13. Roth, H., Lu, L., Liu, J., Yao, J., Seff, A., Kevin, C., Kim, L., Summers, R.M.: Improving computer-aided detection using convolutional neural networks and random view aggregation. arXiv preprint arXiv:1505.03046 (2015)
14. Kuruvilla, J., Gunavathi, K.: Lung cancer classification using neural networks for CT images. Comput. Methods Programs Biomed. **113**(1), 202–209 (2014)

Left Ventricle Wall Detection
from Ultrasound Images Using Shape
and Appearance Information

Gerardo Tibamoso$^{(\boxtimes)}$, Sylvie Ratté, and Luc Duong

École de technologie supérieure, Montreal, Canada
gerardo.tibamoso-pedraza.1@ens.etsmtl.ca

Abstract. Clinical analysis of heart conditions take into account parameters such as thickness, perimeter, and area of the left ventricle wall. These measurements are normally obtained from manual segmentation in ultrasound images, which depends on operator experience. Supporting this process through automatic segmentation methods is very challenging due to low resolution, missing information, noise, and blurring on these images. In this work, we propose a novel semi-automatic method of left ventricle detection in ultrasound images based on supervised learning. The method combines appearance and shape ventricle information implicitly through ring partitions, following the ventricle shape pattern in axial views. The results show the convenience of the method to deal with noise and missing information.

Keywords: Ultrasound images · Left ventricle · Segmentation · Supervised learning · Feature extraction

1 Introduction

Manual measurement of wall thickness of ventricular myocardium through ultrasound images is a common practice for relevant parameters estimation in clinical protocols [1]. However, the complexity of acquisition and interpretation of these images makes the measurements dependent on operator skills, who has to deal with low resolution, noise, blurring (due to cardiorespiratory motion), and incomplete anatomical information. Indeed, the image intensities not only depend on density differences between adjacent biological tissues, but also on the incidence angle of the ultrasound waves [2]. In addition, the extraction of this information from large sequences is a tedious process.

Automatic and semiautomatic segmentation methods in ultrasound images have been proposed in recent literature to support estimations of ventricular wall thickness [3]. Deformable models are particularly appealing for segmentation in medical images due to their adaptability to the variability of anatomical shapes. Belaid et al. [4] proposed a left ventricle segmentation method in ultrasound images based on Level Sets, which is robust against noise and diffuse boundaries, but it is computationally demanding, hindering its real-time performance.

© Springer International Publishing AG 2017
F. Karray et al. (Eds.): ICIAR 2017, LNCS 10317, pp. 63–70, 2017.
DOI: 10.1007/978-3-319-59876-5_8

On the other hand, Machine Learning techniques have shown to be appropriate for ultrasound image segmentation given their capacity to learn from examples. Lempitsky et al. [5] proposed a 3D miocardium segmentation based on Random Forest, to classify every voxel according to the appearance of its neighbors and its spatial coordinates. This method is computationally efficient but the voxel position dependence could affects its performance in cases where the heart has an unusual orientation or when the probe position has changed. In both cases, the validation has been limited to a small group of data, possibly due to the difficulty of gathering plenty of manual segmentations from experts.

In this work we propose a novel ultrasound left ventricle (LV) detection method, which combines appearance and shape information implicitly based on supervised learning. The information is extracted from ring shape partitions in images, following the ventricle shape pattern in axial views. A Quadratic Bayes classifier separates rings, whose initial parameters from training data are complemented with information from a manual wall ventricle segmentation in the first frame in each ultrasound sequence of the test data set. Finally, the greater region of adjacent rings classified as part of the ventricle in each frame was preserved, while the rest were removed.

The results show that the combination of intensities and shape information of the LV wall complement each other well in order to overcome noise and missing information. This document is organized as follow: Sect. 2 presents the database, features extraction and selection, and the classifier design. Results and discussion are presented in Sect. 3. Finally, the conclusion is presented in Sect. 4.

2   Materials and Methods

The proposed method regards the LV wall detection as a classification problem. Hence, based on a training data set, the method classifies rings into two classes: "ventricle wall" (1 or white) and "background" (0 or black), generating binary responses as it is represented in Fig. 1. These rings are generated by means of a regular partition in every image, where the features are extracted. Subsequently, the key features are selected for the classification process. Finally, rings classified as "ventricle wall" are filtered for every frame, such that the ones that form the larger region are preserved while the rests are discarded.

2.1   Database

The database was built from axial slices of LV extracted from 3D Ultrasound sequences of a cardiac cycle (CC). Slice positions were the same for all frames in the sequence, where the LV wall had the shape of a closed ring. The reference segmentations (binary images) were generated with the intersection between every slice with reference surfaces that enclose the LV wall in every stage of the CC.

Ultrasound and segmentation sequences used in this project were obtained from an open access database built for the Motion Tracking Challenge, held at

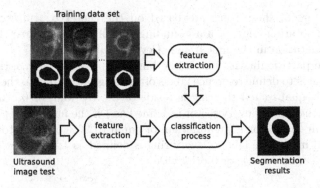

Training data set

feature extraction

classification process

Ultrasound image test

feature extraction

Segmentation results

Fig. 1. General description of the proposed classification method for left ventricle segmentation in ultrasound images.

MICCAI workshops of 2011 [6]. For this database, 3D ultrasounds of left ventricle were acquired in full-volume mode, during breath-hold, and from apical view. Every sequence captures a CC (between 10 and 15 frames) of fifteen healthy volunteers. Every frame is composed by $208 \times 224 \times 208$ voxels, where each voxel represents $0.82 \times 0.84 \times 0.73 \, \text{mm}^3$. The reference surfaces were defined through the LV segmentation in MRI and registered in the ultrasound images of the same patient.

2.2 Feature Extraction and Selection

In ultrasound images, the LV wall is represented by different intensity ranges with different local maximums, as can be seen in some histograms shown in Fig. 2. However, based on preliminary observations, it can be assumed that if this region is divided into small zones, the range of intensities in each one is close to a Gaussian distribution, whose parameters (mean μ and standard deviation σ) can be used as representative features of the wall. In our experimentation we chose μ as the only representative feature of every sector.

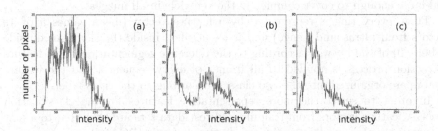

Fig. 2. Histograms of LV wall in ultrasound images from different patients.

In this way, the feature extraction is done through a regular partition per frame, defined as a sequence of concentric rings divided into sectors (see Fig. 3).

The parameters of these rings are based on the center c, *width* and *height* relation, and rotation angle θ of a bounding box of minimum area that delimit the LV segmentation in the first frame of every sequence.

The same partitioning is done in the binary images (see Fig. 3, right hand), where the goal is to define reference labels of the rings according to their position, namely label equal to 1 if the ring is inside of the LV wall, or 0 otherwise. To achieve this, the label of every sector is defined as 1 if the number of pixels inside the LV wall in the sector is greater than or equal to 25% of all sector pixels, or 0 otherwise. Finally, the label of every ring is defined as 1 if at least 75% of the labels of its sectors were 1, or 0 otherwise.

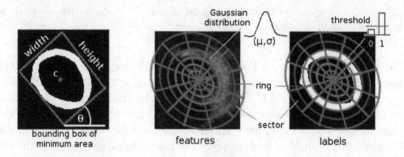

Fig. 3. Representation of the feature extraction and labeling process for each ultrasound image and reference segmentation of LV wall.

The area of every ring is set constant to facilitate the direct comparisons between features. The curves that form every ring have elliptical shape. The shortest radius of the external curve of every ring i is defined as $r_i = \sqrt{i} * r_{min}$, where r_{min} is a constant value experimentally defined, while the largest one is defined as $R_i = p * r_i$ where $p = width/height$ if $width/height \geq 1$ or $p = height/width$ in other case. The rings center for all the frames in the sequence is the same and equal to c. The total number of rings per frame is defined constant, and large enough to cover completely the ventricle in all images.

Then, every ring is represented by a feature vector (mean values μ of its sectors from ultrasound image) and it is classified inside (label $= 1$) or outside (label $= 0$) of the LV wall according to the reference segmentations. The feature extraction process is applied to all frames of every sequence, and the feature vectors are organized into the two classes, according to the rings labels.

In order to select the more representative features for the classification process, feature vectors from training and test data sets are projected to a new feature space through Principal Components Analysis (PCA). The eigenvectors and eigenvalues required to this projection are estimated from the training data. After this projection, the new features are sorted in descending order according to their variance values (eigenvalues). Greater variance, greater probability of separation between classes.

2.3 Classification Process

As a first approximation to the classification problem, it is assumed that features in each class follow a Gaussian distribution. Hence, a Quadratic classifier is designed due to its simplicity and flexibility of parameter definitions. Then the results give an insight into data distribution.

A manual LV segmentation on the first frame for every sequence of the test data set is required to define the ring parameters and extract the features. These features are used to define a test mean vector $\mu_{j,te}$ and a test covariance matrix $\Sigma_{j,te}$ for every class j to bias the classification process toward the specific sequence. Therefore, a mean vector (μ_j) and a covariance matrix (Σ_j) that define the Gaussian distribution for every j class are estimated from training and test data set to feed the discriminant function of Eq. 1. This function assigns every ring ($\mathbf{x}$) to class w_i, if $g_i > g_j$, for all $j \neq i$, where

$$g_j(\mathbf{x}) = log(p(w_j)) - \frac{1}{2}log(|\Sigma_j|) - \frac{1}{2}(\mathbf{x} - \mu_j)\Sigma_j^{-1}(\mathbf{x} - \mu_j)^T, \tag{1}$$

and $p(w_j)$ is the a priori probability of each class from training data set. The classification parameters (μ_j, Σ_j) are modulated with $\mu_{j,te}$ and $\Sigma_{j,te}$, such that

$$\begin{aligned} \mu_j &= \alpha * \mu_{j,tr} + (1 - \alpha) * \mu_{j,te} \\ \Sigma_j &= \alpha * \Sigma_{j,tr} + (1 - \alpha) * \Sigma_{j,te} \end{aligned} \tag{2}$$

where $\mu_{j,tr}$ and $\Sigma_{j,tr}$ are respectively the mean vector and covariance matrix of every class j from the training data, and α is constant that controls the weights of every component.

3 Results and Discussion

For feature extraction, the axial slice number 104 (in the middle of the volume) was obtained from all frames in all sequences, due to it is the image in the middle of the ventricle. From these, we selected the sequences with closed ring ventricles. Totally, eleven sequences were chosen and divided randomly such that eight were devoted for training ($\approx 75\%$, sequences V6–V9, V11–V13, and V16) and three for test ($\approx 25\%$, sequences V2, V4, and V5).

3.1 Results of Feature Extraction and Selection

Twenty-one rings per frame were obtained. The number of sectors per ring was 15, and $r_{min} = 15$ pixels. These experimental parameters were defined to ensure that the ventricle wall was split into several rings and a small area per sector but at the same time no so small to avoid noise deviations. In total, the training data set was shaped with 2604 rings $\times$ 15 features, where 2235 $\times$ 15 were of class 0, and 369 $\times$ 15 of class 1. Hence, $p(w_0) \approx 0.86$ and $p(w_1) \approx 0.14$. Likewise, the test data set was shaped with 882 rings $\times$ 15 features, where 764 $\times$ 15 were of class 0, and 118 $\times$ 15 of class 1.

Measurements error of LV wall and internal cavity areas between the ring representation and the reference segmentations, in each frame for test data set, are presented in Fig. 4. From these results, it is noted that the error distribution of every sequence is highly concentrated around the median values, and its extreme values did not exceed the 25%. The higher deviations can be explained by the irregular shape of the ventricle.

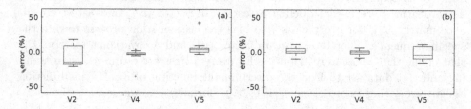

Fig. 4. Differences between reference segmentations and rings representation of test data set, frame by frame in every sequence. (a) error of segmentation area, (b) error left ventricle internal area.

3.2   Results of the Classification Process

Results of the classifier were evaluated by decreasing the number of features (removing the ones with minor variance), and trying with α equal to 0.0, 0.5, and 1.0. The classifier showed to be more effective with 13 features, to preserve the 99% of the total variance, and $\alpha = 0.5$, which reduces the number of misclassified rings. Confusion matrices of the classification results for test data set are shown in Table 1. The global results in the test dataset showed that the class 0 is highly recognized, with a sensitivity of 93%, while the identification of class 1 is harder, with a sensitivity of 69% (specificity class 0), being the accuracy 90%.

Table 1. Confusion matrices for V2, V4, and V5 sequences (test data set), with 13 features and $\alpha = 0.5$.

V2				V4				V5		
	True class				True class				True class	
	0	1			0	1			0	1
Predicted 0	278	19		Predicted 0	191	13		Predicted 0	241	5
class 1	19	20		class 1	8	19		class 1	27	42

In order to validate our method, we measured the similarity between the results of the segmentations and the reference segmentations using Dice coefficients. Additionally, we applied an edge-based level set technique for the ventricle segmentation [7]. The similarity results of both our method and the level set technique are presented in Fig. 5. From the results, our method is more robust

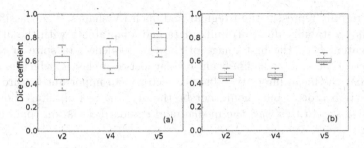

Fig. 5. Results of the comparison between segmentations: (a) reference and our method, and (b) reference and level sets, for the test dataset using the Dice coefficient.

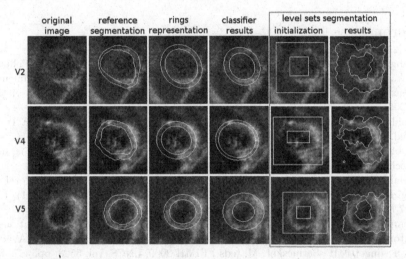

Fig. 6. Images from test data set (V2, V4, and V5). Columns from left to right: original ultrasound image, reference segmentation, rings representation, classification/segmentation results ($\alpha = 0.5$ and 13 features), and level sets segmentation (initialization and results).

than the level set technique, which was initialized with internal and external curves for expansion and shrinkage, to extract the ventricle wall. However, the irregular LV shape and its intensity variability hinders the rings classification as can be seen in Fig. 6.

4   Conclusion

We have proposed a semi-automatic method of LV detection in axial views of 3D ultrasound, through supervised learning. We considered that a ring partitioning is a promising hypothesis allowing to become the segmentation problem into a classification problem. This partition simplified the process of feature representation considering the anatomical ventricle shape, but rings more flexible are required to fully represent the irregular and open LV shapes. The classification results shown stability due to training data and adaptability with the inclusion of information from the first image of the test sequence. In spite of that, it was noticed that there is a significant overlap between classes, which makes us believe that the mean intensity value per sector is an important feature in the discrimination process but should not be the one due to the diffuse boundary between classes. In this way, the median and standard deviation could also be considered.

Acknowledgements. The authors thank Atefeh Abdolmanafi, Rémi Martin, Jigen Bouchtiba, Binh Tran, Ruben Dorado, and Cleiton Martins de Souza for helpful comments and improvement ideas. This research was funded by NSERC Discovery grant.

References

1. Lang, R.M., Bierig, M., Devereux, R.B., Flachskampf, F.A., Foster, E., Pellikka, P.A., Picard, M.H., Roman, M.J., Seward, J., Shanewise, J., et al.: Recommendations for chamber quantification. Eur. Heart J.-Cardiovasc. Imaging **7**(2), 79–108 (2006)
2. Noble, A.J., Boukerroui, D.: Ultrasound image segmentation: a survey. IEEE Trans. Med. Imaging **25**(8), 987–1010 (2006)
3. Noble, J.A.: Ultrasound image segmentation and tissue characterization. Proc. Inst. Mech. Eng. Part H: J. Eng. Med. **224**(2), 307–316 (2010)
4. Belaid, A., Boukerroui, D., Maingourd, Y., Lerallut, J.-F.: Phase-based level set segmentation of ultrasound images. IEEE Trans. Inf. Technol. Biomed. **15**(1), 138–147 (2011)
5. Lempitsky, V., Verhoek, M., Noble, J.A., Blake, A.: Random forest classification for automatic delineation of myocardium in real-time 3D echocardiography. In: Ayache, N., Delingette, H., Sermesant, M. (eds.) FIMH 2009. LNCS, vol. 5528, pp. 447–456. Springer, Heidelberg (2009). doi:10.1007/978-3-642-01932-6_48
6. Tobon-Gomez, C., De Craene, M., Mcleod, K., Tautz, L., Shi, W., Hennemuth, A., Prakosa, A., Wang, H., Carr-White, G., Kapetanakis, S., et al.: Benchmarking framework for myocardial tracking and deformation algorithms: an open access database. Med. Image Anal. **17**(6), 632–648 (2013)
7. Li, C., Xu, C., Gui, C., Fox, M.D.: Distance regularized level set evolution and its application to image segmentation. IEEE Trans. Image Process. **19**(12), 3243–3254 (2010)

Probabilistic Segmentation of Brain White Matter Lesions Using Texture-Based Classification

Mariana Bento[1,3]([✉]), Yan Sym[1], Richard Frayne[2,3,4], Roberto Lotufo[1], and Letícia Rittner[1]

[1] Faculty of Electrical and Computer Engineering,
University of Campinas, Campinas, SP, Brazil
marianapbento@gmail.com
[2] Radiology and Clinical Neuroscience, Hotchkiss Brain Institute,
University of Calgary, Calgary, AB, Canada
[3] Calgary Image Processing and Analysis Centre,
Foothills Medical Centre, Calgary, AB, Canada
[4] Seaman Family MR Research Centre,
Foothills Medical Centre, Calgary, AB, Canada

Abstract. Lesions in brain white matter can cause significant functional deficits, and are often associated with neurological disease. The quantitative analysis of these lesions is typically performed manually by physicians on magnetic resonance images and represents a non-trivial, time-consuming and subjective task. The proposed method automatically segments white matter lesions using a probabilistic texture-based classification approach. It requires no parameters to be set, assumes nothing about lesion location, shape or size, and demonstrates better results (Dice coefficient of 0.84) when compared with other, similar published methods.

Keywords: White matter lesion (WML) · Magnetic resonance (MR) imaging · Brain · Segmentation · Texture features

1 Introduction

Lesions in the brain white matter (white matter lesions, WMLs) are commonly observed on magnetic resonance (MR) images. WMLs are caused by a variety of pathologies including immune-mediated and cerebrovascular diseases [1] and are present in both symptomatic and asymptomatic patients [2].

The manual analysis of WMLs is a time-consuming task, and also susceptible to intra- and inter-operator variability. To overcome these problems, automated, computerized methods have been proposed to identify WMLs on MR images. These tools aid physicians in diagnosis and treatment monitoring by reducing the subjectivity of the procedure and by making it more robust and agile [3–5]. They usually combine techniques from different areas, such as texture analysis [6,7]

© Springer International Publishing AG 2017
F. Karray et al. (Eds.): ICIAR 2017, LNCS 10317, pp. 71–78, 2017.
DOI: 10.1007/978-3-319-59876-5_9

and classification [8,9], commonly use *a priori* information about WML shape, size and location, but often present operational issues, *e.g.*, requiring multiple imaging modalities and the setting/tuning of many parameters, limiting their wide-spread usability.

Contributions. Different from related works presented in the literature, the proposed automated WML segmentation method requires no *a priori* information and performs a texture-based classification of pixels within the brain white matter. Texture descriptors extracted in a pixel-by-pixel basis are based on both statistical and geometrical properties. The main goal was to compute a probability map of pixels being lesion. Automatic segmentation of WMLs can be performed using this probability map. Evaluation was conducted using a publicly available MR dataset.

Paper Organization. Section 2 describes the MR dataset. Section 3 details the proposed method. Section 4 presents our results and compares their performance against other methods. Section 5 summarizes our conclusions.

2 MR Dataset

Experiments were conducted using a public dataset from the 18th International Conference on Medical Image Computing and Computer Assisted Intervention held in 2015 (MICCAI-15). The data were assembled for the Medical Image Segmentation Challenge on Ischemic Stroke Lesion Segmentation.[1] The challenge had two different goals: automatic segmentation of ischemic stroke lesion and stroke outcome prediction [10]. In order to perform the segmentation task, a dataset, consisting of 28 patients diagnosed with stroke, was made available that contained T1-weighted, T2-weighted, diffusion-weighted (DWI, $b = 1000\,\text{s/mm}^2$) and fluid attenuated inversion recovery (FLAIR) images acquired on a 3 T MR scanner (Magnetom Trio; Siemens, Erlangen, Germany). Also included were expert segmentation of the stroke lesions (Fig. 1).

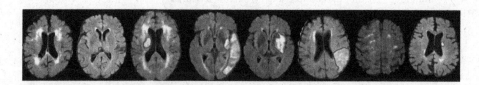

Fig. 1. Samples of the ischemic stroke lesion dataset from eight MICCAI-15 subjects: FLAIR images with ground truth WML outlined in red [10]. (Color figure online)

Prior to inclusion in the MICCAI-15 dataset, all MR images were identified, skull-stripped and resampled to an isotropic spacing of 1 mm^3 using a b-spline algorithm [11]. No other pre-processing, such as co-registration, was performed.

---

[1] www.isles-challenge.org/.

3   Method

Our proposed pixel-based classification method used the FLAIR images and had four main steps: (1) pre-processing, (2) attribute extraction, (3) classification and (4) post-processing (Fig. 2).

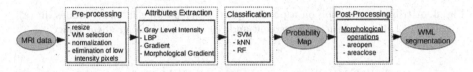

Fig. 2. Key steps of the proposed method: pre-processing, texture attribute extraction, probabilistic classification, and post-processing. The main outputs were a WML probability map and a WML segmentation mask.

Pre-processing. Automatically resized the FLAIR image to a fixed shape (256 × 256 per anatomical slice), segmented the white matter (WM) using a semi-automated region growing tool [12], and normalized image intensities to lie between 0 and 255. In order to reduce the computational effort, the mean intensity of pixels within the WM was calculated, and pixels with intensities lower than this mean value were discarded (as WML are hyperintensity in FLAIR images). This step requires no user intervention.

Attribute Extraction. Computed the local binary pattern (LBP), structural and morphological gradients for each image (Fig. 3). A total of seven attributes were extracted for each pixel: (1) intensity value, intensity value of the same pixel in the adjacent (2) inferior and (3) superior images, the intensity value in the (4) LBP, (5) structural gradient and (6) morphological gradient images and the (7) WM intensity mean value over the image.

Besides the texture attributes given by the LBP (representing the relationship between a pixel and its neighborhood), and gradient images, the local pixel intensity was a relevant attribute since WML appears as a bright region in FLAIR images. The intensity values of the pixels in the adjacent images were included because WMLs usually appear across multiple adjacent images. A similar spatial continuity constraint is often used in manual analysis.

Classification. Distinguishing normal appearing WM (NAWM) from WML pixels by using three classifiers (k nearest neighbor (kNN), random forest (RF) algorithms and support vector machine (SVM) methods).

The training phase was performed by using 43% of the available data (*i.e.*, 12 of the 28 subjects) and included images with WML of various sizes, shapes and location. A grid-search technique was used to select optimal parameters for SVM and kNN algorithms. The training data were unbalanced and had more NAWM than WML pixels. Several approaches were tested to balance the

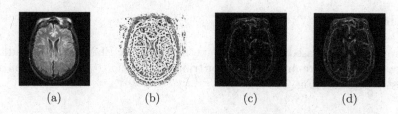

(a)   (b)   (c)   (d)

Fig. 3. Attribute extraction on a pixel-by-pixel basis showing intensity values in: (a) original, (b) LBP, (c) structural gradient and (d) morphological gradient images.

training samples [13]: (WM_1) randomly select NAWM pixels to match the number of WML pixels; (WM_2) randomly select NAWM pixels to match the number of WML pixels avoiding transitions between WML and NAWM; (WM_3) randomly select NAWM pixels to match the number of WML pixels avoiding transitions between WM and gray matter; or (WM_4) the intersection of approaches WM_2 and WM_3 (Table 1).

Table 1. Evaluated approaches for balancing the number of NAWM pixels in the training data. Selected NAWM pixels were randomly chosen from the identified search region to match the number of WML pixels.

Approach	NAWM search region
WM_1	Only avoiding lesioned pixels
WM_2	WM_1 and avoiding transitions between NAWM and WML
WM_3	WM_1 and avoiding transitions between white and gray matter
WM_4	Intersection of WM_2 and WM_3

In the testing phase, each pixel within the testing data (*i.e.*, the remaining 16 subjects within the dataset) was evaluated by the trained classifiers. Instead of outputting a class label (such as 0 = normal; 1 = lesion), a probabilistic implementation was used [14], providing a continuous class membership probability (*i.e.*, the probability of representing lesion). A WML probability map was generated for each slice (Fig. 4b).

Post-processing. WML segmentation was achieved by post-processing the WML probability map using a series of morphological operators (Fig. 4c). Connected components (CC) in the WML probability map first were identified. CC were determined using eight pixel connectivity, *i.e.* all pixels within a CC are connected with each other [15]. Post-processing then eliminated all CC with an area <3 pixels (area open operation) and closed residual gaps in CC (area close operation). A binary mask representing the results of the WML segmentation was obtained by then thresholding the CC.

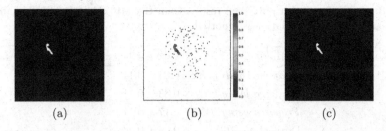

(a)   (b)   (c)

Fig. 4. WML probability map example: (a) WML ground truth for comparison purposes; (b) probability map visualization using a color scale; (c) post-processing result. (Color figure online)

4   Results and Discussion

The data were randomly subdivided into training and testing datasets. Using approach WM_4 to balance training samples (Table 1), the highest accuracy prior to post-processing (0.87) was achieved by SVM using a linear kernel with $C = 10$. The highest Dice coefficient after post-processing (0.84) was achieved by the RF classifier (Table 2).

Interestingly, the SVM results with the highest accuracy (SVM), achieved the lowest Dice coefficient likely due to lower sensitivity (Table 2). The low sensitivity indicates the number of WMLs that were not properly classified as lesions and thus were not properly segmented.

Table 2. Mean accuracy, sensitivity and specificity before post-processing and Dice coefficient after post-processing by classifier. These results used approach WM_4 to balance the number of training samples (see Table 1).

Classifier	Accuracy	Sensitivity	Specificity	Dice coefficient
RF	0.74	0.87	0.61	0.84
kNN	0.79	0.86	0.71	0.83
SVM	0.87	0.79	0.94	0.77

Our results were compared against published results obtained using the MICCAI-15 dataset. These papers presented different approaches to perform experiments and to subdivide data into training and testing sets.

All experimented classifiers (RF, kNN and SVM) presented higher Dice coefficient than the best Dice coefficient found in the challenge (0.69) [16] (Table 3).[2]

We also conducted an exploratory analysis of the Dice coefficient on a subject-by-subject basis (Fig. 5). The Dice coefficient was measured in each image making it possible to compute statistics across all images for each subject.

---

[2] http://www.virtualskeleton.ch/ISLES/Start2015.

Table 3. Comparison with published results from the MICCAI 2015 SISS challenge [10]. Mean ± standard deviation is reported.

Method	Dice coefficient
Proposed using RF	0.84 ± 0.11
Proposed using kNN	0.83 ± 0.11
Proposed using SVM	0.77 ± 0.17
Dutil *et al.* [16]	0.69 ± 0.30
Kamnitsas *et al.* [17]	0.66 ± 0.24
Chen *et al.* [18]	0.65 ± 0.33
Feng *et al.* [19]	0.63 ± 0.28
Halme *et al.* [20]	0.61 ± 0.24

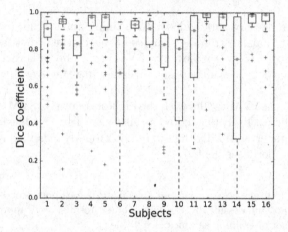

Fig. 5. Box plot of Dice coefficients by subject using the RF classifier. Red circles represents the median (or second quartile) value, the blue boxes extend from the first to third quartile, and the flier blue points are those past boxes to show the range of the data. Subjects were ordered according to decreasing number of WML pixels (Subject 1 had 26 WMLs with total volume of 18226 pixels, Subject 16 had 2 WMLs with total volume of 30 pixels). (Color figure online)

The median Dice coefficient varied from 0.60 to 0.95 across the subjects. Dice coefficient variation across each subject (as represented by box size, *i.e.*, the first-to-third quartile range [15]) was highly variable and independent of the total number of WML pixels.

The proposed method is limited to properly segment WML of >10 pixels, since it uses texture attributes that are based on changes across a local neighborhood. Most other methods have similar limitation which fundamentally is related to the WML minimum size that can be correctly segmented.

5   Conclusions

This paper presented a method to automatically segment WML using pixel-based probabilistic classification approaches. It does not use any *a priori* information about the WMLs, such as location, size and shape, and combined texture attributes with probabilistic classifiers to distinguish normal from lesioned pixels. This proposed WML segmentation method was compared with similar results from the literature and presented higher Dice coefficients. However, it is unable to correctly segment small WML (<10 pixels). The method may also be useful in multi-center datasets because pre-processing techniques are applied to standardize the images prior to attribute extraction and classification.

Acknowledgments. The authors thank FAPESP, CAPES and CNPQ for their financial support.

References

1. Appenzeller, S., Li, L.M., Faria, A.V., Costallat, L.T., Cendes, F.: Quantitative magnetic resonance imaging analyses and clinical significance of hyperintense white matter lesions in systemic lupus erythematosus patients. Ann. Neurol. **64**(6), 635–643 (2008)
2. Vernooij, M.W., Arfan Ikram, M., Tanghe, H.L., Vincent, A.J.P.E., Hofman, A., Krestin, G.P., Niessen, W.J., Breteler, M.M.B., Lugt, A.: Incidental findings on brain MRI in the general population. New Engl. J. Med. **357**(18), 1821–1828 (2007)
3. Despotovic, I., Goossens, B., Philips, W.: MRI segmentation of the human brain: challenges, methods, and applications. Comput. Math. Methods in Med. **2015**(1), 1–23 (2015)
4. Roura, E., Oliver, A., Cabezas, M., Valverde, S., Pareto, D., Vilanova, J., Ramió-Torrentà, L., Rovira, A., Lladó, X.: A toolbox for multiple sclerosis lesion segmentation. Neuroradiology **57**(10), 1031–1043 (2015)
5. Oppedal, K., Eftestol, T., Engan, K., Beyer, M., Aarsland, D.: Classifying dementia using local binary patterns from different regions in magnetic resonance images. Int. J. Biomed. Imaging **1–14**, 2015 (2015)
6. Loizou, C., Pantziaris, M., Pattichis, C., Seimenis, I.: Brain MR image normalization in texture analysis of multiple sclerosis. J. Biomed. Graph. Comput. **3**(1), 20 (2013)
7. Kloppel, S., Abdulkadir, A., Hadjidemetriou, S., Issleib, S., Frings, L., Thanh, T.N., Mader, I., Teipel, S.J., Hull, M., Ronnebeger, O.: A comparison of different automated methods for the detection of white matter lesions in MRI data. Neuroimage **57**(2), 416–422 (2011)
8. Steenwijk, M., Pouwels, P., Daams, M., Dalen, J., Caan, M., Richard, E., Barkhof, F., Vrenken, H.: Accurate white matter lesion segmentation by k nearest neighbor classification with tissue type priors (kNN-TTPs). NeuroImage: Clin. **3**, 462–469 (2013)
9. Anbeek, P., Vincken, K.L., Osch, M.J.P., Bisschops, R.H.C., Grond, J.: Probabilistic segmentation of white matter lesions in MR imaging. NeuroImage **21**(3), 1037–1044 (2004)

10. Maier, O., Menze, B.H., von der Gablentz, J., Häni, L., Heinrich, M.P., Liebrand, M., Winzeck, S., Basit, A., Bentley, P., Chen, L., Christiaens, D., Dutil, F., Egger, K., Feng, C., Glocker, B., Götz, M., Haeck, T., Halme, H.L., Havaei, M., Iftekharuddin, K.M., Jodoin, P.M., Kamnitsas, K., Kellner, E., Korvenoja, A., Larochelle, H., Ledig, C., Lee, J.H., Maes, F., Mahmood, Q., Maier-Hein, K.H., McKinley, R., Muschelli, J., Pal, C., Pei, L., Rangarajan, J.R., Reza, S.M., Robben, D., Rueckert, D., Salli, E., Suetens, P., Wang, C.W., Wilms, M., Kirschke, J.S., Krämer, U.M., Münte, T.F., Schramm, P., Wiest, R., Handels, H., Reyes, M.: ISLES 2015 - a public evaluation benchmark for ischemic stroke lesion segmentation from multi-spectral MRI. Med. Image Anal. **35**, 250–269 (2017)

11. Rueckert, D., Sonoda, L., Hayes, C., Hill, D., Leach, M., Hawkes, D.: Nonrigid registration using free-form deformations: application to breast MR images. IEEE Trans. Med. Imaging **18**(8), 712–721 (1999)

12. Lu, Q., Gobbi, D., Frayne. R., Salluzzi, M.: Cerebra-WML: a stand-alone application for quantification of white matter lesion. In: Proceedings of Imaging Network Ontario Symposium (2014)

13. Griffanti, L., Zamboni, G., Khan, A., Li, L., Bonifacio, G., Sundaresan, V., Schulz, U., Kuker, W., Battaglini, M., Rothwell, P., Jenkinson, M.: BIANCA (Brain Intensity Abnormality Classification Algorithm): a new tool for automated segmentation of white matter hyperintensities. NeuroImage **141**, 191–205 (2016)

14. Pedregosa, F., Varoquaux, G., Gramfort, A., Michel, V., Thirion, B., Grisel, O., Blondel, M., Prettenhofer, P., Weiss, R., Dubourg, V., Vanderplas, J., Passos, A., Cournapeau, D., Brucher, M., Perrot, M., Duchesnay, E.: Scikit-learn: machine learning in python. J. Mach. Learn. Res. **12**(1), 2825–2830 (2011)

15. Woods, R., Gonzalez, R.C.: Digital Image Processing. Edgard Blucher, São Paulo (2000)

16. Havaei, M., Dutil, F., Pal, C., Larochelle, H., Jodoin, P.-M.: A convolutional neural network approach to brain tumor segmentation. In: Crimi, A., Menze, B., Maier, O., Reyes, M., Handels, H. (eds.) BrainLes 2015. LNCS, vol. 9556, pp. 195–208. Springer, Cham (2016). doi:10.1007/978-3-319-30858-6_17

17. Kamnitsas, K., Chen, L., Ledig, C., Rueckert, D., Glocker, B.: Multi-scale 3D convolutional neural networks for lesion segmentation in brain MRI. In: Proceedings of Ischemic Stroke Lesion Segmentation Challenge, Held in Conjunction with International Conference on Medical Image Computing and Computer Assisted Intervention 2015 (2015)

18. Chen, L., Bentley, P., Rueckert, D.: A novel framework for sub-acute stroke lesion segmentation based on random forest. In: Proceedings of Ischemic Stroke Lesion Segmentation Challenge, Held in Conjunction with International Conference on Medical Image Computing and Computer Assisted Intervention 2015 (2015)

19. Feng, C., Zhao, D., Huang, M.: Segmentation of stroke lesions in multi-spectral MR images using bias correction embedded FCM and three phase level set. In: Proceedings of Ischemic Stroke Lesion Segmentation Challenge, Held in Conjunction with International Conference on Medical Image Computing and Computer Assisted Intervention 2015 (2015)

20. Halme, H.-L., Korvenoja, A., Salli, E.: ISLES (SISS) challenge 2015: segmentation of stroke lesions using spatial normalization, random forest classification and contextual clustering. In: Crimi, A., Menze, B., Maier, O., Reyes, M., Handels, H. (eds.) BrainLes 2015. LNCS, vol. 9556, pp. 211–221. Springer, Cham (2016). doi:10.1007/978-3-319-30858-6_18

A Machine Learning-Driven Approach to Computational Physiological Modeling of Skin Cancer

Daniel S. Cho[1], Farzad Khalvati[2(✉)], David A. Clausi[1],
and Alexander Wong[1]

[1] Systems Design Engineering, University of Waterloo, Waterloo, Canada
[2] Medical Imaging, Sunnybrook Research Institute, University of Toronto,
Toronto, Canada
farzad.khalvati@sri.utoronto.ca

Abstract. Melanoma is the most lethal form of skin cancer in the world. To improve the accuracy of diagnosis, quantitative imaging approaches have been investigated. While most quantitative methods focus on the surface of skin lesions via hand-crafted imaging features, in this work, we take a machine-learning approach where abstract quantitative imaging features are learned to model physiological traits. In doing so, we investigate skin cancer detection via computational modeling of two major physiological features of melanoma namely eumelanin and hemoglobin concentrations from dermal images. This was done via employing a non-linear random forest regression model to leverage the plethora of quantitative features from dermal images to build the model. The proposed method was validated by separability test applied to clinical images. The results showed that the proposed method outperforms state-of-the-art techniques on predicting the concentrations of the skin cancer physiological features in dermal images (i.e., eumelanin and hemoglobin).

Keywords: Melanoma detection · Dermal radiomics · Random forests

1 Introduction

Skin cancer is the most common form of cancer in the world [1]. It is generally categorized into two types: melanoma and non-melanoma where the latter is known to be malignant with poor prognosis [2]. Melanoma occurs when melanocytes produce excessive melanin uncontrollably [3] with the potential to aggressively move further into skin and to blood vessels causing metastasis. Therefore, early detection is the key to improve prognosis of melanoma [4].

A. Wong—This research was undertaken, in part, thanks to funding from the Canada Research Chairs program. The study was also funded by the Natural Sciences and Engineering Research Council of Canada (NSERC).

© Springer International Publishing AG 2017
F. Karray et al. (Eds.): ICIAR 2017, LNCS 10317, pp. 79–86, 2017.
DOI: 10.1007/978-3-319-59876-5_10

Initial diagnosis of melanoma is traditionally conducted by dermatologists or clinicians to visually assess any abnormal skin lesions [5]. Given a suspicious lesion, a dermatologist examines it to determine whether the lesion is melanoma, based on the ABCDE criteria [6], which stands for structural asymmetry (A), border irregularity (B), colour variegation (C), diameter greater than 6 mm (D), and changes over time, namely, evolution (E). The major limitation of this method is that the diagnosis of melanoma solely depends on the expertise of the doctor, and thus it may be subjective and may lead to a biased conclusion due to lack of quantitative measurements [7].

In recent years, radiomics has emerged to utilize medical images via advanced image analysis to extract a large amount of quantitative features for characterizing tumor phenotype [8]. The extracted features, which are typically more than a few hundreds, are processed to infer genomic and proteomic patterns, and eventually are employed for the improved diagnostic decisions. Currently, radiomics is actively investigated in lung, prostate, and head-and-neck cancer and the results have shown significant improvements on detection and prognosis of cancer [9–11].

Existing methods for radiomics-driven computer-aided melanoma diagnostic system have focused on quantifying ABCDE rubric, and mimic clinician's decision making process with the collected features (*i.e.,* classification). Celebi *et al.* extracted features from a skin lesion based on shape, color, and texture [12]. The limitation is that feature extractions were performed by purely mathematical formulation and thus, the features themselves do not provide any clinical meaning. As an alternative, a new set of features, high level intuitive features (HLIF), was proposed to provide clinicians with more intuitive meaning for each feature [13]. For example, one of HLIFs is an asymmetry score, and the score reflects the extent of the asymmetry of the lesion.

While the radiomics-driven melanoma detection techniques have successfully extracted and quantified skin features used in diagnosing melanoma, the chief limitation still lies in that only the surface of the skin lesion is utilized. Moreover, these cancer detection algorithms rely on hand-crafted imaging features, which limit the ability of such algorithms in capturing the true underlying physiological characteristics of skin cancer. Melanoma is generated from melanocytes, which are located in the epidermal layer of skin, and therefore, tracking the activity of melanocytes and their products could provide valuable diagnostic information. In this paper, we propose a novel machine learning-driven approach to learn and construct a computational model of the physiological features of melanoma from dermal images via employing a random forest regression model. The proposed computational model can be used to extract physiological skin biomarkers suitable for radiomics-driven computer-aided diagnosis tool for skin cancer. To accommodate the complex skin-light interaction of a skin lesion, the extraction model is designed in a non-linear fashion via random forest regression.

2   Methods

This section presents a modeling framework for physiological features of melanoma.

2.1   Physiological Biomarkers

While most skin feature quantification techniques are based on hand-crafted imaging features extracted from the surface of the lesion, our approach utilizes the underneath physiological biomarkers, which provides information not available to the naked eye. In this paper, we take a machine-learning approach where abstract imaging features are learned to model physiological traits. This is done by constructing a computational model of the physiological features of melanoma from dermal images via a non-linear random forest regression model. The key physiological biomarkers which could be used for melanoma diagnosis are melanin and hemoglobin. In a malignant lesion, the melanin levels are expected to be higher compared to lesion's surrounding. There are two main classes of melanin: eumelanin and pheomelanin. For concentrations of these melanins in melanomic lesions, eumelanin amounts tend to increase while the level of pheomelanin decreases [14]. Moreover, studies showed that the concentration of hemoglobin is also expected to increase [15].

In this work, we focus on modeling two major physiological features of melanoma namely eumelanin and hemoglobin. Here, we describe the problem formulation about extracting these physiological biomarkers, which can be used to generate radiomics sequences for diagnosis of melanoma.

2.2   Physiological Model

When light hits the surface of skin, it undergoes various interaction such as scattering, absorption, and reflection based on the optical properties of skin, and finally remitted. The remitted light can be decomposed into three visible bands: red, green and blue, and they are captured by a camera or dermoscopy [16]. The image acquisition process can be divided into two part, and the first step can be described as $r = s(p)$ where p is the concentration of physiological features such as eumelanin and λ is reflectance values generated through skin-light interaction.

The second step is followed to generate the standard camera images from the reflectance values as $i = t(r)$ where i represents the intensity in the three spectral bands, which is computed via non-linear conversion from reflectance values. In the matrix form, the whole process can be combined as $I = FP$, where $p = [p_1, p_2, \ldots p_n]^T$, $I = [(i_{11}, i_{12}, i_{13}), (i_{21}, i_{22}, i_{23}), \ldots, (i_{m1}, i_{m2}, i_{m3})]^T$, and F is the forward model that generates intensities in RGB color space from the different physiological features. Each component of the forward model, F, is discussed in the following sections in detail.

Generating Reflectance Values from Subcutaneous Concentrations of Physiological Features. To observe the relationship between reflectance values and concentrations of physiological features, the accurate measurements on the concentrations are necessary. However, it is clinically not feasible to obtain concentrations of the features at the pixel level. Therefore, biophysically-based spectral model of light interaction with human skin (*BioSpec*), was employed.

BioSpec model simulates the reflectance values from skin feature concentrations by adapting the light propagation in various skin layers [17] (i.e., stratum corneum, epidermis, papillary dermis, reticular dermis, and hypodermis). In order to simulate the light propagation, the *BioSpec* model takes into account surface reflectance, subsurface reflectance and transmittance. Based on the prior information of optical properties of physiological features (e.g., eumelanin), the model predicts the reflectance values of given combination of features.

Generating Standard Camera Images from Reflectance Values. While the RGB spectral bands make up standard camera image, the mapping from the reflectance values to RGB color space involves an intermediate step, which is the color tristimulus values, XYZ. XYZ color space was first introduced by the International Commission on Illumination (CIE) in 1931 to describe the color space mathematically. This color space was derived from the RGB model and expanded beyond the RGB color space. The XYZ can be calculated from reflectance values by the additive law of color matching [18].

$$X = N \sum_{\lambda} R(\lambda)S(\lambda)\overline{x}(\lambda)\Delta\lambda, \tag{1}$$

$$Y = N \sum_{\lambda} R(\lambda)S(\lambda)\overline{y}(\lambda)\Delta\lambda, \tag{2}$$

$$Z = N \sum_{\lambda} R(\lambda)S(\lambda)\overline{z}(\lambda)\Delta\lambda, \tag{3}$$

where $S(\lambda)$ is the relative spectral power distribution of the illuminant; $R(\lambda)$ is the reflectance functions, which were modeled from the *BioSpec* model; $\overline{x}(\lambda), \overline{y}(\lambda)$ and $\overline{z}(\lambda)$ are the spectral sensitivity function, and $\Delta\lambda$ is the wavelength interval. For our experiment, CIE standard Illuminant D65 was used for $S(\lambda)$, and the CIE 1931 sensitivity functions of the standard observer for 2° and $\Delta\lambda = 5\,\text{nm}$ were employed for the spectral sensitivity function, $\overline{x}(\lambda), \overline{y}(\lambda)$ and $\overline{z}(\lambda)$, and wavelength intervals, respectively. The constant N was defined as

$$N = \sum_{\lambda} S(\lambda)\overline{y}(\lambda)\Delta\lambda \tag{4}$$

Once tristimulus values XYZ had been obtained, they were converted to the RGB space.

$$R' = 3.2410X - 1.5374Y - 0.4986Z, \tag{5}$$

$$G' = -0.9692X + 1.8760Y + 0.0416, \tag{6}$$

$$B' = 0.0556X - 0.2040Y + 1.0570Z, \tag{7}$$

RGB' represents standard RGB color space and to obtain the RGB colors in the correct range, an additional non-linear transform is performed. If any values of R', G', B' are less than 0.0031308,

$$\overline{RGB} = 12.92RGB' \tag{8}$$

otherwise,
$$\overline{RGB} = 1.055RGB'^{(1/2.4)} - 0.055 \qquad (9)$$

The obtained $\overline{RGB}$ are normalized, and the correct 8-bit values are restored by multiplying $\overline{RGB}$ by 255.

2.3   Inverse Model

Based on the formulation above, the estimation of the concentrations of the physiological features is to find an inverse function:

$$P = F^{-1}(I) \qquad (10)$$

F which computes individual concentration of physiological features from the acquired skin lesion image, was constructed in non-linear fashion due to the complexity of skin-light interaction as well as color space conversion using random forest regression modeling [19]. Random forest regression model is an ensemble technique that uses many decision trees and aggregates them to elicit the final answer. The uniqueness of random forest regression is that it randomly chooses predictors at each node of the tree.

In this study, for each physiological feature, at least 1500 decision trees were generated to estimate concentrations. For the choice of predictors, we expand from the three spectral bands (red, green, and blue band for standard camera and dermoscopic images), and generated different colour spaces that was converted from RGB. Consequently, for each pixel, a total of 15 predictors (3 from RGB colour space, 3 from XYZ, 3 from L*a*b*, 3 from L*u*v*, and 3 from xyz colour space). An individual random forest model was constructed for each physiological features (i.e., eumelanin and hemoglobin) as shown in Fig. 1.

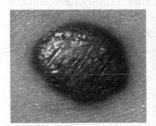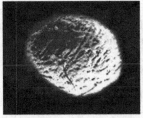

Fig. 1. (Left) melanomic lesion, different physiological features were extracted using random forest regression model: (Centre) eumelanin, and (Right) hemoglobin

3   Experimental Design and Results

The training data for the proposed model requires the different compositions of physiological features and their corresponding colour intensities in visual spectrum. As aforementioned, it is extremely difficult to collect those data in *in vivo*, thus *BioSpec* model was employed for data generation.

The parameters for *BioSpec* model have been set up as following: the concentrations of eumelanin and hemoglobin has been set from 20 to 300 g/L and 120 to 184 g/L with step of 4 g/L, respectively, while other parameters for the forward model were kept at default values.

For validation, we used separability test to distinguish the prediction of concentration of the two physiological features (eumelanin and hemoglobin) that belong to two classes of melanoma, non-melanoma, via non-linear random forest regression model (RF). For comparison, the same data was also trained by the Cavalcanti's nearest neighbour model (CNN), the only state-of-the-art technique for extracting physiological features in non-linear fashion [20].

The rationale behind separability test is that the physiological features are extracted to ultimately classify malignant lesions against benign ones. The separability test measures the strength of each feature for how well a given feature can discriminate between two classes (benign and malignant melanoma). Among different classification algorithm, Fisher's linear discriminant analysis (FDA) was chosen [21]. The mathematical formulation of FDA is formulated as follows:

$$J(w) = \frac{|m_1 - m_2|^2}{s_1^2 + s_2^2} \tag{11}$$

where m is mean, s is standard deviation, and the subscript represents a class. A dataset of 206 clinical images (119 melanoma, 87 non-melanoma) from DermIS [22] and DermQuest [23] was adapted for this study, and RF and CNN were employed to extract eumelanin and hemoglobin at each pixel of the segmented lesion in the dataset. The concentrations at each pixel were treated as a sample and FDA was performed for comparison.

To examine the proposed extraction technique in the clinical setting, each physiological features underwent the separability test. From a total of 206 clinical images (87 benign and 119 malignant), each physiological features were extracted using RF and CNN, and the Fisher separability measure was computed, as shown in Table 1. The visualization of the concentrations of each physiological feature is shown in Fig. 2.

Table 1. Fisher information of eumelanin and hemoglobin predicted by RF and CNN

Features	Eumelanin	Hemoglobin
RF	**0.083**	**0.0069**
CNN	0.059	0.0032

The separability test is designed to examine the accuracy and robustness of the proposed algorithm performed on clinical images. The corresponding concentration on every pixel of lesion was treated as a sample, and all of the samples were aggregated and underwent Fisher's linear discriminant for each biomarker. Fisher score shows the ability of each biomarker to differentiate benign and malignant lesion. As shown in Table 1, the biomarkers extracted from the

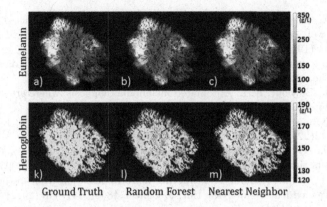

Ground Truth   Random Forest   Nearest Neighbor

Fig. 2. Color maps of the concentrations of physiological features predicted by random forest (RF) and Cavalcanti's nearest neighbor model (CNN) (Color figure online)

proposed method outperformed the ones from CNN, which is the only state-of-the-art technique. Although these results do not provide direct comparison on the feature extraction accuracy, they certainly infer the performance of each algorithm when dealing with actual clinical lesions, indicating the proposed method is preferable.

4   Conclusion

In this work, we proposed a novel method to extract physiological features of skin cancer, which can be used for automated detection of melanoma. In contrast to existing radiomics-driven melanoma detection algorithms that rely only on hand-crafted imaging features extracted from the skin surface, we construct a computational model of the physiological features of melanoma from dermal images via employing a non-linear random forest regression model to predict the concentration of melanin and hemoglobin. A separability analysis was conducted and the results showed that the proposed method outperformed over the state-of-the-art techniques. The proposed computational model can be used to extract physiological skin biomarkers suitable for radiomics-driven computer-aided diagnosis tool for skin cancer. We believe that adapting physiological features into computer-aided diagnostic system may provide additional information as well as improved accuracy, and ultimately help clinicians and dermatologists making better diagnosis for melanoma and other types of skin cancer.

References

1. Rogers, H.W., et al.: Incidence estimate of nonmelanoma skin cancer in the United States, 2006. Arch. Dermatol. **146**(3), 283 (2010)
2. Ascierto, P., et al.: The role of spectrophotometry in the diagnosis of melanoma. BMC Dermatol. **10**(1), 5 (2010)

3. Thody, A., Higgins, E., Wakamatsu, K., Ito, S., Burchill, S., Marks, J.: Pheomelanin as well as eumelanin is present in human epidermis. J. Invest. Dermatol. **97**(2), 344 (1991)

4. Psaty, E.L., et al.: Current and emerging technologies in melanoma diagnosis: the state of the art. Clin. Dermatol. **27**(1), 35–45 (2009)

5. Day, G.R., Barbour, R.H.: Automated melanoma diagnosis: where are we at? Skin Res. Technol. **6**(1), 1–5 (2000)

6. Abbasi, N.R., et al.: Early diagnosis of cutaneous melanoma. JAMA: J. Am. Med. Assoc. **292**(22), 2771–2776 (2004)

7. Argenziano, G., et al.: Dermoscopy of pigmented skin lesions: results of a consensus meeting via the internet. J. Am. Acad. Dermatol. **48**(5), 679–693 (2003)

8. Lambin, P., et al.: Radiomics: extracting more information from medical images using advanced feature analysis. Eur. J. Cancer **48**(4), 441–446 (2012)

9. Aerts, H.J., et al.: Decoding tumour phenotype by noninvasive imaging using a quantitative radiomics approach. Nat. Commun. **5** (2014). Article no. 4006. doi:10.1038/ncomms5006

10. Khalvati, F., Wong, A., Haider, M.A.: Automated prostate cancer detection via comprehensive multi-parametric magnetic resonance imaging texture feature models. BMC Med. Imaging **15**(1), 27 (2015)

11. Cameron, A., Khalvati, F., Haider, M., Wong, A.: MAPS: a quantitative radiomics approach for prostate cancer detection. IEEE Trans. Bio-med. Eng. **63**(6), 1145–1156 (2016)

12. Celebi, M.E., et al.: A methodological approach to the classification of dermoscopy images. Comput. Med. Imaging Graph. **31**(6), 362–373 (2007)

13. Amelard, R., Glaister, J., Wong, A., Clausi, D.A., et al.: High-level intuitive features (HLIFs) for intuitive skin lesion description. IEEE Trans. Biomed. Eng. **62**(3), 820–831 (2015)

14. Zonios, G., Dimou, A., Carrara, M., Marchesini, R.: In vivo optical properties of melanocytic skin lesions: common nevi, dysplastic nevi and malignant melanoma. Photochem. Photobiol. **86**(1), 236–240 (2010)

15. Garcia-Uribe, A., et al.: In-vivo characterization of optical properties of pigmented skin lesions including melanoma using oblique incidence diffuse reflectance spectrometry. J. Biomed. Optics **16**(2), 020501 (2011)

16. Menzies, S.W.: Automated epiluminescence microscopy: human vs machine in the diagnosis of melanoma. Arch. Dermatol. **135**(12), 1538 (1999)

17. Baranoski, G.V., Krishnaswamy, A.: Light and Skin Interactions: Simulations for Computer Graphics Applications. Morgan Kaufmann, Burlington (2010)

18. Lee, H.-C.: Color Imaging Science. Cambridge University Press, Cambridge (2005)

19. Breiman, L.: Random forests. Mach. Learn. **45**(1), 5–32 (2001)

20. Cavalcanti, P.G., Scharcanski, J., Baranoski, G.V.: A two-stage approach for discriminating melanocytic skin lesions using standard cameras. Expert Syst. Appl. **40**(10), 4054–4064 (2013)

21. Stigler, S.M.: Statistics on the Table: The History of Statistical Concepts and Methods. Harvard University Press, Cambridge (2002)

22. DermIS, November 2014. http://www.dermis.net

23. DermQuest, November 2014. http://www.dermquest.com

Ejection Fraction Estimation Using a Wide Convolutional Neural Network

AbdulWahab Kabani and Mahmoud R. El-Sakka[✉]

Department of Computer Science, The University of Western Ontario,
London, ON, Canada
{akabani5,melsakka}@uwo.ca

Abstract. We introduce a method that can be used to estimate the ejection fraction and volume of the left ventricle. The method relies on a deep and wide convolutional neural network to localize the left ventricle from MRI images. Then, the systole and diastole images can be determined based on the size of the localized left ventricle. Next, the network is used in order to segment the region of interest from the diastole and systole images. The end systolic and diastolic volumes are computed and then used in order to compute the ejection fraction. By using a localization network before segmentation, we are able to achieve results that are on par with the state-of-the-art and by annotating only 25 training subjects (5% of the available training subjects).

Keywords: Localization · Detection · Recognition · Artificial neural networks · Deep learning · Convolutional neural network · Image classification · Cardiac MRI · Left ventricle · Automatic ejection fraction estimation

1 Introduction

The ejection fraction is a measure of the outbound blood pumped from the heart after each heartbeat. The ejection fraction, and the systolic and diastolic volumes are important measures that can help cardiologists assess how healthy the heart is. Manually estimating the end-systolic volume, the end-diastolic volume, and the ejection fraction from cardiac MRI images is a time consuming process. A cardiologist or an expert has to manually segment all slices before the volume can be calculated. We propose a deep and wide convolutional neural network that can be used to localize the left ventricle from an MRI image. This network strikes a nice balance between performance and hardware requirement. Then, the systole and diastole images can be determined based on the size of the localized left ventricle. Next, the same network can be used to segment the cavity in the left ventricle. Using the DICOM meta fields, we can compute the volume size.

Left ventricle volume estimation has been an active area of research [4,9,14,15]. Recently CNNs were used in order to tackle this problem [7,13,27]. For instance, in [27] a U-Net [19] network was used in order to segment the left ventricle.

© Springer International Publishing AG 2017
F. Karray et al. (Eds.): ICIAR 2017, LNCS 10317, pp. 87–96, 2017.
DOI: 10.1007/978-3-319-59876-5_11

A Convolutional Neural Network (CNN) is a neural network that contains some layers with restricted connectivity. CNNs were introduced in [3] and achieved good results on the MNIST data set [12]. The success of CNNs on the MNIST [12] was reproduced on image classification problems. CNNs can now produce state of the art performance in many classification tasks and on challenging datasets such as [1,20]. The success of CNNs is due to many reasons including large training data sets [1,20], powerful hardware, regularization techniques such as Dropout [6,24], initialization methods [5], ReLU activations [17], and data augmentation. Since 2012, many networks that can perform image recognition were introduced [11,22,25]. Recently, CNNs have been used to perform detection and localization [2,19,21,23,26].

CNNs typically require large training data sets in order to be successful. However, cleaning and standardizing the data can help a lot in alleviating this requirement. We use a dataset of MRI images provided by the National Heart, Lung, and Blood Institute (NHLBI) [8]. The dataset [8] contains 500 training studies, 200 validation studies, and 440 testing studies. This dataset includes the MRI images along with the diastole and systole volumes. The dataset does not contain the left ventricle segmentation. Therefore, the left ventricle in the training data should be manually segmented and annotated to be able to train a CNN that can segment left ventricle. Manually annotating the training data is a time consuming process. Therefore, we first train a model to localize the left ventricle which is a much easier task. To perform this step, one image from each subject is annotated and used to train a localization network. Once trained, this localization network is used to localize the left ventricle in all the images. Once the left ventricle is localized, the task of training a model to segment the left ventricle becomes much easier and requires a very small amount of annotated training data. For the segmentation, we annotated only 25 training studies out of 500 training studies (5% of the number of studies). Finally, after segmenting the left ventricle, the systolic and diastolic volumes can be easily calculated. The results are reported on the whole testing set, which includes 440 subjects. The test set was never manually annotated in any way.

In this paper, we divide the task into several steps in order to ensure that the training process is successful. First, the left ventricle is localized (Sect. 2). Then, the systole and diastole images for each slice are determined. Next, the left ventricle segmentation process is described in Sect. 3. Finally, the volume of the left ventricle is estimated (Sect. 4). In Sect. 5, we present the results. We conclude our work in Sect. 6. An overview of our proposed solution is shown in Fig. 1.

2   Localization

The dataset contains (500 training subjects, 200 validation subjects, and 440 testing subjects). Each subject includes 8–16 short axis views (or slices). These slices provide a view of the heart at different levels. Each slice contains a set of (30 images or less) spanning one cardiac cycle. We are interested in only

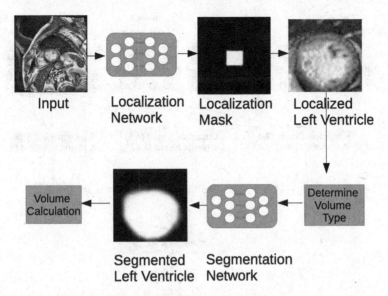

Fig. 1. Overview: a set of training images (1 per subject) are used to train the localization network. Then, the network is used to predict the localization mask. The localization mask is thresholded using Otsu [18]. Once the left ventricle is localized, we can determine the systole and diastole images. The left ventricle is segmented by training a network on only 25 training subjects. After thresholding, the area of the left ventricle corresponds to the sum of pixel values that are not 0. The volume can be calculated by multiplying each area by the slice height and summing up the volumes.

two images in each slice (the end-systolic and end-diastolic images). In each slice, these two images show us when the left ventricle is fully contracted and expanded, respectively. In order to determine these two phases, we first need to localize the left ventricle.

We trained a localization network on a set of randomly sampled images from the training set. We made sure that one image is sampled from each patient. The architecture of the localization network is described in Fig. 2. As shown in Fig. 2, the input image is re-sized to (128, 128). Then, it is passed to the network and it goes through a set of convolutional layers. Next, the last layers from each block -except for the first one- are upsampled and merged. Three types of merging is done: type_1 involves merging (blocks_2, blocks_3, blocks_4, blocks_5), type_2 involves merging (blocks_3, blocks_4, blocks_5), and finally type_3 merging (block_4, and block_5). It is worth noting that type_1 carries more high resolution information while type_3 carries highly abstract information but with less resolution (due to downsampling).

Each convolutional layer is followed by rectified linear units (ReLU) activation [17]. The output layer has a sigmoid activation to ensure that each pixel in the output mask ranges between 0 (black) to 1 (white). Maxpooling is used in order to extract more abstract features. We opted for upsampling by repeating

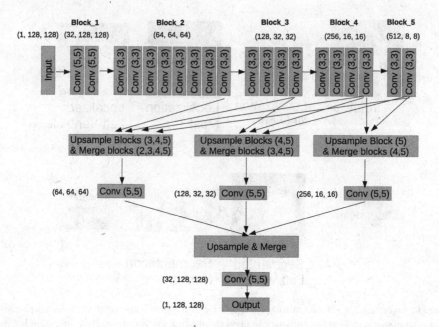

Fig. 2. Network architecture: this figure shows the architecture of one of the networks that were used for localization and segmentation of the left ventricle. The input goes through a set of convolutional layers. Then, the last layer from each block (except for block_1) are upsampled and merged in a hierarchical fashion in order to get the output mask in the original resolution.

the units rather than through parametrized upsampling in order to reduce the training time and GPU RAM consumption. This network strikes a nice balance between performance and hardware requirement. Indeed, this network was also tested on a laptop with around 3.5 GB of available GPU RAM.

The only input pre-processing that we did was subtracting the mean input of each channel and then dividing by the standard deviation. It is worth mentioning that in order for the left ventricle area to be determined correctly, we resized all images using the pixel spacing DICOM field. This ensures that each pixel represents 1 mm. However, when training the localization network, we resized the input image to (128, 128) because the network can only accept fixed sizes. Once the network is used to predict the mask, the mask is resized to the original size where 1 $pixel = 1$ mm. The following equations show how the pixel spacing is used to resize the image:

$$w_{new} = w_{old} \times \Delta s \tag{1}$$

$$h_{new} = h_{old} \times \Delta s \tag{2}$$

where w_{old} is the old image width, w_{new} is the new image width, h_{old} is the old image height, h_{new} is the new image height, and Δs is the pixel spacing, respectively.

The output of this step is shown in Fig. 3.

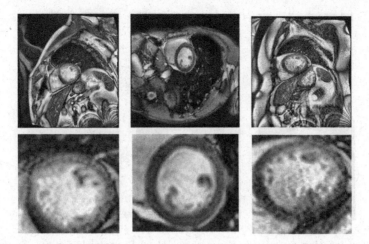

Fig. 3. Localization sample: first row shows the original images. Second row shows the localized left ventricle.

3 Determining Images of Interest and Segmentation

Because the size of the left ventricle varies between the end-systolic and end-diastolic phases, we can determine these phases by the size of the white area in the masks. For each slice, we consider the image with the smallest mask as end-systolic and the one with the largest mask as end-diastolic. A neural network with the same architecture as shown in Fig. 2 can be used to segment the left ventricle. The network is trained only on the end-sysolic and end-diastolic images because these are the images that are needed to compute the end-systolic and end-diastolic volumes.

It should be noted that while the network has a similar architecture to the localization network, the input size for the segmentation network is smaller (80 × 80) vs (128 × 128) for the localization network. This is because the input images are much smaller than the original images. Consequently, the segmentation network is trained much faster than the localization network. The segmentation network was trained only on 540 images representing 25 subjects. The training set contains 500 training subjects. We could not use the whole training set because the process of performing manual segmentation annotation is very time consuming. Because of the localization process, training the segmentation network on such a small data is possible as most of the irrelevant pixels are removed before segmentation.

Figure 4 shows a sample of left ventricle images and the predicted segmentation masks. These masks were generated by training an ensemble of 10 segmentation networks. The ensemble contains networks similar to the one described in Fig. 2 but with varying the number of layers and kernel size in each block. We found that training an ensemble helps a lot since the number of training images is very low.

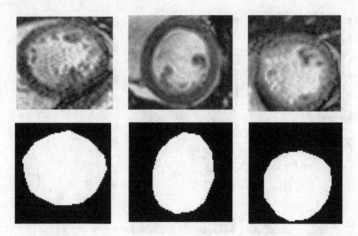

Fig. 4. Segmentation sample: first row shows the left ventricle images. Second row shows the segmentation mask.

4   Volume Estimation

At this stage, we can calculate the volume because the left ventricle is segmented. We calculated the volume using the following equation:

$$V = \sum_{i=0}^{N} SA_i \times SH_i \tag{3}$$

where SA_i and SH_i are the area and height of the slice i, respectively. The area of the slice is calculated by taking the sum of all active pixels (nonzero pixels) in the segmentation mask. SH_i is the result of taking the absolute difference of slice locations from two consecutive slices.

Equation 4 shows how to compute the ejection fraction which is a measure of the outbound blood pumped from the heart after each heartbeat.

$$EF = 100 \times \frac{(V_D - V_S)}{V_D} \tag{4}$$

where V_D and V_S are the end diastolic and end systolic volumes, respectively.

Because we did not train on the whole training set, we train four linear regression and random forest models to improve the results of the volume calculation. The features that are used to train these models include the systole and diastole volumes along with other features such as average slice height, average slice area, etc. Next section, we will report the results that we achieved.

5   Results

We trained the model on a computer with titan X (12 GB RAM). It is worth mentioning that the networks that we described in this paper were also tested on a laptop with a 3.5 GB of available GPU RAM.

Table 1. Diastole RMSE comparison with other state of the art solutions. The state of the art solutions were obtained from [16]

Method	Diastolic RMSE (ml)
Tencia Woshialex [13]	12.02
Ours	**13.37**
JuliandeWit [27]	13.63
Kunsthart [10]	13.65
ShowMeTheMoney	13.2

Table 2. Systole RMSE comparison with other state of the art solutions. The state of the art solutions were obtained from [16]

Method	Systolic RMSE (ml)
ShowMeTheMoney	9.31
Tencia Woshialex [13]	10.19
JuliandeWit [27]	10.32
Kunsthart [10]	10.43
Ours	**12.1**

Table 3. Ejection fraction RMSE comparison with other state of the art solutions. The state of the art solutions were obtained from [16]

Method	Ejection fraction RMSE (%)
ShowMeTheMoney	4.69
Tencia Woshialex [13]	4.88
JuliandeWit [27]	5.04
Ours	**5.97**
Kunsthart [10]	6.99

Training the localization network for 500 epochs with learning rate 0.001 can take around 4.1 h. On the other hand, training the segmentation network for 100 epochs takes only 28 min. Since we trained an ensemble of 10 segmentation networks, the segmentation training process takes around 5 h. The mean absolute value error (MAE) we were able to achieve for the end-diastolic, end-systolic, and ejection fraction are 9.94 ml, 8.42 ml, and 4.47%, respectively. On the other hand, the RMSE errors are slightly higher (13.37 ml for the diastolic volume, 12.1 for the systolic volume and 5.97% for the ejection fraction). According to [16], these values are comparable to the performance of humans in estimating the volumes. The differences between two humans in estimating the end diastolic, end systolic, and ejection fraction are: 13 ml, 14 ml, and 6%, respectively. Tables 1, 2, and 3 show a comparison between the results we achieved and the ones achieved

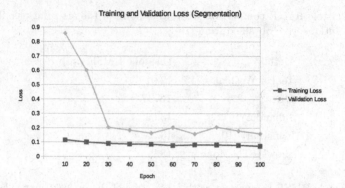

Fig. 5. Segmentation loss: the training and validation loss while training one of the segmentation models. There's a relatively big gap between the training and validation loss. This indicates that the model is likely overfitting the data.

by other methods. These results are very good considering the fact that the localization and segmentation CNNs were trained only on a small subset of the training set. Figure 5 shows the training and validation losses while training one of the segmentation models. Because of the gap between the two losses, it is very likely that the model is overfitting the data. This issue is likely because we only annotated 25 training subjects and used it to train the model. The results are likely to be improved if we annotate more subjects and train the model again.

6   Conclusion

We introduced a network that can be used to localize a region of interest in cardiac MRI images. Once the region of interest (left ventricle) is localized, the systole and diastole images are determined. Next, we segmented the left ventricle using the CNN we used for localization. Finally, we performed volume and ejection fraction estimation using the DICOM fields.

Acknowledgements. This research is partially funded by the Natural Sciences and Engineering Research Council of Canada (NSERC). This support is greatly appreciated. We would also like to thank kaggle, Booz Allen Hamilton, and the National Heart, Lung, and Blood Institute (NHLBI) for providing the MRI images.

References

1. Deng, J., Dong, W., Socher, R., Li, L.J., Li, K., Fei-Fei, L.: ImageNet: a large-scale hierarchical image database. In: IEEE Conference on Computer Vision and Pattern Recognition, CVPR 2009, pp. 248–255. IEEE (2009)
2. Erhan, D., Szegedy, C., Toshev, A., Anguelov, D.: Scalable object detection using deep neural networks. In: Proceedings of the IEEE Conference on Computer Vision and Pattern Recognition, pp. 2147–2154 (2014)

3. Fukushima, K.: Neocognitron: a self-organizing neural network model for a mechanism of pattern recognition unaffected by shift in position. Biol. Cybern. **36**(4), 193–202 (1980)
4. Germano, G., Kiat, H., Kavanagh, P.B., Moriel, M., Mazzanti, M., Su, H.T., Train, K.F.V., Berman, D.S.: Automatic quantification of ejection fraction from gated myocardial perfusion spect. J. Nucl. Med. **36**(11), 2138 (1995)
5. Glorot, X., Bengio, Y.: Understanding the difficulty of training deep feedforward neural networks. In: International Conference on Artificial Intelligence and Statistics, pp. 249–256 (2010)
6. Hinton, G.E., Srivastava, N., Krizhevsky, A., Sutskever, I., Salakhutdinov, R.R.: Improving neural networks by preventing co-adaptation of feature detectors. arXiv preprint arXiv:1207.0580 (2012)
7. Kabani, A.W., El-Sakka, M.R.: Estimating ejection fraction and left ventricle volume using deep convolutional networks. In: Campilho, A., Karray, F. (eds.) ICIAR 2016. LNCS, vol. 9730, pp. 678–686. Springer, Cham (2016). doi:10.1007/978-3-319-41501-7_76
8. Kaggle: Data science bowl cardiac challenge data. https://www.kaggle.com/c/second-annual-data-science-bowl. Accessed 19 Mar 2016
9. Kaus, M.R., von Berg, J., Weese, J., Niessen, W., Pekar, V.: Automated segmentation of the left ventricle in cardiac MRI. Med. Image Anal. **8**(3), 245–254 (2004)
10. Korshunova, I.: Diagnosing heart diseases with deep neural networks. http://irakorshunova.github.io/2016/03/15/heart.html. Accessed 01 Feb 2017
11. Krizhevsky, A., Sutskever, I., Hinton, G.E.: Imagenet classification with deep convolutional neural networks. In: Advances in neural information processing systems, pp. 1097–1105 (2012)
12. LeCun, Y., Bottou, L., Bengio, Y., Haffner, P.: Gradient-based learning applied to document recognition. Proc. IEEE **86**(11), 2278–2324 (1998)
13. Lee, T., Liu, Q.: Solution to win the second annual data science bowl. https://github.com/woshialex/diagnose-heart. Accessed 01 Feb 2017
14. Lin, X., Cowan, B.R., Young, A.A.: Automated detection of left ventricle in 4D MR images: experience from a large study. In: Larsen, R., Nielsen, M., Sporring, J. (eds.) MICCAI 2006. LNCS, vol. 4190, pp. 728–735. Springer, Heidelberg (2006). doi:10.1007/11866565_89
15. Lynch, M., Ghita, O., Whelan, P.F.: Automatic segmentation of the left ventricle cavity and myocardium in MRI data. Comput. Biol. Med. **36**(4), 389–407 (2006)
16. Mulholland, J.: Leading and winning team submissions analysis. http://www.datasciencebowl.com/leading-and-winning-team-submissions-analysis/. Accessed 04 Aug 2016
17. Nair, V., Hinton, G.E.: Rectified linear units improve restricted Boltzmann machines. In: Proceedings of the 27th International Conference on Machine Learning (ICML 2010), pp. 807–814 (2010)
18. Otsu, N.: A threshold selection method from gray-level histograms. Automatica **11**(285–296), 23–27 (1975)
19. Ronneberger, O., Fischer, P., Brox, T.: U-net: convolutional networks for biomedical image segmentation. In: Navab, N., Hornegger, J., Wells, W.M., Frangi, A.F. (eds.) MICCAI 2015. LNCS, vol. 9351, pp. 234–241. Springer, Cham (2015). doi:10.1007/978-3-319-24574-4_28
20. Russakovsky, O., Deng, J., Su, H., Krause, J., Satheesh, S., Ma, S., Huang, Z., Karpathy, A., Khosla, A., Bernstein, M., et al.: Imagenet large scale visual recognition challenge. Int. J. Comput. Vis. **115**(3), 211–252 (2015)

21. Sermanet, P., Eigen, D., Zhang, X., Mathieu, M., Fergus, R., LeCun, Y.: OverFeat: integrated recognition, localization and detection using convolutional networks. arXiv preprint arXiv:1312.6229 (2013)
22. Simonyan, K., Zisserman, A.: Very deep convolutional networks for large-scale image recognition. arXiv preprint arXiv:1409.1556 (2014)
23. Song, H.O., Girshick, R., Jegelka, S., Mairal, J., Harchaoui, Z., Darrell, T.: On learning to localize objects with minimal supervision. arXiv preprint arXiv:1403.1024 (2014)
24. Srivastava, N., Hinton, G., Krizhevsky, A., Sutskever, I., Salakhutdinov, R.: Dropout: a simple way to prevent neural networks from overfitting. J. Mach. Learn. Res. **15**(1), 1929–1958 (2014)
25. Szegedy, C., Liu, W., Jia, Y., Sermanet, P., Reed, S., Anguelov, D., Erhan, D., Vanhoucke, V., Rabinovich, A.: Going deeper with convolutions. arXiv preprint arXiv:1409.4842 (2014)
26. Szegedy, C., Toshev, A., Erhan, D.: Deep neural networks for object detection. In: Advances in Neural Information Processing Systems, pp. 2553–2561 (2013)
27. de Wit, J.: Third place solution for the second kaggle national datascience bowl. https://github.com/juliandewit/kaggle_ndsb2. Accessed 01 Feb 2017

Fully Deep Convolutional Neural Networks for Segmentation of the Prostate Gland in Diffusion-Weighted MR Images

Tyler Clark[1], Alexander Wong[2], Masoom A. Haider[1], and Farzad Khalvati[1(✉)]

[1] Medical Imaging, Sunnybrook Research Institute, University of Toronto, Toronto, Canada
farzad.khalvati@sri.utoronto.ca
[2] Systems Design Engineering, University of Waterloo, Waterloo, Canada

Abstract. Prostate cancer is a leading cause of mortality among men. Diffusion-weighted magnetic resonance imaging (DW-MRI) has shown to be successful at monitoring and detecting prostate tumors. The clinical guidelines to interpret DW-MRI for prostate cancer requires the segmentation of the prostate gland into different zones. Moreover, computer-aided detection tools which are designed to detect prostate cancer automatically, usually require the segmentation of prostate gland as a pre-processing step. In this paper, we present a segmentation algorithm for delineation of the prostate gland in DW-MRI via fully convolutional neural network. The segmentation algorithm was applied to images of 30 (testing) and 104 (training) patients and a median Dice Similarity Coefficient of 0.89 was achieved. This method is faster and returns similar results compared to registration based methods; although it has the potential to produce improved results given a larger training set.

Keywords: Prostate segmentation · Convolutional neural networks

1 Introduction

Artificial intelligence is increasingly affecting different fields in medicine including radiology. Although still not fully mature, computer vision techniques are proving to have potential to be as good or better than human experts at many of medical image analysis tasks once thought to be out of computers' grasp.

The main workhorse behind this innovation is the deep convolutional neural network (CNN) introduced by [1]. A deep CNN consists of many consecutive layers that transform the original image until it is of the dimension of the desired output. The first layer can be seen as an n-dimensional filter sliding across the image creating a new matrix of values that correspond to the filter's interactions with the image. The second layer is another group of filters interacting with the matrix created by the first layer. There can be hundreds of layers in a standard

© Springer International Publishing AG 2017
F. Karray et al. (Eds.): ICIAR 2017, LNCS 10317, pp. 97–104, 2017.
DOI: 10.1007/978-3-319-59876-5_12

CNN. The filters consist of weights, and the accuracy of the network is measured using a loss function. The loss created by each weight is determined, and progressively the weights are updated in order to minimize the loss.

In men, prostate cancer is the leading cause of cancer death. Importantly, the prognosis of a prostate tumor does not often change during its development [2]. The measure of its prognosis in technical terms is known as the Gleason score. This means that a low Gleason score tumor that has a low probability of being dangerous does not often become a dangerous one rapidly. This has lead to a course of action among low Gleason score tumors known as active monitoring or surveillance. Instead of surgery, radiation or other radical treatment, the tumor is monitored using various imaging techniques to ensure there are no changes. This means that the buck is passed to radiologists who are already overwhelmed due to the rise of diagnostic images being requested.

Active monitoring of a prostate tumor is usually done using Diffusion-weighted Magnetic Resonance Imaging (DW-MRI). The European Society of Urogenital Radiology (ESUR) has developed a scoring system (PI-RADS: Prostate Imaging Reporting and Data System) for accurate interpretations of prostate cancer by radiologists [3]. To reduce inter-observer variability of readers, PI-RADS scoring partitions the prostate gland into 10 different zones. Prostate segmentation is a necessary step for PI-RADS zoning and automating this process will significantly improve the accuracy and consistency of prostate cancer diagnosis. Otherwise, the radiologists must go through the slices of 3D volume and manually segment (draw boundaries around) the prostate gland as whole, and the transitional zone of the prostate. For a radiologist, this could mean around 30 s per slice; which equates to 10 min per patient (e.g., 20 slices) [4]. Several computer-aided detection (CAD) algorithms have been proposed for automated detection of prostate cancer with the goal of improved diagnosis accuracy and consistency [5,6]. These CAD algorithms require the prostate gland be segmented as a preprocessing step. Otherwise, the CAD algorithms would perform poorly due to the contamination of data from the surrounding of the prostate gland.

Most algorithms that have been proposed for prostate segmentation [7–9] rely on users drawing bounding boxes around the prostate or manually contouring the prostate gland in several slices of the volume (e.g., 5). They also rely on image registration algorithms which are computationally expensive. The heavy user interaction requirement and the time consuming nature of image registration-based algorithms highlight the need for fully automatic segmentation algorithms with pseudo-real time performance. On the other hand, deep CNNs have shown great promise in computer vision problems. In this work, a deep CNN as the U-net [10] was adopted, modified, and configured to segment the prostate gland using DW-MRI data, as discussed in the following sections.

2   Methods

In this section, the data used for the experiments, the deep CNN used for segmentation, and the results are presented.

2.1 Data and Pre-processing

DW-MRI data of 104 patients was acquired using a Philips Achieva 3.0T machine at Sunnybrook Health Sciences Centre. Institutional research ethics board approval and written informed consent was waived. Each patient had a 3D volume divided into 12–26 slices for which, the prostate was manually segmented by an expert for every slice on b0 images of DW-MRI (b = 0 s/mm^2).

Data Augmentation. To increase the training data size, it was artificially increased by altering the images in various ways using Keras functions [11]. This increased training data size from 1498 to 7490. The test set remained unchanged.

- The image is flipped on both X and Y axis
- Random rotation: The image is rotated randomly between 0–7°
- Random channel shift: Channel shifting is the process of taking pixels in an image and applying those values to pixels in different positions on the image.

2.2 Architecture

First, a brief introduction to the main parts of a deep CNN is presented.

Convolutions. The convolution in the first layer can be seen as an interaction of a filter with the matrix representation of the image. There is a smaller $n \times n$ matrix or kernel that performs a $n \times n$ dimensional dot product with the image starting on the top left corner of the image. The kernel is spatially moved one or more pixels to the right, and then another dot product is performed. When it is at the farthest right pixel, it slides down one or more rows then begins again.

A new matrix is created by the dot product of the kernel and the image. The number of filters in the first layer is 32, which then increases at each layer by a factor of 2. The values in the filters are called weights and these are the values that are changed in the learning process by gradient descent.

Activation Functions. After most convolutional layers, there is a nonlinear function that is used as a decision boundary for which values in the matrix will be passed on to the next layer. The activation function used until the last layer is the rectified linear unit or *relu*.

$$relu = max(0, x) \tag{1}$$

The activation function used on the last layer to return the image to a binary mask was the sigmoid function.

$$sigmoid = \frac{1}{1 + e^{-x}} \tag{2}$$

Max-pooling. Max pooling is a method to reduce the dimensions of the matrix by picking the largest value within a group of pixels.

Upsampling. Upsampling or deconvolution is a way of increasing the dimension of a matrix to bring it to the original size [12].

U-Net. The U-net (Fig. 1) is first composed of a series of four convolution and max-pooling operations that greatly reduce the dimension of the input image; followed by four convolution and up-sampling operations. During the up-sampling, the weights of the max-pooling/convolutions are concatenated with their reciprocal down-sampled counterparts. The use of skip connections such as this have shown to be effective at increasing detail in segmentation maps [10].

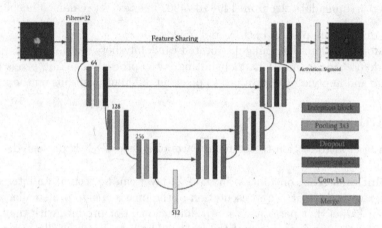

Fig. 1. Convolutional network architecture used for the segmentation prostate

In this work, we have made several modifications to the original U-net architecture. First, the regular convolution blocks in the U-net (3×3 with N filters) were replaced with inception reduction blocks with N filters (Fig. 2).

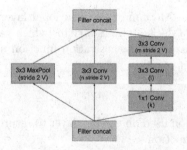

Fig. 2. Reduction block. Taken from [13]

Inception reduction blocks use parallel convolutional filters to reduce computational cost and improve performance [13]. Second, a spatial Drop-out [14]

was implemented to reduce over fitting, which is often a problem with small data sets. For standard dropout, during training, each neuron is "dropped out" meaning set to zero with a probability p, independent of all other neurons. Spatial dropout instead will drop entire feature maps with a probability p. For example, if there is a $n \times h \times w$ feature tensor, each of the n $h \times w$ matrices will be dropped with a probability p.

Table 1 shows the dimensions of the pooling, convolution, dropout, and upsampling in each of the 10 layers of the proposed network.

Table 1. U-net architecture detail

Layer	Max pooling	Convolution	Dropout	Upsampling
1	3×3	$3 \times 3 \times 32$	0.6	None
2	3×3	$3 \times 3 \times 64$	0.6	None
3	3×3	$3 \times 3 \times 128$	0.6	None
4	3×3	$3 \times 3 \times 256$	0.6	None
5	None	$3 \times 3 \times 512$	None	None
6	None	$3 \times 3 \times 256$	0.6	2×2
7	None	$3 \times 3 \times 128$	0.6	2×2
8	None	$3 \times 3 \times 64$	0.6	2×2
9	None	$3 \times 3 \times 32$	0.6	2×2
10	None	$1 \times 1 \times 1$	None	None

The optimizer used for gradient descent was Adam [15], which is currently industry standard. The loss function was a modified Dice Similarity Coefficient (DSC). The DSC essentially measures the overlap between the predicted mask and ground truth.

$$DSC = \frac{2|P \cap G|}{|P| + |G|} \tag{3}$$

where P is the predicted mask, and G is the ground truth.

The proposed U-net architecture in this paper adds additional loss to the function only when the sum of the pixels of the predicted mask is larger then the sum of the ground truth. The reason for this is that the structure of the prostate gland has an uneven proportion of slices with different cross sectional areas. This means the ends of the prostate gland which are smaller end up having a predicted mask that is too large, due to under sampling. This addition increased the average test DSC by 3%. The loss function used for our architecture is shown below.

$$Loss = -DSC(1 - max(0, sum(P) - sum(G)/\mu_{pixels}) \tag{4}$$

where $\mu_{pixels} = 6000$ is approximately the average amount of pixels contained within a mask. It was a hyper parameter used in order to scale the max function properly to create optimal results.

The implementation of the proposed architecture was initially based on [16], which had the U-net structure but significant changes and modifications were made to the original code to implement our proposed architecture.

2.3 Computer System

The processor used for the proposed algorithm was a Nvidia GEFORCE GTX GPU with Cuda parallel processing. The language used was python with a Keras wrapper on top of a Theano back end. In total there have been 42 iterations of the algorithm, which translates to 378 training hours.

3 Results

The proposed architecture was tested using a 3-fold validation: The data was randomly divided into 9 equal sub-samples, then a single sub-sample was retained as the validation set for testing the model and the remaining 8 sub-samples were used as training data. The validation process was then repeated 3 times, with 3 of the sub-samples used once as validation data. This means that 30% of the images (30 patients) were used for validation. The size of each of the training and validation (test) sets were 1498/149 slices respectively. The metric used for evaluation was DSC as mentioned above. The results obtained by this procedure are illustrated in Fig. 3 for each cross validation separately. The median DSC after the 3-fold validation was 0.89.

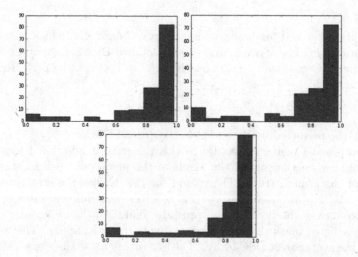

Fig. 3. DSC values of the 3-fold cross validation for segmentation of the prostate gland with median values of top-left: 0.896, top-right: 0.875, and bottom: 0.901

Figure 4 shows sample prostate images along with the segmentation of the prostate gland (manual and the proposed algorithm).

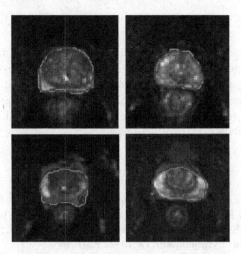

Fig. 4. Examples of successful results. The ground truth (green) and predicted mask (blue) are overlaid on the DWI (b0) (Color figure online)

The results achieved (DSC 0.89) by the proposed architecture for the segmentation of the prostate gland for DW-MRI images show the potential for further improvement of this algorithm. When considering clinical usage, the time required to segment images for a radiologist is an important parameter. Most of related work has been on segmenting the prostate gland in T2-weighted MRI (T2w). For example, a registration-based segmentation algorithm was proposed in [7] for T2w which produced slightly better results (DSC of 0.92) while requiring the user to manually contour 3 slices of the prostate volume. Another work produced DSC of 0.85–0.93 depending on how many slices were required to be manually contoured by the user (approximately 5 to 10 slices per 3D volume, respectively). In a recent work, an atlas-based segmentation was introduced by [9] for the prostate gland in DWI which yielded average DSC of 0.80–0.85 depending on the number of bounding boxes drawn by the user around the prostate gland (3 or 5 slices).

These algorithms require intensive user interaction and image registration algorithms. For example, the method in [9] is based on image registration, which is a time consuming task (e.g., 4.25 s per slice [7]). Moreover, there are 5 bounding boxes required to be drawn by the radiologist for the algorithm to work efficiently. An estimate of the time required for this action is at least 60 s. Thus, even ignoring the processing time for image registration, the algorithm in [9] would require roughly 2 min of manual intervention to segment a 3D image with 15 slices. The method proposed in this paper requires 0.310 s per slice to complete. Thus, for a patient with 15 images, this only requires 4.5 s, which is approximately a 13× speedup compared to [9] with superior accuracy. Thus, the proposed segmentation algorithm in this paper can produce the results in near real time, which is crucial for successful integration into clinical workflow.

4   Conclusion

In this paper we introduced a new algorithm for segmenting the prostate gland using DWI (b0) images. The proposed algorithm is based on U-net architecture, a novel deep CNN algorithm. The U-net architecture was modified via integrating novel inception reduction and dropout blocks as well as a loss function specific to the problem at hand. The proposed architecture was applied to b0 images of 30 patients and a median DSC of 0.89 was achieved.

References

1. Krizhevsky, A., et al.: ImageNet classification with deep convolutional neural networks. In: Neural Information and Processing Systems (NIPS), pp. 1–9 (2012)
2. Mcmaster, M.L., Kristinsson, S.Y., Turesson, I., Bjorkholm, M., Landgren, O.: NIH public access. Clin. Lymphoma **9**(1), 19–22 (2010)
3. Barentsz, J.O., et al.: ESUR prostate MR guidelines 2012. Eur. Radiol. **22**(4), 746–757 (2012)
4. Martin, S., et al.: A multiphase validation of atlas-based automatic and semiautomatic segmentation strategies for prostate MRI. Int. J. Radiat. Oncol. Biol. Phys. **85**(1), 95–100 (2013)
5. Khalvati, F., et al.: Automated prostate cancer detection via comprehensive multiparametric magnetic resonance imaging texture feature models. BMC Med. Imaging **15**, 1–14 (2015). doi:10.1186/s12880-015-0069-9
6. Cameron, A., Khalvati, F., Haider, M., Wong, A.: MAPS: a quantitative radiomics approach for prostate cancer detection. IEEE Trans. Bio-Med. Eng. **63**(6), 1145–1156 (2016)
7. Khalvati, F., et al.: Inter-slice bidirectional registration-based segmentation of the prostate gland in MR and CT image sequences. Med. Phys. **40**(12), 123503 (2013)
8. Khalvati, F., et al.: Sequential registration-based segmentation of prostate gland in MR image volumes. J. Digit. Imaging **29**, 254–263 (2016)
9. Zhang, J., Baig, S., Wong, A., Haider, M.A., Khalvati, F.: A local ROI-specific atlas-based segmentation of prostate gland and transitional zone in diffusion MRI. J. Comput. Vis. Imaging Syst. **2**(1) (2016)
10. Ronneberger, O., Fischer, P., Brox, T.: U-net: convolutional networks for biomedical image segmentation. In: Navab, N., Hornegger, J., Wells, W.M., Frangi, A.F. (eds.) MICCAI 2015. LNCS, vol. 9351, pp. 234–241. Springer, Cham (2015). doi:10.1007/978-3-319-24574-4_28
11. Keras: Deep learning library for theano and tensorflow (2015). https://keras.io/
12. Long, J., Shelhamer, E., Darrell, T.: Fully convolutional networks for semantic segmentation. In: Proceedings of the IEEE Computer Society Conference on Computer Vision and Pattern Recognition, vol. 07, pp. 3431–3440, 12 June 2015
13. Szegedy, C., Ioffe, S., Vanhoucke, V.: INCEPTION - V4, INCEPTION - R and the impact of residual connections on learning. In: International Conference on Learning Representations (ICLR) (2016)
14. Goroshin, R., et al.: Efficient object localization using convolutional networks. In: IEEE Conference on Computer Vision and Pattern Recognition (CVPR) (2015)
15. Kingma, D.P., Ba, J.L.: Adam: a method for stochastic optimization. In: International Conference on Learning Representations 2015 (2015)
16. Marko, J.D.: https://github.com/jocicmarko (2015)

Image Enhancement and Reconstruction

Compensated Row-Column Ultrasound Imaging System Using Three Dimensional Random Fields

Ibrahim Ben Daya$^{(\boxtimes)}$, Albert I.H. Chen, Mohammad Javad Shafiee, Alexander Wong, and John T.W. Yeow

Vision and Image Processing Research Group and Advanced Micro-/Nano- Devices Research Group, Department of Systems Design Engineering, University of Waterloo, 200 University Avenue W, Waterloo, ON N2T 3G1, Canada
ibendaya@uwaterloo.ca

Abstract. The row-column method received a lot of attention for 3-D ultrasound imaging. This simplification technique reduces the number of connections required to address a 2-D array and therefore reduces the amount of data to handle. However, Row-column ultrasound imaging still has its limitations: the issues of data sparsity, speckle noise, and the spatially varying point spread function with edge artifacts must all be taken into account when building a reconstruction framework. In this work, we introduce a compensated row-column ultrasound imaging system, termed 3D-CRC, that leverages 3-D information within an extended 3-D random field model to compensate for the intrinsic limitations of row-column method. Tests on 3D-CRC and previously published row-column ultrasound imaging systems show the potential of our proposed system as an effective tool for enhancing 3-D row-column imaging.

Keywords: Ultrasound imaging · Row-column · Random fields · Image reconstruction · Statistical estimation

1 Introduction

3-D ultrasound imaging offers unique opportunities for applications ranging from material science to medical diagnostics. It is possible to generate certain material properties more accurately when given volumetric data [1], and it is easier to get a volumetric scan of an area than trying to find the same 2-D slice for the purpose of follow up studies in medical diagnostics [2].

To get a 3-D scan from a fixed transducer, a 2-D array is required. However, a fully addressed $N \times N$ array requires N^2 connections, which poses a challenge both in terms of addressing individual connections as well as the large amount of data to handle. One simplification technique for 2-D arrays that received a lot of attention is the row-column method.

In the row-column method, two sets of orthogonally positioned 1-D transducers are used. One set will be responsible for transmit beamforming, and the other for receive beamforming (Fig. 1). Using this method, the number of connections

© Springer International Publishing AG 2017
F. Karray et al. (Eds.): ICIAR 2017, LNCS 10317, pp. 107–116, 2017.
DOI: 10.1007/978-3-319-59876-5_13

required for an $N \times N$ array is $N + N$. There are a few limitations for this method and ultrasound imaging in general: data sparsity, speckle noise, and a spatially dependant point spread function (PSF) that suffers from edge artifacts.

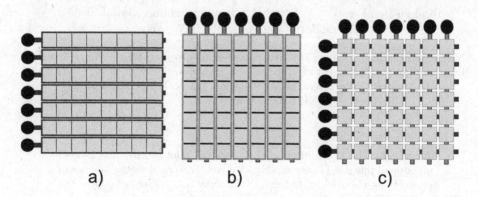

Fig. 1. 1-D arrays arranged as (a) a set of rows and (b) a set of columns to form (c) a row-column array

There are a few proposed row-column ultrasound imaging systems in literature that attempt to address some of these limitations. Chen *et al.* [4] proposed a real time row-column ultrasound imaging system (henceforth referred to as baseline RC). Another system presented in Rasmussen *et al.* [3] and Christiansen *et al.* [5] proposed an integrate apodization system that attempts to address edge waves through transducer design. Daya *et al.* [6] proposed CRC-UIS a compensated ultrasound imaging system that uses a physical model for the point spread function in a reconstruction framework to algorithmically address the limitations of the row-column method. However, their reconstruction framework only works for 2-D images. They address 3-D volumes by working in each 2-D slice separately and then combining all slices to form a 3-D image. Incorporating information in 3-D should give a better reconstructed image, and this is the motivation behind this work: we propose 3D-CRC, a compensated row-column ultrasound imaging system where we leverage 3-D information within an extended 3-D random field model.

The rest of the paper will be organized as follows: Sect. 2 will outline the methodology of the proposed system, Sect. 3 will detail experimental setup and results, and Sect. 4 will discuss results and future work.

2   Methodology

The proposed system has two main stages: characterization and compensation. For the characterization stage, an image formation model is used to help account for data sparsity, a noise model is defined to understand speckle noise, and a PSF model is use to generate row-column's PSF at different spatial positions. In the compensation stage, the characteristics of the system are taken into account in a reconstruction framework. Both stages will be discussed in detail.

2.1 Characterization

Characterization of the row-column system includes an image formation model, a noise model, and a PSF model.

Image Formation Model. Equation 1 describes image formation of a volumetric row-column scan:

$$g_r(x,y,z) = M(x,y,z)[f(x,y,z) * h(x,y,z) + u(x,y,z)]. \tag{1}$$

where x, y, and z are the Cartesian coordinates. The term $g_r(x,y,z)$ is the observed RF image, $M(x,y,z)$ is the data acquisition unit's sampling function, $f(x,y,z)$ represents the tissue reflectivity function, the operator '$*$' denotes the convolution operation, $h(x,y,z)$ represents the PSF function (PSF), and $u(x,y,z)$ is the noise component.

The observed RF image $g_r(x,y,z)$ can be visualized as a set of parallel 2-D plates of a series of fan-beams of 'readings', where the fan beams of each plate originate from the 1-D transducers responsible for receive beamforming. This is visualized in Fig. 2. These fan-beams are set using the sampling function $M(x,y,z)$. $h(x,y,z)$ is the ultrasound system's spatially dependant PSF. $u(x,y,z)$ describes both the measurement noise as well as the physical phenomena which are not accounted for by the convolution model. The noise model will now be discussed.

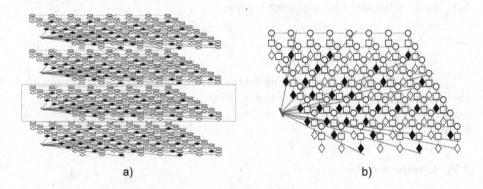

a)   b)

Fig. 2. (a) For every 1-D array responsible for receive, there will be a plane with 'fan beams' of readings, (b) shows one plain in more detail, black diamonds indicate available readings while white diamonds indicate missing readings

Noise Model. Noise in ultrasound images can be modeled as:

$$g_e(x,y,z) = f(x,y,z)\xi_m(x,y,z) + \xi_a(x,y,z) \tag{2}$$

where $g_e(x,y,z)$ is the observed envelope image, $f(x,y,z)$ is the noise-free image, $\xi_m(x,y,z)$ is the multiplicative speckle noise component, and $\xi_a(x,y,z)$ is the additive speckle noise component [25].

There were a few distributions proposed to model speckle [25], empirical tests done on the envelop data captured by the data acquisition unit show that the Generalized Gamma distribution has the best fit. The noise samples of the logarithmic transformed multiplicative noise can be modeled with the FisherTippett distribution given by:

$$p(I(x,y,z)) = 2 \exp\left[(2I(x,y,z) - \ln 2\sigma^2) - \exp[2I(x,y,z) - \ln 2\sigma^2]\right] \tag{3}$$

where voxel intensity at point (x,y,z) is denoted by $I(x,y,z)$, and their standard deviation is denoted by σ. The PSF model will now be discussed.

Point Spread Function Model. One of the most commonly used models for the point spread function of ultrasound systems is the one based on the Tupholme-Stephanisshen model for spatial impulse response, which was further derived for the pulse echo case by Jensen [8]. In this model, the point spread function of a row-column system at point r_1 with transducers at point r_2 and geometry S is given by:

$$H_{pe}(r_1, r_2, t) = h(r_1, r_2, t) * h(r_2, r_1, t) \tag{4}$$

where the convolution is over time and not space, $\delta(.)$ is the Dirac delta function and c is the speed of sound at a homogeneous medium of density ρ_0, and $h(r_1, r_2, t)$ is the one way impulse response:

$$h(r_1, r_2, t) = \int_S \frac{\delta\left(t - \frac{|r_1 - r_2|}{c}\right)}{2\pi|r_1 - r_2|} dS \tag{5}$$

The sparsity due to the sampling function from the image formation model, the speckle noise which follows the Fisher-Tippett model, and the spatially varying PSF with edge artifacts can all be incorporated in a 3-D compensation framework.

2.2   Compensation

The goal of an ultrasound image reconstruction framework is to estimate the tissue reflectivity function f, thus an inverse problem of Eq. 1 must be solved. The reconstruction problem can be formulated as a Maximum a Posteriori (MAP) problem, where an estimate can be obtained by maximizing the posterior distribution $P(F|G)$:

$$F^* = \overline{F} argmax\{P(F|G)\} \tag{6}$$

where the MAP solution, the possible results set, and the observation are denoted by $F^*, \bar{F}$, and G respectively.

Conditional random field (CRF), first proposed by Lafferty *et al.* [12], is a discriminative modeling framework that allows for a direct model of the conditional probability $P(F|G)$ through potential functions instead of specifying a prior model [13]:

$$P(F|G) = \frac{1}{Z(G)} \exp\left(-\psi(F, G)\right) \tag{7}$$

where Z is the partition function and $\psi(\cdot)$ is the potential function [12–17], which can be a combination of unary $\psi_u(\cdot)$ and pairwise $\psi_p(\cdot)$ potential functions:

$$\psi(F, G) = \sum_{i=1}^{n} \psi_u(f_i, G) + \sum_{c \in C} \psi_p(f_c, G) \tag{8}$$

where C is a set of a clique structure for each node where random variable interactions are involved in modeling.

With regular CRFs, local cliques structures are adopted and are considered with the same degree of certainty; observations are assumed to be complete, and data sparsity is not taken into account [6]. However, this poses a problem as data sparsity is a challenges we aims to address. Multilayered conditional random fields (MCRFs) addresses this issue by introducing an uncertainty layer to the CRF model, where each observation is linked with a value to specify whether an observation exists or not. Equation 7 can be rewritten as:

$$P(F|Cr, G) = \frac{1}{Z(G)} \exp\left(-\psi(F|Cr, G)\right) \tag{9}$$

where the uncertainty layer is denoted by Cr. This layer can be visualized as a zero-one plane where $Cr = 1$ indicates missing observations and $Cr = 0$ indicates availability of observations.

The unary potential function is a data driven function. It incorporates into the MCRF model information corresponding to the observation. This function is modeled after the Fisher-Tippett noise model shown in 3, since we believe that is how the observation degraded [6].

The pairwise potential function incorporates spatial information into the MCRF model. Through this function, small noises are removed, consistent labels are provided, and areas with no prior data are estimated. Information is incorporated from a subset of random variables determined by a clique structure, where relations between random variables are defined according to set penalty functions. Two penalty functions are used in this work: spatial proximity penalty term w_{sp}, which is based on the assumption that far away voxels do not belong to the same label, and First Order Variation (FOV) of intensity values w_{fov}, which uses difference in intensity between voxels to outline tissue transitions [6].

Taking into account the MCRF represented in Eq. 9 and potential function in Eq. 8, the energy function that drives the MAP model can be formulated as:

$$E(F, G, Cr) = \sum_{i=1}^{n} \psi_u(f_i, G, Cr_i) + \sum_{c \in C} \psi_p(f_c, G). \tag{10}$$

The MAP can now be reformulated as:

$$F^* = \overline{F}argmin\{E(F,G,Cr)\}. \tag{11}$$

This is an optimisation problem that can be solved using gradient descent. The gradient descent for possible solution F^* can be expressed as:

$$F^{*t+1} = F^{*t} + \frac{\nabla E(F,G,Cr)}{\nabla F} \tag{12}$$

where $\frac{\nabla E(F,G,Cr)}{\nabla F}$ is the energy gradient with respect to F and F^{*t} is the estimated solution at iteration t.

Given that the pairwise potential function takes data sparsity into account, and the unary potential function takes speckle noise as well as the spatially dependant PSF into account. This compensation framework is capable of addressing the limitations of the row-column method. We will now test our proposed system.

3   Results

Simulated ultrasound scans were used to evaluate our proposed system. The simulated scans were compared against CRC-UIS [6] to highlight the value of incorporating information in 3-D over 2-D.

3.1   Simulation

Row-column ultrasound beamforming was simulated using the open source MATLAB toolkit Field II [18]. In our tests, envelope data from the simulated scans was mapped into a regular 3-D lattice through linear interpolation before passing it to the compensation stage.

The phantom used in our tests was a cysts 6 mm in diameter placed in a 20 mm × 20 mm × 30 mm volume, with the center of the cyst placed at $[x, y, z] = (0, 0, 25)$ mm. The simulation was done with 100,000 scatterers, with a 32 × 32 row column setup, with each element width and height set to 4.8 mm.

3.2   Metrics

For quantitative evaluation of our system, Peak Signal to Noise Ratio (PSNR), Effective Number of Looks (ENL), and Coefficient of Correlation (CoC) were used as metrics to evaluate the performance of our system. These metrics were defined according to recent literature [6, 19–25].

Table 1. Quantitative evaluation for the cysts phantom. The proposed 3D-CRC outperforms all systems in literature as well as a fully addressed 2-D array when it comes to PSNR, CoC, and ENL

System	PSNR (dB)	CoC	ENL
3D-CRC	**24.8557**	**0.1268**	**14.6329**
CRC-UIS [6]	19.2640	0.0386	11.1775
Integrated apodization [26]	24.2688	0.0077	0.8900
Baseline RC [4]	19.8386	0.0011	0.8900
Fully addressed 2-D	19.2640	0.0007	1.4489

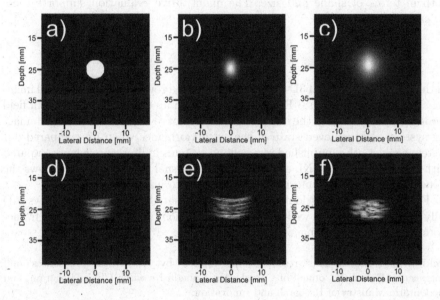

Fig. 3. Visual evaluation for the output image of our proposed system as opposed to other systems in literature. The phantom image is shown in (a). 3D-CRC system is shown in (b). CRC-UIS [6] is shown in (c). Integrated apodization [26] is shown in (d). Baseline RC [4]is shown in (e). Fully addressed 2-D array is shown in (f).

3.3 Quantitative Evaluation

The results of the proposed 3D-CRC were compared against other systems in literature. They are summarized in Table 1.

Quantitative analysis of the results show that the proposed 3D-CRC shows the highest PSNR, outperforming other row-column systems in literature as well as a fully addressed 2-D array. 3D-CRC also had the highest CoC, indicating better preserving edges. 3D-CRC had the highest ENL score, indicating it was the closest at reconstructing the homogeneous region of the phantom image.

3.4 Visual Evaluation

Figure 3 shows the reconstruction of the phantom image for 3D-CRC and other row-column systems in literature as well as a fully addressed 2-D array.

3D-CRC shows the closest reconstruction of the ideal phantom image in terms of shape and size. 3D-CRC and CRC-UIS show the smoothest regions that more closely resemble the smooth inner part of the cyst in the phantom image. CRC-UIS as well as the fully addressed 2-D array were capable of reconstructing the circular shape of the cyst, but the reconstructed cysts was larger than the phantom image. For the integrated apodization, the reconstructed cysts was less circular, although the size was close. The baseline RC had poor reconstruction both in terms of shape and size. The quantitative evaluation supports these observations.

4 Conclusion

In this work, we proposed 3D-CRC: a compensated row-column ultrasound imaging system that leverages 3-D information within an extended 3-D random field model to compensate for the intrinsic limitations of row-column ultrasound imaging systems. We achieved state of the art performance when we compared our system against other published row-column systems in literature when compared with PSNR, CoC, and ENL as well as through visual evaluation. This shows the value of leveraging 3-D information as opposed to just 2-D.

Future work includes looking into a compensated system that employs 3-D fully connected CRF, as well as an implementation of the compensation framework on real phantom images.

Acknowledgement. This research was funded by the Natural Sciences and Engineering Research Council of Canada, the Canada Research Chairs Program, and the Ontario Ministry of Research and Innovation.

References

1. Smith, R.A., Nelson, L.J.: 2D transmission imaging with a crossed-array configuration for defect detection. Insight J. Br. Inst. NDT **51**, 82–87 (2009)
2. Szabo, T.L.: Diagnostic Ultrasound Imaging: Inside Out. Elsevier Academic Press, Cambridge (2004)
3. Rasmussen, M., Christiansen, T., Thomsem, E., Jensen, J.: 3-D imaging using row-column-addressed arrays with integrated apodization - part I: apodization design and line element beamforming. IEEE Trans. Ultrason. Ferroelectr. Freq. Control **62**(5), 947–958 (2015)
4. Chen, A., Wong, L., Logan, A., Yeow, J.T.W.: A CMUT-based real-time volumetric ultrasound imaging system with row-column addressing. IEEE Int. Ultrason. Symp. **1**, 1755–1758 (2011)
5. Christiansen, T., et al.: 3-D imaging using row-column-addressed arrays with integrated apodization - part II: transducer fabrication and experimental results. IEEE Trans. Ultrason. Ferroelectr. Freq. Control **62**(5), 959–971 (2015)

6. Daya, I.B., Chen, A.I.H., Shafiee, M.J., Wong, A., Yeow, J.T.W.: Compensated row-column ultrasound imaging system using fisher tippett multilayered conditional random field model. PLoS One **10**(12), e0142817 (2015)
7. Michailovich, O., Tannenbaum, A.: Despeckling of medical ultrasound images. IEEE Trans. Ultrason. Ferroelectr. Freq. Control **53**(1), 64–78 (2006)
8. Jensen, J.A.: Linear descriptions of ultrasound imaging systems. Technical University of Denmark, DK-2800 Lyngby, Denmark (1999)
9. Black, A., Kohli, P., Rother, C.: Markov Random Fields for Vision and Image Processing. The MIT Press, Cambridge (2011)
10. Dolui, S.: Variable splitting as a key to efficient image reconstruction. Ph.D. thesis, University of Waterloo (2012)
11. Sanches, J., Bioucas-Dias, J., Marques, J.: Minimum total variation in 3D-ultrasound reconstruction. IEEE Int. Conf. Image Process. **3**, 597–600 (2005)
12. Lafferty, J.D., McCallum, A., Pereira, F.C.N.: Conditional random fields: probabilistic models for segmenting and labeling sequence data. In: Proceedings of the Eighteenth International Conference on Machine Learning, pp. 282–289 (2001)
13. Kazemzadeh, F., Shafiee, M.J., Wong, A., Clausi, D.A.: Reconstruction of compressive multispectral sensing data using a multilayered conditional random field approach. In: SPIE Proceedings, vol. 9217 (2014)
14. Shafiee, M.J., Wong, A., Siva, P., Fieguth, P.: Efficient Bayesian inference using fully connected conditional random fields with stochastic cliques. In: IEEE International Conference on Image Processing, pp. 4289–4293 (2014)
15. Broomand, A., et al.: Multi-penalty conditional random field approach to super-resolved reconstruction of optical coherence tomography images. Biomed. Optics Express **4**(10), 2032–2050 (2013)
16. Tanaka, K., Kataoka, S., Yasuda, M.: Statistical performance analysis by loopy belief propagation in Bayesian image modeling. J. Phys: Conf. Ser. **233**(1), 012013 (2010)
17. Yao, F., Qian, Y., Hu, Z., Li, J.: A novel hyperspace remote sensing images classification using Gaussian processes with conditional random fields. In: International Conference on Intelligent Systems and Knowledge Engineering, pp. 197–202 (2010)
18. Jensen, J.: FIELD: a program for simulating ultrasound systems. In: 10th Nordic-Baltic Conference on Biomedical Imaging Published in Medical Biological Engineering Computing, vol. 34, pp. 351–353 (1996)
19. Achim, A., Bezerianos, A., Tsakalides, P.: Novel Bayesian multiscale method for speckle removal in medical ultrasound images. IEEE Trans. Med. Imaging **20**(8), 772–783 (2001)
20. Shruthi, G., Usha, B.S., Sandya, S.: A novel approach for speckle reduction and enhancement of ultrasound images. Int. J. Comput. Appl. **45**(20), 14–20 (2012)
21. Wu, S., Zhu, Q., Xie, Y.: Evaluation of various speckle reduction filters on medical ultrasound images. In: Engineering in Medicine and Biology Society, pp. 1148–1151, July 2013
22. Sivakumar, R., Gayathri, M.K., Nedumaran, D.: Speckle filtering of ultrasound B-scan images- a comparative study between spatial and diffusion filters. In: IEEE Conference on Open Systems, pp. 80–85, December 2010
23. Nageswari, C., Prabha, K.: Despeckle process in ultrasound fetal image using hybrid spatial filters. In: International Conference on Green Computing, Communication and Conservation of Energy, pp. 174–179, December 2013
24. Srivastava, R., Gupta, J., Parthasarthy, H.: Comparison of PDE based on other techniques for speckle reduction from digitally reconstructed holographic images. Opt. Lasers Eng. **48**(5), 626–635 (2010)

25. Michailovich, O., Tannenbaum, A.: Blind deconvolution of medical ultrasound images: a parametric inverse filtering approach. IEEE Trans. Image Process. **16**(12), 3005–3019 (2007)
26. Rasmussen, M., Jensen, J.: 3-D ultrasound imaging performance of a row-column addressed 2-D array transducer: a measurement study. In: IEEE International Ultrasonics Symposium, pp. 1460–1463 (2013)
27. Xu, L., et al.: Oil spill candidate detection from SAR imagery using a thresholding-guided stochastically fully-connected conditional random field model. In: 2015 IEEE Conference on Computer Vision and Pattern Recognition Workshops (CVPRW), pp. 79–86 (2015)
28. Shafiee, M.J., Chung A.G., Wong, A., Fieguth P.: Improved fine structure modeling via guided stochastic clique formation in fully connected conditional random fields. In: 2015 IEEE International Conference on Image Processing (ICIP), pp. 3260–3264 (2015)

Curvelet-Based Bayesian Estimator for Speckle Suppression in Ultrasound Imaging

Rafat Damseh[1]($\boxtimes$) and M. Omair Ahmad[2]

[1] Institut de génie biomédical, École Polytechnique de Montréal,
Montreal, QC H3T 1J4, Canada
rafat.damseh@polymtl.ca
[2] Department of Electrical and Computer Engineering,
Concordia University, Montreal, QC H3G 1M8, Canada
omair@encs.concordia.ca

Abstract. Ultrasound images are inherently affected by speckle noise, and thus reducing this noise is crucial for successful post-processing. One powerful approach for noise suppression in digital images is Bayesian estimation. In the Bayesian-based despeckling schemes, the choice of suitable statistical models and the development of a shrinkage function for estimation of the noise-free signal are the major concerns. In this paper, a novel curvelet-based Bayesian estimator for speckle removal in ultrasound images is developed. The curvelet coefficients of the degradation model of the noisy ultrasound image are decomposed into two components, namely noise-free signal and signal-dependent noise. The Cauchy and two-sided exponential distributions are assumed to be statistical models for the two components, respectively, and an efficient low-complexity realization of the Bayesian estimator is proposed. The experimental results demonstrate the superiority of the proposed despeckling scheme in achieving significant speckle suppression and preserving image details.

Keywords: Ultrasound imaging · Curvelet transform · Speckle noise · Bayesian estimation · Statistical modeling

1 Introduction

Ultrasound imaging is important for medical diagnosis and has the advantages of cost effectiveness, port-ability, acceptability and safety [1]. However, ultrasound images are of relatively poor quality due to its contamination by the speckle noise, which considerably degrades the image quality and leads to a negative impact on the diagnostic task. Thus, reducing speckle noise while preserving anatomic information is necessary to better delineate the regions of interest in ultrasound images.

This work was supported in part by the Natural Sciences and Engineering Research Council (NSERC) of Canada and in part by the Regroupement Strategique en Micro-electronique du Quebec (ReSMiQ).

© Springer International Publishing AG 2017
F. Karray et al. (Eds.): ICIAR 2017, LNCS 10317, pp. 117–124, 2017.
DOI: 10.1007/978-3-319-59876-5_14

In the work of speckle suppression in ultrasound images, many spatial-based techniques that employ either single-scale or multi-scale filtering have been developed in the literature [2–4]. Early-developed single-scale spatial filtering [2] are limited in their capability for significantly reducing the speckle noise. More promising spatial single-scale techniques such as those using bilateral filtering [4] and nonlocal filtering [3] have been recently proposed. In general, the performances of these techniques depend on the size of the filter window, and hence, for a satisfactory speckle suppression, they require large computational time. Alternatively, multi-scale spatial techniques [5], based on partial differential equations, have been investigated in the literature. These techniques are iterative and can produce smooth images with preserved edges. However, important structural details are unfortunately degraded during the iteration process. As an appropriate alternative to spatial-based speckle suppression in ultrasound images, many other despeckling techniques based on different transform domains, such as the ones of wavelet, contourlet, and curvelet, have been recently proposed in the literature [6–8]. Wavelet transform has a good reputation as a tool for noise reduction but has the drawback of poor directionality, which makes its usage limited in many applications. Using contourlet transform provides an improved noise reduction performance due to its property of flexible directional decomposability. However, curvelet transform offers a higher directional sensitivity than that of contourlet transform and is more efficient in representing the curve-like details in images.

For the development of despeckling techniques based on transform domains, thresholding [7] has been presented as a technique to build linear estimators of the noise-free signal coefficients. However, the main drawback of this thresholding technique is in the difficulty of determining a suitable threshold value. To circumvent this problem, non-linear estimators [6] have been statistically developed in a Bayesian estimation framework.

The suitability of the probabilistic models assumed for the transform domain coefficients is a major concern in the development of an efficient Bayesian despeckling scheme. Also, when investigating suitable probabilistic models, the computational complexity of the Bayesian estimator should be taken into account. In this paper, to achieve a satisfactory performance for despeckling of ultrasound images at a lower computational effort, a new curvelet-based Bayesian scheme is proposed. The multiplicative degradation model representing an observed ultrasound image is decomposed into an additive model consisting of noise-free and signal-dependent noise components. Two-sided exponential distribution is used to model the curvelet coefficients of the signal-dependant noise component. This model is employed along with the Cauchy distribution to construct a Bayesian shrinkage function of low-complexity. The efficiency of the proposed despeckling scheme is examined on both synthetically-speckled and real ultrasound images, and its performance is compared with that of other existing despeckling schemes in the literature.

2   Modeling of Curvelet Coefficients

A noisy ultrasound image $g(i,j)$ in the spatial domain is modeled as follows

$$g(i,j) = v(i,j)s(i,j) \tag{1}$$

where $v(i,j)$ and $s(i,j)$ are, respectively, the noise-free signal and the speckle noise. However, This noisy observation can be decomposed into an additive model:

$$g(i,j) = v(i,j) + (s(i,j) - 1)v(i,j)$$
$$= v(i,j) + u(i,j) \tag{2}$$

where $u(i,j)$ is a signal-dependant noise. By applying the curvelet transform on (2) at level l, we have

$$y^{[l,d]}(i,j) = x^{[l,d]}(i,j) + n^{[l,d]}(i,j) \tag{3}$$

where $y^{[l,d]}(i,j)$, $x^{[l,d]}(i,j)$ and $n^{[l,d]}(i,j)$ represent, respectively, the (i,j)th curvelet coefficient of the noisy ultrasound image, the corresponding noise free counterpart and the corresponding signal-dependant noise at direction $d = 1, 2, 3, \cdots, D$. Henceforth, the superscripts l and d and the index (i,j) will be dropped for simplicity. In this work, a Bayesian estimator for suppressing speckle inherited in ultrasound images is developed by exploiting the statistical characteristics of the curvelet coefficients in (3). For this purpose, prior probabilistic models need to be assumed for the curvelet coefficients x and n. In [9], it has been shown that the curvelet coefficients of noise-free images can be suitably modeled by the Cauchy distribution given by

$$p_x(x) = (\gamma/\pi)(x^2 + \gamma^2) \quad . \tag{4}$$

where γ is the dispersion parameter that can be estimated by minimizing the function

$$\int \left| \hat{\phi}_y(t) - \phi_y(t) \right| e^{-t^2} dt \tag{5}$$

In (5), $\hat{\phi}_y(t)$ is the empirical characteristic function corresponding to the curvelet coefficients y of the noisy image, $\phi_y(t) = \phi_x(t)\phi_\epsilon(t)$, $\phi_x(t) = e^{-\gamma|t|}$, and $\phi_\epsilon(t) = e^{-(\sigma_\epsilon^2/2)|t|^2}$ with the standard deviation σ_ϵ obtained as

$$\sigma_\epsilon = \frac{\text{MAD}(y(i,j))}{0.6745} \tag{6}$$

where MAD is the median absolute deviation operation. Now, to construct the Bayesian estimator, a prior statistical distribution for the curvelet coefficients n should also be assumed. From experimental observation, it is noticed that the tail part of the empirical distribution of n decays at a low rate. Hence, the use a two-sided exponential (TSE) distribution given by

$$p_n(n) = \frac{1}{2\beta} e^{-|n|/\beta} \tag{7}$$

where β is a scale parameter, is proposed. The parameter β can be estimated by using the following expression [10]:

$$\tilde{\beta} = \exp\left[\left\{\frac{1}{N_1 N_2} \sum_{i=1}^{N_1} \sum_{j=1}^{N_2} \log(y(i,j))\right\} + \xi\right] \tag{8}$$

where ξ is the Euler-Mascheroni constant and N_1 and N_2 define the size $N_1 \times N_2$ of the curvelet subband considered.

3 Bayesian Estimator

Due to the fact that each of the Cauchy and TSE distributions has only one parameter, one could expect the process of Bayesian estimation to be of lower complexity. The estimates of the noise-free curvelet coefficients $\hat{x}$ obtained in the minimum mean square error (MMSE) Bayesian framework are given by the following shrinkage function

$$\hat{x}(y) = \frac{\int p_n(y-x) p_x(x) x \, dx}{\int p_n(y-x) p_x(x)} \tag{9}$$

The bayseian estimates in the above expression are computed by replacing the associated integrals with infinite series as suggested in [10]. Accordingly, the Bayesian estimates are calculated by the shrinkage function:

$$\hat{x}(y) = \frac{e^{-y/\beta}\left[f_{11}(y) - \zeta_1\right] + e^{y/\beta}\left[-f_{12}(y) + \zeta_1\right]}{e^{-y/\beta}\left[f_{21}(y) + \zeta_2\right] + e^{y/\beta}\left[-f_{22}(y) + \zeta_2\right]} \tag{10}$$

where

$$f_{11}(y) = f_{12}(-y) = \sin(\gamma/\beta)\left[\text{Im}\left\{E_1(\frac{-y+j\gamma}{\beta})\right\} - \text{Si}(\gamma/\beta) + \frac{\pi}{2}\right]$$
$$-\cos(\gamma/\beta)\left[\text{Re}\left\{E_1(\frac{-y+j\gamma}{\beta})\right\} + \text{Ci}(\gamma/\beta)\right], \tag{11}$$

$$f_{21}(y) = -f_{22}(-y) = -\frac{1}{\gamma}\sin(\gamma/\beta)\left[\text{Re}\left\{E_1(\frac{-y+j\gamma}{\beta})\right\} + \text{Ci}(\gamma/\beta)\right]$$
$$-\frac{1}{\gamma}\cos(\gamma/\beta)\left[\text{Im}\left\{E_1(\frac{-y+j\gamma}{\beta})\right\} - \text{Si}(\gamma/\beta) + \frac{\pi}{2}\right], \tag{12}$$

$$\zeta_1 = \lim_{y\to\infty} f_{12}(y) = \sin(\gamma/\beta)\left[-\text{Si}(\gamma/\beta) + \frac{\pi}{2}\right] - \cos(\gamma/\beta)\text{Ci}(\gamma/\beta), \text{ and} \tag{13}$$

$$\zeta_2 = \lim_{y\to\infty} f_{22}(y) = \frac{1}{\gamma}\sin(\gamma/\beta)\text{Ci}(\gamma/\beta) + \frac{1}{\gamma}\cos(\gamma/\beta)\left[-\text{Si}(\gamma/\beta) + \frac{\pi}{2}\right] \tag{14}$$

In the equations above, $j = \sqrt{-1}$, $\text{Im}\{\cdot\}$ and $\text{Re}\{\cdot\}$ are, respectively, the imaginary and real parts of their corresponding argument, and $E_1(\cdot)$, $\text{Si}(\cdot)$ and $\text{Ci}(\cdot)$ are the exponential, sine and cosine integral functions, respectively, obtained as in [10].

4   Experimental Results

In this section, experimentations are conducted to assess the performance of the proposed curvelet-based despeckling scheme. The results are compared with that of existing despeckling schemes that use improved-Lee filtering [2], adaptive-wavelet shrinkage [6], and contourlet thresholding [7]. Performance evaluation is done on synthetically-speckled and real ultrasound images. In the implementation of the proposed speckling scheme, the 5-level decomposition of the curvelet transform is applied. From the experimental observation, applying a higher level of decomposition of the curvelet transform does not lead to any improvement in the despeckling performance. The curvelet transform is shift-variant, thus the cycle spinning [11] procedure is applied on the observed noisy image to avoid pseudo-Gibbs artifacts around discontinuities. It is to be noted that, in the proposed despeckling scheme, only the detail curvelet coefficients are despeckled using the developed shrinkage function in (10).

Peak signal-to-noise ratio (PSNR) and structural similarity index (SSIM) are used as quantitative measures to assess the despeckling performance of the various schemes when applied on synthetically-speckled images. Table 1 gives the

Table 1. The PSNR/SSIM values obtained when applying the various despeckling schemes on *Lena* and *Boat* images contaminated by speckle noise at different levels.

Despeckling scheme	Standard deviation of noise						
	0.1	0.2	0.3	0.4	0.5	0.7	1
	Lena						
[2]	31.42/0.750	26.63/0.627	24.96/0.542	23.71/0.488	21.95/0.447	19.76/0.436	18.16/0.355
[6]	32.11/0.757	27.74/0.636	25.97/0.567	24.63/0.492	22.07/0.452	20.39/0.441	19.04/0.361
[7]	31.75/0.751	27.97/0.655	26.19/0.557	25.24/0.517	23.85/0.465	22.02/0.439	20.53/0.362
Proposed	**32.24/0.760**	**28.7/0.684**	**27.31/0.569**	**26.13/0.528**	**25.14/0.481**	**23.15/0.448**	**21.91/0.390**
	Boat						
[2]	28.45/0.630	23.59/0.612	21.81/0.468	20.74/0.433	19.88/0.381	18.07/0.342	16.07/0.306
[6]	29.11/0.659	24.53/0.604	22.49/0.501	21.55/0.438	20.75/0.403	19.21/0.357	17.41/0.316
[7]	29.75/0.679	25.15/0.613	23.61/0.514	22.68/0.464	21.85/0.392	20.44/0.350	18.95/0.319
Proposed	**30.38/0.681**	**25.64/0.616**	**24.42/0.520**	**23.61/0.488**	**23.03/0.429**	**21.79/0.384**	**19.87/0.342**

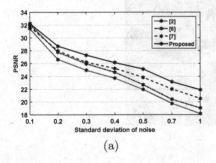
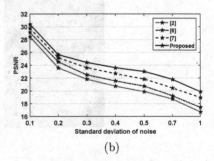

(a)   (b)

Fig. 1. Quantitative comparison between the various despeckling schemes in terms of PSNR values: (a) *Lena* image; (b) *Boat* image.

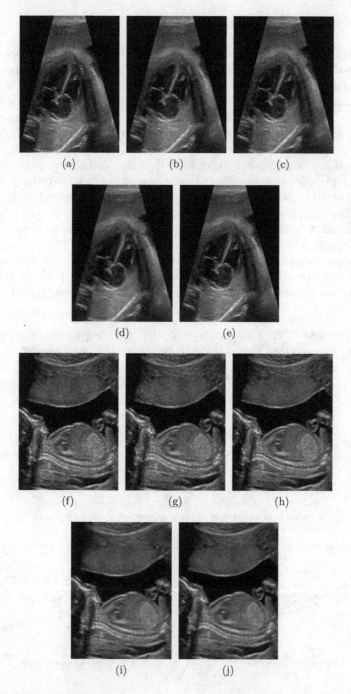

Fig. 2. Qualitative comparison between the various despeckling schemes. (a), (f) Noisy ultrasound images. Despeckled images obtained by applying the schemes in (b), (g) [2], (c), (h) [6], (d), (i) [7] and (e), (j) the proposed scheme.

PSNR and SSIM values obtained after despeckling two synthetically-speckled images of size 512×512, namely, *Lena* and *Boat*. It is obviously seen from this table that, in all cases, the proposed despeckling scheme provides higher values of PSNR and SSIM compared to that provided by the other schemes. To have a better insight on the despeckling performance of the various schemes, the PNSR results in Table 1 are visualized in Fig. 1. It is clearly seen in this figure that the superiority of the proposed scheme is more evident when a higher level of speckle noise is introduced to the test images. In order to study the despeckling performance on real ultrasound images, two images obtained from [12] and shown in Fig. 2 are used. Since the noise-free images cannot be made available, one can only give a subjective evaluation of the performance of the various despeckling schemes. From Fig. 2, it is clearly seen that the schemes in [2] and [6] provide despeckled images that suffer from the presence of visually noticeable speckle noise. On the other hand, the scheme in [7] severely over-smooth the noisy images thus providing despeckled images in which some of the texture details are lost. However, the proposed despeckling scheme results in images with not only a significant suppression of speckle noise but also a good preservation of the textures of the original images.

5   Conclusion

In this paper, a new curvelet-based Bayesian estimator for despeckling of ultra-sound images has been developed. First, the observed ultrasound image is decomposed into two additive components, namely, noise-free signal and signal-dependant noise. The Cauchy and two-sided exponential distributions have been selected as prior probabilistic models for the curvelet coefficients of the two decomposed components, respectively. The proposed probabilistic models of the curvelet coefficients have been employed in building a Bayesian shrinkage function to provide estimates of the noise-free curvelet coefficients. To demonstrate the performance of the proposed scheme, experiments have been conducted on synthetically-speckled and real ultrasound images. In comparison with some other existing despeckling schemes, the results have shown that the proposed scheme provides higher PSNR/SSIM values and gives well-despeckled images with better diagnostic details.

References

1. Dhawan, A.P.: Medical Image Analysis, vol. 31. Wiley, Hoboken (2011)
2. Loupas, T., McDicken, W., Allan, P.: An adaptive weighted median filter for speckle suppression in medical ultrasonic images. IEEE Trans. Circuits Syst. **36**(1), 129–135 (1989)
3. Coupé, P., Hellier, P., Kervrann, C., Barillot, C.: Nonlocal means-based speckle filtering for ultrasound images. IEEE Trans. Image Process. **18**(10), 2221–2229 (2009)

4. Sridhar, B., Reddy, K., Prasad, A.: An unsupervisory qualitative image enhancement using adaptive morphological bilateral filter for medical images. Int. J. Comput. Appl. **10**(2i), 1 (2014)
5. Abd-Elmoniem, K.Z., Youssef, A.B., Kadah, Y.M.: Real-time speckle reduction and coherence enhancement in ultrasound imaging via nonlinear anisotropic diffusion. IEEE Trans. Biomed. Eng. **49**(9), 997–1014 (2002)
6. Swamy, M., Bhuiyan, M., Ahmad, M.: Spatially adaptive thresholding in wavelet domain for despeckling of ultrasound images. IET Image Process. **3**(3), 147–162 (2009)
7. Hiremath, P., Akkasaligar, P.T., Badiger, S.: Speckle reducing contourlet transform for medical ultrasound images. Int. J. Compt. Inf. Eng. **4**(4), 284–291 (2010)
8. Jian, Z., Yu, Z., Yu, L., Rao, B., Chen, Z., Tromberg, B.J.: Speckle attenuation in optical coherence tomography by curvelet shrinkage. Opt. Lett. **34**(10), 1516–1518 (2009)
9. Deng, C., Wang, S., Sun, H., Cao, H.: Multiplicative spread spectrum watermarks detection performance analysis in curvelet domain. In: 2009 International Conference on E-Business and Information System Security (2009)
10. Damseh, R.R., Ahmad, M.O.: A low-complexity MMSE Bayesian estimator for suppression of speckle in SAR images. In: 2016 IEEE International Symposium on Circuits and Systems (ISCAS), pp. 1002–1005. IEEE (2016)
11. Temizel, A., Vlachos, T., Visioprime, W.: Wavelet domain image resolution enhancement using cycle-spinning. Electron. Lett. **41**(3), 119–121 (2005)
12. Siemens Healthineers. https://www.healthcare.siemens.com/ultrasound. Accessed 06 Jan 2017

Object Boundary Based
Denoising for Depth Images

Mayoore S. Jaiswal[1(✉)], Yu-Ying Wang[2], and Ming-Ting Sun[1]

[1] University of Washington, Seattle, WA 98195, USA
mayoore@uw.edu
[2] National Taiwan University, Taipei, Taiwan

Abstract. Economical RGB-D cameras such as Kinect can produce both RGB and depth (RGB-D) images in real-time. The accuracy of various RGB-D related applications suffers from depth image noise. This paper proposes a solution to the problem by estimating depth edges that correspond to the object boundaries and using them as priors in the hole filling process. This method exhibits quantitative and qualitative improvements over the current state-of-the-art methods.

Keywords: RGB-D camera · Depth image · Hole-filling · Edge detection

1 Introduction

Popular commercial RGB-D cameras can capture both color and depth images at an economical price, and have been used widely in many applications. However, the quality of the depth images is degraded by various holes and noises. A Kinect depth image and its associated color image is shown in Fig. 1. Shadow holes caused by the displacement between the IR projector and IR sensor, generally occurs along object boundaries in depth images as highlighted in red in Fig. 1. Additionally, depth sensors often are unable to accurately estimate the depth values of smooth, shiny, and reflective surfaces because IR rays reflected from these surfaces are weak or scattered. This results in random noises in depth images as highlighted in green in Fig. 1.

Depth hole filling and denoising is challenging because it needs to accurately estimate depth while preserving depth discontinuities around object boundaries. Many approaches for denoising and hole-filling for depth images have been proposed. They fall into one of the following broad categories: (1) filtering based [1, 2, 6–8], (2) plane-fitting based [3, 4], and (3) probabilistic methods [5]. Filtering based approaches use variations of bilateral filters to denoise the depth image. However, due to the false edges from the texture of color images, illumination of the scene, or weak edges due to the similar color of the object and the background, the bilateral filters or the variations of thereof may not produce the best result. Plane fitting based methods fit planes to neighborhood pixels. However, these methods fail when there are object boundaries inside the holes and in complex scenes consisting of multiple objects with complex geometries. For an example of the probabilistic methods, an autoregressive model with color images is proposed in [5]. Though this method produces qualitatively good results in large structures, it blurs smaller object boundaries.

© Springer International Publishing AG 2017
F. Karray et al. (Eds.): ICIAR 2017, LNCS 10317, pp. 125–133, 2017.
DOI: 10.1007/978-3-319-59876-5_15

Fig. 1. A depth image and its corresponding color image obtained from a Kinect camera. (Color figure online)

The main contribution of this paper is that we propose a new approach which focuses on constructing edges which correspond to the true object boundaries and using them to guide the hole filling process. We propose a novel depth edge estimation algorithm which estimates depth edges using information from the color edges and noisy depth edges by observing the statistics of the neighborhood depth pixel values. We use adaptive thresholds in detecting color edges to overcome the problem when objects have similar color as background. The refined depth edge image can be used in other applications, such as depth denoising and object segmentation. In the hole-filling process, we use a trilateral filter with an adaptive window size to prevent pixels from the opposite side of an object boundary being used as reference pixels. Experimental results show the effectiveness of our proposed algorithm.

The rest of this paper is organized as follows. Section 2 describes the proposed method in detail. Section 3 shows the quantitative and qualitative effectiveness of the proposed method using various datasets. Finally, Sect. 4 lists the conclusions.

2   Proposed Method

Our proposed approach is shown in Fig. 2 and described in the following subsections.

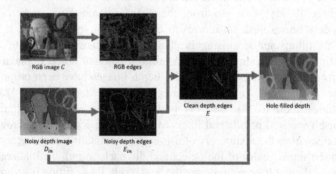

Fig. 2. Illustration of the proposed method.

2.1   Color Edge and Noisy Depth Edge Extraction

The objective of the color edge extraction module is to be able to extract all possible edges even when the foreground and background have similar colors. The edges from the color image were extracted using Canny edge detector [9]. To reduce the localization error, we use a small standard deviation of 0.1 for the Gaussian filter. Also, in order to detect all possible edges, we use 0.01 and 0.1 for the lower and upper hysteresis thresholds respectively. These choices could yield many edges caused by texture and illumination changes in the scene which will be rejected later. Many previous algorithms have problems when the foreground object has a similar color as the background, since the color edges will be faint and may not be detected by the edge detector. To overcome this problem, we use a threshold value proportional to the variance of the pixel color values inside the hole region. With this adaptive thresholding strategy, the faint color edges can be detected inside the hole regions, without generating many false edges in other regions.

Small random perturbations in the noisy depth image are filtered out using a 5×5 median filter before the Canny edge detector is applied to the depth image.

2.2   Generating Clean Depth Edges

The noisy depth edge image E_{in} obtained from the previous step has false edges caused by random noise and holes, and true edges produced by object boundaries. The clean depth edges are produced in two stages. First, edges coinciding with object boundaries are identified and retained. Other false edges are suppressed. Second, object boundaries in the hole regions of the raw depth image are extracted from the color edges.

To identify edges associated with boundary edges, it is observed that if there are less hole-pixels in the input depth image D_{in} in the neighborhood of an edge pixel location $E_{in}(p)$, this edge pixel is likely caused by the presence of an object boundary. Otherwise it is probably caused by erroneous depth values in the input depth image. Let I be a binary mask and I_q is the value of the mask at pixel q. $I_q = 1$ if q is a hole-pixel in the input depth image D_{in}, otherwise $I_q = 0$. Equation (1) is used to find edges caused by the presence of object boundaries.

$$E_1(p) = \begin{cases} 1, & if \sum_{q \in N_m(p)} G_q \cdot I_q < th, and\, E_{in}(p) = 1; \\ 0; & otherwise. \end{cases} \tag{1}$$

where $E_{in}(p) = 1$ means p is a pixel on an edge in the noisy depth edge image E_{in}, th is a threshold, $N_m(p)$ are pixels in an $m \times m$ local neighborhood centered at p. G_q is the weight at location q with an $m \times m$ Gaussian kernel G centered at p. $E_1(p) = 1$ indicates that p is on an edge caused by an object boundary. Gaussian kernel G with σ_m is used to spatially weigh the depth pixel locations within the $m \times m$ local neighborhood. By using a Gaussian kernel, a distant hole pixel from p would weigh less than a hole pixel close to p within the neighborhood.

Next, false edges produced by random noise and holes in the depth image are removed. The variation of depth pixel values in the locale of false edges caused by random noise in the depth image is small; whereas in the neighborhood of false edges caused by holes in the depth image, the variation of depth pixel values is generally higher when there is an object boundary within the hole. The variation of depth values is evaluated by fitting a plane to an $h \times h$ local neighborhood of a false edge pixel in the raw depth image and computing the distances from each depth pixel in this local neighborhood to the fitted plane. If the variance of the distances is larger than a threshold, the false edge pixel was likely caused by a hole with an object boundary. These false edges are replaced by edges in the color image. The window size, h is adaptively chosen such that at least 80% of the pixels in the $h \times h$ neighborhood are non-hole depth pixels. The adaptive window size ensures that sufficient number of non-hole depth pixels are taken into account. The initial window size of h is set to 21.

To find object boundaries in the hole regions of the raw depth image D_{in}, we elicit the edges from the color image. The depth hole regions with likely object boundaries are dilated with a 3×3 structuring element with all the elements equal to 1 and the origin at the center. The intersection of the dilated region and the color edge image is the depth edge image corresponding to object boundaries. The clean depth edge image E is the combination of this depth edge image and $E_1(p)$ from Eq. (1).

2.3   Depth Hole Filling

Once the clean depth edges are recovered, we use a trilateral filter as described in Algorithm 1 to construct the recovered depth image D_{rec}. An edge and color aware adaptive trilateral filter in Eq. (2) is used to fill the hole-pixels progressively from the perimeter of the hole to its center. The non-hole pixels outside the perimeter of the hole is used to estimate the hole depth pixels. The size of the local neighborhood $n \times n$ used to fill a depth hole-pixel p varies adaptively to ensure that at least 80% of pixels in the neighborhood are non-hole pixels with initial $n = 7$. With this approach, more non-hole pixels are used in the trilateral filter than using a fixed neighborhood size.

Algorithm 1: Depth hole-filling with edge and color aware adaptive trilateral filter

---

Input: Recovered depth edge image E, input raw depth image D_{in}
Output: Recovered depth D_{rec}
Initialization: Push every hole-pixel in input depth image $D_{in}(p)$ into a priority queue Q with decreasing ordered by the percentage of non-hole pixels in a $n \times n$ window centered at p, initial $n = 7$
While Q is not empty: $D_{in}(p) \leftarrow$ pop Q
   Update $D_{in}(p)$:
     $num \leftarrow$ percentage of non-hole pixels in $n \times n$ window centered at $D_{rec}(p)$
     If $num < 80\%$: $n \leftarrow n + 2$
        Push $D_{in}(p)$ into Q
     Else: Compute $D_{rec}(p)$ using Equation (2)
     End
End
Return D_{rec}

The depth hole-pixels $D_{rec}(p)$ can be found as:

$$D_{rec}(p) = \frac{1}{k_p} \sum_{q \in N_n(p)} D_{in}(q) \cdot f_s(\|q - p\|) \cdot f_r(E, q, p) \cdot f_i(C, q, p) \tag{2}$$

where $f_s(\cdot)$ is a zero mean spatial Gaussian kernel with a standard deviation σ_d. k_p is a normalizing factor. $\|\cdot\|$ is the Euclidean distance measure. The range kernel $f_r(\cdot)$ is a binary indicator defined as:

$$f_r(E,q,p) = \begin{cases} 1, \textit{if the shortest path connecting pixels } q \textit{ and } p \\ \quad (\textit{excluding pixels } q \textit{ and } p) \\ \quad \textit{does not intersect an edge in } E; \\ 0, \textit{ otherwise.} \end{cases} \tag{3}$$

The range kernel f_r assures that only pixels on the same side of an edge are averaged. The intensity kernel $f_i(\cdot)$ is a binary indicator defined as:

$$f_i(C,q,p) = \begin{cases} 1, & \textit{if } \|C(q) - C(p)\| \\ 0, & \textit{otherwise.} \end{cases} \tag{4}$$

where C is the color image with the R, G and B components normalized to $[0, 1]$ and ε is a threshold. The intensity kernel $f_i(\cdot)$ ensures that neighborhood depth pixels with similar color in the corresponding color image are used to fill a hole depth pixel. When p is an edge pixel, it is ambiguous which side of the edge does p belong to. In this case, a surrogate pixel p' which is a pixel in the 8 connected neighborhood of p that is not an edge pixel and has the closest color to p in the corresponding color image is found using Eq. (5). The surrogate pixel p' is used to determine which side of an edge, pixel p belongs to.

$$p' = \min_{u \in N'(p)} \|C(p) - C(u)\| \tag{5}$$

where u is a pixel in the neighbourhood $N'(p)$ such that it is one of the 8 connected pixels of p but is not an edge pixel. $C(p)$ is the RGB color value of pixel p.

3   Experimental Results

In this section, we present results from evaluating our method with respect to other state-of-the-art methods using Mean Absolute Error (MAE) as in Eq. (6).

$$MAE = \frac{1}{N} \sum_{p \in \{P | D_{syn}=0 \cap D_{gt} \neq 0\}} |D_{rec}(p) - D_{gt}(p)| \tag{6}$$

where D_{rec}, D_{gt} and D_{syn} are the recovered depth, ground truth depth and synthetically degraded depth, respectively. N is the number of hole-pixels in the synthetically degraded depth that are not hole-pixels in the ground truth depth. MAE is calculated only in the synthetically degraded hole-pixels in D_{syn}, which means $D_{syn}(p) = 0$ but the ground truth depth image D_{gt} has valid depth pixel values. This enables us to quantitatively evaluate the algorithms' ability to recover synthetic hole-pixels in object boundary and smooth regions. The parameters settings are: $m = 9$, $\sigma_m = 1$, $th = 0.055$, $t = 0.09$, $\sigma_d = 1$, and ε is 0.1,

Yang et al. introduced a synthetically degraded dataset in [5] which consists of a subset of the Middlebury dataset [10] with random noise and structural holes created on the depth images to imitate the holes in depth images produced by a Kinect. Figure 3 shows a sample of the images in the Yang dataset. To mimic Kinect depth images, structural missing pixels are created along depth discontinuities, and random missing pixels are generated in flat areas.

The proposed method is compared to bicubic interpolation and four state-of-the-art methods JBF [11], Guide [1], CLMF [2] and AAR [5] to produce recovered depth images. The structure similarity index (SSIM) is calculated as described in [12]. Table 1 tabulates the results. The proposed method gets the least *MAE* score on all the images. On average the method reduces the MAE error by 14.02%. Our method also gets better or comparable SSIM values on the dataset.

Fig. 3. Thumbnails from the synthetically degraded dataset [5]. The upper row shows the color images. Middle row shows the corresponding depth images degraded by random noise and structural holes. The bottom row shows the ground truth depth images. (Color figure online)

Table 1. Table lists the quantitative results comparing our method to the state-of-the-art methods tested on the synthetically degraded dataset [5].

Algorithm	Art		Book		Dolls		Laundry		Moebius		Reindeer	
	MAE	SSIM	MAE	SSIM	MAE	SSIM	MAE	SSIM	MAE	SSIM	MAE	SSIM
AAR	4.78	0.987	1.99	0.991	1.91	0.988	3.14	0.982	2.40	0.989	2.12	0.990
Bicubic	7.45	0.984	2.30	0.991	2.12	0.986	3.81	0.980	2.65	0.988	3.05	0.987
CLMF	6.40	0.988	2.07	0.992	1.92	0.988	3.51	0.982	2.38	0.990	2.80	0.988
Guide	6.94	0.985	2.10	0.992	1.94	**0.989**	3.42	**0.985**	2.42	0.990	2.81	0.989
JBF	6.97	0.983	2.36	0.989	2.13	0.985	3.70	0.979	2.76	0.986	2.95	0.984
Ours	**4.67**	**0.988**	**1.61**	**0.993**	**1.79**	0.988	**2.50**	0.984	**1.89**	**0.991**	**1.81**	**0.991**

To test the proposed method's ability to reconstruct depth values in hole regions, we repeated the above experiment by calculating the mean absolute error using:

$$MAE_{ob} = \frac{1}{N} \sum_{p \in \{P \mid D_{syn}=0 \cap D_{gt} \neq 0 \cap J'=1\}} \left| D_{rec}(p) - D_{gt}(p) \right| \tag{7}$$

where J is a mask indicating the depth hole regions which have object boundaries. N is the number of hole-pixels in the synthetically degraded depth that are not hole-pixels in the ground truth but in J. MAE_{ob} is calculated only in the synthetically degraded hole-pixels in D_{syn} which have object boundaries inside the hole region, which means $D_{syn}(p) = 0$ and $M(p) = 1$ but the ground truth depth image D_{gt} has valid depth pixel values. Table 2 tabulates the results. The proposed method gets the best results on all the images. On average the method reduces the MAE error by 36.78%.

Table 2. Table lists the MAE_{ob} results comparing our method to the state-of-the-art methods tested on the synthetically degraded dataset [5].

Algorithm	Art	Book	Dolls	Laundry	Moebius	Reindeer
AAR	10.50	5.93	4.62	8.06	5.55	6.17
Bicubic	17.45	7.71	5.67	10.65	6.55	10.80
CLMF	14.57	6.84	5.08	9.85	5.80	9.93
Guide	16.08	6.96	5.13	9.53	5.93	9.91
JBF	16.35	8.00	5.69	10.15	6.92	10.17
Ours	**10.31**	**3.80**	**3.87**	**5.58**	**3.54**	**4.34**

Figure 4 further illustrates the effectiveness of the proposed method in preserving depth edges. They show cropped and zoomed portions of the 'Book' images in the synthetically degraded dataset. Bicubic interpolation blurs the depth edge. Our method produces the least distorted edges. Guided and CLMF blur the object boundaries similar to bicubic. JBF and AAR produce jagged artefacts near the edges. Our method is able to fill holes while producing clean edges at object boundaries. Quantitative and qualitative experiments prove the strength of the proposed method.

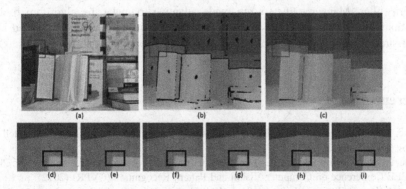

Fig. 4. Visual comparison results on the 'Book' image. (a), (b) and (c) are the color, synthetically degraded depth and the ground-truth depth respectively. (d) – (i) show zoomed in view in the red rectangle region. (d) bicubic (e) JBF (f) Guide (g) CLMF (h) AAR and (i) ours. Note that our method preserves the edge structures the best, there are no "orange" colored pixels in the colored recovered depth image as illustrated in the boxes highlighted in black. (Color figure online)

Next we test the proposed method using color and depth data from a Kinect camera. For this purpose, we use the Kinect data provided by [5]. The depth image was denoised using AAR, Guide, JBF and our method. Due to the lack of ground truth, we present only a qualitative comparison of the results. Figure 5 compares the AAR, Guide, JBF, and our method. The cropped and zoomed images show that the proposed method preserves the depth edges better than other methods. JBF, AAR and Guide blur and distort the object boundaries in the depth image as highlighted in Fig. 5.

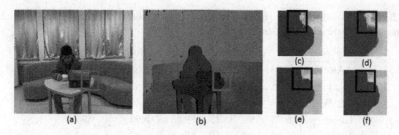

Fig. 5. A visual comparison of results from the state-of-the-art methods on a set of Kinect images. The thumbnails are zoomed in portions of the area highlighted in red the input Kinect images. (a) color image from Kinect, (b) depth from Kinect, (c) AAR, (d) Guide, (e) JBF, and (f) Our method. (Color figure online)

4   Conclusion

In this paper, we propose a new approach to denoise and fill the holes in depth images. The proposed algorithm first extracts an initial depth edge image from the raw depth image which is noisy due to the inaccuracies present in the raw depth image. This initial edge image is refined by using the edges from the color images. The clean depth edge image is used to fill the holes in the depth image by using an edge and color aware trilateral filter. Many parameters are made adaptive to make the algorithm robust. Quantitative and qualitative experimental results demonstrate that the proposed hole filling strategy can generate more accurate depth images than existing methods.

References

1. He, K., Sun, J., Tang, X.: Guided image filtering. IEEE Trans. Pattern Anal. Mach. Intell. **35** (6), 1397–1409 (2013)
2. Lu, J., Shi, K., Min, D., Lin, L., Do, M.N.: Cross-based local multipoint filtering. In: 2012 IEEE Conference on Computer Vision and Pattern Recognition (CVPR) (2012)
3. Xu, L., Au, O.C., Sun, W., Li, Y., Li, J.: Hybrid plane fitting for depth estimation. In: Signal & Information Processing Association Annual Summit and Conference (APSIPA ASC) (2012)
4. Matsumoto, K., De Sorbier, F., Saito, H.: Plane fitting and depth variance based upsampling for noisy depth map from 3D-ToF cameras in real-time. In: SciTePress (2015)

5. Yang, J., Ye, X., Li, K., Hou, C.: Depth recovery using an adaptive color-guided auto-regressive model. In: Fitzgibbon, A., Lazebnik, S., Perona, P., Sato, Y., Schmid, C. (eds.) ECCV 2012. LNCS, vol. 7576, pp. 158–171. Springer, Heidelberg (2012). doi:10. 1007/978-3-642-33715-4_12
6. Camplani, M., Salgado, L.: Efficient spatio-temporal hole filling strategy for Kinect depth maps. In: International Society for Optics and Photonics, IS&T SPIE Electronic Imaging (2012)
7. Liu, J., Gong, X., Liu, J.: Guided inpainting and filtering for Kinect depth maps. In: 2012 21st International Conference on Pattern Recognition (ICPR) (2012)
8. Wang, Z., Hu, J., Wang, S., Lu, T.: Trilateral constrained sparse representation for Kinect depth hole filling. Pattern Recogn. Lett. **65**, 95–102 (2015)
9. Canny, J.: A computational approach to edge detection. IEEE Trans. Pattern Anal. Mach. Intell. **6**, 679–698 (1986)
10. Hirschmuller, H., Scharstein, D.: Evaluation of cost functions for stereo matching. In: Computer Vision and Pattern Recognition (2007)
11. Riemens, O., Gangwal, O., Barenbrug, B., Berretty, R.-P.: Multistep joint bilateral depth upsampling. In: International Society for Optics and Photonics, IS&T SPIE Electronic Imaging (2009)
12. Wang, Z., Bovik, A.C., Sheikh, H.R., Simoncelli, E.P.: Image quality assessment: from error visibility to structural similarity. IEEE Trans. Image Process. **13**(4), 600–612 (2004)

A Note on Boosting Algorithms for Image Denoising

Cory Falconer, C. Sean Bohun, and Mehran Ebrahimi[✉]

Faculty of Science, University of Ontario Institute of Technology,
2000 Simcoe Street North, Oshawa, ON L1H 7K4, Canada
{cory.falconer,sean.bohun,mehran.ebrahimi}@uoit.ca

Abstract. In recent years, non-local methods have been among most efficient tools to address the classical problem of image denoising. Recently, Romano et al. have proposed a novel algorithm aimed at "boosting" of a number of non-local denoising algorithms as a "black-box." In this manuscript, we consider this algorithm and derive an analytical expression corresponding to successive applications of their proposed "boosting scheme." Mathematically, we prove that such successive application does not always enhance the input image and is equivalent to a re-parameterization of the original "boosting" algorithm. We perform a set of computational experiments on test images to support this claim. Finally, we conclude that considering the blind application of such boosting methods as a general remedy for all denoising schemes is questionable.

Keywords: Image denoising · Nonlocal methods · Boosting

1 Introduction

Regardless of the acquisition process, digital images always contain undesired variation of pixel intensity, causing an image degradation in the form of noise, unavoidably reducing the quality of the image. Given a corrupted image $\mathbf{y}$, the goal of image denoising algorithms is to recover the original signal $\mathbf{x}$. Treating the noise to be additive, the image degradation takes the following form

$$\mathbf{y} = \mathbf{x} + \mathbf{n} \tag{1}$$

in which we assume $\mathbf{n}$ is a zero-mean additive white noise that shares no dependence on $\mathbf{x}$. A wide variety of powerful algorithms have been proposed to address the classical problem of image denoising. This includes TV denoising [1,2], bilateral filtering [3], non-Local means (NLM) [4], and block matching 3D (BM3D) [5] to list a few. It is well known that patch based denoising algorithms are quite successful removing noise, though these methods also may remove relevant image content. Naturally, multiple methods have been proposed with intent to enhance the ability of these algorithms to remove noise while retaining image content.

© Springer International Publishing AG 2017
F. Karray et al. (Eds.): ICIAR 2017, LNCS 10317, pp. 134–142, 2017.
DOI: 10.1007/978-3-319-59876-5_16

More specifically, we focus on algorithm enhancement schemes that utilize a sequential approach such as [6–8], classified by their usage of "post-processing."

The first class of methods entails recycling of noisy residuals with intent to import any "stolen" image content back into the estimation. The latter strengthens the approximation using a "cleaned" signal $\hat{\mathbf{x}}^k$, bypassing any unnecessary reintroduction of noise back into our image [7].

In [7], the authors demonstrate the effectiveness of their "boosting" method with a clear improvement in both SNR and visual clarity regardless of image structure or noise. In this paper, we raise the question if it is possible to apply this boosting algorithm sequentially, with a goal to iteratively improve the overall image quality at each iteration. We focus specifically on the Non-Local means algorithm for two reasons; (i) it is a fundamental algorithm utilizing a patch based approach and (ii) the NLM was validated as a method in which the "Boosting" scheme was applied successfully [7].

2 Non-local Means Denoising and Its Associated Boosting Algorithm

This section focusses on two separate aspects. In the first part, a review of the NLM denoising algorithm [4] is presented and in the second part, the concept of "boosting" in image denoising as introduced in [7] is reviewed.

2.1 Non-local Means Denoising

The NLM algorithm [4] exploits the innate redundancy of natural images, that replaces every pixel in the image with a weighted average of all pixels in the image. The weights are determined using neighborhood similarity of the pixels. In the process "similar" neighborhoods are assigned a high weight and dissimilar neighborhoods take low weights. Following the formation of the NLM algorithm, the neighborhood of size d is defined for $\forall\, x \in \Omega$, i.e., every point in the discrete image domain Ω, as

$$\mathcal{N}^d\{x\} = \{x + r|\, \|r\|_\infty \le d\}.$$

The NLM algorithm denoted by $\mathbf{f}(\cdot)$ denoises the intensity of every pixel $x \in \Omega$ of an image $\mathbf{y}$ via

$$\mathbf{f}\left(\mathbf{y}(x)\right) = \frac{1}{C(x)} \sum_{y \in \Omega} w(x, y)\mathbf{y}(y), \tag{2}$$

where the weight $w(x, y)$ and normalization $C(x)$ are defined as

$$w(x, y) = \exp\left(-\frac{1}{h^2} \left\|\mathbf{y}(\mathcal{N}^d\{x\}) - \mathbf{y}(\mathcal{N}^d\{y\})\right\|_{2,a}^2\right), \quad C(x) = \sum_{y \in \Omega} w(x, y)$$

respectively. Note that $\|.\|_{2,a}$ denotes the Gaussian-weighted-semi norm, for any patch $\mathbf{P}$

$$\|\mathbf{P}\|_{2,a} = \|G \star \mathbf{P}\|_{2,a}$$

where G is a Gaussian kernel with a variance of a^2, and $\star$ is the convolution operator.

2.2   Boosting of Image Denoising Algorithms

As proposed in [7], the so called "SOS boosting" of an image denoising algorithm $\mathbf{f}$, to recover a superior approximation of the image $\mathbf{x}$ given a noisy realization $\mathbf{y}$, consists of iterations of the following three steps: **S**trengthen, **O**perate, and **S**ubtract. In detail for a single iteration:

1. **S**trengthen the signal by adding the previously denoised image and original noisy input;
2. **O**perate an image denoising algorithm on the new enhanced signal;
3. **S**ubtract the previously denoised image from the result of 2.

Treating the denoising operator $\mathbf{f}(\cdot)$ as a "black-box," the generalized boosting method proposed in [7] is formulated as

$$\hat{\mathbf{x}}^{k+1} = \tau\mathbf{f}(\mathbf{y} + \rho\hat{\mathbf{x}}^k) - (\tau\rho + \tau - 1)\hat{\mathbf{x}}^k \tag{3}$$

where $\hat{\mathbf{x}}^k$ denotes the approximate solution at iteration k, and $\hat{\mathbf{x}}^0$ is an initial approximation of $\mathbf{x}$. It is necessary to mention that parameter ρ determines the steady state solution and τ controls the rate of convergence. Since the boosting operation is sequential and fixed for all iterations, we pose the assumption $\hat{\mathbf{x}}^{k+1} = \hat{\mathbf{x}}^k = \hat{\mathbf{x}}^*$ to find a steady-state solution satisfying

$$\hat{\mathbf{x}}^* = ((\rho + 1)\mathbf{I} - \rho\mathbf{f}))^{-1}\mathbf{f} \cdot \mathbf{y}. \tag{4}$$

Note that here we considered a linearization of the discretized denoising operator $\mathbf{f}$ as $\mathbf{f}(\mathbf{y}) = \mathbf{f} \cdot \mathbf{y}$, as in [7], and $\mathbf{I}$ denotes the identity matrix. As proposed by [7] the solution given by (4) is the "boosted" image that yields a superior approximation compared to the $\mathbf{f}(\mathbf{y})$, i.e., denoising of $\mathbf{y}$ using the denoising operator $\mathbf{f}$.

From the structure of (4) we define $\mathcal{B}(\mathbf{f})$ as

$$\mathcal{B}(\mathbf{f}) = ((\rho + 1)\mathbf{I} - \rho\mathbf{f}))^{-1}\mathbf{f} \tag{5}$$

which is referred to as the "boosted" form of $\mathbf{f}$ and is envisioned as a "new" denoising algorithm compared to $\mathbf{f}$. The question is raised if it would be possible to further enhance the "boosting" by considering it as a new operator and successively applying the "boosting" process.

3   Iterative "Boosting" of "Boosting" Operators

Since we are trying to apply a boosting operation onto itself, we begin the derivation by substitution $\mathcal{B}(\mathbf{f})$ in for the denoising filter matrix $\mathbf{f}$ and repeat the process for a fixed parameter ρ. The following theorem summarizes the result of n consecutive boosting operations.

Theorem 1. *Given some denoising operator $\mathbf{f}$, and some fixed parameter $\rho, \forall n \in \mathbb{N}$*

$$\mathcal{B}^n(\mathbf{f}) = ((\rho+1)^n(\mathbf{I} - \mathbf{f}) + \mathbf{f})^{-1} \mathbf{f}. \tag{6}$$

Proof. By induction, taking $n = 1$,

$$\mathcal{B}^1(\mathbf{f}) = ((\rho+1)^1(\mathbf{I} - \mathbf{f}) + \mathbf{f})^{-1} \mathbf{f} = ((\rho+1)\mathbf{I} - \rho\mathbf{f} - \mathbf{f} + \mathbf{f})^{-1} \mathbf{f}$$
$$= ((\rho+1)\mathbf{I} - \rho\mathbf{f})^{-1} \mathbf{f}.$$

Hence, $\mathcal{B}^1(\mathbf{f}) = \mathcal{B}(\mathbf{f}) = ((\rho+1)\mathbf{I} - \rho\mathbf{f})^{-1} \mathbf{f}$. Now, assuming Eq. (6) holds for some positive integer $n = k$, that is

$$\mathcal{B}^k(\mathbf{f}) = ((\rho+1)^k (\mathbf{I} - \mathbf{f}) + \mathbf{f})^{-1} \mathbf{f},$$

we find that

$$\mathcal{B}^{k+1}(\mathbf{f}) = \mathcal{B}(\mathcal{B}^k(\mathbf{f}))$$
$$= ((1+\rho)\mathbf{I} - \rho\mathcal{B}^k(\mathbf{f}))^{-1} \mathcal{B}^k(\mathbf{f})$$
$$= \left((1+\rho)\mathbf{I} - \rho\left((1+\rho)^k(\mathbf{I} - \mathbf{f}) + \mathbf{f}\right)^{-1}\mathbf{f}\right)^{-1} \left((1+\rho)^k(\mathbf{I} - \mathbf{f}) + \mathbf{f}\right)^{-1} \mathbf{f}$$
$$= \left(\left((1+\rho)^k(\mathbf{I} - \mathbf{f}) + \mathbf{f}\right)\left((1+\rho)\mathbf{I} - \rho\left((1+\rho)^k(\mathbf{I} - \mathbf{f}) + \mathbf{f}\right)^{-1}\mathbf{f}\right)\right)^{-1} \mathbf{f}$$
$$= ((1+\rho)^{k+1}(\mathbf{I} - \mathbf{f}) + (1+\rho)\mathbf{f} - \rho\mathbf{f})^{-1} \mathbf{f}.$$

This means that the theorem holds for $n = k + 1$, and this completes the induction. ∎

As illustrated by Theorem 1, a sequential application of $\mathcal{B}$ onto denoising filter matrix $\mathbf{f}$, could be performed in one step. Such sequential application is nothing but a re-parametrization of the expression. That is $(\rho + 1) \mapsto (\rho + 1)^n$, or equivalently $\rho \mapsto (\rho + 1)^n - 1$. We can also observe that if $\mathbf{I} - \mathbf{f}$ is invertible and $\rho > 0$, then

$$\lim_{n\to\infty} \mathcal{B}^n(\mathbf{f}) = \lim_{n\to\infty} ((\rho+1)^n(\mathbf{I} - \mathbf{f}) + \mathbf{f})^{-1}\mathbf{f} = \lim_{n\to\infty} ((\rho+1)^n(\mathbf{I} - \mathbf{f}))^{-1}\mathbf{f} = 0.$$

4   Experiments

In this section we investigate the effect of boosting of the NLM operator as well as its consecutive boosting on test images.

Fig. 1. (a) Ground truth $\mathbf{x}$, (b) Corrupted with AWGN ($\sigma = 0.01$), $\mathbf{y}$, (c) NLM restored $\mathbf{f} \cdot \mathbf{y}$, (d) Ground truth $\mathbf{x}$, (e) Corrupted with AWGN ($\sigma = 0.01$), $\mathbf{y}$, (f) NLM restored $\mathbf{f} \cdot \mathbf{y}$

We implemented a sparse representation of the NLM algorithm as the denoising operator $\mathbf{f}$, with a search window of size 5×5 and smoothing parameter $h = 10\sigma$. Note that σ is the standard deviation of the additive Gaussian white noise corrupting the ideal image $\mathbf{x}$.

As it can be seen in Fig. 1 after an application of the NLM on noisy Cameraman and Saturn images with Additive White Gaussian Noise (AWGN) of $\sigma = 0.01$, an increase in overall quality is achieved. To quantitatively measure the performance at each application of the boosting $\mathcal{B}^n(\mathbf{f})$, we measure the SNR (signal-to-noise ratio) as a function of boosting iteration count n. We also define $\mathcal{B}^0(\mathbf{f}) = \mathbf{f}$ i.e., the NLM restoration operator.

Curves relating to the computed SNR of $\mathcal{B}^n(\mathbf{f}) \cdot \mathbf{y}$ are given in Figs. 2 and 3 respectively for Cameraman and Saturn. In each figure two different values of noise standard deviation, namely $\sigma = 0.01$ and $\sigma = 0.05$, are considered. The parameter ρ is varied in the range of 0.1 to 1.5 for these curves.

It is interesting to note that for both images, when the noise standard deviation is $\sigma = 0.05$, no boosting can outperform the original NLM algorithm. This fact can be verified by considering the family of curves in Figs. 2(a) and 3(a) that are all decreasing.

For the smaller noise standard deviation $\sigma = 0.01$, the peak of the curves is observed for Figs. 2(b) and 3(b). A zoomed-in version of both images for $\sigma = 0.01$ can be seen on Figs. 2(c) and 3(c). It can be observed that $\rho = 1.3$ and

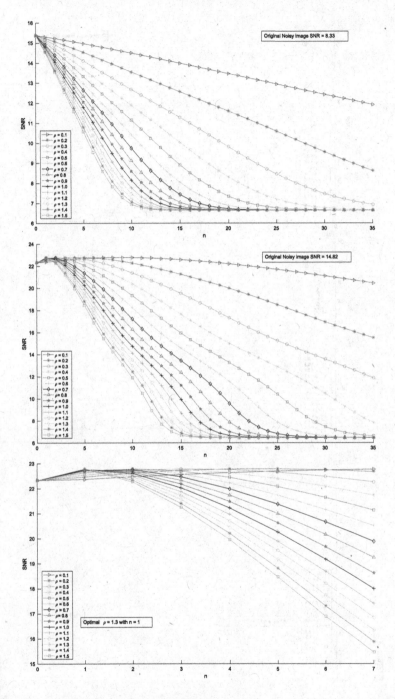

Fig. 2. Sequential boosting of NLM algorithm under operator $\mathcal{B}^n(\mathbf{f})$ applied on noisy Cameraman (a) AWGN ($\sigma = 0.05$), (b) AWGN ($\sigma = 0.01$), (c) Zoomed AWGN ($\sigma = 0.01$) and corresponding SNR

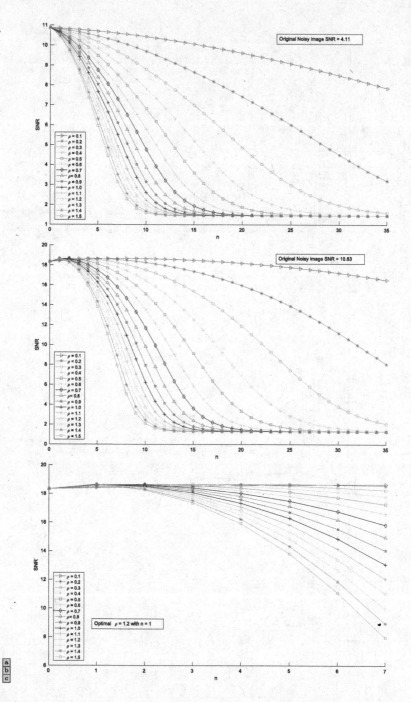

Fig. 3. Sequential boosting of NLM algorithm under operator $\mathcal{B}^n(\mathbf{f})$ applied to noisy Saturn (a) AWGN ($\sigma = 0.05$), (b) AWGN ($\sigma = 0.01$), (c) Zoomed AWGN ($\sigma = 0.01$) and corresponding SNR

Fig. 4. Resulting $\mathcal{B}^n(\mathbf{f}) \cdot \mathbf{y}$ on noisy image $\mathbf{y}$ with AWGN ($\sigma = 0.01$) for (a) $n = 1$, (b) $n = 10$, (c) $n = 20$, (d) $n = 40$, (e) $n = 50$, (f) $n = 60$, corresponding to the optimal $\rho = 1.3$, (g) $n = 1$, (h) $n = 10$, (i) $n = 20$, (j) $n = 40$, (k) $n = 50$, (l) $n = 60$, corresponding to the optimal $\rho = 1.2$

$\rho = 1.2$ for $n = 1$ correspond to the peak of SNR corresponding the two images of Cameraman and Saturn for the noise standard deviation of $\sigma = 0.01$. Finally, Fig. 4 shows the resulting images $\mathcal{B}^n(\mathbf{f}) \cdot \mathbf{y}$ for a range of values of n fixing the corresponding peak ρ values of $\rho = 1.3$ and $\rho = 1.2$.

5 Discussion and Conclusions

Based on our experiments, choosing an optimal value of ρ can be a challenging task for the SOS boosting algorithms. We conclude that for some higher noise levels, SOS boosting does not improve the NLM algorithm, regardless of the choice of ρ. Assuming SOS boosting as a new denoising operator, we asked the question whether we can improve its performance by its iterative applications. We proved that iterative applications of the SOS boosting is equivalent to a re-parametrization of ρ. In addition, we showed that iterative applications of the SOS boosting for a fixed positive ρ converges to a zero image. This proves that considering the blind application of the SOS boosting as a general remedy for all denoising schemes is incorrect. We believe more research is required to further study the domain of applicability of the proposed boosting operator.

Acknowledgments. This research was supported in part by the Natural Sciences and Engineering Research Council of Canada (NSERC).

References

1. Rudin, L.I., Osher, S., Fatemi, E.: Nonlinear total variation based noise removal algorithms. Phys. D: Nonlinear Phenom. **60**(1–4), 259–268 (1992)
2. Selesnick, I.: Total variation denoising (an MM algorithm). NYU Polytechnic School of Engineering Lecture Notes (2012)
3. Zhang, M., Gunturk, B.K.: Multiresolution bilateral filtering for image denoising. IEEE Trans. Image Process. **17**(12), 2324–2333 (2008)
4. Buades, A., Coll, B., Morel, J.-M.: A non-local algorithm for image denoising. In: IEEE Computer Society Conference on Computer Vision and Pattern Recognition, CVPR 2005, vol. 2, pp. 60–65. IEEE (2005)
5. Dabov, K., Foi, A., Katkovnik, V., Egiazarian, K.: Image denoising by sparse 3-D transform-domain collaborative filtering. IEEE Trans. Image Process. **16**(8), 2080–2095 (2007)
6. Charest, M.R., Elad, M., Milanfar, P.: A general iterative regularization framework for image denoising. In: 2006 40th Annual Conference on Information Sciences and Systems, pp. 452–457. IEEE (2006)
7. Romano, Y., Elad, M.: Boosting of image denoising algorithms. SIAM J. Imaging Sci. **8**(2), 1187–1219 (2015)
8. Talebi, H., Zhu, X., Milanfar, P.: How to SAIF-ly boost denoising performance. IEEE Trans. Image Process. **22**(4), 1470–1485 (2013)

Image Segmentation

K-Autoregressive Clustering: Application on Terahertz Image Analysis

Mohamed Walid Ayech[✉] and Djemel Ziou

Department of Computer Science, University of Sherbrooke,
Sherbrooke, QC J1K 2R1, Canada
{walid.ayech,djemel.ziou}@usherbrooke.ca

Abstract. In this paper, we propose to segment Terahertz (THz) images and introduce a new family of clustering based regression techniques suitable to time series. In particular, we propose a novel approach called K-Autoregressive (K-AR) model in which we assume that the time series depicting the pixels were generated by AR models. The K-AR model consists to minimize a new objective function for recovering the original K autoregressive models describing each cluster of time series. The corresponding pixels are then assigned to the clusters having the best AR model fitting. The order of K-AR model is automatically estimated using a model selection criterion. Our algorithm is tested on two real THz images. Comparison with existing clustering algorithms shows the efficiency of the proposed approach.

Keywords: Segmentation · Terahertz imaging · Time series · Autoregressive model · Model selection criterion

1 Introduction

Terahertz imaging is a novel modality of imaging which occupy a portion of the electromagnetic radiations between microwave and infrared bands [6]. Compared to other technologies, the analysis and the interpretation of THz images are yet in its infancy. However, with the rapid progression of the sources and the detectors of THz radiations, recent advances in acquisition technologies have made the electromagnetic THz region accessible for exploitation in various applications, such as security, quality control, and medical diagnosis [1,8,12]. Terahertz images are measured by acquisition of sequences of THz pulses, called signals or time series, transmitted through or reflected from a sample. Each time series can be represented by diverse temporal features or bands (e.g. 1000 features) behind every pixel. The huge amount of raw features can be a hurdle to analyze this type of data. Moreover, some features can be uninformative, redundant or noisy for further processing.

Several works have been proposed to overcoming these problems and analyzing more efficiently the Terahertz images. Some ones are summarized in terms of the space of features and the algorithms of classification or clustering. The

© Springer International Publishing AG 2017
F. Karray et al. (Eds.): ICIAR 2017, LNCS 10317, pp. 145–152, 2017.
DOI: 10.1007/978-3-319-59876-5_17

main space of features is the raw set of time series of Terahertz images [5]. Some works reduce the feature space by using time series models, such as autoregressive models or autoregressive moving average models [2, 11]. Traditional clustering algorithms, such as K-means techniques, were largely used for THz image segmentation [3–5]. These techniques were generally used with clusters defined by measures of the central tendency, called arithmetic means or centers, and pixels described by the whole feature space. However, the relevance problem of the huge amount of features can be a hurdle to analyze THz images based time series. Moreover, arithmetic means are not robust statistics, i.e. they are strongly influenced by outliers because data distributions are not usually normal [9]. So, it is interesting to use statistical methods that are more adapted to time series and not excessively affected by outliers.

In THz image segmentation, it is known that the different regions can be shown as clusters of time series and each time series is efficiently represented by an autoregressive model [2, 11]. We believe that autoregressive modeling leads to model each cluster of time series. In this paper, we introduce a new approach of clustering called K-Autoregressive model (K-AR) which consists to classify the set of THz time series into a fixed number of clusters, each one is represented by its predictive prototype. We assume that each pixel of the THz image is a stationary time series generated from an autoregressive (AR) model where the order is unknown. A new objective function is proposed for recovering the original AR models and then assigns the corresponding pixels to the clusters according to their predictive prototypes. The order of the K-AR approach is automatically estimated using a model selection criterion. There are three major differences between our work and the literature. First, our work constitutes a new family of clustering based regression techniques which produces clusters represented by its predictive prototypes. Second, the time series constituting the THz image are assigned to clusters having the best AR model fitting. Third, the automatic estimation of the K-AR order by using a model selection criterion.

The rest of the paper is organized as follows: in Sect. 2 we introduce the AR for univariate signal. Section 3 presents an original clustering approach, called K-AR, for the THz image segmentation and shows an automatic estimation of the order by using a model selection criterion. In Sect. 4, the results are presented and discussed.

2 Autoregressive Modeling

Univariate autoregressive (AR) models consist to model the current value of the time series variable as a weighted linear sum of its previous values plus an error, considered as a centered Gaussian random variable of variance σ_t^2. Because the time series is stationary, the σ does not depend on t. The model order is the number of preceding observations used, and the weights (also called coefficients) characterize the time series. Consider a time series $x = \{x_t; t = 1, \ldots, T\}$ and a model order P. The AR(P) model predicts the next value x_t in the time series as a linear combination of the P previous values. The AR coefficients

$w = (w_1, \ldots, w_P)'$ will be computed by fitting the model to the training time series observations. This can be realized by minimizing an error function that measures the misfitting between the predicted model, for any given values of AR coefficients, and the time series observations. Let us consider the vector $\phi_t = (x_{t-1}, x_{t-2}, \ldots, x_{t-P})'$ which represents the P previous observations of the time series. The most used error function is given by the quadratic errors between the target values x_t and the corresponding predicted values $\phi_t' w$. Thus, we minimize the following error function

$$J_{AR} = \sum_{t=P+1}^{T} (x_t - \phi_t' w)^2 \tag{1}$$

This fitting problem is determined by selecting the value of w for which the error function is as small as possible. J_{AR} is a quadratic function of the weights w, so its derivatives with respect to the weights are linear in the components of w. The minimization of J_{AR} with respect to w has an unique solution which can be found in closed form. We can deduce then the least squares solution for the AR coefficients as follows

$$w = \left(\sum_{t=P+1}^{T} \phi_t \phi_t' \right)^{-1} \left(\sum_{t=P+1}^{T} \phi_t x_t \right) \tag{2}$$

By using the estimated autoregressive coefficients w, the noise variance σ^2 is computed by $\frac{1}{T-P} J_{AR}$.

In fact, each univariate time series x was assumed originally generated by an AR model. In this section, we see how to recover the original model by estimating their coefficients w. Let us recall that each pixel of the THz image is generated from an AR model. In the next section, we will formulate the segmentation of THz images by using the clustering of AR models.

3 K-Autoregressive Clustering

Standard clustering techniques, such as K-means [7], were generally used with a fixed number of clusters defined by their arithmetic means (i.e. centers) and pixels described by the whole feature space and classified using distance measures. Nevertheless, this is undesirable in Terahertz imaging where the pixels are depicted by time series represented by a huge number of raw features. Besides, distance measures between time series are generally not suitable in the case of large feature space due to uninformative features. The relevance problem of the huge amount of features can be a hurdle to analyze THz images. Moreover, arithmetic means are not robust statistics [9] and other statistical methods more adapted to time series can improve the clustering process.

In THz image segmentation, it is known that the different regions can be shown as clusters of time series [4] and each time series is efficiently represented by an autoregressive (AR) model [2,11]. We believe that autoregressive modeling leads to model each cluster of time series. In this section, we propose a new clustering approach called K-Autoregressive (K-AR) model which consists to classify the set of THz time series into a fixed number of clusters, each one is represented by its predictive prototype. We assume that each pixel of the THz image is a stationary time series generated from an AR model where the order is unknown. The K-AR approach consists to minimize a suitable objective function for recovering the original K autoregressive models describing each cluster of time series. Let us consider $x_n = \{x_{nt}; t = 1, \ldots, T\}$ a time series of the nth pixel, P is the AR order, the weights $w_k = (w_{k1}, \ldots, w_{kP})'$ are the AR coefficients which characterize the kth time series cluster and $\phi_{nt} = (x_{n(t-1)}, x_{n(t-2)}, \ldots, x_{n(t-P)})'$ is the vector of the P previous observations of x_{nt}. For each cluster, the fitting error of an AR(P) is $\sum_{n \in C_k} \sum_{t=P+1}^{T} (x_{nt} - \phi_{nt}' w_k)^2$, where C_k is the set of time series of the kth cluster. For all clusters, this error is equal to the sum of K AR(P) fitting errors. We need to find coefficients $\mathbf{w} = (w_1, \ldots, w_K)$, which minimize the following error, for all pixels:

$$J_{KAR} = \sum_{k=1}^{K} \sum_{n=1}^{N} \sum_{t=P+1}^{T} u_{nk} \left(x_{nt} - \phi_{nt}' w_k \right)^2 \tag{3}$$

Note that in this equation, the vector $\mathbf{w}$ is the same for all pixels. This equation is resolved by selecting the values of w_k for which J_{KAR} is as small as possible. The J_{KAR} is a quadratic function of the weights w_k, its derivatives with respect to the weights are linear in the components of w_k. Thus, the minimization of the objective function J_{KAR} in Eq. 3 gives the following expression

$$w_k = \left(\sum_{n=1}^{N} \sum_{t=P+1}^{T} u_{nk} \phi_{nt} \phi_{nt}' \right)^{-1} \left(\sum_{n=1}^{N} \sum_{t=P+1}^{T} u_{nk} \phi_{nt} x_{nt} \right) \tag{4}$$

The time series are assigned to their closest cluster by computing the membership degrees u_{nk}. The membership degrees must ascertain the constraints $\{u_{nk} \mid u_{nk} \in \{0, 1\}$ and $\sum_{k=1}^{K} u_{nk} = 1\}$. The necessary condition for minimizing the objective function J_{KAR} gives the following expression

$$u_{nk} = \begin{cases} 1, & \text{if } k = \arg \min_l \{\mathcal{D}_{nl}\} \\ 0, & \text{otherwise} \end{cases} \tag{5}$$

where $\mathcal{D}_{nk} = \sum_{t=P+1}^{T} (x_{nt} - \phi_{nt}' w_k)^2$ represents the sum of the squares of the errors between the target values x_{nt} and the corresponding predicted values $\phi_{nt}' w_k$. We assume that the number K of clusters is known. For clustering, an AR model is assigned with the cluster k if $u_{nk} = 1$. The K-AR consists to classify the set of time series data into K clusters; each cluster is represented by one AR coefficients w_k, where $k = 1, \ldots, K$. The clustering process is therefore realized by iterating between two phases, updating the AR coefficients of clusters and the membership degrees of the time series, until the objective function values do not change. We remark that AR coefficients are computed according to their clustering importance and the time series observations. To summarize, let us consider a parameter P less to T. The K-AR algorithm can be summarized as follows:

K-AR algorithm
1. Initialize w_k by random values
2. **Do**
  Update membership degrees u_{nk} using Eq. 5
  Update AR weights w_k using Eq. 4
 Until $|J_{KAR}^{(t)} - J_{KAR}^{(t-1)}|$ <threshold
3. Return U.

The pixels represented by the time series are then assigned to the clusters having the best AR model fitting. The obtained clusters are determined by the final membership degrees of the time series. For all time series, the residual variance of the error between time series observations and the corresponding predicted values is given by $\frac{J_{KAR}}{N(T-P)}$. Using the error variance, the best K-AR order is selected using Schwartz information criterion (SIC) which aims to determine the best compromise between model fitting and model complexity [10].

4 Experimental Results

In this section, K-AR, W-K-means, K-means, KHM and GMM are tested on moth and chemical Terahertz images. Pixels of moth and chemical Terahertz images are constituted respectively by 894 and 1052 bands in the temporal domain. We present in the left of Figs. 1 and 2 the objects acquired in the visible light for the validation. The chemical image ground truth and the 570th band of the moth wings THz image are shown in the right. The chemical image comprises 4 compounds, L-Valine (0.200 g), Proline (0.200 g), L-Tryptophan (0.200 g) and L-Tryptophan (0.100 g), divided into 4 false colored zones, whereas the moth THz image mainly comprises two wings. The weights of K-AR approach and the centers of KHM, W-K-means, GMM and K-means were initialized by random values. The segmentation of chemical and moth images was tested respectively with $K = 4$ and $K = 5$ clusters.

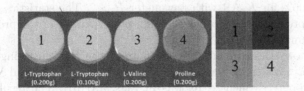

Fig. 1. In the left, a visible image of 4 chemical compounds. In the right, the THz image ground truth. The false colors green, yellow, red and blue correspond respectively to the compounds L-Valine (0.200 g), Proline (0.200 g), L-Tryptophan (0.200 g) and L-Tryptophan (0.100 g). (Color figure online)

Fig. 2. In the left, a visible image of the moth. In the right, the band number 570 of the Terahertz image.

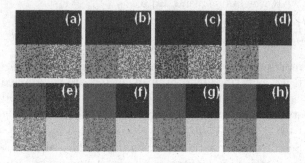

Fig. 3. The segmentation of chemical Terahertz image for (a) K-means, (b) KHM, (c) GMM, (d) W-K-means and K-AR for (e) $P = 2$, (f) 5, (g) 10, (h) 15.

Figure 3 shows the segmentation of the chemical Terahertz image for different techniques. In Figs. 3(a)–(d), W-K-means, GMM, KHM and K-means give as output over-segmented regions. The clusters of L-Tryptophan (0.100 g) and (0.200 g) are combined together in the case of KHM, K-means and GMM which clearly shows their segmentation shortcomings. Also, L-Tryptophan (0.200 g) cluster is largely affected by noisy points in the case of the W-K-means. Figure 3 from (e) to (h) show the output clusters of K-AR for different values of P (2, 5, 10 and 15). For $P = 2$, the K-AR produces as output over-segmented images. The best image segmentations are obtained when P surpasses 2 in Figs. 3(f)–(h), the different compounds become well segmented, with the exception of few pixels of L-Valine (0.200 g) are misclassified. The statistics of the five algorithms are shown in Fig. 4. The performances of the clustering have not exceed 70% for KHM, GMM and k-means, 80% for W-k-means, 80% for K-AR with $P = 2$ and near to 100% for K-AR with parameter $P > 2$. The statistics confirm the results found previously and prove the high performances of the proposed clustering algorithm.

Figure 5 shows the segmentation outputs of the five clustering techniques on the Terahertz image of the moth. K-means, KHM, GMM and W-K-means generate a wrongly segmented regions in Figs. 5(a)–(d). The resulted regions illustrate the shortcomings of the four approaches to provide clear inner structure of moth wings. Figure 5 from (e) to (h) show the output regions of K-AR for P equal to 2, 4, 6, and 10. The structure of the moth is well identified for different values of P and the wings of the moth are preserved.

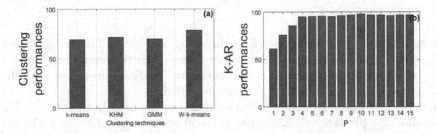

Fig. 4. The performances of the clustering algorithms on chemical Terahertz image for (a) W-*K*-means, GMM, KHM and *K*-means and (b) *K*-AR for divers values of *P*.

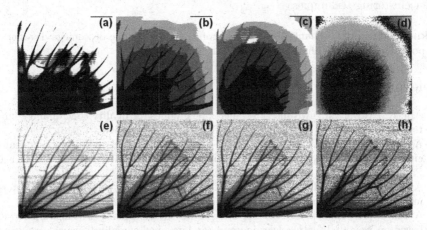

Fig. 5. Moth THz image segmentation for *K*-means (a), KHM (b), GMM (c), W-*K*-means (d) and *K*-AR for $P = 2$ (e), 4 (f), 6 (g), 10 (h).

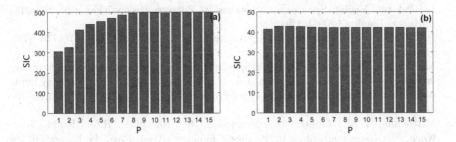

Fig. 6. Schwartz information criterion (SIC) for both (a) chemical and (b) moth images

Figures 6(a) and (b), show plots of SIC criterion for divers orders of *K*-AR approach on both THz images. In the case of the chemical image, the best orders are around 10. While, in the case of the moth image, the best order is equal to 2. The statistics confirm the results found previously in Figs. 3 and 5.

5   Conclusion

We have proposed a new approach of clustering suitable to THz images based time series. The K-AR consists to regroup a set of THz time series into clusters represented by their prediction prototypes. The K-AR assumes that the time series depicting the pixels were generated by AR models and consists to recover the original K autoregressive models describing each cluster of time series. The corresponding pixels are then assigned to the clusters having the best AR model fitting. The order of K-AR model is automatically estimated using a modified information criterion. Our approach is tested on two THz images. Experimental results show that K-AR approach produces more accurate segmentation than other clustering techniques.

Acknowledgments. Thanks are due to Prof. Thomas Tongue of Zomega Terahertz Corporation for the Terahertz measurements.

References

1. Ahi, K., Anwar, M.: Advanced Terahertz techniques for quality control and counterfeit detection. In: Proceedings of SPIE, vol. 9856, p. 98560G (2016)
2. Ayech, M.W., Ziou, D.: Terahertz image segmentation based on k-harmonic-means clustering and statistical feature extraction modeling. In: International Conference Pattern Recognition, pp. 222–225. IEEE, Tsukuba (2012)
3. Ayech, M.W., Ziou, D.: Segmentation of Terahertz imaging using k-means clustering based on ranked set sampling. Expert Syst. Appl. **42**, 2959–2974 (2015)
4. Ayech, M.W., Ziou, D.: Terahertz image segmentation using k-means clustering based on weighted feature learning and random pixel sampling. Neurocomputing **175**(Part A), 243–264 (2016)
5. Berry, E., Boyle, R.D., Fitzgerald, A.J., Handley, J.W.: Time frequency analysis in Terahertz pulsed imaging. In: Bhanu, B., Pavlidis, I. (eds.) Computer Vision beyond the Visible Spectrum. Advances in Pattern Recognition, pp. 271–311. Springer, London (2005). Chap. 9
6. Hu, B.B., Nuss, M.C.: Imaging with Terahertz waves. Opt. Lett. **20**(16), 1716–1718 (1995)
7. MacQueen, J.B.: Some methods for classification and analysis of multivariate observations. In: Cam, L.M.L., Neyman, J. (eds.) Proceeding of the Fifth Berkeley Symposium on Mathematical Statistics and Probability, vol. 1, pp. 281–297. University of California Press (1967)
8. Mittleman, D.M., Gupta, M., Neelamani, R., Baraniuk, R.G., Rudd, J.V., Koch, M.: Recent advances in Terahertz imaging. Appl. Phys. B: Lasers Optics **68**, 1085–1094 (1999)
9. Rousseeuw, P.J., Leroy, A.M.: Robust Regression and Outlier Detection. Wiley, New York (1987)
10. Schwarz, G.: Estimating the dimension of a model. Ann. Stat. **6**(2), 461–464 (1978)
11. Yin, X.X., Ng, B.W.H., Ferguson, B., Mickan, S., Abbott, D.: Statistical model for the classification of the wavelet transforms of T-ray pulses. In: 18th International Conference Pattern Recognition, vol. 3, pp. 236–239 (2006)
12. Zhang, X.C.: Terahertz wave imaging: horizons and hurdles. Phys. Med. Biol. **47**(21), 3667–3677 (2002)

Scale and Rotation Invariant Character Segmentation from Coins

Ali K. Hmood$^{(\boxtimes)}$, Tamarafinide V. Dittimi, and Ching Y. Suen

Center for Pattern Recognition and Machine Intelligence (CENPARMI),
Computer Science and Software Engineering Department,
Concordia University, Montreal, Canada
{a_alfraj, t_dittim, suen}@encs.concordia.ca

Abstract. This paper presents a robust method for character segmentation from coin images. While many papers studied character segmentation and recognition from structured and unstructured documents. Several methods proposed that vary, in terms of targeted documents, from complex (degraded) into different languages. This is the first paper to study and propose a solution for character segmentation from coins. Character segmentation plays a crucial role in coin recognition, grading and authentication systems. Scaling and rotating the coins are challenging in character segmentation due to the circular nature of coins. In this paper, we transform the coin from circular into rectangular shape and then perform morphological operations to compute the horizontal and vertical projection profiles and apply dynamic adaptive mask to extract characters. Our method is evaluated on several coins from diverse countries with different image background complexity. The proposed method achieved precision and recall rates as high as 93.5% and 94.8% respectively demonstrating the effectiveness of the proposed method.

Keywords: Computer vision · Character extraction · Rotation invariant character segmentation

1 Introduction

Character segmentation and word recognition is the focus of many researchers around the world. There are several papers demonstrating research conducted on different kinds of documents for different languages using various features and approaches. Compared to structured and hand-written documents, character recognition on coins is more challenging because the variance between the character stroke and the background is minimum. Also the size, font, spacing and locations of characters may vary on the same coin and between different coins.

Character segmentation from coins is an essential step in coin recognition, grading and authentication systems. Unlike documents, coins are (1) subject to severe degradation due to circulation in our daily use; (2) no prior knowledge of character orientation; (3) bump height of characters minted on the coin's surface varies which affect the stroke sharpness and width; (4) different lighting sources highly affect the character

© Springer International Publishing AG 2017
F. Karray et al. (Eds.): ICIAR 2017, LNCS 10317, pp. 153–162, 2017.
DOI: 10.1007/978-3-319-59876-5_18

appearance due to highlight and shadow variation; (5) identical color of the characters and the background. Therefore, the traditional OCR and other segmentation methods are unsuitable for coin images [1].

While much research have been conducted on character recognition on documents [2], very few researchers have tried to recognize characters on coins surface [1, 4, 5]. However, only few existing work [6, 7] have studied character segmentation to recognize coins based on extracting features of the segmented characters appears on the surface. In this research, we propose a method for character segmentation from coin images using a dynamic adaptive mask. The mask is dynamic in terms of size based on the character width and height of different coins. The proposed method starts by preprocessing steps and straightening the coin images; a set of morphological operations are performed to find the vertical and horizontal profiles, then the adaptive mask is applied to extract the characters. The binarized image is used to locate the Region of Interest (ROI) and finally the characters are extracted. The proposed method uses a dynamic adaptive mask to fit all characters regardless of their height, width, or language. Character segmentation and character extraction will be used interchangeably in the rest of the paper.

2   Related Work

Numerous studies have been presented in the literature to extract characters from printed or hand-written documents and natural scene images. Different methods have been proposed to combine different feature extraction and machine learning algorithms for different languages. Natural scene images added additional challenges to character extraction due to illumination conditions, combination of different colors, geometric distortion, and different font styles [8]. On the other hand, several coin recognition, grading, and authentication systems have been proposed based on different features such as edge-based statistical features, local image features, and texture features. Some of these systems considered the characters on the coin in terms of character locations and their edges only without extracting them which limits their proposed systems. We will study related work in terms of character extraction methods and coin recognition and authentication systems.

2.1   Character Extraction

Character extraction and recognition have received intensive attention since the mid-1940s and new methods are still being introduced to recognize characters within more complex and challenging images. The character extraction and recognition methods, sorted and presented in [3], are appropriate for printed and hand-written documents but not for natural scene and coin images [4]. Several improved methods have been studied for character or line segmentation including contour tracing, morphological techniques, grayscale feature combination, graph-based techniques, and vertical and horizontal projection profiles.

However, the above mentioned and majority of character segmentation research are not suitable for natural scene and coin images [1]. Character extraction from coin images is an essential step in several coin based systems such as coin recognition which aims at recognizing words and forming a textual document from textual images.

2.2 Coin Recognition and Authentication

The increased demands on automated coin recognition, grading, and authentication processes has led to the continuous improvements in image processing and machine intelligence methods for coin images. Several papers have been reported in the literature for coin grading [9], recognition [1, 4, 5], and authentication [10, 11]. However, the coin grading systems have not considered characters or their features in grading the coin quality. Though, character segmentation and analysis add advantages to the coin grading methods. Unlike coin authentication where the work of [10, 11] rely heavily on character properties to identify genuine coins and reject counterfeit ones. Yet, none of these systems considered character segmentation for accurate results based on character analysis. Sun et al. [10] proposed a counterfeit coin detection algorithm based on the character properties such as the stroke width, edge smoothness, height and width of the character, as well as, relative distances and angles between characters. However, the researchers have only located the character on the coin and no character segmentation have been reported. Khazaee et al. [11] also presented a counterfeit coin detection for 3D coin images. They examined the outer circle of the coin where all characters and numbers occur using the height and depth information obtained by the 3D scanner to identify genuine coins. Hence, character segmentation method for coins is needed in coin authentication systems.

Moreover, several coin recognition methods have been proposed which aim at sorting large numbers of coins based on their country or denomination. A few researchers worked on extracting features from characters location on the coin [1, 4, 5]. Nonetheless, the researchers extracted SIFT features from either a full word or characters after locating the ROI on the coin. However, no automated character segmentation has not been reported in those papers. Unlike previous work, Arandjelović [6] was the first to present character extraction from coins. Number of small windows are placed on the coin image and for each window the HOG-like descriptor is computed and compared against a manually annotated dataset to recognize characters. The recognition accuracy rate of the 25 coin of ancient Roman Denarii dataset was not reported in this paper. On the other hand, Pan and Tougne [7] proposed a solution for coin date detection. The proposed system should have a priori knowledge about the coin to locate the date. The histogram is computed for gradients of each candidate location and is then compared to mean histogram of synthetic date model. The system was evaluated on the US coins with 900 images of individual date numbers. The recognition rate was 44% at its best on the real coin dataset. In this paper, we propose a fully automated and dynamic character segmentation method for coin images. The proposed method returns a set of characters on the coin surface and, due to coin straightening algorithm, the characters are all rotated and aligned. The method fits several coins regardless of their image size and language. The segmented characters can be used in several coin systems.

3 Character Segmentation

The increased focus on text regions contained in coins to extract their features and the properties of their characters (e.g. stroke width) to achieve higher accuracies have raised the need for a reliable character segmentation method. The traditional character segmentation methods for printed/handwritten documents and natural scenes have some assumptions that are not valid for coin images. Due to these facts, we propose a new method for character segmentation based on vertical and horizontal projection profiles and dynamic adaptive mask. The method starts by scaling the coin image to fit the whole image by removing the left, right, top, and bottom margins and then straightening the circular shape of the coin into rectangular shape; then the connected components are identified. The horizontal and vertical projection profiles are calculated to define the average height and width of characters, then a dynamic mask is applied to cover each character.

3.1 Coin Scaling

Coin scaling is an essential step in coin recognition, grading and authentication systems. Many researchers have ignored this step and simply applied a threshold to discriminate the coin from the background pixels due to the large variance between the image background and the coin colors. However, this is not always the case and we consider coin scaling to fit the entire image. The algorithm starts by detecting all edges E in the image $I(x, y)$ using Canny edge detection and the edge image I_E is then dilated using a morphological structuring element (e.g. circular shape). The algorithm starts identifying all circular shapes in the image and defines a set of center point $c(x, y) \in I(x, y)$ and radii R for each circle. The largest radius $r \in R$ and its corresponding center point $c_r(x_r, y_r)$ are selected as a candidate circle of the coin. Dynamic mask dm of the coin's size is created to cover the whole coin and the coin is cropped to fit the whole image. Due to the circular shape of the coin, there are still areas where the background still appear. Hence during the masking operation, the margins outside the actual coin's boarder is transformed into black pixels as shown in Fig. 1.

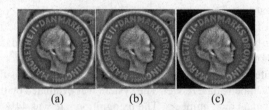

(a)   (b)   (c)

Fig. 1. (a) The original image of Danish coin, (b) locating the coin borders within the image, and (c) masked and scaled Danish coin image

3.2   Straightening Algorithm

The aim of this algorithm is to overcome the orientation problem when extracting the characters. Transforming the circular shape of the coin into rectangular shape facilitates the character segmentation process using projection profiles. Since the focus of this research is on characters on the coin surface, the straightening algorithm will consider the outer circle of the coin where the characters are located. The straightening algorithm reads the image pixels in diagonal style and writes each pixel value into a new rectangular matrix.

Given the circle equation, assume $p(x, y)$ is the pixel of the original image and $p'(x', y')$ is the new pixel of the straighten image calculated from Eq. (1):

$$x' = n * sin(z) + (w/2) \quad \begin{cases} \pi < z < -\pi \\ b < n < c \end{cases} \qquad (1)$$
$$y' = n * cos(z) + (h/2)$$

where w and h are the width and height of the original image respectively. b is half the total number of columns from the original image and c is the number of rows of the outer circle of the image containing the characters. Straightening the coin adds a great value not only during the segmentation but also after a set of rotated characters are obtained. Figure 2(a) and (b) show the straightened images of US and Danish coins.

(a)
(b)

Fig. 2. (a) and (b) represent a straightened US and Danish coins, used in our experiments described in Sect. 4.

3.3   Coin Analysis

Coin analysis allows the segmentation algorithm to learn the properties of the coin and refine the parameters and thresholds for character segmentation.

A 3×3 Gaussian filter is applied to remove noise from the coin and Otsu binarization algorithm [12] is utilized to obtain the binary image. Then all connected components (blobs) are identified on the coin.

The vertical projection profile $f(x, p(x))$ is calculated to find the characters distribution on the coin, and average width w_{avg} of connected components is found to adjust the dynamic mask dm. The horizontal projection profile $g(y, p(y))$ is also found to set the average height h_{avg} of connected components. The slopes between each peak and the next valley in the vertical and horizontal projection profiles are considered to determine the spaces between connected components. The slopes in the vertical projection profile define the spaces between characters; whereas the slopes in the horizontal profile define the distances between characters and other objects on the coin. The slopes are found by Eq. (2) below:

$$slope_{vertical} = \frac{p(x_{peak}) - p(x_{valley})}{x_{peak} - x_{valley}} \tag{2}$$

where $p(x)$ is the total number of white pixels in column x, and x_{peak} and x_{valley} are the coordinates of columns where the peak and valley are located.

Figure 3 illustrates the vertical and horizontal projection profiles for the Canadian coin. The h_{avg}, w_{avg}, $slope_{vertical}$, and $slope_{horizantal}$ are defined and passed to character segmentation algorithm.

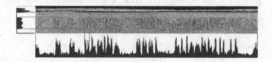

Fig. 3. Horizontal and vertical projection profiles of a Canadian coin.

3.4　Character Segmentation

The four parameters that we obtained from the coin analysis are used to initiate the character segmentation algorithm together with the binarized coin image. The algorithm checks for every connected component lm that falls within:

$$h_{avg} + c_1 > h_{lm} \geq h_{avg} - c_1$$
$$w_{avg} * c_2 > w_{lm} \geq w_{avg}/c_2$$

where c_1 and c_2 are constant thresholds, and c_1 is the difference between h_{avg} and h_{max}, where h_{max} is the maximum height of all characters and is obtained from the horizontal projection profile; and $c_2 = 4$. Note that the use of different arithmetic operations for defining the range of the height and the width is due to the variation in height is much less than the variation in width from one character to another.

The algorithm starts segmenting characters using an adaptive mask around each connected component that has the height and width within the range and discard the ones that exceed the range. For improving character segmentation results, all connected components lm that occur below the width and height ranges but below the maximum height h_{max} of all characters that we obtain from the horizontal projection profile. These lm commonly occur due to degradations on the coin surface which results in broken characters. For each $lm_i \subseteq lm$ check the distance d between $C(lm_i)$ and $C(lm_j)$ where $lm_j \subseteq lm$ that occurs within the immediate mask see Fig. 4. If d is less than h_{max}/α for lm_j above or below lm_i or is less than w_{max}/α for lm_j to the left or right lm_i, where α (equals 2 for our dataset) is the threshold for acceptable distance between two lm_i, $lm_j \subseteq lm$. The two blobs are then combined and checked again against the original range of acceptable character width and height. However, the size of lm_i should not be less than h_{avg}/β and w_{avg}/β where β equals 4 for height and width respectively.

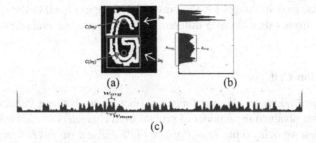

(a)   (b)

(c)

Fig. 4. (a) Illustrates bounding box lm_i and lm_j and their centers, d is the distance between C (lm_i) and $C(lm_j)$, (b) the maximum and the average height found from the horizontal projection profile, and (c) the maximum and the average width found from the vertical projection profile

4 Experimental Results

Our solution was tested on 6 datasets (Sect. 4.1) that contain 6 different coins from four different countries. The algorithm showed an efficient performance in segmenting characters for all coins. The final output for our method was a single character in binary and grayscale formats. Figure 5 shows examples on the results of character segmented from Canadian and Chinese coins using the proposed method. Furthermore, the use of connected components (blobs) for character segmentation would be appropriate for all kinds of coins regardless of the language used on the coin.

Fig. 5. Examples of segmented characters of Canadian and Chinese coins using our proposed method. Each character is segmented from the binary and grayscale images of the coins.

4.1 Dataset

We evaluated the performance of our method on 6 datasets from 4 different countries, Canadian, Danish, Chinese, and US coins. The Canadian and Chinese coin images were collected and scanned at the Center for Pattern Recognition and Machine Intelligence (CENPARMI). The Canadian coins dataset consists of 2 versions on the C\$2, before and after 2002. This dataset has a total of 100 coins each with 2 sides (obverse and reverse sides) and in total we had 200 images. While the Chinese coin dataset contains 4 images, we modified these images by rotating the coins and adding some noise to have 28 images. The Danish coins dataset as provided by the Danish authorities, and it contains a set of fake coins and the aim of the Danish authorities is to build a system that is able to detect counterfeit coins. However, in this research, we used 40 genuine Danish coins (obverse side) since counterfeit coins fall out of scope. Finally, the publicly available US coin dataset were obtained from the Professional Coin Grading Service (PCGS) website that contains several US coin images. In this research, we obtained 40 images for \$1 coins and 40 images of half dollar coins. A total of 348 images from 4 datasets of 6 different coins were used to evaluate the proposed method.

4.2 Evaluation Criteria

We used the *precision*, *recall*, and *f-measure* metrics to evaluate the results of our method. We first obtained the number of ground truth characters in each coin as shown in Table 1. Then we defined the *True Positive (TP), False positive (FP)*, and the *False Negative (FN)*. The $precision = \frac{TP}{TP+FP}$, $recall = \frac{TP}{TP+FN}$, and $f - measure = \frac{(1+\beta)\times precision \times recall}{\beta^2 \times (precision + recall)}$ are calculated where β is set equal to 1 to equalize the importance of *recall* and the *precision* when finding the *f-measure*.

Additionally, the performance of character segmentation method is evaluated by the *segmentation error rate*. Assume C_i is the number of mismatched segments and T_j is total number of ground truth segments. Then the *segmentation error rate* is computed where $Err_{seg} = \frac{C_i \times 100}{T_j}$.

If a ground truth segment is not included in the resulted segments, the ground truth segment is considered as a mismatched segment. Thus, the segmentation error rate is a complement to the *recall*.

Table 1. Recall, Precision, and f-measure values for character segmentation of the 6 different datasets

		Ground truth characters	Recall	Precision	f-measure
Canadian (Old version)	Obverse	23	0.93	0.896	0.913
	Reverse	14	0.905	0.876	0.89
Canadian (New version)	Obverse	23	0.935	0.904	0.919
	Reverse	14	0.91	0.874	0.892
Danish	Obverse	27	0.953	0.935	0.944
	Reverse	–	–	–	–
Chinese	Obverse	10	0.886	0.849	0.867
	Reverse	9	0.873	0.88	0.876
US (One Dollar Coins)	Obverse	7	0.921	0.866	0.893
	Reverse	30	0.948	0.89	0.918
US (Half Dollar Coins)	Obverse	11	0.914	0.878	0.895
	Reverse	31	0.929	0.894	0.911

4.3 Results and Discussion

We evaluated the segmentation results based on 6626 ground truth segments belonging to 348 images. The results of *recall, precision,* and *f-measure* are represented in Table 1. We observe that the highest *f-value* obtained on segmentation of Danish coins due to the high quality images obtained by the specialized scanner. The images of

Danish coins are very clear and all the edges on the image are sharp. The lowest *f-value* obtained from the Chinese segments. The main reasons are: (1) The Chinese coins that we worked on were over 100 year old and the quality was lower than other dataset coins. (2) Some of the Chinese characters are separated (detached) by nature which makes it harder to segment them as a single character.

The overall *f-measure* values were between 0.867 and 0.944 which reflects the effectiveness of the proposed character segmentation method. The segmented characters are of different sizes based on the input straight coin image size and the character height and width. The *segmentation error rate* was also reported to determine the rate of unsegmented characters from the coin image.

Figure 6 depicts the *segmentation error rate* which is complement to *recall* rate. The *x-axis* represents the 6 different datasets used in our experiments in Table 1. whereas the *y-axis* depicts the *segmentation error rate* percentage. The highest segmentation error rate also was reported for the Chinese coins dataset for the reasons stated earlier. The average *segmentation error rate* is 8%, where the highest is 12.6% and lowest is 4.6%. Whereas, the lowest rate was also reported for the Danish coins dataset where the highest quality images were obtained.

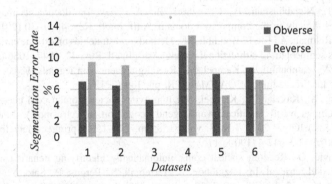

Fig. 6. Segmentation error rate for the 6 datasets

The main observations drawn by the experimental results are: (1) the use of dynamic mask improves the segmentation performance for different coins with various character sizes, (2) the coin image quality has the highest effect on segmentation results, (3) the more characters on the coin, the lower segmentation error. Since the character height and width ranges would increase and it could increase *false positive* segments, (4) the language used in coins has very little effect on the segmentation accuracy where languages that have detached characters are lowering the segmentation accuracy, and (5) the thresholds selected in this research and described in Sect. 3.4 are the optimal selection of multiple threshold studied in this research; changing the thresholds would considerably reduce the segmentation accuracy. The segmentation results would remarkably decrease when applied to degraded and scratched coin image. Thus, the proposed method can be extended in the future to include the attached characters to the outer edge or other stamp symbols.

5 Conclusion

In this paper, we studied the problem of character segmentation from coin images. Character segmentation is an essential step for various systems that work on coins such as coin recognition, grading, and authentication. The challenges of this work are the character orientation, heavily degraded coin quality, and character appearance due to highlight and shadow variations caused by different lighting sources. We proposed a reliable solution for segmenting single characters and return a binary and grayscale images of each character separately. The proposed solution was experimentally proved to handle different languages and extract characters accurately. Experimental results suggest the image quality of the coin can have the highest impact on the segmentation results while the number of characters on the coin has the second highest impact on the segmentation accuracy.

References

1. Kavelar, A., Zambanini, S., Kampel, M.: Word detection applied to images of ancient roman coins. In: Proceedings of 18th International Conference on Virtual Systems and Multimedia (VSMM), pp. 577–580. IEEE (2012)
2. Javed, M., Nagabhushan, P., Chaudhuri, B.: A review on document image analysis techniques directly in the compressed domain. Artif. Intell. Rev. **48**, 1–30 (2017)
3. Saba, T., Rehman, A., Elarbi-Boudihir, M.: Methods and strategies on off-line cursive touched characters segmentation: a directional review. Artif. Intell. Rev. **42**, 1047–1066 (2014)
4. Kavelar, A., Zambanini, S., Kampel, M.: Reading ancient coin legends: object recognition vs. OCR. In: Proceedings of OAGM/AAPR, pp. 1–9 (2013)
5. Zambanini, S., Kavelar, A., Kampel, M.: Improving ancient roman coin classification by fusing exemplar-based classification and legend recognition. In: Petrosino, A., Maddalena, L., Pala, P. (eds.) ICIAP 2013. LNCS, vol. 8158, pp. 149–158. Springer, Heidelberg (2013). doi:10.1007/978-3-642-41190-8_17
6. Arandjelović, O.: Reading ancient coins: automatically identifying denarii using obverse legend seeded retrieval. In: Fitzgibbon, A., Lazebnik, S., Perona, P., Sato, Y., Schmid, C. (eds.) ECCV 2012. LNCS, vol. 7575, pp. 317–330. Springer, Heidelberg (2012). doi:10.1007/978-3-642-33765-9_23
7. Pan, X., Tougne, L.: Topology-based character recognition method for coin date detection. Int. J. Comput. Electr. Autom. Control Inf. Eng. **10**(10), 1752–1757 (2016)
8. de Campos, T.E., Babu, B.R., Varma, M.: Character recognition in natural images. In: VISAPP, vol. 8, no. 2, pp. 273–280 (2009)
9. Bassett, R., Gallivan, P., Gao, X., Heinen, E., Sakalaspur, A.: Development of an automated coin grader: a progress report. In: Proceedings of the 8th Annual Mid-Atlantic Student Workshop on Programming Languages and Systems, pp. 15.1–15.10 (2002)
10. Sun, K., Feng, B.-Y., Atighechian, P., Levesque, S., Sinnott, B., Suen, C.Y.: Detection of counterfeit coins based on shape and letterings features. In: The Proceedings of (CAINE 2015), pp. 165–170 (2015)
11. Khazaee, S., Sharifi Rad, M., Suen, C.Y.: Detection of counterfeit coins based on modeling and restoration of 3D images. In: Barneva, R., Brimkov, V., Tavares, J. (eds.) CompIMAGE 2016. LNCS, vol. 10149, pp. 178–193. Springer, Cham (2017). doi:10.1007/978-3-319-54609-4_13
12. Otsu, N.: A threshold selection method from gray-level histograms. Automatica **11**(285–296), 23–27 (1975)

Image Segmentation Based on Solving the Flow in the Mesh with the Connections of Limited Capacities

Michael Holuša[1]([✉]), Andrey Sukhanov[1,2], and Eduard Sojka[1]

[1] VŠB Technical University of Ostrava, Ostrava, Czech Republic
{michael.holusa,eduard.sojka}@vsb.cz
[2] Rostov State Transport University, Rostov-on-Don, Russia
a.suhanov@rfniias.ru

Abstract. This paper presents a novel seeded segmentation technique inspired by the flowing of a liquid in a mesh of pipes. The method can be likened to the anisotropic diffusion algorithm. On the other hand, some substantial changes in the relation of how the diffusion works are included. The method is based on the spreading of liquid from the foreground seeds to the neighboring image points that represent basins with an initial amount of liquid. The background seeds drain the liquid from the neighboring basins. If a basin is full or empty, the corresponding pixel becomes a new source or sink. The algorithm runs until all pixels become either sources or sinks. The properties of the method are illustrated on the image segmentation of synthetic images. The comparison with other segmentation techniques is presented on real-life images. The experiments show promising results of the new method.

Keywords: Image segmentation · Seeded segmentation · Anisotropic diffusion based segmentation · Flow in mesh · Liquid transmission

1 Introduction

Image segmentation is a technique of computer vision for partitioning image into various segments. The segmented areas can be used for an analysis of their content. Image segmentation can be divided into two main branches: automatic and seeded segmentation. The automatic segmentation consists in a clustering of image pixels into segments with a similar brightness, the number of clusters is usually estimated automatically. The seeded segmentation requires a priori user-provided seeds scribbled into particular segment areas. The goal is to divide all unseeded points into the particular segments initially represented by the seeds. The number of segments is given by the number of distinct seeds. Many of the seeded segmentation approaches are inspired by spreading the energy from the seeds to other image points. This behavior can be likened to the diffusion process. In 1990, Perona and Malik [11] presented an anisotropic diffusion used

F. Karray et al. (Eds.): ICIAR 2017, LNCS 10317, pp. 163–170, 2017.
DOI: 10.1007/978-3-319-59876-5_19

for edge detection. The authors in [14] presented the seeded segmentation algorithm based on heat diffusion. The well-known random walker algorithm, introduced by Grady [5], is also a special case of the diffusion process. This algorithm computes the probabilities that a random walk from an image point reaches one of the seed points. The random walker method was applied in many tasks of computer vision, and modifications of this method have been introduced [3,4,6,7,12]. The authors in [13] combined the random walker technique with another segmentation method called graph cuts [2]. The authors in [10] proposed diffusion-based watershed. The watershed method [1,9] is a segmentation technique based on region growing.

In this paper, we propose a new method that is inspired by flowing a liquid in a mesh of pipes. It can be regarded as a similar approach as diffusion, but substantial differences are introduced. The sources of liquid with an infinite capacity are placed in the foreground seeds, the background seeds are used as the sinks where the liquid is leaving the mesh, the other pixels represent basins that may contain an initial amount of liquid. The process runs until all the basins are either full or empty. The paper is organized as follows. The description of our method is in Sect. 2. Section 3 contains the experiments and comparisons with the random walker and the watershed segmentation techniques since their way of segmentation is similar to the proposed method. Section 4 is a conclusion.

2   The Proposed Method

The main idea of the technique is inspired by flowing a liquid in a mesh of pipes. At the beginning, the seeds of objects are regarded as the sources of liquid with an infinite capacity, the seeds of the background are regarded as the sinks, whose capacity is infinite too. The remaining pixels are considered as the basins that are filled with a certain initial liquid amount $\mu_0 \in [0, 1]$. In the mesh, the neighboring pixels are connected with the pipes. The capacity of pipes is determined on the basis of local contrast in image. Big contrast corresponds to a small capacity and vice versa. From the sources, the liquid can flow into the neighboring pixels (basins). Similarly, the sinks drain the liquid from the neighboring pixels. The amount of liquid at pixel x at time t is denoted by $\mu(x, t), \mu(x, 0) = \mu_0$. If a basin at a certain time t reaches the state $\mu(x, t) = 1$ (or $\mu(x, t) = 0$), it becomes a new source (or sink).

Let c be the source in the neighborhood of an observed pixel x at time t which is neither another source nor a sink. The following equation is used for determining $\mu(x, t + \Delta t)$ which is observed at the next time step $t + \Delta t$.

$$\mu(x, t + \Delta t) = \begin{cases} \min(\mu(x, t) + \Delta t \cdot \delta(c, x), 1) & \text{if } \mu(c, t_0) = 1 \\ \max(\mu(x, t) - \Delta t \cdot \delta(c, x), 0) & \text{if } \mu(c, t_0) = 0 \end{cases}, \quad (1)$$

where $\delta(c, x)$ is a capacity of the pipe connecting c and x. The capacity can be computed using the Gaussian function, which is of the form

$$\delta(c, x) = e^{\frac{-d^2(c,x)}{2\sigma^2}}, \quad (2)$$

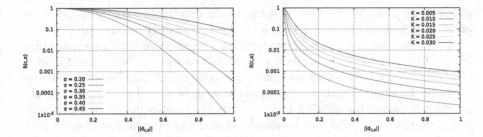

Fig. 1. Comparison between Gaussian (*left*) and Perona&Malik (*right*) functions.

or using the Perona&Malik formula [11]

$$\delta(c, x) = \frac{1}{1 + \dfrac{\|d(c, x)\|^2}{K^2}}. \tag{3}$$

In both the formulae, $d(c, x)$ is the brightness difference between the points c and x. We observed that the Perona&Malik formula convergences faster than the Gaussian function because of its smoother character (see Fig. 1). Moreover, the Perona&Malik formula is less sensitive to the constant value K (compared to σ in the case of the Gaussian function).

The algorithm runs until all pixels become either sources or sinks. To avoid the case of infinite execution, if the uncertainty is present because of neighboring sink and source, the algorithm considers that the filling process is a little bit "stronger" than the draining one. In other words, if an observed pixel x has both types of seeds in its neighborhood, it becomes source.

It should be noted that the above described process substantially differs from diffusion. The amount of fluid running through the pipe between two pixels does not depend on the difference of potentials in the pixels as it is in diffusion (i.e. it does not depend on the difference between the levels of liquid in the corresponding basins). The amount is only limited by the capacity of pipe. The fluid either flows according to the full capacity of pipe or does not flow at all. In diffusion, the flow is linearly dependent on the difference of potentials, the capacity itself is not limited. In our approach, the amount of fluid running through a pipe is a step function. In diffusion, it is a linear function.

In the known diffusion approaches, the boundary conditions are kept constant during the whole time interval. In our approach, the boundary conditions change. It is due to the fact that the undecided pixels may fall into the category of sources or sinks, which, in fact, is a change of boundary conditions.

In the presented approach, a certain kind of decision is involved, which is not the case of the diffusion approach. In diffusion, the decision is made by analyzing the value of potential, which is computed after the certain time interval, whose value highly influences the result. In our approach, the pixels are sorted into the groups of object and background pixels directly during solving the above equations, and the effectiveness is not decreased when the time of flowing converges to infinity. It resembles to something like region growing (a recognized technique of image segmentation) that is driven by the flow analogy in this case.

For illustrating how the method works, we present a synthetic grayscale noisy image of the size 100×100 pixels that contains two segments with a unit brightness step separated by an edge in the image center (Fig. 2a). The noise has the Gaussian distribution with $\sigma = 0.33$. Let the seeds be placed as shown in Fig. 2b, i.e. the left part of the image is considered as the background (black seeds with $\mu(x,0) = 0$) and the right part is considered as the foreground (white seeds with $\mu(x,0) = 1$), the other points have an initial value of $\mu_0 = 0.5$. The capacities of the pipes are computed using Eq. (3) with $K = 0.04, \Delta t = 1$ (Eq. (1)). The situation after 8 iterations is illustrated in Fig. 2c. It is seen that many new points became new seeds (their basins are either full or empty) and, as it is seen in the image detail, the liquid is flowing from the foreground seeds to the neighboring points, which causes that their amount of liquid is increasing (their brightness is higher). The same process runs around the background seeds that are draining the liquid from their neighbors. In Fig. 2d, we show the situation after 32 iterations, when the foreground and background areas touch each other. As it was written in the method description, the boundary points are filled with the liquid and are labeled as new sources. The final segmentation is shown in Fig. 2e.

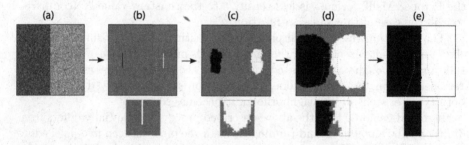

Fig. 2. Visualization of the proposed segmentation method; the input image (a), the initial segmentation (b) (object and background seeds), the state after 8 iterations (c), after 32 iterations (d), and the final segmentation after 55 iterations (e). See the text for further explanation.

3   Experiments

In this section, we evaluate the proposed segmentation technique in the area of seeded image segmentation and compare our results with other methods that have a similar segmentation approach.

In the first experiment, we show how the parameters of the method influence the segmentation result. Two synthetic images were used for this purpose; the image with two segments separated by an edge in the image center (Fig. 3a) (the seeds are placed as in Fig. 3b), and the image with a rectangular object in the image center (Fig. 3c) with the seeds placed as in Fig. 3d. Both images contain a unit brightness step between the segments, and both are influenced by the Gaussian noise with $\sigma = 0.33$. Equation (3) is used for computing the capacities of pipes. The segmentation error is evaluated as a relative number of incorrectly labeled pixels. The method was tested on an Intel Xeon 2.2 GHz computer. The average error, the average number of iterations, and the average running time are shown in Table 1. It can be seen that a lower value of K (see Eq. (3)) leads to a better segmentation result. On the other hand, a low value of K reduces the capacity of pipes and the method converges slower.

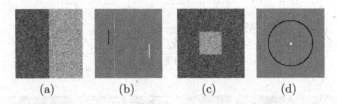

| (a) | (b) | (c) | (d) |

Fig. 3. The synthetic images (a) and (c) with the seeds placed as in (b) and (d), respectively. The segmentation errors of our method are shown in Table 1.

Table 1. The average error rate, the average number of iterations, and the average running time of the new method for various values of K. The method was tested on the images shown in Fig. 3. The values were computed from 10^5 samples.

K	Figure 3(a)			Figure 3(c)		
	Error [%]	Iter	Time [ms]	Error [%]	Iter	Time [ms]
0.002	1.51	5313	308.2	0.64	3757	295.8
0.005	1.59	903	57.2	0.66	632	55.0
0.010	1.89	271	18.1	0.72	185	17.7
0.020	2.76	109	6.9	0.96	72	5.6
0.050	4.22	57	2.8	1.56	37	2.3
0.075	4.38	50	2.3	1.83	32	1.9
0.100	4.46	50	2.3	2.08	29	1.5

We also experimented with the initial amount of liquid in the basins (μ_0). For this purpose, we used Fig. 3a and we were changing the horizontal position of seeds within the segments to achieve different distances from the seeds to the segment boundary (see Table 2, row 1). The parameters were set to $K = 0.01$ and, again, the Gaussian noise ($\sigma = 0.33$) was added. The average error rate, and the average number of iterations are shown in Table 2 (the average time can be derived from the number of iterations, see Table 1). It is seen that the choice of μ_0 influences the segmentation result. If the liquid sources (white seeds) are closer to the boundary (Table 2, column 1) than the liquid sinks (black seeds), a lower μ_0 value is required (and vice versa). If $\mu_0 = 0.5$ and if both segments

Table 2. The average error rate and the average number of iterations of the new method for various values of μ_0. The method was tested on Fig. 2a with the noise amount $\sigma = 0.33$, and with the various combinations of seed positions (shown in the *first row*). The values were computed from 10^5 samples.

μ_0	Error [%]	Iter	Error [%]	Iter	Error [%]	Iter
0.3	7.04	331	13.02	297	22.57	267
0.4	**0.99**	299	2.79	274	11.62	262
0.5	2.97	262	**0.33**	249	3.71	265
0.6	10.59	263	2.32	276	**1.13**	302
0.7	21.60	270	12.00	301	6.71	335

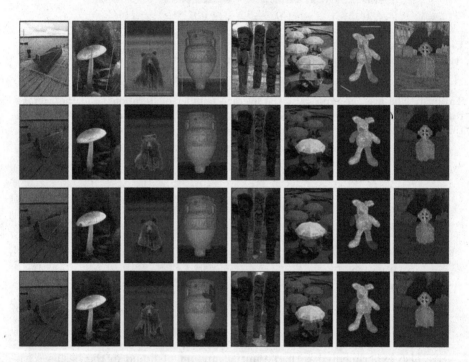

Fig. 4. The comparison of segmentations of real-life images from the Berkeley dataset [8]. Red and green strokes in the input images (*row* 1) represent the foreground and background seeds, respectively. The results of the proposed method (*row* 2), the random walker (*row* 3), and the watershed method (*row* 4) are shown. The segmentation errors are presented in Table 3. (Color figure online)

have similar capacities of pipes (i.e. the brightness variability in both segments is similar), the liquid is filled and drained in a similar pace. If the source is closer to the segment boundary than the sink, the liquid reaches this boundary earlier

Table 3. The segmentation errors [%] of the images shown in Fig. 4.

Method	Boat	Mushroom	Bear	Vase	Totems	Umbrella	Teddy	Grave
Our method	**2.59**	**0.67**	**4.42**	**2.01**	**0.65**	2.59	**0.84**	**2.08**
Rand. Walk.	3.29	0.96	5.77	2.12	0.78	**0.63**	3.89	2.39
Watershed	3.32	0.94	7.04	6.23	4.22	3.32	3.17	**2.08**

than it is drained, and a certain amount of liquid could overflow to the second segment. Therefore, in such cases, it is better to use a lower value of μ_0 because the basins require more liquid to be filled, which slows down the filling procedure. If the source seeds are farther from the boundary than the sink seeds, the filling should be faster than draining and the initial amount of liquid is required to be higher (Table 2, column 3). This observation is also applicable in the images where one segment has different brightness variability than the second segment, which usually happens in real-life images. In such cases, it can be useful to set μ_0 to a different value than 0.5.

Finally, we tested the method on selected real-life images from the Berkeley dataset [8]. The segmentation outputs of our method, the random walker [5], and the watershed method [9] are compared and shown in Fig. 4, the percentage of incorrectly labeled pixels with respect to the image ground truths is in Table 3. The parameters K and μ_0 (our method) and β (random walker that computes the weights as $w_{i,j} = e^{-\beta(g_i - g_j)^2}$, g_i is the brightness at pixel i) were set such that each method gives the best possible result. It is seen that our method achieved the lowest segmentation error in most cases, other methods gave similar or worse results (depending on the specific image).

4 Conclusion

In this paper, we proposed a method for binary seeded image segmentation that is inspired by the anisotropic diffusion process. The principle of the method can be likened to the flowing of liquid in a mesh of pipes. The sources and the sinks of liquid correspond to the foreground and the background seeds, respectively, other image points correspond to the basins with an initial amount of liquid. The capacities of pipes, which connect the neighboring basins, are given by the similarity of pixel brightness. The initial amount of liquid and the coefficient for computing the pipe capacities are the two parameters that are set manually. The properties of the method were illustrated on the segmentation of synthetic images. The comparison with other known segmentation techniques is presented for a set of real-life images. In many cases, the new method achieved better results than other tested techniques.

Acknowledgements. This work is partially supported by Grant of SGS No. SP2017/61 of VŠB - TU Ostrava, and by RFBR Grant No. 16-07-00032 A.

References

1. Bleau, A., Leon, L.: Watershed-based segmentation and region merging. Comput. Vis. Image Underst. **77**(3), 317–370 (2000)
2. Boykov, Y., Jolly, M.P.: Interactive graph cuts for optimal boundary & region segmentation of objects in N-D images. In: Proceedings of the 8th IEEE International Conference on Computer Vision, ICCV 2001, vol. 1, pp. 105–112 (2001)
3. Collins, M.D., Xu, J., Grady, L., Singh, V.: Random walks based multi-image segmentation: quasiconvexity results and GPU-based solutions. In: Proceedings of the IEEE Conference on Computer Vision and Pattern Recognition (CVPR) (2012)
4. Gopalakrishnan, V., Hu, Y., Rajan, D.: Random walks on graphs for salient object detection in images. IEEE Trans. Image Process. **19**(12), 3232–3242 (2010)
5. Grady, L.: Random walks for image segmentation. IEEE Trans. Pattern Anal. Mach. Intell. **28**(11), 1768–1783 (2006)
6. Lee, C., Jang, W.D., Sim, J.Y., Kim, C.S.: Multiple random walkers and their application to image cosegmentation. In: 2015 IEEE Conference on Computer Vision and Pattern Recognition (CVPR), pp. 3837–3845 (2015)
7. Lee, S.H., Jang, W.D., Park, B.K., Kim, C.S.: RGB-D image segmentation based on multiple random walkers. In: 2016 IEEE International Conference on Image Processing (ICIP), pp. 2549–2553 (2016)
8. Martin, D., Fowlkes, C., Tal, D., Malik, J.: A database of human segmented natural images and its application to evaluating segmentation algorithms and measuring ecological statistics. In: Proceedings of the 8th International Conference on Computer Vision, vol. 2, pp. 416–423 (2001)
9. Meyer, F.: Color image segmentation. In: 1992 International Conference on Image Processing and its Applications, pp. 303–306 (1992)
10. Nguyen, H.T., Ji, Q.: Improved watershed segmentation using water diffusion and local shape priors. In: 2006 IEEE Computer Society Conference on Computer Vision and Pattern Recognition (CVPR 2006), vol. 1, pp. 985–992 (2006)
11. Perona, P., Malik, J.: Scale-space and edge detection using anisotropic diffusion. IEEE Trans. Pattern Anal. Mach. Intell. **12**(7), 629–639 (1990)
12. Shen, J., Du, Y., Wang, W., Li, X.: Lazy random walks for superpixel segmentation. IEEE Trans. Image Process. **23**(4), 1451–1462 (2014)
13. Sinop, A.K., Grady, L.: A seeded image segmentation framework unifying graph cuts and random walker which yields a new algorithm. In: 2007 IEEE 11th International Conference on Computer Vision, pp. 1–8 (2007)
14. Zhang, J., Zheng, J., Cai, J.: A diffusion approach to seeded image segmentation. In: 2010 IEEE Computer Society Conference on Computer Vision and Pattern Recognition, pp. 2125–2132 (2010)

Motion and Tracking

Exploiting Semantic Segmentation for Robust Camera Motion Classification

François-Xavier Derue[✉], Mohamed Dahmane, Marc Lalonde, and Samuel Foucher

Computer Research Institute of Montreal,
405, avenue Ogilvy, bureau 101, Montreal, QC H3N 1M3, Canada
{deruefx,dahmanmo,lalondma,fouchesa}@crim.ca

Abstract. The goal of camera motion classification is to identify how the camera moves during a shot (Zoom, Pan, Tilt, etc.). This dynamic information about a video is valuable for many applications such as video indexing and retrieval. For that purpose, we propose an optical flow-based SVM classification to identify 9 types of motion. Numerous methods fail when large moving foreground objects are present in the scene, and we address this problem by combining semantic segmentation with our feature extraction in order to select only relevant motion vectors. We conducted an evaluation that shows promising results as our method reaches 90% correct classifications on a large set of varied video samples.

1 Introduction

Film directors often leverage camera movement techniques to make a shot more interesting. For example, a *zoom in* on a face emphasizes a subject's feeling. To set up a context, a long *panning* is usually used to reveal an environment through the character's eyes. Even camera *rotation* could be exploited to show disorder in a scene.

Identifying these types of movement is the goal of camera motion classification, also referred to as global motion classification. As Duan et al. [6] pointed out, unlike content-based visual queries of still images, motion information is an important tool for content-based video retrieval. Video summarization also benefits from it, as for example in [13], where a long panning on a landscape could be summarized into a single panoramic keyframe. Camera motion characterization is also applied as a preprocessing step in [15] for better video stabilization, and in the domain of action recognition [5,17,18] to subtract camera motion from subject motion.

In the literature, we can distinguish two different approaches in camera motion analysis. The first one is based on global motion estimation, which attempts to find the parameters of a homographic transformation between two images, then infer a motion class label from these parameters [2,4,9]. While such variables are necessary for video stabilization and 3D reconstruction, the computations required are often too expensive for simple camera movement classification. The other approach, rather considered as a qualitative estimation,

© Springer International Publishing AG 2017
F. Karray et al. (Eds.): ICIAR 2017, LNCS 10317, pp. 173–181, 2017.
DOI: 10.1007/978-3-319-59876-5_20

analyzes the motion vector field in a statistical way, such as representing the motion distribution using histograms and deducing a class label from it.

Motion vector field is a common denominator for most of the motion analysis methods. Motion vectors are either provided directly from MPEG video or computed by techniques such as block matching and optical flow. Lee and Hayes [10] use 6 templates of motion vector field, each one characterizing a specific camera movement. To determine the type of motion, they select the closest template to the motion vector field of the current frame. While quite simple, this method does not deal with rotation and is limited to MPEG-1 videos. Another way to classify motion vectors is to use machine learning techniques. Okade et al. [15] chose Multilayer Perceptron (MP) while Liu et al. [11] preferred Support Vector Machine (SVM). In their method, the motion descriptor is simply a concatenation of every motion vectors. However, this high dimensional feature tends to make the classifier overfit. In addition, it is usually very noisy, as no strategy is proposed to filter motion belonging to foreground objects. Geng et al. [8] also use SVM but only to distinguish between translation and non-translation motion. They attempt to tackle the issue of noisy vector motion by assuming two types of noise: Abnormal motion noise is caused by mistakes in the computation of the motion vectors, and foreground noise is due to the presence of foreground objects whose motion is different from the background. Assuming that a subject of interest should be in the center region of a frame, they build a motion-based saliency map that is binarized in an adaptive way. Although the technique would be good for videos that follow general filming guidelines, it would not handle videos with many moving objects distributed in the entire frame.

Instead of motion fields, keypoints are matched in [2] to compute a homography between two images. By extracting a single feature from a series of homographies, the approach classifies a whole shot instead of two successive frames. The complexity of this method makes it slow and it does not suggest any solution for filtering foreground object motion. A quite different approach to process shots is also proposed by Ngo et al. [14]. They introduced the concept of a *shot slice*, which gathers temporal information in a single picture. The distinct trails moving in specific directions allow them to determine the global camera motion. Although original, this method is affected by cluttered background with many moving objects, as they create visual patterns on the slice that are hard to distinguish.

Similarly to [11], our method relies on multiple SVM classifiers. However, we provide a more accurate classification in 9 categories, including camera translation in four directions, zoom in and out, as well as clockwise and counterclockwise rotation. Moreover, our feature is computationally inexpensive, as only the orientation of the motion vectors is considered. Finally, to the best of our knowledge, we apply for the first time semantic segmentation for noisy foreground motion removal, leading to more accurate classifications. The experimental results presented at the end of this paper show that our method is able to correctly classify camera motion for a large set of different videos.

2   Proposed Approach

The different steps of the proposed method for camera motion classification are shown in Fig. 1. At each frame, we first compute the motion vectors via dense optical flow [3]. In order to remove noisy motion vectors belonging to the foreground, we extract a background mask via semantic segmentation applied to the current frame. Then, motion vectors belonging to that mask are used to create an histogram-based feature, which will finally be classified by a multi-SVM classifier.

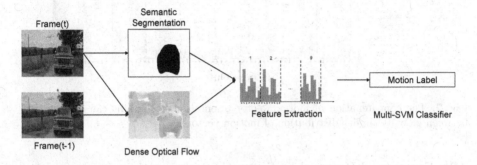

Fig. 1. Overview of our strategy for camera motion classification.

As said previously, we consider 9 types of motion: translation *right* and *left* which correspond to panning and trucking in cinematographic terms; translation *up* and *down* for tilting and pedestal; *in* and *out* for zoom and dolly shots; and *static* when the camera does not move.

2.1   Feature Extraction

Given the motion vectors computed at each pixel, we only keep the orientation component to compute our feature. Indeed, if every vector had the same magnitude, they could still be differentiated by their orientation. A naive approach would gather all these vectors into a single orientation histogram. However, the resulting feature would not be discriminative enough. Since using a histogram would eliminate the spatial information about the location of each vector, a histogram for a clockwise rotation would be exactly the same as a counterclockwise rotation, a zoom in, or even a zoom out. Therefore, in order to preserve spatial localization, we split the optical flow image into 9 blocks as represented in Fig. 2. For each block, a N-bin histogram is computed with the vector orientations inside that block. As a result, the final descriptor is a concatenation of 9 histograms. Now, if we compare a zoom in (Fig. 2a) and a zoom out (Fig. 2b), we observe that the two descriptors are very distinct.

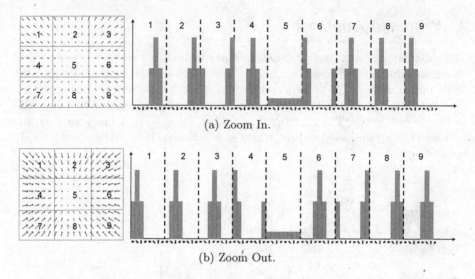

(a) Zoom In.

(b) Zoom Out.

Fig. 2. The concatenation of motion vector orientation histograms can discriminate between globally similar distribution of motion vectors.

2.2 Foreground Removal by Semantic Segmentation

In this section, we show how we leverage semantic segmentation to remove noisy motion vectors for better classification.

Imagine a static camera filming a moving object. In the optical flow image, there would be two different groups of motion vectors: those corresponding to the background which should be close to zero, and those that belong to the moving object which are the noisy foreground motion vectors that will affect the classification. If the moving object fills most of the frame, it is very likely that a camera motion will be detected while it is supposed to be static. Ideally, only the background vectors should be considered for the classification.

In the literature, the methods that propose a solution to deal with this issue usually fail when the subject fills more than one half of the frame. Wang et al.'s method [18] can handle large foreground, yet it only takes into consideration humans as foreground objects since their main goal was action recognition. We propose to detect any kind of foreground in order to remove their respective motion vectors.

Semantic segmentation assigns to each pixel a label corresponding to the object or region it belongs to. In our case, we only look for background areas. For that purpose, we use the method of Long et al. [12] based on a fully convolutional Network (FCN), which has been trained on PASCAL VOC 2011 segmentation dataset [7] including a background class. Therefore, since we are able to label every background pixel, the remaining pixels considered as foreground could belong to any moving object. This method is efficient since it takes less that 50 ms to process a typical image while achieving state-of-the-art segmentation results.

Figure 3 shows the benefits of using semantic segmentation as background mask to remove the ambiguity due to foreground motion vectors. The car is moving towards the camera during a panning left. The optical flow image clearly shows two groups of motion vectors. Computed on that image, our feature is ambiguous and will be misclassified. With the mask obtained from the FCN, only the vectors inside that mask are used to make the feature, resulting in a noise-free feature easy to classify. The difference is visible in the blocks 5, 6, 8 and 9. Note that we use the color code specified by [1] to represent the optical flow at each pixel. Vector magnitude and angle correspond to saturation and hue respectively.

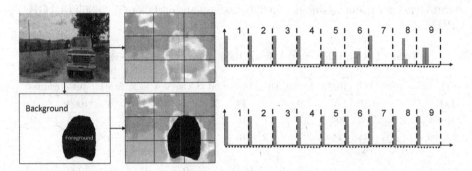

Fig. 3. Semantic segmentation identifies pixels belonging to the background. Used as a mask, only relevant motion vectors are selected to make a clean descriptor (below) compared to a noisy one (above).

2.3   Multi-SVM Classifier

Support Vector Machines (SVM) [16] is a learning algorithm which has many advantages. Compared to other methods, it requires fewer parameters and it can achieve high generalization performance with a limited sample set. SVMs are typically used as binary classifiers within a supervised learning framework: They require positive and negative samples so that they can build a hyperplane to separate these two groups in the feature space. By use of kernels, it is even capable of splitting non-linearly separable data classes.

In our case, we face a multiclass problem. Therefore, we train a *one-against-all* SVM classifier for each of the motion types. For example, when training the classifier *Panning Left*, negative samples include all motion samples except those corresponding to a left pan motion. Instead of a binary classification, we use SVM as regressor, which outputs a continuous value representing the confidence that the input sample belongs to the given class. Thus, an unknown sample is given a score from each classifier and the one that gets the highest one is the winner. Although there are actually 9 different motion classes, we only build 8 classifiers: The *static* class is not associated with any particular direction, so an unknown camera motion is classified as static if the median of the magnitude of the optical flow is under a threshold, otherwise it is classified by our multi-SVM classifier.

3   Results

For our experimentation, each orientation histogram has 32 bins, resulting in a feature of $32 * 9 = 288$ dimensions. Each SVM uses a linear kernel as it empirically gives the best results for our data. The threshold on the optical flow magnitude to filter out static motion has been set to 0.01. While the dense optical flow algorithm provided by OpenCV has been used for our evaluation, more complex optical flow techniques could further improve our method. Our training set consists of 100 samples for each class, and we perform a 10-fold cross validation to optimize the internal parameters. As for the semantic segmentation, we use the implementation provided by [12] available online[1]. Computation takes about 60 ms per frame of dimension 320×240 on an Intel Core i7-2600k (3.4 GHz) CPU.

Table 1. Confusion matrix from 1000 samples tested for each class.

GT/Estimation	Left	Right	Down	Up	In	Out	CClock	Clock	Static	Total frames
Left	980	0	0	0	14	2	0	1	3	1000
Right	42	904	7	11	16	10	9	1	0	1000
Down	18	21	939	2	13	7	0	0	0	1000
Up	26	5	0	920	12	18	0	16	3	1000
In	14	0	6	11	944	0	0	0	25	1000
Out	22	0	36	4	0	853	0	0	85	1000
CClock	2	105	0	9	8	4	870	0	2	1000
Clock	139	3	0	3	12	3	0	818	22	1000
Static	36	44	38	13	23	18	4	2	822	1000

As no standard dataset is available online to evaluate camera motion classification, we build our own test set (different from the training set) using various videos collected from YouTube that we manually annotated.

In total, 1000 samples have been tested for each class. Table 1 shows the confusion matrix with the rows corresponding to the ground truth and the columns to our predictions. From that matrix, we compute the precision $P = \frac{TP}{TP+FP}$ and recall $R = \frac{TP}{TP+FN}$ where TP stands for true positive count, FP for false positive count and FN for false negative count. Table 2 summarizes the results for each motion class. On average, we reach a precision and a recall of 0.9 and 0.89 respectively. Compared to a basic classification without semantic segmentation, the improvement is about 8% for precision and 6% for recall. Notice that the most affected class is the *Static* class. Indeed, if there is a moving object in a static shot, it will definitely mislead the classification, unless it is removed from the background. Note that the lower precision of the class *Left* could be due to the presence of shaky and composite motion, such as a zoom and a pan in the same shot. Moreover, as our method relies on optical flow, it inherently

---

[1] https://github.com/shelhamer/fcn.berkeleyvision.org.

Table 2. Precision and recall: improvement by using semantic segmentation.

Class	With segmentation		Without segmentation	
	Precision	Recall	Precision	Recall
Left	0.77	0.98	0.72	0.97
Right	0.84	0.91	0.82	0.90
Down	0.92	0.94	0.80	0.91
Up	0.95	0.92	0.85	0.84
In	0.91	0.94	0.80	0.97
Out	0.93	0.85	0.94	0.86
CClock	0.99	0.87	0.95	0.97
Clock	0.98	0.82	0.87	0.98
Static	0.85	0.82	0.63	0.09
Average	0.9	0.89	0.82	0.83

suffers from non-textured scene where motion vectors cannot be found. Those issues will be considered for further research.

As a comparison to the state-of-the-art, we gather in Table 3 the overall results of four similar methods. Since neither code source nor video data set is available, those values are directly imported from their respective paper. Compared to the SVM-based classification methods [6,8,11], ours shows a real improvement in both precision and recall, which shows the benefit of our foreground removal strategy. While Okade et al. [15] reports the highest precision and recall, their test set only includes 100 samples per class.

Table 3. Comparison to similar approaches.

	Okade et al. [15]	Duan et al. [6]	Liu et al. [11]	Geng et al. [8]	Ours
Precision	0.94	0.80	0.82	0.85	0.9
Recall	0.93	0.85	0.83	0.84	0.89

4   Conclusion

In order to classify camera motion, we developed a simple yet effective method based on multiple SVM classifiers and optical flow-based orientation histograms. In addition, we leveraged the high quality semantic segmentation produced by a CNN to remove noisy foreground motion vectors, therefore improving the classification. Our evaluation on a large set of video samples shows that our method is accurate to more than 90% for most of the classes. Finally, as a future work, we plan to extend our method to the classification of composite camera movements.

Acknowledgements. This work was supported by the Ministère de l'Économie, des Sciences et de l'Innovation (MESI) of the province of Québec.

References

1. Baker, S., Scharstein, D., Lewis, J.P., Roth, S., Black, M.J., Szeliski, R.: A database and evaluation methodology for optical flow. Int. J. Comput. Vis. **92**(1), 1–31 (2011)
2. Bhattacharya, S., Mehran, R., Sukthankar, R., Shah, M.: Classification of cinematographic shots using lie algebra and its application to complex event recognition. IEEE Trans. Multimed. **16**(3), 686–696 (2014)
3. Bouguet, J.Y.: Pyramidal implementation of the Lucas Kanade feature tracker. Intel Corporation, Microprocessor Research Labs (2000)
4. Bouthemy, P., Gelgon, M., Ganansia, F.: A unified approach to shot change detection and camera motion characterization. IEEE Trans. Circuits Syst. Video Technol. **9**(7), 1030–1044 (1999)
5. Chen, Q.Q., Liu, F., Li, X., Liu, B.D., Zhang, Y.J.: Saliency-context two-stream convnets for action recognition. In: 2016 IEEE International Conference on Image Processing (ICIP), pp. 3076–3080, September 2016
6. Duan, L.Y., Jin, J.S., Tian, Q., Xu, C.S.: Nonparametric motion characterization for robust classification of camera motion patterns. IEEE Trans. Multimed. **8**(2), 323–340 (2006)
7. Everingham, M., Van Gool, L., Williams, C.K.I., Winn, J., Zisserman, A.: The PASCAL Visual Object Classes Challenge 2011 (VOC2011) Results
8. Geng, Y., Xu, D., Feng, S., Yuan, J.: A robust and hierarchical approach for camera motion classification. In: Yeung, D.-Y., Kwok, J.T., Fred, A., Roli, F., Ridder, D. (eds.) SSPR/SPR 2006. LNCS, vol. 4109, pp. 340–348. Springer, Heidelberg (2006). doi:10.1007/11815921_37
9. Guironnet, M., Pellerin, D., Rombaut, M.: Camera motion classification based on transferable belief model. In: 2006 14th European Signal Processing Conference, pp. 1–5, September 2006
10. Lee, S., Hayes, M.H.: Real-time camera motion classification for content-based indexing and retrieval using templates. In: 2002 IEEE International Conference on Acoustics, Speech, and Signal Processing, vol. 4, pp. IV-3664–IV-3667, May 2002
11. Liu, L., Zhang, R., Fan, L.: Camera motion classification based on SVM. In: 2010 3rd International Congress on Image and Signal Processing, vol. 1, pp. 392–394 (2010)
12. Long, J., Shelhamer, E., Darrell, T.: Fully convolutional networks for semantic segmentation. In: 2015 IEEE Conference on Computer Vision and Pattern Recognition (CVPR), pp. 3431–3440, June 2015
13. Mei, T., Tang, L.X., Tang, J., Hua, X.S.: Near-lossless semantic video summarization and its applications to video analysis. ACM Trans. Multimed. Comput. Commun. Appl. **9**(3), 16:1–16:23 (2013)
14. Ngo, C.W., Pong, T.C., Zhang, H.J.: Motion analysis and segmentation through spatio-temporal slices processing. IEEE Trans. Image Process. **12**(3), 341–355 (2003)
15. Okade, M., Patel, G., Biswas, P.K.: Robust learning-based camera motion characterization scheme with applications to video stabilization. IEEE Trans. Circuits Syst. Video Technol. **26**(3), 453–466 (2016)

16. Vapnik, V.N.: The Nature of Statistical Learning Theory. Springer-Verlag New York Inc., New York (1995)
17. Wang, H., Kläser, A., Schmid, C., Liu, C.L.: Action recognition by dense trajectories. In: CVPR 2011, pp. 3169–3176, June 2011
18. Wang, H., Schmid, C.: Action recognition with improved trajectories. In: 2013 IEEE International Conference on Computer Vision, pp. 3551–3558, December 2013

An Event-Based Optical Flow Algorithm for Dynamic Vision Sensors

Iffatur Ridwan and Howard Cheng[✉]

Department of Mathematics and Computer Science, University of Lethbridge,
Lethbridge, Canada
{iffatur.ridwan,howard.cheng}@uleth.ca

Abstract. We present an event-based optical flow algorithm for the Davis Dynamic Vision Sensor (DVS). The algorithm is based on the Reichardt motion detector inspired by the fly visual system, and has a very low computational requirement for each event received from the DVS.

1 Introduction

Motion detection is a common task in many areas of video processing and computer vision, and optical flow computation is one method of performing such detections. In applications such as autonomous vehicle or robot navigation [5], these computations must be done in real-time, using devices that may have limitations on power consumption as well as computational power. We propose a fast algorithm to compute optical flow that is useful on such restricted platforms, using a camera that has been inspired by biological retinas.

There are existing works on the computation of optical flow for videos obtained from conventional frame-based cameras (for example, [6,8,9]). Since consecutive frames are highly correlated, these cameras often capture frames with redundant data which are later removed in the processing. Computational time and electrical power are wasted in capturing and processing this data.

The Davis Dynamic Vision Sensor (DVS) is a camera that is modelled upon the human retina [7]. The DVS is an asynchronous device that only transmits events indicating changes in brightness in individual pixels. If there is no change, this system does not give any output. The DVS has lower power and computational requirements, as well as faster reaction times. Algorithms for the DVS must be designed so that it works with "sparse" input in order to take advantage of the unique properties of the DVS.

In this paper, we propose an algorithm for the DVS based on the Reichardt detector—a simple correlation-based movement detection model inspired by the visual system of flies [4,10]. This model cannot be used directly for the DVS but we will show that our algorithm can be considered as a variation of the

H. Cheng—Supported by a Discovery Grant from the Natural Sciences and Engineering Research Council (NSERC) of Canada.

© Springer International Publishing AG 2017
F. Karray et al. (Eds.): ICIAR 2017, LNCS 10317, pp. 182–189, 2017.
DOI: 10.1007/978-3-319-59876-5_21

Reichardt detector on conventional frame-based video input. We will show that our algorithm requires relatively little processing for each event received from the DVS and therefore maintains the advantage of using the DVS. While there are some other works on optical flow algorithms for this type of cameras [2,3], our approach is different in that it is based on the Reichardt detector.

The paper is organized as follows. Section 2 reviews some of the previous works our algorithm is based on. Section 3 describes our algorithm, and experimental results and analysis are given in Sect. 4.

2   Preliminaries

2.1   Reichardt Detector

The Reichardt detector is a model of how neurons detect motion from photoreceptors, and it is inspired by the visual system of flies [4,10]. The Reichardt detector consists of two mirror symmetric sub-units (Fig. 1). In each sub-unit, the luminance values as measured in two adjacent image locations (one of them is delayed by a low pass filter) are multiplied together. The product can be viewed as the correlation of the two locations at different times. The resulting output signal is the difference between the multipliers of the two sub-units.

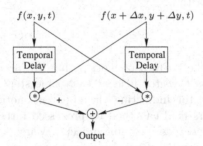

Fig. 1. The Reichardt detector.

More formally, let $f(x, y, t)$ is the luminance value at location (x, y) at time t. We also let Δx and Δy be the offset between two adjacent sub-units, and Δt be the time delay. Then the output of the Reichardt Detector is

$$RD(f, x, y, t, t') = f(x', y', t) \cdot f(x, y, t') - f(x, y, t) \cdot f(x', y', t'), \qquad (1)$$

where $x' = x + \Delta x, y' = y + \Delta y$, and $t' = t + \Delta t$. If $|RD(f, x, y, t, t')|$ exceeds a threshold T_{RD}, motion is detected along the direction $\pm(\Delta x, \Delta y)$, and the sign of $RD(f, x, y, t, t')$ indicates the actual direction. Adjacent pixels in stationary objects will not be detected because the difference is close to zero.

Each Reichardt detector can only detect motion at a particular location with velocities of $\pm(\Delta x/\Delta t, \Delta y/\Delta t)$. Different velocities can be detected by varying

$\Delta x, \Delta y$, and Δt. In practice, this is accomplished by a grid of Reichardt detectors on the pixels of successive frames, so that motion can be detected at all locations in a number of pre-defined directions (e.g. the 8 compass directions).

2.2   The Davis Dynamic Vision Sensor

The Davis Dynamic Vision Sensor (DVS) is a type of "neuromorphic camera" that is inspired by biological retinas [7]. Each pixel is an independent sensor from the others. When the log-luminance level at a particular pixel changes by more than a predefined threshold T_{DVS}, an event is reported indicating the location and its polarity (positive or negative). The events are reported asynchronously as soon as they occur. As a result, it is possible to react to local changes quickly without waiting for a "frame" to be collected. Since only significant changes are reported, redundant data are not reported and do not need to be processed. It may also result in lower power requirement. Of course, the algorithm to process this data must not "convert" these events into a frame-based video or the advantages of the sparseness of the data will be lost.

The events generated by the DVS are communicated using the Address Event Representation (AER) protocol. Conceptually, each of the events we are interested in contains the following information: timestamp, location (x, y), and polarity $(\pm)$. The DVS may also generate other types of events but they are ignored by our optical flow algorithm.

2.3   Optical Flow

Optical flow refers to the pattern of motion that are present in a scene. It is generally represented as a vector field at each time step, in which each pixel is associated with a vector indicating the apparent motion for that pixel at that time. These vectors field can then be processed further to detect specific types of motion of interest (e.g. incoming objects). Many optical flow algorithms for conventional frame-based cameras have been proposed and studied (see, for example, [1]).

3   Proposed Algorithm

In this section, our algorithm to compute optical flow for the Dynamic Vision Sensor (DVS) is described. The connection between the algorithm and Reichardt detectors will also be shown.

The input to our algorithm is an event stream in the AER format. Moreover, the output is also an event stream indicating when motion is detected. When motion is detected at a particular pixel, the algorithm generates an event specifying a timestamp, the location (x, y), as well as the direction of the motion. In our algorithm, we only detect one of the eight compass directions, which we denote by the vectors $v_1 = (-1, -1), v_2 = (-1, 0), v_3 = (-1, 1), v_4 = (0, -1), v_5 = (0, 1), v_6 = (1, -1), v_7 = (1, 0), v_8 = (1, 1)$. Thus, the output of our algorithm is an event stream indicating the nonzero vectors in the optical flow at specific times. The direction vectors are fixed and only these 8 directions can be reported, but it is possible that multiple directions are reported at the same location and time. If desired, the multiple directions at a location can be combined (e.g. by taking the "average" of the detected directions).

For each pixel location (x, y), we maintain the most recently received event $e_{(x,y)}$. Each event is represented by the timestamp and its polarity (t, p). As each event arrives, we search for a recent event (occurring no more than some threshold T before the current event) that has the same polarity. If a match is found, an event indicating detected motion from the neighbour to the current pixel is reported. This is described in Algorithm 1. The parameter T is used to control how "recent" a neighbouring event is considered a match to the current event received. Note that only the direction of the motion is reported. If desired, the magnitude of the motion can be reported by comparing the timestamps of the two matching events to determine the velocity of the movement.

Algorithm 1. Optical flow computation with DVS for a single event received.

Input: an event from the DVS consisting of timestamp t, location (x, y), and polarity $p \in \{+, -\}$; a threshold T.
Output: if motion is detected, event(s) each consisting of timestamp t, location (x, y), and direction v.

$e_{(x,y)} \leftarrow (t, p)$;
for $v \in \{v_1, \ldots v_8\}$ **do**
  Let $(x', y') = (x, y) - v$;
  Let $(t', p') = e_{(x',y')}$;
  if $0 < t - t' \leq T$ and $p = p'$ **then**
    Output event (t, x, y, v) ;
end

In terms of computational complexity, each input event requires only a small constant number of operations proportional to the number of directions. The number of pixels in the image is irrelevant. This is important because the output of the DVS (the input of our algorithm) is sparse and the complexity of our algorithm is directly proportional to the number of events in the input. Thus, the advantage of the DVS is preserved by our algorithm. Other existing approaches [2,3] require more complicated calculations for each event and have a higher computational costs, but the output of these algorithms are more general and are not restricted to the 8 directions as in our algorithm.

3.1 Relationship to Reichardt Detectors

Although the main approach in Algorithm 1 can be considered as "event matching," the algorithm is in fact closely related to the Reichardt detector. A single

Reichardt detector along the direction $v = (\Delta x, \Delta y)$ for a conventional frame-based camera is described by (1). Recall that the output of (1) is compared to the threshold T_{RD} to determine if motion is detected along the direction v.

We first show that Algorithm 1 can be considered an application of the Reichardt detector on the output of the DVS (instead of the original scene). An event is generated by the DVS at location (x, y) when the change in log-luminance is greater than T_{DVS}. If we denote this change $\Delta f(x, y, t)$, then

$$|\Delta f(x,y,t)| = |\log f(x,y,t') - \log f(x,y,t)| > T_{DVS}, \qquad (2)$$

where $t' = t + \Delta t$. When two events are matched in Algorithm 1, each of these events corresponds to a log-luminance change exceeding T_{DVS} at two times t_1 and t_2 with $t_1 < t_2$. In the algorithm, the current event at (x, y) at time t_2 is matched with a previous neighbouring event of the same polarity at time t_1. To simplify notation, we let $t_2 = t_1 + \Delta t_1$, and let $t_3 = t_2 + \Delta t_2$. Applying (1) to the output of the DVS, we have

$$RD(\Delta f, x, y, t_1, t_3) = \Delta f(x',y',t_1) \cdot \Delta f(x,y,t_2) - \Delta f(x,y,t_1) \cdot \Delta f(x',y',t_2). \quad (3)$$

A match in polarity of two events at (x, y, t_2) and (x', y', t_1) means that the product $\Delta f(x,y,t_2) \cdot \Delta f(x',y',t_1)$ is positive and greater than $(T_{DVS})^2$, which can be considered as the threshold T_{RD} that is used in (3) for detecting motion. The second term of (3) may be assumed to be 0 as there are no events at (x', y') at time t_2. Thus, our algorithm can be viewed as applying the Reichardt detector to DVS events.

Furthermore, the application of Reichardt detector to the DVS output can be thought of as a combination of the outputs of different Reichardt detectors on the original scene at closely related times. Simple algebraic manipulation shows that

$$\begin{aligned} RD(\Delta f, x, y, t_1, t_3) &= RD(\log f, x, y, t_1, t_2) + RD(\log f, x, y, t_2, t_3) \\ &\quad - RD(\log f, x, y, t_1, t_3). \end{aligned} \qquad (4)$$

Thus, we have shown that Algorithm 1 can be considered to be a variation of the Reichardt detector on the original scene.

4   Experimental Results

Our algorithm has been designed and implemented for the DVS. However, to demonstrate the effectiveness of Algorithm 1 in this paper, we use a DVS simulation algorithm to process input videos from conventional frame-based cameras to produce an event stream, which is then processed by Algorithm 1. We did not compare our results to these previous methods [2,3] as the output of our algorithm is not directly comparable.

Two test videos are used in our experiments on Algorithm 1. The first video consists of a single circular object that moves around in a dark background. The second video contains a person moving his head and body. The camera is not steady so both the foreground and the background of the scene are moving. The properties of the two videos are shown in Table 1. In the first video, events are generated by the DVS only on the boundary of the circular object, resulting in a significant reduction in the amount of data sent compared to conventional frame-based cameras. This is true even in the second video—the number of events generated is less than a tenth of the total number of pixels among all frames. Two example frames from the second video are shown in Fig. 2, and the DVS events generated corresponding to these frames are shown in Fig. 3.

Table 1. Properties of the test videos.

	Size	Number of frames	Number of DVS events
Video 1	1280×720	60	32,940
Video 2	1920×1080	45	8,953,097

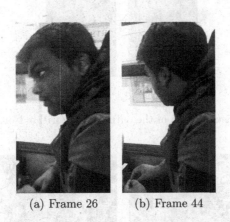

(a) Frame 26   (b) Frame 44

Fig. 2. Two example frames from the second test video.

Figures 4 and 5 show some of the optical flow computed by Algorithm 1 on Video 1 and Video 2, respectively. To visualize the results, motion events generated are collected and those occurring at the same time are displayed as individual images. To make it easier to visualize, not all vectors reported by our algorithm are shown—only one vector from a group of closely located vectors are shown. We can visually observe that the optical flow computed reflect the actual motion present in the videos, though there are very rarely extraneous motion detected due to noise (e.g. the motion vector detected in Frame 30 in Fig. 4).

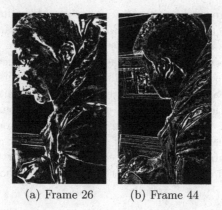

(a) Frame 26   (b) Frame 44

Fig. 3. DVS events corresponding to the example frames in Fig. 2. Pixels with DVS events are shown in white.

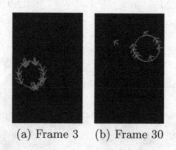

(a) Frame 3   (b) Frame 30

Fig. 4. A visualization of the output of Algorithm 1 on Video 1.

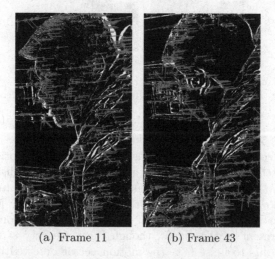

(a) Frame 11   (b) Frame 43

Fig. 5. A visualization of the output of Algorithm 1 on Video 2.

5   Conclusion

In this paper, we described a new event-based approach to perform optical flow calculations for the Davis Dynamic Vision Sensor based on the Reichardt detector. The proposed algorithm is efficient and requires only a small number of operations for each event received from the DVS. Thus, the advantages of the DVS is maintained.

We are working to incorporate the output of our optical flow algorithm in other applications such as object tracking and looming detection. We believe that the restricted optical flow output will be beneficial for these tasks.

Acknowledgement. The authors would like to acknowledge Dr. Matthew Tata for providing access to the Davis Dynamic Vision Sensor for our work, and Cody Barnson for implementing some of our algorithms.

References

1. Beauchemin, S.S., Barron, J.L.: The computation of optical flow. ACM Comput. Surv. **27**(3), 433–466 (1995)
2. Benosman, R., Ieng, S.H., Clercq, C., Bartolozzi, C., Srinivasan, M.: Asynchronous frameless event-based optical flow. Neural Netw. **27**, 32–37 (2012)
3. Brosch, T., Tschechne, S., Neumann, H.: On event-based optical flow detection. Front. Neurosci. **9**, 137 (2015)
4. Egelhaaf, M., Reichardt, W.: Dynamic response properties of movement detectors: theoretical analysis and electrophysiological investigation in the visual system of the fly. Biol. Cybern. **56**(2–3), 69–87 (1987)
5. Fortun, D., Bouthemy, P., Kervrann, C.: Optical flow modeling and computation. Comput. Vis. Image Underst. **134**(C), 1–21 (2015). http://dx.doi.org/10.1016/j.cviu.2015.02.008
6. Fülöp, T., Zarandy, A.: Bio-inspired looming object detector algorithm on the Eye-RIS focal plane-processor system. In: 2010 12th International Workshop on Cellular Nanoscale Networks and their Applications (CNNA 2010) (2010)
7. Lichtsteiner, P., Posch, C., Delbruck, T.: A 128×128 120 dB 15 μs latency asynchronous temporal contrast vision sensor. IEEE J. Solid-State Circuits **43**(2), 566–576 (2008)
8. Pantilie, D., Nedevschi, S.: Real-time obstacle detection in complex scenarios using dense stereo vision and optical flow. In: 13th International IEEE Conference on Intelligent Transportation Systems (2010)
9. Park, S.S., Sowmya, A.: Autonomous robot navigation by active visual motion analysis and understanding. In: Proceedings of IAPR Workshop on Machine Vision Applications (1998)
10. Reichardt, W., Egelhaaf, M.: Properties of individual movement detectors as derived from behavioural experiments on the visual system of the fly. Biol. Cybern. **58**(5), 287–294 (1988)

People's Re-identification Across Multiple Non-overlapping Cameras by Local Discriminative Patch Matching

Rabah Iguernaissi[✉], Djamal Merad, and Pierre Drap

Aix-Marseille University, LSIS - UMR CNRS 7296,
163 Avenue of Luminy, 13288 Marseille Cedex 9, France
rabah.iguernaissi@lsis.org

Abstract. People's tracking in multi-camera systems is one of the most important parts for the study of human's behavior. In this work, we propose a re-identification method for associating people across non-overlapping cameras for tracking purposes. The proposed method is based on the use of discriminatives patches (salient regions). Our method is based on the proposal of a new framework that is used for selecting the most discriminative patches for each tracked individual. This framework is based on exploiting both appearance and spatial information to find the most discriminative salient regions. In this framework, each individual is represented by a set of values representing a rough description for several local patches extracted from the given individual. Then, this representation is used to select some interest patches that most represent the individual of interest compared to other individuals. At the end, these patches are used for associating new detected individuals to tracked ones.

Keywords: People re-identification · Multi-camera tracking · Salient regions

1 Introduction

In the last few years, the use of automatic tools in the study of human's behavior is one of the most active research fields in computer vision. One of the most addressed subjects in this field is the people's tracking in multi-camera systems which is considered as the basic step in many applications that are designed for the study of human behavior such as the study of customers' behavior in a sale area. Generally tracking systems are composed of two main steps a mono-camera tracking step and a re-identification step. The mono-camera tracking consists in the design of algorithms that enable tracking people within one camera field of view for a short period of time to reconstruct trajectory segments. Then, the second step consists in the design of re-identification strategies that allow associating individuals detected within the fields of view of different cameras. In this work, we propose a re-identification strategy that is used to associate

© Springer International Publishing AG 2017
F. Karray et al. (Eds.): ICIAR 2017, LNCS 10317, pp. 190–197, 2017.
DOI: 10.1007/978-3-319-59876-5_22

individuals across non-overlapping cameras. It is realized within the MAGNUM project which is funded by the French government in the context of behavioral marketing analysis.

Several works were done and many methods were proposed to solve the problem of associating people across non-overlapping cameras. This problem is commonly known as people re-identification. The proposed works are generally based on the use of one of three main approaches the discriminative signature based methods, the feature transformation based methods, and the metric learning based methods.

The most used approach is the discriminative signature based methods. These methods focus on finding the most discriminative signature that can be used to differentiate individuals. In this category, we can mention methods that are based on learning a classifier that is used to recover the identity of individuals such as the ones proposed in [1,2], and the methods that are based on directly finding the most discriminative signature either by segmenting individuals into different body parts such as the ones proposed in [3,4] or accumulating several local features in a single signature as done in [5] or based on saliency calculations as done in [6,7].

The second category of methods are those based on the feature transformation approach. These re-identification methods aim to find the transformation that relates features within a certain camera to their corresponding ones in another camera. Several works were done in this category for either finding the direct transformation that relates a certain descriptor in one camera to the corresponding one in another camera [8–10] or learning a classifier that differentiates between pair of images that represent the same individual and those that doesn't represent the same individual [11,12].

The last category of methods are those based on the metric learning approach. These methods aim to learn optimal signature distance metric that is used to compare the feature vectors. These methods try to improve the performance of re-identification by proposing an adequate metric for the used descriptors [13–16]. For this category of methods, the proposed metric is generally used with a selected discriminative signature to improve the re-identification performance.

Independently of the adopted approach, most methods focus on solving the same main issues. In general, proposed re-identification systems deal with illumination variations, viewpoint changes, and similar appearances between different individuals.

In this work, we propose a method for person re-identification across non-overlapping cameras. Our method is based on the use of discriminative patches for re-identifying individuals. The main contribution of this work is the design of a novel framework for salient regions selection. This framework is based on the use of a rough color description of several local patches for person representation in the framework.

The rest of this paper is organized in two main parts. The first part is dedicated to the proposal of our re-identification strategy and the second part is used to show some results obtained using this proposed re-identification method. This paper ends up with a conclusion.

2 Proposed Method

In this work, we propose a re-identification method for associating individuals observed across non-overlapping cameras for tracking purposes in a multi-camera tracking system. The proposed method is based on exploiting local differences between tracked individuals by using the most discriminative local patches for re-identification. Our re-identification strategy is made up of two main phases the learning and the matching steps. The first one is used to identify the most discriminative patches for each tracked individual by exploiting both appearance and spatial information from a set of detections for tracked individuals. Then, the second phase consists in using these discriminative local patches to associate these individuals.

2.1 Learning Process

This part of our algorithm is used to select the most discriminative local patches for each tracked individual. These local patches will be used later on to associate individuals in different cameras. It is based on the use of an online learning strategy. The patch selection is done in two main steps the first step is the framework construction in which each individual is represented by some graphs obtained from the appearance and spatial information obtained using a set of detection of the individual of interest. Then, the second step consists in using this framework to select the most discriminative patches for each individual.

Proposed Framework. This work is done in the context of multi-camera tracking of people to associate tracked Individuals from different cameras. Thus, we proposed a multiple-shot strategy to exploit the tracklets obtained from a mono-camera tracking system. The tracklets are made up of set of successive detections for each individual. These detections are used to construct a framework in which each individual is represented by some graphs that are obtained by exploiting both spatial and appearance information.

In this algorithm, the framework construction is done in two main steps. The first step consists in estimating of the central vertical axis for each detection. To estimate this axis, we start by using a background subtraction method to detect foreground objects. For this, we used a background subtraction method that is based on the use of running average for estimating the background. Then, we use frame differencing between current frame and estimated background to obtain the difference image $diff_t$. The $diff_t$ is used to obtain the foreground mask B_t using a threshold th_{min}. This is done using Eqs. (1) and (2).

$$diff_t(x,y) = |frame_t - background|. \tag{1}$$

$$B_t(x,y) = \begin{cases} 1 & if \quad th_{min} < diff_t(x,y) \\ 0 & otherwise \end{cases} \tag{2}$$

Ones the foreground segmented into blobs representing single individuals, the next step consists in estimating the central vertical axis for each detection. The central vertical axis is considered to be the axis of symmetry that separates the detected blob among the $x - axis$ into two parts that have the same number of foreground pixels. Then, the second step consists in the calculation of descriptors that exploit both spatial and appearance characteristics for each detection to construct the framework. For this, detected persons are represented by rectangles with width w and height h. Then, a set of rectangular local patches $p(y, wp, hp)$ are propagated around the central vertical axis. These patches are centered at the point (x, y) that belong to the central vertical axis and have different widths and heights. This is illustrated in Fig. 1.

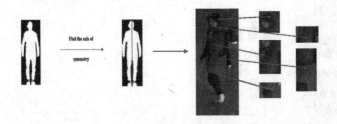

Fig. 1. Central vertical axis estimation and patches propagation.

The next step in framework construction consists in calculating a rough color description for each patch. The used descriptor is the surface under the cumulative color histogram where each channel is taken apart. This is illustrated in Fig. 2. The surface s_i^n that represent the description of patch i for candidate n is calculated based on Eq. (3). This ends up with a framework in which each individual is represented by a set of detections and each detection is represented by a set of values representing a rough description for different patches.

$$s_i^n = \int H_i^n \tag{3}$$

Where H_i^n is the cumulative HSV color histogram of patch i for candidate n.

Patch Selection. The second step in our learning procedure consists in the selection of the most representative patches for each individual. From the previous step, each individual is represented in the framework by a set of detections and each detection is represented by a set of values representing a rough description for different patches form the used detection. This is illustrated in Fig. 3.

Then, the next step consists in selecting patches that both minimize the distance between detections belonging to the same individual and maximize the distance between detections that represent different individuals. For each individual, we try to find the most stable patches by selecting patches with lowest standard deviation among different detections of the individual of interest. Then, among these stable patches, we select those maximizing the minimum distance between the individual of interest and the other individuals.

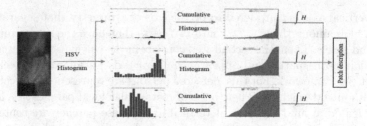

Fig. 2. Descriptor calculation.

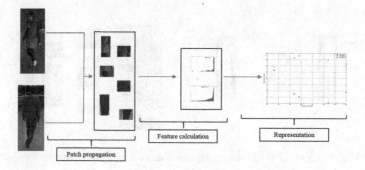

Fig. 3. Representation of individuals in the framework (Example of two individuals by taking 5 patches in the H channel on the HSV color space).

2.2 Matching Process

In this step, we extract descriptors that describe each selected patch for each individual in order to use it in the association of tracked individuals. We create a set that gathers all the selected patches from different individuals. This set is considered to be the candidates set. Then, during the matching process we compare each patch to its corresponding ones that have the same position and size. To perform matching, we used a color based re-identification that is based on the use of HSV color histograms. We used color histograms in HSV color space as descriptors due to their simplicity, effectiveness, and relative stability to illumination changes. For histograms comparison, we use the earth movers distance (EMD) which shows better match than bin-to-bin distances.

Our people association strategy is based on the use of the K-nearest neighbors' algorithm to select the most similar appearances. Then, for each new detected individual (query individual), we calculate the EMD distance between all members of the candidates set that gathers all the selected patches from the previous step and the corresponding patches for the query individual. These distances are used to select the first k nearest neighboring patches.

At the end, the set of neighboring patches is used to confirm re-identification. This is done by calculating the number of patches that belong to each candidate individual in the set of neighboring patches. Then, we confirm a re-identification if the number of patches belonging to the same individual is greater than a certain threshold (more than 50%) and the total distance between different patches is less than another threshold. Figure 4 illustrates some results of patch matching.

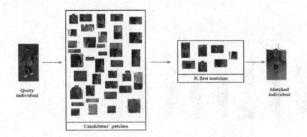

Fig. 4. Example for patch matching.

3   Evaluation

For evaluating our re-identification strategy in the context of people tracking in a multi-camera system with non-overlapping cameras. We used a dataset that gathers 9 individuals (shown in Fig. 5). This dataset is obtained by recording these individuals while they are moving within the fields of view of three disjoint cameras.

Fig. 5. Used dataset.

The performance of the re-identification system is evaluated on the described dataset by calculating the accuracy (precision)-recall metric based on Eqs. (4) and (5).

$$precision = \frac{correct\_matches}{correct\_matches + false\_matches} \tag{4}$$

$$recall = \frac{correct\_matches}{queries\_number} \tag{5}$$

Tests were performed by formulating 100 queries. Then, for each individual the accuracy and recall were calculated. The resulting accuracy and recall for each individual are shown in Fig. 6(a). Whereas Fig. 6(b) represents the cumulative matching characteristic curve over the same number of queries (100 queries).

From these figures we can notice that our re-identification strategy gives a good results for associating individuals across non-overlapping cameras. We can also see that the CMC curve converges quickly to 1. This convergence is considered to be very important in re-identification systems.

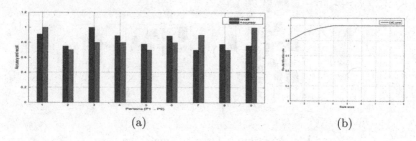

(a)   (b)

Fig. 6. Re-identification results: (a) precision and recall curves, (b) cumulative matching characteristic.

4   Conclusion

In this paper, we proposed a people's re-identification strategy that is used in the context of people's tracking in a multi-camera system. This method is designed to associate tracked individuals across non-overlapping cameras to recover the global trajectories of tracked individuals while moving within the cameras network.

The proposed method is based on the selection of most discriminative patches for each tracked individual. These patches are selected based on a learning procedure that is done in two main steps. The first step consists in the construction of a novel framework in which all individuals are described by a set of values that represents the surfaces under curves representing the cumulative color histogram of several local patches. This patches are centered on the central vertical axis to reduce the influence of background pixels. Then, this representations are used to select the most discriminative patches for each individual. At the end, the selected patches are used to match new detected individuals in other cameras with learnt candidates. This re-identification strategy gives a simple and performant way for exploiting both the spatial and appearance information. As future work, we plan to extend our method by integrating other descriptor in addition to used color histograms to improve its performance.

References

1. Corvee, E., Bremond, F., Thonnat, M., et al.: Person re-identification using Haar-based and DCD-based signature. In: 2010 Seventh IEEE International Conference on Advanced Video and Signal Based Surveillance (AVSS), pp. 1–8. IEEE (2010)

2. Zhao, R., Ouyang, W., Wang, X.: Learning mid-level filters for person re-identification. In: Proceedings of the IEEE Conference on Computer Vision and Pattern Recognition, pp. 144–151 (2014)
3. Aziz, K.-E., Merad, D., Fertil, B.: People re-identification across multiple non-overlapping cameras system by appearance classification and silhouette part segmentation. In: 2011 8th IEEE International Conference on Advanced Video and Signal-Based Surveillance (AVSS), pp. 303–308. IEEE (2011)
4. Martinel, N., Foresti, G.L.: Multi-signature based person re-identification. Electron. Lett. **48**(13), 765–767 (2012)
5. Bazzani, L., Cristani, M., Perina, A., Farenzena, M., Murino, V.: Multiple-shot person re-identification by HPE signature. In: 2010 20th International Conference on Pattern Recognition (ICPR), pp. 1413–1416. IEEE (2010)
6. Martinel, N., Micheloni, C., Foresti, G.L.: Kernelized saliency-based person re-identification through multiple metric learning. IEEE Trans. Image Process. **24**(12), 5645–5658 (2015)
7. Zhao, R., Ouyang, W., Wang, X.: Person re-identification by salience matching. In: Proceedings of the IEEE International Conference on Computer Vision, pp. 2528–2535 (2013)
8. Datta, A., Brown, L.M., Feris, R., Pankanti, S.: Appearance modeling for person re-identification using weighted brightness transfer functions. In: 2012 21st International Conference on Pattern Recognition (ICPR), pp. 2367–2370. IEEE (2012)
9. Colombo, A., Orwell, J., Velastin, S.: Colour constancy techniques for re-recognition of pedestrians from multiple surveillance cameras. In: Workshop on Multi-camera and Multi-modal Sensor Fusion Algorithms and Applications-M2SFA2 2008 (2008)
10. Javed, O., Shaque, K., Rasheed, Z., Shah, M.: Modeling inter-camera space-time and appearance relationships for tracking across non-overlapping views. Comput. Vis. Image Underst. **109**(2), 146–162 (2008)
11. Martinel, N., Micheloni, C., Foresti, G.L.: A pool of multiple person re-identification experts. Pattern Recogn. Lett. **71**, 23–30 (2016)
12. Avraham, T., Lindenbaum, M.: Learning appearance transfer for person re-identification. In: Gong, S., Cristani, M., Yan, S., Loy, C.C. (eds.) Person Re-Identification. ACVPR, pp. 231–246. Springer, London (2014). doi:10.1007/978-1-4471-6296-4_11
13. Li, W., Wu, Y., Mukunoki, M., Minoh, M.: Common-near-neighbor analysis for person re-identification. In: 19th IEEE International Conference on Image Processing, pp. 1621–1624. IEEE (2012)
14. Zheng, W.-S., Gong, S., Xiang, T.: Re-identification by relative distance comparison. IEEE Trans. Pattern Anal. Mach. Intell. **35**(3), 653–668 (2013)
15. Liao, S., Hu, Y., Zhu, X., Li, S.Z.: Person re-identification by local maximal occurrence representation and metric learning. In: Proceedings of the IEEE Conference on Computer Vision and Pattern Recognition, pp. 2197–2206 (2015)
16. Pedagadi, S., Orwell, J., Velastin, S., Boghossian, B.: Local fisher discriminant analysis for pedestrian re-identification. In: Proceedings of the IEEE Conference on Computer Vision and Pattern Recognition, pp. 3318–3325 (2013)

3D Computer Vision

Hybrid Multi-modal Fusion for Human Action Recognition

Bassem Seddik[1,2(✉)], Sami Gazzah[1], and Najoua Essoukri Ben Amara[1]

[1] LATIS Laboratory, National Engineering School of Sousse,
University of Sousse, Sousse, Tunisia
bassem.seddik@gmail.com, sami.gazzah@gmail.com,
najoua.benamara@eniso.rnu.tn
[2] National Engineering School of Sfax, University of Sfax, Sfax, Tunisia

Abstract. We introduce in this paper a hybrid fusion approach allowing the efficient combination of the Kinect modalities within the feature, representation and decision levels. Our contributions are three-fold: (i) We propose an efficient concatenation of complementary per-modality descriptors that rely on the joint modality as a high-level information. (ii) We apply a multi-resolution analysis that combines the local frame-wise decisions with the global BoVW ones. We rely in this context on the scalability of the Fisher vector representation in order to handle large-scale data and apply additional concatenation of its output. (iii) We also propose an efficient score merging scheme by generating multiple weighting-coefficients that combine the strength of different SVM classifiers with a given action label. By evaluating our approach on the Cornell activity dataset, state-of-the-art performances are obtained.

Keywords: Human action recognition · Hybrid fusion · Fisher vectors · Kinect modalities · Cornell activity dataset

1 Introduction

Different types of modalities have been considered for human action recognition. Works focusing on RGB, depth or skeletal streams [1] have made available real-world applications ranging from human-machine interfaces to entertainment and automated surveillance [2]. The Kinect sensor accelerated the progress in this context by of offering synchronized streams of the afore-mentioned modalities. The recent advances in RGB-D-based 3D human-action recognition [3] prove that the combination of these inputs is still an attractive research field.

We focus in this paper on the efficient combination of RGB, depth and joint modalities for the goal of human-action recognition. We rely here on a joint-based efficient description as a high-level information and the Fisher Vectors (FV) effectiveness with large-scale datasets [8] and contribute by combining the global BoVW representations with the local frame-wise descriptors in a hybrid multi-modal fusion scheme. Compared to our previous work [10,11], our improvements

© Springer International Publishing AG 2017
F. Karray et al. (Eds.): ICIAR 2017, LNCS 10317, pp. 201–209, 2017.
DOI: 10.1007/978-3-319-59876-5_23

are: (i) Compared to [11], we propose a new hybrid fusion scheme and improve our weighting coefficients. (ii) Compared to [10], we increase our RGB/depth descriptors sizes and improve our joint normalization. A new dataset is also used: The Cornell Activity Dataset (CAD-60) [6].

In what follows, we present a literature review for feature extraction, BoVW representation and multi-modal fusion. Later, we propose our contribution in Sect. 3. Our experimentations and results are discussed in Sect. 4. Paper summary and future actions are given in the last section.

2   Related Work

We review here the advances in feature extraction, BoVW representation and multi-modal fusion, then present our literature analysis.

2.1   Local Feature Extraction

Hereafter, we review the literature features for separate Kinect modalities.

RGB: Image descriptors have also reached remarkable performances. As a start, the Spatio-Temporal Interest Points (STIP) and their extension to 3D temporal volumes [4] have been widely used. Multiple multi-modal works relied on the HOG and HOF low-level features [6,7,12]. Dense extractors dominated afterwards by adding the motion boundaries and trajectories descriptors [19].

Depth: Out of shape landmarks, shape-context derived descriptors have been proved useful to produce distance and orientation histograms [1]. The depth temporal differences have been also widely used. They proved their efficiency in [10] and brought better performances than the RGB-based displacements [7]. Other approaches rely on depth-normals [20], STIP-based features [21] and depth-based point cloud analysis [22].

Joint: This modality proved to be very efficient within many contributions. A number of works rely first on skeletal normalization as a pre-processing stage [14]. Then, different measures relative to the joint positions [15], first and second order gradients [13,16], inter-distances [9,10,17] and rotation angles [9,14,18] are computed. Advanced methods rely on dedicated representations such as the lie-curves [14] in combination with specific scores.

2.2   Global BoVW Feature Representation

The core of the BoVW approaches is a meaningful grouping of the local features into more coherent global representations. This is commonly applied by finding the K most-action-representative local descriptors (codebook centroids) and then attributing the less-important ones to them. To find these important codebook centroids and use them to encode all the features, a number of increasingly efficient methods have been used [23]. While early approaches relied on a hard-assignment strategy using K-means clustering to find the K centroids and then applying Vector Quantization (VQ) for encoding [12]. Following improvements used soft-assignments through the probability distributions of the Gaussian Mixture Models (GMM) [23]. Other works rather focused on the use of over-complete codebooks dictionaries. More recently, high-dimensional feature representations such as the Fisher vectors have been successfully used with large-scale datasets [8].

2.3 Multi-modal Fusion Strategies

Multi-modal fusion allows gaining the modalities' complementarity [2]. The concatenation of the low-level descriptors, can be made using full or conditional combinations. Such strategies were successfully utilized in [19] to fuse multiple features coming from the same RGB modality. They were also used in [12] with both the RGB and depth modalities to reduce feature variability. The use of the joint positions to extract local features from the RGB or depth space has been also commonly used [13]. The fusion of multiple BoVW representations is a very efficient source of performance improvement. The obtained feature representations can be then concatenated [24] for better performances. At the score level, the fusion is made out of modality-relative scores. The fusion of modal-weighted scores provides better performances than the separate ones [2]. The weighted fusion of multiple SVM classifiers' outputs proved their efficiency with the CAD-60 dataset in [18]. A fourth family of hybrid-fusion methods make call to combinations of the afore-mentioned fusions. Peng *et al.* [23] proposed a hybrid super-vector approach combining multiple BoVW architectures decisions.

2.4 Literature Analysis and Approach Proposition

From literature, the joint features have proved to bring gains around 19 to 29% [5] compared to RGB and depth. For this reason, we create rich skeletal-based descriptors and design complimentary displacement and appearance features using the depth and RGB inputs. Referring to the super-vector representation efficiency [24], we have opted for concatenating the FV representations obtained from the separate modalities. Our proposed approach can be considered as a hybrid fusion method as we concatenate different multi-modal descriptors, different FV representations and later combine the different decision scores generated. More details are presented in the next section.

3 Approach Details

We present hereafter our local feature extraction, FV-based representation and fusion scheme improvements compared to our previous work [10,11].

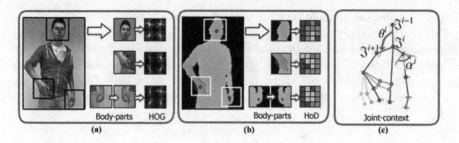

(a)　　　　　　　　　　　(b)　　　　　　　　　　　(c)

Fig. 1. Considered features: a-RGB HOG; b-depth HoD; c-joint context

3.1 Per-modality Feature Extraction

As shown in Fig. 1, we define a complementarity feature setup from the joints, RGB and depth modalities. Relying on the joint 2D positions, we apply a dense sampling every t frames and extract localized hands and face descriptors.

RGB Modality: The RGB modality is designed to save the localized appearance of the face and hands (Fig. 1a). Based on previous research, we consider windows of 48×48 pixels and extract per-frame HOG descriptors of 8 orientations and 4×4 cells using the joint 2D projected positions for the face and both hands. The selected configuration generates a total of 384 features.

Depth Modality: The depth stream is mainly useful for measuring the displacements of the silhouette. We extract local Histograms of Displacements (HoD) [11] from 3 sub-windows relative to the face, the right and left hands obtained using the 2D joint projections (Fig. 1b). The retained configuration applies a downs-scaling to grids of 8×8 pixels for a total of 192 features.

Joint Modality: The joint modality saves the general posture of the body. We enhance our joint normalization stage proposed in [10] by resetting the body rotation while preserving the inter-joint distances. Using the joint positions $J_{x,y,z}^i$, $i = 1..N$, we have predefined skeletal bone-sizes as in [14] and generated more stable joint streams $\mathfrak{J}_{x,y,z}^i$ with reference to the hip-center (see Fig. 1c). We have also computed the joint accelerations and velocities, then generated the joint pair-wise distances. We have used quaternion angles for the joint α^i inter-bone angles θ^i [10]. The concatenated feature vector has a size of 371.

3.2 Fisher Vector Representation

Fisher vectors are a special case of the Fisher kernels combining the strengths of both generative and discriminative models [2]. For each of the sample action features $S = [x_1, \ldots, x_t]$ of size D, produced for all t available frames, we first apply PCA with whitening enabled [23] to de-correlate and reduce our data size. After that, we generate a GMM of K centroids associated with their π_k, μ_k and Σ_k parameters respectively relative to the prior probabilities, the means and the diagonal covariance matrices [11]. Each action sequence is described a

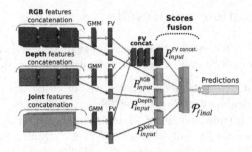

Fig. 2. Hybrid fusion approach proposed

large sparse vector composed of the mean and covariance partial derivatives of all x_t features. The generated FV has a size of $2 \times K \times D$ features composed of the mean and covariance's partial derivatives. Reader may refer to [8] for the FV derivatives definition suitable for large-scale datasets. One last important step for the FV performance is a double normalization stage using first the function $f(x) = sign(x) \times \sqrt{|x|}$ and then the L^2-normalization.

3.3 Fusion and Recognition Strategy

Figure 2 gives an insight about our hybrid fusion pipeline. As input, we apply a selective frame-wise feature concatenation. In addition, the local descriptors in Fig. 2 follow a double path: (i) The first one whitens the features, applies PCA reduction [23] then produces local SVM scores. (ii) The second one applies FV representation and produces global SVM scores. As output, we merge the decision scores relative to the different multi-modal specialized classifiers. For each considered modality, we first combine the local and global resolutions probabilities. Then merge the per-modality probabilities P_{modal} to produce the predictions $\mathcal{P}_{pred}$. To rapidly generate all required models, we pre-compute linear classification kernels from the learning and test sets as implemented in [23]. Figure 2 shows the modality-based and FV-concatenation scores combination.

Our weighted fusion scheme is flexible and profits from the individuals classifiers strengths. For separate action-label, an exhaustive search is applied to find the best weights, using the function $f(P)$ given by Eq. 1:

$$\mathcal{P}_{pred} = f(P_{modal}) = \max_{C_{modal}} \left(\sum_{j=1}^{m} C_{modal}^T P_{modal}^{(j)} \right) \qquad (1)$$

where P_{modal} denotes the probabilities obtained after the first fusion iteration and $P_{modal} = f(P_{L/G})$. $P_{L/G}$ is relative to the initial SVM local/global paths probabilities. $C_{modal} = \{c_{modal}^1, \ldots, c_{modal}^m\} \in [0,1]$ are the per-label weighting coefficients generated during the training. By generating different weighting coefficients per action-label, we are able to reduce the impact of any weakly-contributing modality and to combine the distinctive classifiers capabilities (e.g. classifier-1 finds all 'still' labels while classifier-2 finds all 'brush-teenth' labels).

4   Experimentation and Results

We present hereafter our considered dataset and evaluation metrics. The concatenated FV parameter choices and the obtained performances are also discussed.

Dataset Details: Our experimental choice was oriented towards the CAD-60 [6]. It has the particularity of presenting 68 action samples that can last for long periods (over 1.5 min) resulting in hight inter-variability within a given action-label. It concerns 14 daily-life actions (e.g. 'relaxing on couch', 'cooking', 'working on computer'). Four actors are called to produce RGB and depth image frames, in addition to the 15 body-joint streams. As the performances are reported using cross-validation between the different actors, the hardest case is when learning from 3 right-handed actors and testing on the left-handed one. In our experiments, we have duplicated the learning-set with its horizontal mirror and generated the user mask by segmenting the nearest depth region within a central sub-window. An illustration of the left/right-handed actors and their relative modalities are given in Fig. 3.

For literature comparison, we followed the 'new person' configuration as in [6]: Each cross-validation iteration, we learn from 3 different actors then test over the lasting one. By computing the precision and recall scores we obtain the F_β measure given by Eq. 2:

$$F_\beta = (1 + \beta^2) \times \frac{precision \times recall}{(\beta^2 \times precision) + recall} \qquad (2)$$

Fusion Evaluation: Using the modality-relative local paths, we are able to reach 70.4% for the left-handed actor-3 (hardest case) detailed in Table 1. From another side, the FV-concatenated global representation depends on two parameters: The PCA-reduced feature dimensionality (D) and the number (K) of GMM used for dictionary construction. Figure 4a evaluates the per-modality

Fig. 3. Left-handed and right-handed performances and their relative modalities

Table 1. Modality-based fusion 4-fold cross-validation: Actor 3 detailed

	Actor 1	Actor 2	Actor 3	Actor 4	Average
Precision ($\mathcal{P}_{pred}$)	89.9	96.3	70.4	91.0	89.0

	RGB		Depth		Joint	
	Loc	FV	Loc	FV	Loc	FV
Precision ($\mathcal{P}_{L/G}$)	53.2	30.8	30.5	30.8	56.3	53.9
Precision ($\mathcal{P}_{modal}$)	57.7		46.4		61	

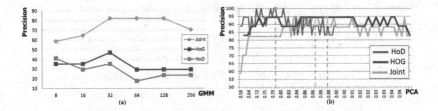

Fig. 4. FV parameters: a-GMM number; b-PCA dimensionality

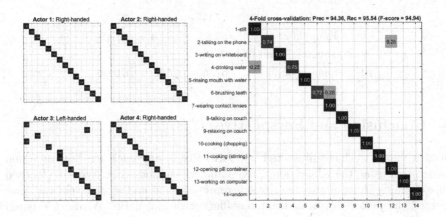

Fig. 5. CAD-60 cross-validation confusion matrices

accuracies using different GMM numbers for CAD-60 actor-1 samples. It shows that reduced amounts of $K = 8$ and 32 mixtures are sufficient to produce the best performances for the HoD and HOG/joint features, respectively. Figure 4b overlays the cumulative impacts of the joint, RGB and HoD PCA energies on the FV-concatenation performance. Using factors around 0.86 and 0.88 of the joint and HOG respective energies, their concatenated FV representations has an accuracy of 95%. By adding a factor of 0.79 from the HoD feature energy, the accuracy oscillates between 95% and 100% on actor-1 samples.

As expected, the fusion of the FV-concatenation scores with those relative to the local modality-based ones results in raising the average precision from 89% in Table 1 to 94.4% as in Fig. 5. An additional grouping is applied by searching for the most occurring label within the central frames $[\frac{N}{6}, \frac{5N}{6}]$ where N is the sequence frame number. This operation allows the 100% precisions reported in Fig. 5. Table 2 proves the competitiveness of our approach on the CAD-60 dataset. Our solution positions first with an F_1-score measure of 94.9% and demonstrates a robust precision/recall ratio. Note that the labels grouping stage allows additional performance gain by reducing the noisy frames impact.

Table 2. Our approach positioning within literature on CAD-60 dataset

Rank	Approach	Precision	Recall	F_1-score
1	**Our approach**	**94.4%**	**95.5%**	**94.9%**
2	Shan *et al.* [16]	93.8%	94.5%	94.2%
3	Cippitelli *et al.* [18]	93.9%	93.5%	93.7%
4	Faria *et al.* [17]	91.1%	91.9%	91.5%
5	Parisi *et al.* [22]	91.9%	90.2%	91.0%
6	Zhu *et al.* [21]	93.2%	84.6%	88.7%
7	Gaglio *et al.* [15]	77.3%	76.7%	77.0%
8	Koppula *et al.* [13]	80.8%	71.4%	75.8%
9	Yang and Tian [25]	71.9%	66.6%	69.2%
10	Sung *et al.* [6]	67.9%	55.5%	61.1%

5   Conclusion and Perspectives

In this paper we have presented a hybrid fusion approach for human-action recognition using the Kinect joint, RGB and depth modalities. We have focused on combining them at the description, representation and score levels and have merged the performances of local classifiers with global BoVW and FV-based ones. Using the CAD-60 dataset, we have experimented the balance between the GMM number and the feature dimensionality parameters and produced state-of-the-art performances.

As we have designed our own fine-tuned descriptors in this work, other perspectives include using different well-established features for each separate modality type: For instance CNN with the RGB/depth modalities [9] and curve-lie-groups with the joint [14]. The facial expressions and inertial sensors are also interesting as additional modalities for action recognition.

References

1. Aggarwal, J.K., Xia, L.: Human activity recognition from 3D data: a review. Pattern Recognit. Lett. **48**, 70–80 (2014)
2. Vrigkas, M., Nikou, C., Kakadiaris, I.A.: A review of human activity recognition methods. Frontiers Robot. AI **2**, 28 (2015)
3. Hadfield, S., Lebeda, K., Bowden, R.: Hollywood 3D: what are the best 3D features for action recognition? ICCV **121**, 95–110 (2017)
4. Laptev, I., Marszalek, M., Schmid, C., Rozenfeld, B.: Learning realistic human actions from movies. In: CVPR, pp. 1–8 (2008)
5. Jhuang, H., Gall, J., Zuffi, S., Schmid, C., Black, M.J.: Towards understanding action recognition. In: ICCV, pp. 3192–3199 (2013)
6. Sung, J., Ponce, C., Selman, B., Saxena, A.: Unstructured human activity detection from RGBD images. In: ICRA, pp. 842–849 (2012)

7. Guyon, I., Athitsos, V., Jangyodsuk, P., Escalante, H.J.: The ChaLearn gesture dataset (CGD 2011). Mach. Vis. Appl. **25**(8), 1929–1951 (2014)
8. Perronnin, F., Sánchez, J., Mensink, T.: Improving the fisher kernel for large-scale image classification. In: Daniilidis, K., Maragos, P., Paragios, N. (eds.) ECCV 2010. LNCS, vol. 6314, pp. 143–156. Springer, Heidelberg (2010). doi:10.1007/978-3-642-15561-1_11
9. Neverova, N., Wolf, C., Taylor, G., Nebout, F.: Moddrop: adaptive multi-modal gesture recognition. IEEE Trans. Pattern Anal. Mach. Intell. **38**(8), 1692–1706 (2016)
10. Seddik, B., Gazzah, S., Amara, N.E.B.: Hands, face and joints for multi-modal human-action temporal segmentation and recognition. In: EUSIPCO, pp. 1143–1147 (2015)
11. Seddik, B., Gazzah, S., Amara, N.E.B.: Modalities combination for italian sign language extraction and recognition. In: Murino, V., Puppo, E. (eds.) ICIAP 2015. LNCS, vol. 9280, pp. 710–721. Springer, Cham (2015). doi:10.1007/978-3-319-23234-8_65
12. Wan, J., Ruan, Q., Li, W., Deng, S.: One-shot learning gesture recognition from RGB-D data using bag of features. J. Mach. Learn. Res. **14**, 2549–2582 (2013)
13. Koppula, H.S., Gupta, R., Saxena, A.: Learning human activities and object affordances from RGB-D videos. Int. J. Rob. Res. **32**(8), 951–970 (2013)
14. Vemulapalli, R., Arrate, F., Chellappa, R.: R3DG features: relative 3D geometry-based skeletal representations for human action recognition. Comp. Vis. Image Underst. **152**, 155–166 (2016)
15. Gaglio, S., Re, G.L., Morana, M.: Human activity recognition process using 3-D posture data. IEEE Trans. Hum.-Mach. Syst. **45**(5), 586–597 (2015)
16. Shan, J., Akella, S.: 3D human action segmentation and recognition using pose kinetic energy. In: ARSO, pp. 69–75 (2014)
17. Faria, D.R., Premebida, C., Nunes, U.: A probabilistic approach for human everyday activities recognition using body motion from RGB-D images. In: RO-MAN, pp. 732–737 (2014)
18. Cippitelli, E., Gasparrini, S., Gambi, E., Spinsante, S.: A human activity recognition system using skeleton data from RGBD sensors. Comput. Intell. Neurosci. **2016**, 1–14 (2016)
19. Wang, H., Schmid, C.: Action recognition with improved trajectories. In: ICCV, pp. 3551–3558 (2013)
20. Oreifej, O., Liu, Z.: HON4D: histogram of oriented 4D normals for activity recognition from depth sequences. In: CVPR, pp. 716–723 (2013)
21. Zhu, Y., Chen, W., Guo, G.: Evaluating spatiotemporal interest point features for depth based action recognition. Image Vis. Comput. **32**(8), 453–464 (2014)
22. Parisi, G.I., Weber, C., Wermter, S.: Self-organizing neural integration of pose-motion features for human action recognition. Frontiers Neurorobotics **9**, 3 (2015)
23. Peng, X., Wang, L., Wang, X., Qiao, Y.: Bag of visual words and fusion methods for action recognition: comprehensive study and good practice. Comput. Vis. Image Underst. **150**, 109–125 (2016)
24. Peng, X., Wang, L., Cai, Z., Qiao, Y.: Action and gesture temporal spotting with super vector representation. In: Agapito, L., Bronstein, M.M., Rother, C. (eds.) ECCV 2014. LNCS, vol. 8925, pp. 518–527. Springer, Cham (2015). doi:10.1007/978-3-319-16178-5_36
25. Yang, X., Tian, Y.: Effective 3D action recognition using EigenJoints. J. Vis. Commun. Image Represent. **25**, 2–11 (2014)

Change Detection in Urban Streets by a Real Time Lidar Scanner and MLS Reference Data

Bence Gálai[1] and Csaba Benedek[1,2(✉)]

[1] Machine Perception Research Laboratory,
Institute for Computer Science and Control,
H-1111 Kende u. 13-17, Budapest, Hungary
{galai.bence,benedek.csaba}@sztaki.mta.hu
[2] Péter Pázmány Catholic University, 1083 Práter u. 50/A,
Budapest, Hungary

Abstract. In this paper, we introduce a new technique for change detection in urban environment based on the comparison of 3D point clouds with significantly different density characteristics. Our proposed approach extracts moving objects and environmental changes from sparse and inhomogeneous instant 3D (i3D) measurements, using as reference background model dense and regular point clouds captured by mobile laser scanning (MLS) systems. The introduced workflow consist of consecutive steps of point cloud classification, crossmodal measurement registration, Markov Random Field based change extraction in the range image domain and label back projection to 3D. Experimental evaluation is conducted in four different urban scenes, and the advantage of the proposed change detection step is demonstrated against a reference voxel based approach.

Keywords: Change detection · Lidar

1 Introduction

The progress of real time Lidar sensors, such as rotating multi-beam (RMB) Lidar scanners, open several new possibilities in comprehensive environment perception for autonomous vehicles (AV) and mobile city surveillance platforms. On one hand, RMB Lidars directly provide instant 3D (i3D) information facilitating the detection of moving street objects and environmental changes. On the other hand, with registering the i3D measurements to a detailed 3D city map, the detected objects and changes can be accurately localized and mapped to a geo-referred global coordinate system.

Using new generation Geo-Information Systems, several major cities maintain from their entire road network dense and accurate 3D point cloud models obtained by Mobile Laser Scanning (MLS) technology. As a possible future utilization, these MLS point clouds can be efficiently considered by the AV's onboard i3D environment sensing modules as highly detailed reference background models. In this context, *change detection* between the instantly sensed

© Springer International Publishing AG 2017
F. Karray et al. (Eds.): ICIAR 2017, LNCS 10317, pp. 210–220, 2017.
DOI: 10.1007/978-3-319-59876-5_24

RMB Lidar measurements and the MLS based reference environment model appears as a crucial task, which indicates a number of key challenges.

Particularly, there is a significant difference in the quality and the density characteristics of the i3D and MLS point clouds, due to a trade-off between temporal and spatial resolution of the available 3D sensors. RMB Lidar scanners, such as the Velodyne HDL-64 provide sequences of full-view point cloud frames with 10–15 fps, and the size of the transferable data is also limited enabling real time processing. As a consequence the measurements have a low spatial density, which quickly decreases as a function of the distance from the sensor, and the point clouds may exhibit particular patterns typical to sensor characteristic, such as the ring patterns of the Velodyne sensor (see Fig. 1(c)). Although the 3D measurements are quite accurate (up to few cms) in the sensor's local coordinate system, the global positioning error of the vehicles may reach several meters in city regions with poor GPS signal coverage.

Recent MLS system such as the Riegl VMX450 are able to provide dense and accurate point clouds from the environment with homogeneous scanning of the surfaces (Fig. 1(a) and (b)) and a nearly linear increase of points as a function of the distance. The point density of MLS point clouds is with 2–3

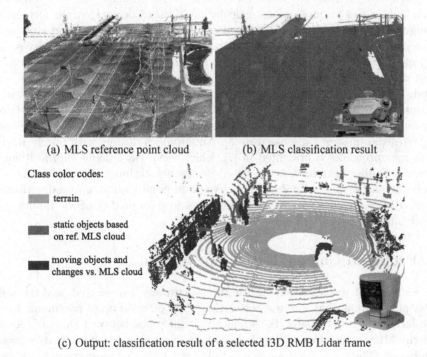

(a) MLS reference point cloud   (b) MLS classification result

Class color codes:

■ terrain

■ static objects based on ref. MLS cloud

■ moving objects and changes vs. MLS cloud

(c) Output: classification result of a selected i3D RMB Lidar frame

Fig. 1. Overview on the proposed approach: based on reference MLS data (a, b), the goal is separation of static scene elements and moving objects/changes on instant RMB Lidar frames (c)

orders of magnitude higher than the density of i3D scans which makes direct point-by-point comparison inefficient. On the other hand, due to the sequential environment scanning process, the result of MLS is a static environment model, which can be updated typically with a period of 1–2 years in large cities. Therefore, apart from the changes caused by moving objects we must expect various differences caused by environmental changes such us altering the buildings and street furniture, or seasonal changes of the tree-crowns or bushes etc.

2   Previous Work

In the recent years various techniques have been published for change detection in point clouds, however, the majority of the approaches rely on dense terrestrial laser scanning (TLS) data recorded from static tripod platforms [6,8]. As explained in [8], classification based on calculation of point-to-point distances may be useful for homogeneous TLS and MLS data, where changes can be detected directly in 3D. However, the point-to-point distance is very sensitive to varying point density, causing degradation in our addressed i3D/MLS cross-platform scenario. Instead, [8] follows a ray tracing and occupancy map based approach with estimated normals for efficient occlusion detection, and point-to-triangle distances for more robust calculation of the changes. Here the Delaunay triangulation step may mean a critical point, especially in noisy and cluttered segments of the MLS point cloud, which are unavoidably present in a city-scale project. [6] uses a nearest neighbor search across segments of scans: for every point of a segment they perform a fixed radius search of 15 cm in the reference cloud. If for a certain percentage of segment points no neighboring points could be found for at least one segment-to-cloud comparison, the object is labeled there as moving entity. A method for change detection between MLS point clouds and 2D terrestrial images is discussed in [5]. An approach dealing with purely RMB Lidar measurements is presented in [7], which use a ray tracing approach with nearest neighbor search. A voxel based occupancy technique is applied in [4], where the authors focus on detecting changes in point clouds captured with different MLS systems. However, the differences in data quality of the inputs are less significant than in our case.

3   Proposed Change Detection Method

We assume that the reference MLS data is accurately geo-referred, and the i3D Lidar platform also has a coarse estimation of its position up to maximum 10 m translational error. Initially, the orientation difference between the car's local and the MLS point cloud's global coordinate systems may be arbitrarily large (see Fig. 2). The proposed approach consists of four main steps: ground removal by point cloud classification, i3D–MLS point cloud registration, change detection in the 2D range image domain, and label backgrojection to the 3D point cloud.

The *ground removal* step separates terrain and obstacle regions using a locally adaptive terrain modeling approach, expecting inhomogeneous RMB Lidar point clouds with typically non-planar ground. First we fit a regular 2D grid with fixed rectangle side length onto the horizontal $P_{z=0}$ plane, using the Lidar sensor's vertical axis as the z direction. We assign each p point of the point cloud to the corresponding cell, which contains the projection of p to $P_{z=0}$. After excluding the sparse grid cells, we use point height information for assigning each cell to the corresponding cell class. All the points in a cell are classified as ground, if the difference of the minimal and maximal point elevations in the cell is smaller than an elevation threshold (used 25 cm), moreover the average of the elevations in neighboring cells does not exceeds an allowed height range. The result of ground segmentation is shown in Fig. 1(b) and (c), which confirms that our technique handles robustly the various i3D and MLS Lidar point cloud types.

For *point cloud registration* we adopt our latest technique [3] for matching point cloud measurements with significantly different density characteristics. The registration process includes three steps. *First*, following the removal of ground points, we search for distinct groups of close points in the remaining obstacles cloud, and assign each group to an abstract object. For handling difficult scenarios with several nearby adjacent objects, we adopted a hierarchical 2-level model [1], which separates first large objects or object groups at a coarse grid level with large cells, then in the refinement it can efficiently separate the individual objects within each group. *Second*, we coarsely align the two point clouds by considering only the center points of the previously extracted abstract objects. We apply here the generalized Hough transform to extract the best similarity transformation in the sense that when applying the transformation to the object centers in the first frame as many of these points as possible overlap with the object centers in the second frame [3]. *Third*, we run a point-level refinement on the above approximate global transform, applying the Normal Distribution Transform (NDT) for all object points. The success of the registration process from an extremely weak initial point cloud alignment is demonstrated in Fig. 2.

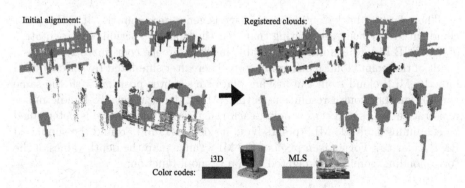

Fig. 2. Demonstration of the proposed point cloud registration step (Deák tér, Budapest). Blue and red points represent the i3D and MLS point clouds, respectively. (Color figure online)

The *change detection* module receives a co-registered pair of i3D and and MLS point clouds, where the terrain is already removed (see Fig. 2 right image). Our proposed solution extracts changes in the range image domain. Creating a range image I_{i3D} from the RMB Lidar's point stream is straightforward as its laser emitter and receiver sensors are vertically aligned, thus every measured point has a predefined vertical position in the image, while consecutive firings of the laser beams define their horizontal position. Geometrically, this mapping is equivalent to projecting the 360° obstacle point cloud to a cylinder surface, whose main axis is equal to the vertical axis of the RMB Lidar scanner. Using Velodyne HDL-64 sensor with 15 Hz rotation frequency, the typical size of this I_{i3D} range image is 64 × 1024. Since the the above projection only concerns the obstacle cloud (without the ground), and several fired laser beams do not produce reflections at all (such as those from the direction of the sky), several pixels of the range map will be assigned to zero (i.e. invalid) depth values. Moreover, such holes may also appear in the range maps due to noise or quantization errors of the rotation angles. On account of this artifact we interpolate the pixel values which have in their 8-neighborhood at least four valid (non-zero) neighboring depth values, as demonstrated in Fig. 3. A sample full-view i3D range image is shown in Fig. 4(a).

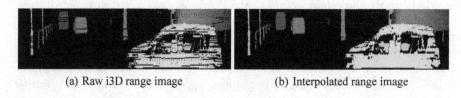

(a) Raw i3D range image   (b) Interpolated range image

Fig. 3. Range image segment from the Velodyne i3D sensor

The reference background range image is generated from the 3D MLS point cloud with ray tracing, exploiting that that the current position and orientation of the RMB Lidar platform are available in the reference coordinate system as a result of the point cloud registration step. Thereafter simulated rays are emitted into the MLS cloud from the moving platform's center position with the same vertical and horizontal resolution as the RMB Lidar scanner. To handle minor registration issues and sensor noise, each range image pixel value is determined by examining multiple MLS points lying inside a pyramid around the simulated RMB Lidar ray. For a given pixel of the MLS range map the depth values of the corresponding points are weighted with a sigmoid function:

$$I_{\mathrm{MLS}}(i,j) = \frac{\sum_{k=1}^{K^{i,j}} w_k^{i,j} D_k^{i,j}}{\sum_{k=1}^{K^{i,j}} w_k^{i,j}}, \quad w_k^{i,j} = \frac{1}{1 + e^{l(D_k^{i,j} - \min_{i,j} D_k^{i,j}) - m}}, \quad (1)$$

where $K^{i,j}$ is the number of MLS points in the (i,j) pyramid, $D_k^{i,j}$ is distance of the k-th point from the ray origin, and the weights $w_k^{i,j}$ are calculated using

(a) Filtered & interpolated Velodyne (i3D) range image

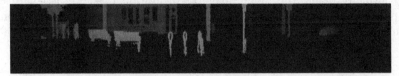

(b) MLS based range image from the actual position of the Velodyne platform

(c) MRF based change mask in the range image domain

Notations:
■ changes
■ background
🏃 see Fig. 5.

(d) Back projection of the change mask to the Velodyne point cloud

Fig. 4. Demonstration of the proposed MRF based change detection process in the range image domain, and result of label back projection to the 3D point cloud

a sigmoid function ($l = 0.5$ and $m = 5$ parameters were empirically set). This calculation formula ensures that the nearest points within the pyramid receive the highest weights, but due to the smoothing effect of weighted averaging, the presence of outlier points, or highly scattered regions (such as vegetation) do not cause significant artifacts. A sample MLS range image generated by the above process is shown in Fig. 4(b).

In the next step, the calculated RMB Lidar-based I_{i3D}, and MLS-based I_{MLS} range images are compared using a Markov Random Field (MRF) model, which classifies each pixel of the range image lattice as foreground (FG) or background (BG). Foreground pixels represent either moving/mobile objects in the RMB Lidar scan, or various environmental changes appeared since the capturing date of the MLS point cloud.

Two sigmoid functions are used to define fitness scores for each class:

$$F_{BG}(i,j) = 1 - \frac{1}{1 + e^{(d^{i,j} - a^{i,j})}}, \quad F_{FG}(i,j) = 1 - \frac{1}{1 + e^{-(d^{i,j} - a^{i,j})}}, \quad (2)$$

where $d^{i,j} = I_{i3D}(i,j)$ and $a^{i,j} = I_{MLS}(i,j)$.

To formally define the range image segmentation task, we assign to each (i,j) pixel of the pixel lattice S a $l_{i,j} \in \{FG, BG\}$ class label so that we aim to minimize the following energy function:

$$E = \sum_{(i,j) \in S} V_D(d^{i,j} | l_{i,j}) + \sum_{(i,j) \in S} \sum_{(m,n) \in N_{i,j}} \beta \cdot 1\{l_{i,j} \neq l_{m,n}\}, \quad (3)$$

where $\beta > 0$ is a smoothness parameter for the label map (used $\beta = 0.5$), and $N_{i,j}$ the four-neighborhood of pixel (i,j). $V_D(d^{i,j} | l_{i,j})$ denotes the data term, derived as:

$$V_D(d^{i,j} | l_{i,j} = BG) = -\log(F_{BG}(i,j)), \quad V_D(d^{i,j} | l_{i,j} = FG) = -\log(F_{FG}(i,j))$$

The MRF energy (3) is minimized via the fast graph-cut based optimization algorithm [2], which process results in a binary change mask in the range image domain, as shown in Fig. 4(c). The final step is *label backprojection* from the range image to the 3D point cloud (see Fig. 4(d)), which can be performed in a straightforward manner, since in our i3D range image formation process, each pixel represents only one Velodyne point.

4   Experiments

We have evaluated the proposed change detection technique in four test scenarios. Each test sequence contains 70 consecutive time-frames from the RMB Lidar sensor, where each i3D frame has a GPS-based coarse location estimation for the point cloud centers, with maximum few meters position error. The MLS reference cloud is accurately geo-referred, and we assume that it only contains the static scene elements such as roads, building facades, and street furniture. For each RMB Lidar frame, we execute the complete workflow of the proposed algorithm.

The Ground Truth (GT) labeling of the RMB Lidar's i3D point clouds was done in a semi-automatic manner. First, using the registered i3D and MLS frames, we applied an automated nearest neighbor classification with a small distance threshold (3 cm), thereafter the labeling of the changed regions was manually revised. As evaluation metrics, we calculated the Precision, Recall and F score values of the detection output at point level, based on comparison to the GT.

Since we have not found any similar i3D-MLS crossmodal change detection approach in the literature, we adopt a voxel based technique [4] as reference, which was originally constructed for already registered MLS/TLS point clouds.

Table 1. Quantitative comparison of the Voxel based (VOX) and the proposed MRF-range image based (MRF) methods on the four test scenes, considering all regions (left), and only the crowded sidewalk areas (right)

Scenes		Overall test set			Sidewalk areas only		
		Precision	Recall	F score	Precision	Recall	F score
Deák	VOX	**0.99**	0.87	0.93	0.81	0.71	0.76
	MRF	**0.99**	**0.90**	**0.94**	**0.87**	**0.89**	**0.88**
Astoria	VOX	**1.00**	0.94	**0.97**	**0.88**	0.81	0.84
	MRF	0.95	**0.98**	**0.97**	0.84	**1.00**	**0.91**
Kálvin	VOX	**1.00**	0.94	0.97	**0.89**	0.96	0.92
	MRF	**1.00**	**0.97**	**0.98**	0.87	**0.99**	**0.93**
Fővám	VOX	**0.98**	0.70	0.82	**0.84**	0.64	0.73
	MRF	0.94	**0.83**	**0.88**	0.81	**0.97**	**0.88**

Therefore by testing both the proposed and the reference models, we apply the same registration workflow introduced in Sect. 3, and only compare the performance of the voxel based and the proposed range image based change detection steps. The reference voxel based technique fits a regular 3D voxel grid to the registered point clouds, thereafter a given RMB Lidar point is classified as foreground if and only if its corresponding voxel does not contain any points in the MLS cloud. We tested this method with multiple w voxel sizes, which parameter naturally affects both the detection performance and the computational time. With larger voxels, we cannot detect some changes in cluttered regions, where the objects can be close to each other and to various street furniture elements. On the other hand, maintaining and processing a fine 3D grid structure with small voxels requires more memory and processing time. The results shown in the upcoming comparative experiments correspond to the voxel size $w = 30$ cm, since we observed with this parametrization approximately the same running speed as using our proposed MRF-range image based model: the change detection step in each frame takes here around 80 msec on a desktop computer, with CPU implementation. Note that by decreasing the w parameter to 20 cm and 10 cm, respectively, the calculation time of the voxel based model starts to rapidly increase (120 msec and 510 msec/frame, resp.), without significant performance improvements.

The comparative results considering the complete dataset are shown in Table 1 (left section), which confirms that the proposed method has an efficient overall performance, and it outperforms the voxel based method in general with 1–6% F scores in the different scenes. We have experienced that the main advantage of the proposed technique is the high accuracy of change detection in cluttered street regions, such as sidewalks with several nearby moving and static objects. As shown in Table 1 (right section), if we restrict the quantitative tests

(a) Voxel based method   (b) Proposed method

Fig. 5. Comparison of the voxel based reference and the proposed range image based approach: a sample bike shed from a magnified image part of the scene in Fig. 4.

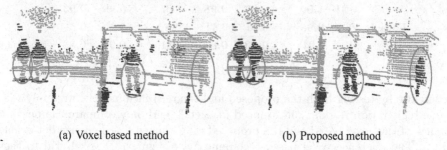

(a) Voxel based method   (b) Proposed method

Fig. 6. Results for a sample region captured at Fővám tér, Budapest, by (a) the voxel based approach, (b) the proposed method. Red and blue points represent the detected background and foreground points respectively. Differences are marked with green ellipses. (Color figure online)

to the sidewalk areas, our method surpasses the voxel approach with 7–15% gaps in three scenes. Similar trends can be observed from the qualitative results of Figs. 5 and 6, which show successful detection samples of small object segments and fine changes with our proposed method, and corresponding limitations of the voxel based approach. As shown in Fig. 6 the voxel based technique results in many falsely ignored moving object segments, in particularly in the regions were people were standing next to static objects. On the other hand, vehicles on the roads with relatively large distances from the street furniture elements can be well separated even with large voxels, therefore the difference between the two methods is less significant in the road regions of the test scenes. Figure 7 shows another test scene.

We display in Fig. 8 synthesized view, visualizing the point clouds of moving objects detected by the i3D RMB Lidar over the geo-referred MLS background data[1].

---

[1] Demo video: http://web.eee.sztaki.hu/i4d/demo_iciar17.html.

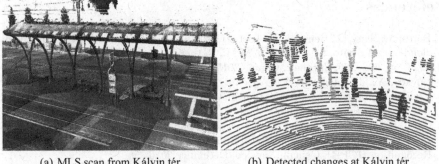

(a) MLS scan from Kálvin tér   (b) Detected changes at Kálvin tér

Fig. 7. *Left:* MLS laser scan of a tram stop in Kálvin tér, Budapest. *Right:* detected changes at the tram stop. Red, blue and green points represent background objects, foreground objects and ground regions, respectively. (Color figure online)

(a)   (b)

Fig. 8. Synthesized view for demonstrating geo-referred moving object detection: object point clouds (tram, car, pedestrians) detected on two subsequent i3D Velodyne frames (marked with blue) are put in and displayed in the MLS reference point cloud (Color figure online)

5   Conclusion and Future Work

We introduced a new method for change detection between different laser scanning measurements captured at street level. The results show that even small and detailed changes can be observed with the proposed method, which cannot be achieved with voxel based techniques. Future work will present a deeper investigation of various background change classes, and tests with lower resolution Lidar sensors.

Acknowledgment. This work was supported by the Hungarian National Research, Development and Innovation Fund (NKFIA #K-120233). C. Benedek also acknowledges the support of the János Bolyai Research Scholarship of the Hungarian Academy of Sciences. MLS test data was provided by the Road Management Department of the City Council of Budapest (Budapest Közút Zrt).

References

1. Börcs, A., Nagy, B., Benedek, C.: Fast 3-D urban object detection on streaming point clouds. In: Agapito, L., Bronstein, M.M., Rother, C. (eds.) ECCV 2014. LNCS, vol. 8926, pp. 628–639. Springer, Cham (2015). doi:10.1007/978-3-319-16181-5_48
2. Boykov, Y., Kolmogorov, V.: An experimental comparison of min-cut/max-flow algorithms for energy minimization in vision. IEEE Trans. Pattern Anal. Mach. Intell. **26**(9), 1124–1137 (2004)
3. Gálai, B., Nagy, B., Benedek, C.: Crossmodal point cloud registration in the Hough space for mobile laser scanning data. In: IEEE International Conference on Pattern Recognition (ICPR). Cancun, Mexico, December 2016
4. Liu, K., Boehm, J., Alis, C.: Change detection of mobile LIDAR data using cloud computing. In: ISPRS International Archives of the Photogrammetry, Remote Sensing and Spatial Information Sciences XLI-B3, pp. 309–313, June 2016
5. Qin, R., Gruen, A.: 3D change detection at street level using mobile laser scanning point clouds and terrestrial images. ISPRS J. Photogram. Remote Sens. **90**, 23–35 (2014)
6. Schlichting, A., Brenner, C.: Vehicle localization by lidar point correlation improved by change detection. In: ISPRS International Archives of the Photogrammetry, Remote Sensing and Spatial Information Sciences XLI-B1, pp. 703–710 (2016)
7. Underwood, J.P., Gillsjö, D., Bailey, T., Vlaskine, V.: Explicit 3D change detection using ray-tracing in spherical coordinates. In: IEEE International Conference on Robotics and Automation. pp. 4735–4741, Karlsruhe, Germany, May 2013
8. Xiao, W., Vallet, B., Brédif, M., Paparoditis, N.: Street environment change detection from mobile laser scanning point clouds. ISPRS J. Photogram. Remote Sens. **107**, 38–49 (2015)

Creating Immersive Virtual Reality Scenes Using a Single RGB-D Camera

Po Kong Lai$^{(\boxtimes)}$ and Robert Laganière

University of Ottawa, Ottawa, ON, Canada
{plai036,laganier}@uottawa.ca

Abstract. We examine the problem of creating immersive virtual reality (VR) scenes using a single moving RGB-D camera. Our approach takes as input a RGB-D video containing one or more actors and constructs a complete 3D background within which human actors are properly embedded. A user can then view the captured video from any viewpoint and interact with the scene. We also provide a manually labeled database of RGB-D video sequences and evaluation metrics.

Keywords: Virtual reality · Scene reconstruction · RGB-D object tracking · RGB-D object segmentation

1   Introduction

The introduction of commodity virtual reality (VR) head mounted displays (HMD) with accurate low latency head tracking [1] and high quality wide field of view (FOV) displays has prompted a resurgence of interest in methods for producing consumable VR content. Creating experiences for VR typically falls along a spectrum with one end containing computer generated imagery (CGI), such as 3D video games, and the other end containing reality based, such as stereo video. CGI often requires a team of specialists to create a virtual 3D world and provide rules for interactions, while reality based content utilizes a combination of visual sensors and algorithms to display or reconstruct a captured scene. We are interested in the latter, specifically the case of a single RGB-D sensor.

One motivation for focusing on the case of a single RGB-D sensor is to explore the possibility of creating VR content through a hand-held device. Realistically speaking, an average consumer may not have the space and/or the resources to setup a multi-sensor environment. Furthermore such an environment has poor mobility making it difficult to record content spontaneously at different locations. Our aim is to provide a framework to empower regular consumers to record their own content to be viewed or interacted with in VR at a later date. Being able to reproduce a dynamic scene through a single sensor is therefore of great interest.

In order for a scene captured using visual sensors to be considered "immersive" in VR, several properties must hold. First, the background of the scene must be displayed to the viewer. While it may be compelling to display just the dynamic objects, without the background there is no context. Second, the

© Springer International Publishing AG 2017
F. Karray et al. (Eds.): ICIAR 2017, LNCS 10317, pp. 221–230, 2017.
DOI: 10.1007/978-3-319-59876-5_25

dynamic objects must be consistent with the captured background. If there is no correlation between the dynamic objects being displayed and the background, then the captured scene does not reflect the viewers expectations. Third, the viewer should retain full rotational and positional freedom while experiencing the captured scene. Without this freedom the viewer will experience an inconsistency between their movements and what they are seeing through the HMD. Fourth, the captured scene, background and dynamic objects, should be as complete as possible. Failure to achieve these properties will decrease the quality of the immersive experience.

The approach presented in this paper aims at producing immersive VR scenes using a single RGB-D camera. An obvious limitation with using a single RGB-D camera is that the fourth property, completely capturing a scene with dynamic objects, is not always possible. A single camera can only record dynamic objects within it's FOV, thus any moving objects out of it's FOV will not be included. Our proposed approach aims to achieve the other three properties by utilizing a combination of static scene reconstruction, skeleton tracking and general visual object tracking to reconstruct the captured scene such that the background and dynamic objects are displayed coherently in VR. The main contributions of this paper are listed below:

1. A framework for creating VR scenes with dynamic objects using a single moving RGB-D camera.
2. A method for extracting dynamic foreground objects from a single moving RGB-D camera.
3. Evaluation metrics and a benchmark dataset with labeled ground truth covering various scenarios.

The remainder of this paper is organized as follows. Section 2 surveys related works and Sect. 3 details our proposed approach for creating VR scenes with dynamic objects using a single RGB-D sensor. Section 4 presents our dataset and evaluation criterion. Section 5 provides the results and Sect. 6 discusses the merits of our approach. Concluding remarks are given in Sect. 7.

2   Related Works

A large quantity of methods exist for producing reality based VR content through visual sensors. We first provide a brief overview of the most dominant categories before highlighting the works most related to ours. When discussing each category we will focus on the degrees of freedom (rotational and positional) allowed for the viewers head as the greater the freedom, the more immersive the VR experience.

Perhaps the oldest, and most studied, approach for producing VR content is the stereo video. The concept is simple: a stereo camera is composed of two cameras which simulates human binocular vision. When the captured video is played back, each camera displays an image for one eye thus allowing a user to feel as if they were at the camera's position. The major drawback is the fact that the viewers head movement, rotational and positional, will not be reflected in the content they are viewing. Multiple monocular or stereo cameras can be combined into one capture rig to produce 360° video [2] or 360° stereo video [3] allowing for rotational head movement. Regardless of the methodology used to produce a 360° video (stereo or otherwise), it still cannot provide the viewers head with positional freedom.

A volumetric or RGB-D video provides viewers with rotational and positional freedom since each frame is a 3D object. Typically, a RGB-D video is captured through active stereo sensors such as Mircosoft's Kinect [4]. Passive stereo systems can also produce a RGB-D video but require additional processing to obtain a depth map [5]. Both active and passive stereo systems have their weaknesses, which are often complementary, and some have tried to combine both to overcome the weaknesses of each [6]. If a RGB-D video does not contain any dynamic objects, then statc scene reconstruction algorithms, such as Kinect-Fusion [7], can be used to accumulate data from multiple viewpoints resulting in a complete 3D model. More recently, DynamicFusion [8] demonstrated dynamic object reconstruction by non-rigidly warping a reference volume to each new RGB-D frame and fusing new depth samples into the model. It should be noted that this work was aimed at reconstructing objects rather than a complete scene and thus the background data is ignored.

Methods using multiple RGB-D sensors have been explored. The approach by Dou et al. [9] is composed of two stages: off-line scanning and on-line running. During the off-line stage a single Kinect sensor scans the target room beforehand to obtain a high quality background model. The on-line stage utilizes 10 Kinect sensors positioned around the target room to record dynamic content. Alignment of the background model and live data is achieved with through SIFT features and planes [10]. More recently, Dou et al. [11] introduced Fusion4D, a pipeline for real-time performance capture that is robust to large frame-to-frame motion and topology. They accomplish this through the use of volumetric key-frames allowing for a more recent model to incorporate new data during quick movement or topological changes. However, similar to DynamicFusion [8], their goal was to capture specific objects within the view of their multiple sensors and thus the background data is not considered.

The works most related to ours are due to Dou et al. [9] and Newcombe et al. [8]. While compelling, multi-view approaches require complex camera setups and thus are not practical in a mobile environment. In contrast, the single sensor approaches for scene reconstruction have either focused on building dynamic objects through precise tracking of the target while ignoring the background [8], or focused on producing static models without dynamic objects [7]. In order to have an immersive VR scene, both the dynamic objects and their corresponding background must be present simultaneously.

3 Proposed Approach

The proposed single RGB-D sensor VR scene capture system consists of three stages: (1) foreground background separation, (2) static background reconstruction and (3) dynamic object overlay. In this section we provide an overview of our approach before covering each stage separately.

3.1   System Overview

Given a set of RGB-D frames, $\mathcal{F}$, we first separate the foreground, $\mathcal{F}_f$, from the background, $\mathcal{F}_b$, using a combination of human pose recognition [12] and visual object tracking [13]. Once the foreground and background have been separated, we feed the background only frames, $\mathcal{F}_b$, into a RGB-D SLAM system to obtain a static background model, $\mathcal{M}$, and a set of camera poses $\mathcal{C}$ such that there is a one-to-one correspondence for each $c_i \in \mathcal{C}$ to each $f_i^b \in \mathcal{F}_b$ and $f_i^f \in \mathcal{F}_f$ where i is the frame number. Since $f_i^b + f_i^f = f_i$ where $f_i \in \mathcal{F}$, then the foreground objects $\mathcal{F}_f$ can be superimposed onto $\mathcal{M}$ via $\mathcal{C}$ allowing for a dynamic scene to be displayed in VR.

3.2   Separating the Foreground and Background

Since people are often the most interesting subject of recorded videos, we first focus on segmenting people from the input RGB-D frames before discussing the case of more general foreground objects. Section 4 provides a comparison of several methods for segmenting the foreground objects.

People Extraction. We utilize Microsoft's Kinect SDK [12] to capture skeletal information and the foreground mask for each detected person. The Kinect SDK is not without its faults and does have skeleton tracking failures (see Fig. 1). To overcome this problem, we deploy a general visual object tracker (VOT) to aid in tracking when the Kinect SDK produces a skeleton tracking failure. Specifically, we use the scalable kernel correlation filter (sKCF) [13] as it is robust and capable of real-time performance. We combine the results of skeletal tracking and the sKCF by examining two events:

Skeletal Tracking Failure (STF). A potential STF is detected when the previous frame contains N skeletons but the current frame has $N - i$ skeletons, where $N \geq 1$ and $N \geq i \geq 1$. The sKCF is initialized using the head positions of the previously detected skeletons and then run for the current frame. The classic flood fill algorithm [14] is applied only to the depth frame to extract a foreground mask. A large overlap between the previous and the current frames foreground masks implies that the person still exists and thus a STF has occurred. Once a STF is detected, the sKCF is given tracking control until a second type of event occurs: tracked person merging.

Tracked Person Merging (TPM). A TPM is required when the previous frame contains N skeletons but the current frame has $N + i$ skeletons, where $N \geq 0$ and $i \geq 1$. In this case we have either rediscovered an existing person who had previously lost skeletal tracking or we have a new person entering the scene. A large overlap between the previous and current foreground masks implies we have rediscovered an existing person. Otherwise a new person has entered the scene and we begin tracking them separately.

Fig. 1. A sampling of frames from the KinectLoop sequence (see Table 1) with (a) color frames and the foreground binary masks from (b) just the Kinect SDK and (c) using Kinect SDK combined with sKCF and segmenting with flood fill.

General Object Extraction. Since we are interested in producing VR scenes from captured RGB-D video, online real-time processing is not strictly required. We present two approaches for general object extraction:

Two-pass filter. In the first pass, similar to Dou et al. [9], we scan the scene without any foreground objects to build a static background model $\mathcal{M}$ (Sect. 3.3). In the second pass, each RGB-D frame f of the input is localized against $\mathcal{M}$. Once localized, f is expanded into a 3D point cloud, P, through the camera intrinsics and the Hausdroff distance is computed between P and $\mathcal{M}$. A threshold on the Hausdroff distance separates the points in P which are too close to $\mathcal{M}$ and the remaining points are considered the foreground.

User selection. In this approach we allow the user to select which objects are considered to be the foreground. To relax the amount of manual labor required, we use sKCF to track the user selected objects.

3.3 Building a Static Background Model

Once the foreground and background are separated (Sect. 3.2), we feed the background only frames into a RGB-D SLAM system [15] with loop closure detection [16,17]. The resulting model can, optionally, be further improved through pose graph refinement and sparse bundle adjustment [18]. To produce a mesh model we use the Poisson surface reconstruction algorithm [19] and perform a transfer of color attributes.

3.4 Superimposing Foreground Frames

Since there is a one-to-one correspondence for each foreground frame to a background frame, we can re-use the camera poses, $\mathcal{C}$, obtained from building a static background model $\mathcal{M}$. More specifically, we use $c_i \in \mathcal{C}$ to transform the foreground frames $f_i^f \in \mathcal{F}$ into the same coordinate system as $\mathcal{M}$. Applying the camera intrinsics results in a properly superimposed 3D object that is consistent with $\mathcal{M}$. A triangular mesh of f_i^f can easily be obtained by exploiting the grid structure of the image.

(a) Kinect SDK only   (b) Kinect SDK + sKCF

Fig. 2. Static background models produced with background masks from the (a) Kinect SDK only (note the unsegmented foreground fused into the background) and (b) Kinect SDK + sKCF.

4   Dataset and Evaluation

Since no evaluation benchmarks exist for single RGB-D sensor VR scene creation, we introduce a dataset (Sect. 4.1) and evaluation metrics (Sect. 4.2).

4.1   Dataset

To test, evaluate and validate our approach we introduce a dataset of 6 RGB-D image sequences along with a manually created binary mask for the foreground objects. The sequences were captured with two different sensors: a structured light and a Kinect V2. Note that our structured light sensor does not provided skeletal information.

4.2   Evaluation Metrics

Two main factors determine the quality of the final result: (1) the accuracy of the foreground background separator (FBSep) and (2) the static background reconstruction method. Since there already exists studies on RGB-D SLAM benchmarking and evaluation [20,21], we focus our evaluation on our FBSep. More specifically, we examine the extracted foreground masks with respect to the manually labeled ground truth. Let GT_i and FG_i be the ground truth and extracted foreground masks at frame i, respectively. Listed below are our evaluation metrics and how they are measured per frame:

- **True Positive (TP):** $|GT_i \cap FG_i|/|GT_i|$
- **False Positive (FP):** $|(\overline{GT_i \cap FG_i}) \cap FG_i|/|FG_i|$
- **False Negative (FN):** $1 - TP$

Intuitively, TP is the percentage of ground truth foreground pixels extracted by the FBSep. The FP and FN are complementary in the sense that they both measure a more subtle visualization element of the final result. The FP measures the amount of background attached to the foreground frames during playback while the FN measures the amount of foreground which was fused into the static background model.

5   Experimental Results

Using our proposed approach, we produced VR scenes for each of the sequences from Sect. 4.1. The results of applying our evaluation metrics from Sect. 4.2 is summarized in Table 1. Figures 3 and 4 illustrate the TP and FP per frame (the FN is not shown as it is simply $1 - TP$) for each sequence in our dataset.

Table 1. Evaluation results using the metrics defined in Sect. 4.2 applied to two methods of foreground background separation (FBSep): (1) use only the Kinect SDK to track and segment dynamic objects and (2) our method where sKCF aids the Kinect SDK in tracking and object segmentation. The TP, FP and FN were computed per frame with the average and median values reported.

Dataset name	FBSep method	Avg.TP	Avg.FP	Avg.FN	Med.TP	Med.FP	Med.FN
KinectTrans	Kinect Only	0.9547	0.2049	0.0453	0.9538	0.2134	0.0462
	Kinect + sKCF	0.9553	0.0215	0.0447	0.9552	0.0023	0.0448
KinectRot	Kinect Only	0.9352	0.0744	0.0648	0.9427	0.0207	0.0573
	Kinect + sKCF	0.9437	0.0060	0.0563	0.9444	0.0010	0.0556
KinectLoop	Kinect Only	0.7352	0.0070	0.2648	0.9640	0.0026	0.0360
	Kinect + sKCF	0.9683	0.0169	0.0317	0.9687	0.0001	0.0313
StructTrans	sKCF Only	0.9150	0.5123	0.0850	0.9138	0.5271	0.0862
StructPath01	sKCF Only	0.9037	0.5155	0.0963	0.9454	0.6179	0.0546
StructPath02	sKCF Only	0.9328	0.3962	0.0672	0.9345	0.4544	0.0655

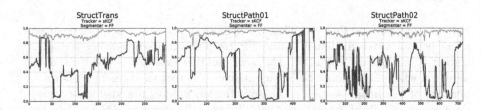

Fig. 3. Per frame TP (green lines) and FP (red lines) results for the sequences captured using a structured light sensor. Vertical and horizontal axis denotes the TP or FP value and frame number, respectively. Columns correspond to a different image sequence from our dataset. Note, since the structured light sensor did not capture body poses, we manually initialized sKCF on the first frame. (Color figure online)

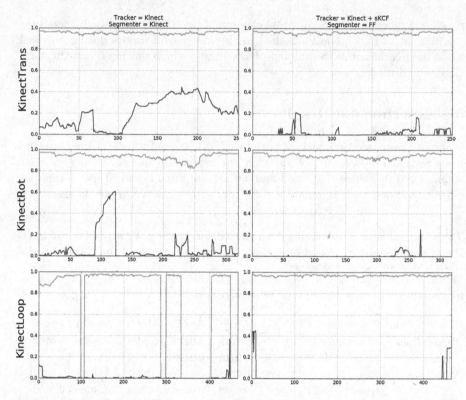

Fig. 4. Per frame TP (green lines) and FP (red lines) results for the sequences captured using a Kinect V2. The left column contains results from using just the Kinect SDK to perform tracking and dynamic foreground object segmentation while the right column combines the sKCF tracker with the Kinect SDK to track and segment the dynamic foreground objects. (Color figure online)

6   Discussion

From Table 1 we note that every image sequence from our dataset observed both a higher and lower median TP and FP (respectively) when using the sKCF tracker to aid the Kinect SDK. The most notable example is the KinectLoop sequence where the Kinect SDK was unable to detect the targets skeleton for portions of the image sequence. These detection failures are visible as gaps in the bottom left graph from Figs. 4 and 2 shows the result of skeletal detection failures.

Our method achieved an average TP and FP of above 0.93 and below 0.03, respectively, for the image sequences captured using a Kinect V2. For the sequences captured with a structured light sensor we achieved an average TP and FP of above 0.90 and below 0.52, respectively. Manual inspection of the structured light sequences revealed that the high FP rate of the structured light sequences was mainly due two factors: (1) the target being at the limits of the sensor range thus resulting in a lower depth quality and (2) a number of static objects cluttering the scene which the classic flood fill algorithm included as part of the foreground. A second pass over the flood fill segmentation should be able to reduce the FP rate significantly and precisely how this is accomplished is left as future work.

7   Conclusion

We have introduced a framework for producing virtual reality scenes which include both a consistent background and dynamic objects using a single moving RGB-D sensor. We tested our framework over a proposed database of RGB-D image sequences and evaluation benchmarks specifically for the specific case of building virtual reality scenes. Open problems include filling in the occlusion of the foreground objects from the single RGB-D sensor and methods for reducing the false positive rate in foreground segmentation. We believe that our work can enable a wider range of consumers to capture virtual reality scenes with both a background and dynamic foreground objects using a hand-held devices.

References

1. LaValle, S.M., Yershova, A., Katsev, M., Antonov, M.: Head tracking for the oculus rift. In: 2014 IEEE International Conference on Robotics and Automation (ICRA), pp. 187–194. IEEE (2014)
2. Yagi, Y., Yachida, M.: Real-time omnidirectional image sensors. Int. J. Comput. Vis. **58**(3), 173–207 (2004)
3. Tzavidas, S., Katsaggelos, A.K.: A multicamera setup for generating stereo panoramic video. IEEE Trans. Multimed. **7**(5), 880–890 (2005)
4. Zhang, Z.: Microsoft kinect sensor and its effect. IEEE Multimed. **19**(2), 4–10 (2012)
5. Tippetts, B., Lee, D.J., Lillywhite, K., Archibald, J.: Review of stereo vision algorithms and their suitability for resource-limited systems. J. Real-Time Image Process. **11**(1), 5–25 (2016)
6. Yang, Q., Tan, K.-H., Culbertson, B., Apostolopoulos, J.: Fusion of active and passive sensors for fast 3D capture. In: 2010 IEEE International Workshop on Multimedia Signal Processing (MMSP), pp. 69–74. IEEE (2010)
7. Newcombe, R.A., Izadi, S., Hilliges, O., Molyneaux, D., Kim, D., Davison, A.J., Kohi, P., Shotton, J., Hodges, S., Fitzgibbon, A.: Kinectfusion: real-time dense surface mapping and tracking. In: 2011 10th IEEE International Symposium on Mixed and Augmented Reality (ISMAR), pp. 127–136. IEEE (2011)
8. Newcombe, R.A., Fox, D., Seitz, S.M.: Dynamicfusion: reconstruction and tracking of non-rigid scenes in real-time. In: Proceedings of the IEEE Conference on Computer Vision and Pattern Recognition, pp. 343–352 (2015)
9. Dou, M., Fuchs, H.: Temporally enhanced 3D capture of room-sized dynamic scenes with commodity depth cameras. In: 2014 IEEE Virtual Reality (VR), pp. 39–44. IEEE (2014)

10. Dou, M., Guan, L., Frahm, J.-M., Fuchs, H.: Exploring high-level plane primitives for indoor 3D reconstruction with a hand-held RGB-D camera. In: Park, J.-I., Kim, J. (eds.) ACCV 2012. LNCS, vol. 7729, pp. 94–108. Springer, Heidelberg (2013). doi:10.1007/978-3-642-37484-5_9

11. Dou, M., Khamis, S., Degtyarev, Y., Davidson, P., Fanello, S.R., Kowdle, A., Escolano, S.O., Rhemann, C., Kim, D., Taylor, J., Kohli, P., Tankovich, V., Izadi, S.: Fusion4D: real-time performance capture of challenging scenes. ACM Trans. Graph. **35**(4), 114:1–114:13 (2016)

12. Shotton, J., Sharp, T., Kipman, A., Fitzgibbon, A., Finocchio, M., Blake, A., Cook, M., Moore, R.: Real-time human pose recognition in parts from single depth images. Commun. ACM **56**(1), 116–124 (2013)

13. Montero, A.S., Lang, J., Laganiere, R.: Scalable kernel correlation filter with sparse feature integration. In: Proceedings of the IEEE International Conference on Computer Vision Workshops, pp. 24–31 (2015)

14. Heckbert, P.S.: A seed fill algorithm. In: Graphics Gems, pp. 275–277. Academic Press Professional Inc (1990)

15. Labbé, M., Michaud, F.: Online global loop closure detection for large-scale multi-session graph-based slam. In: 2014 IEEE/RSJ International Conference on Intelligent Robots and Systems, pp. 2661–2666. IEEE (2014)

16. Labbé, M.,Michaud, F.: Memory management for real-time appearance-based loop closure detection. In: 2011 IEEE/RSJ International Conference on Intelligent Robots and Systems, pp. 1271–1276. IEEE (2011)

17. Labbe, M., Michaud, F.: Appearance-based loop closure detection for online large-scale and long-term operation. IEEE Trans. Robot. **29**(3), 734–745 (2013)

18. Lourakis, M.I.A., Argyros, A.A.: SBA: a software package for generic sparse bundle adjustment. ACM Trans. Math. Softw. (TOMS) **36**(1), 2 (2009)

19. Kazhdan, M., Bolitho, M., Hoppe, H.: Poisson surface reconstruction. In: Proceedings of the Fourth Eurographics Symposium on Geometry Processing vol. 7 (2006)

20. Endres, F., Hess, J., Engelhard, N., Sturm, J., Cremers, D., Burgard, W.: An evaluation of the RGB-D slam system. In: 2012 IEEE International Conference on Robotics and Automation (ICRA), pp. 1691–1696. IEEE (2012)

21. Sturm, J., Engelhard, N., Endres, F., Burgard, W., Cremers, D.: A benchmark for the evaluation of RGB-D slam systems. In: 2012 IEEE/RSJ International Conference on Intelligent Robots and Systems, pp. 573–580. IEEE (2012)

Sunshine Hours and Sunlight Direction Using Shadow Detection in a Video

Palak Bansal$^{(\boxtimes)}$, Chao Sun, and Won-Sook Lee

School of Electrical Engineering and Computer Science,
University of Ottawa, Ottawa, Canada
{pbans035,sc8412,wslee}@uottawa.ca
http://www.eecs.uottawa.ca/~wslee

Abstract. Previous systems used location information like GPS and the Suns location to detect sun light. However how much sunshine an area gets depends on its surround environment too, for instance we seldom get sunshine under a big tree or near a big building. So, we propose estimating sunshine hour just with a video by using image processing. We also calculate sunlight moving direction. One day outdoor video such as backyard, park or forest is processed to measure sunshine hour for every pixel to determine location of sunniest area. Shadow detection based on an algorithm using LAB color space where a difference in the light channel L is compared to neighbours to determine shadow. We improved this common algorithm by using adaptive threshold based on histogram of each frame of the video to overcome difficulty in tree and leaves shadow detection during sunset scene. We have tested 8 videos and the shadow detection rate has been improved to 93.04 from 85.34 by previously published algorithm. Then we use resultant image showing amount of sunlight on each pixel to obtain the sunshine hours. In addition, we calculate a sun direction from these images by using tracking algorithm for shadow movement.

Keywords: Sunshine hours · Sunlight direction · Shadow detection · Shadow tracking · Histogram analysis · Adaptive thresholds

1 Introduction

Sunlight is important in our daily life. It has a significant role in living being's life like humans, animals, and plants. *Sunshine hours* measuring the amount of sunlight on a given spot is one of the ways to estimate the amount of the Sun as energy. In daily life, finding a sunny region or shadowy region helps us to do garden planning, to find the best location for solar panel installation, etc.

Currently available commercial applications such as Sun Surveyor Lite [2], LightTrac application [1], Sun Calc [4] use sunlight tracking. These applications use Global Positioning System (GPS), the position of the camera and the Sun information to provide sunlight direction in a given area. It also needs information about surrounding areas like buildings, large vehicles, and statues to detect

© Springer International Publishing AG 2017
F. Karray et al. (Eds.): ICIAR 2017, LNCS 10317, pp. 231–238, 2017.
DOI: 10.1007/978-3-319-59876-5_26

sunlight at the detection time. However, mostly available videos do not always have information about GPS location, the Sun location or surroundings. This motivates us to estimate sunshine hours just with video scenes.

In this paper, we propose a method to detect sunlight in any given video by using image processing. We measure *Sunshine hours* and estimate a direction of the Sun from shadow detection and tracking. This method does not require any pre-knowledge of location, the direction of the Sun or surroundings.

2   Literature Review

Many commercial applications are present to detect the amount of sunlight and its direction. Sun surveyor [2] gives live view using sun path simulator and shows the direction of the Sun. It requires a precise location of a device such as a camera combined with full network access to give information about sunlight. Sun Seeker shows the path of the sun and a full calendar of sun rises and sets on the location of device [3]. SunCalc is an application that shows the sun movement and sunlight phases during the given day at the given location [4]. All these methods require GPS location and they do not have the capacity to consider 3D information of the surroundings which is related to shadows.

There are many other methods for detecting shadows from images, mostly based on non-variant color features. Jacques et al. [5] developed shadow detection algorithm for gray scale video sequences. Tian et al. [6] used normalized cross-correlation to extract non-shadow region and hence shadow pixels are obtained. Zhang et al. [12] deployed the ratio between the intensities of neighboring pixels, called as ratio edge, to detect indoor and outdoor shadow sequences. Wang et al. [7] detected shadow in indoor sequence using edge spatial and temporal information. Xu et al. [8] detected shadow in grayscale images of an indoor environment using initial change detection masks, canny edge maps, multi-frame integration, edge matching, and conditional dilation. Cucchiara et al. detected shadow by using the color independence property in the HSV color space [13]. They assumed that in shadow pixel, the hue and saturation components of the pixel only change within a certain limit. Salvador et al. proposed cast shadow segmentation algorithm for still and moving images [14].

However, the images and videos used for above paper do not contain tree or leaves shadows which are more difficult to detect. We develop our algorithms focusing on video images with lots of trees and leaves.

3   Calculation of *Sunshine Hours*

We performed experiments on 8 videos (Table 1). These videos are taken from the Internet with a few criteria; (i) cover a whole day containing frames from morning to evening (ii) outdoor scene in sunny day (iii) scenes as diverse as possible such as lawn, park, forest, swimming pool, and backyard. We divided our method into three steps (Fig. 1). In the first step, the video is converted into frames, an improved LAB color space algorithm is used to detect shadows in

Table 1. Videos used for experimenting

Video1	Video2	Video3	Video4	Video5	Video6	Video7	Video8
30 frames	30 frames	30 frames	30 frames	33 frames	27 frames	37 frames	10 frames
Sunset	Forest	Fall	Backyard	Bareland	Grassland	Birdhouse	Swimming pool
Dark with sun in back	Tree shadow	Yellow leaves	Wall shadow	Tree shadows	Thin stick shadow	Clear house shadow	Dark and unclear shadow

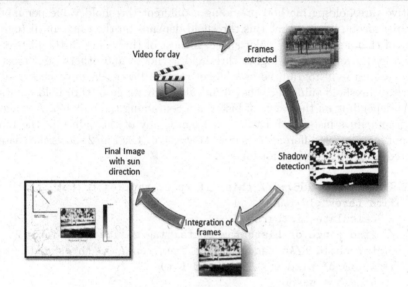

Fig. 1. Steps of method performed

these frames. In the second step, these images are integrated to produce a final image showing amount of sunlight for each pixel. Lastly, sun direction in 2-D is calculated using tracking algorithm.

3.1 Detecting Shadow: Histogram Adaptive Thresholding Method

A full day video is converted into frames at equal intervals. The proposed shadow detecting algorithm is motivated by LAB color space shadow detection algorithm [9]. In this algorithm, the initially RGB image is converted to LAB color space. The lightness channel L ranges from 0 to 100, which gives different shades from black to white. Channels A and B range from -128 to $+127$, giving ratios of red to green color or of yellow to blue respectively.

Initially mean values of the channels A and B of an image are calculated. If mean of channel A + mean of B is less than or equal to 256, then pixels whose L value is less than (mean L standard deviation $L/3$) are classified as shadow pixels. If the sum of mean values is greater than 256, then pixels with value of L and B lower than a threshold value of L and B planes are classified as shadow pixels and others as non-shadow pixels [9]. The threshold value is trial-error based fixed value and varies between a mean of L and B + 2 to a mean of L and B + 25.

Using a fixed threshold value per video, some videos work well while shadows in video 1, 4 and 8 (containing leaves and tree shadows) are not detected properly. These videos had darker surroundings and the suitable brightness, which make shadow detection at a fixed threshold difficult. Therefore, we propose an adaptive thresholding method providing a different threshold value per image. By experiments, we and that this threshold depends on the range of histogram values of channel V of HSV color space. If most of the values in the histogram lie in a lower range then a higher threshold is chosen and otherwise if most of the values in the histogram lie in a higher range a lower threshold chosen. Our empirical threshold values are shown in Table 2, A range of threshold values is chosen depending on the range of histogram containing the maxima. A range of histogram values instead of a value is taken as any of the values in the range does not make much difference in shadow detection. The Fig. 2 shows the shadow detected using this method for images,

```
if (mean L/A > 256): // this 'if' part applies to both L and B
    find threshold L/A:
        calculate histogram V
        find range of histogram containing maxima (use Table 2
        threshold L/A= random value from range of threshold
    if ( L/A of pixels < Threshold L/A)
        pixels = shadow
else: // this 'else' part applies to only L
    if (L of pixels < (mean L - 1/3 standard deviation L)
        pixels = shadow
```

Table 2. Table representing range of threshold values (mean + value) depending on range of histogram of V where maxima lies

Histogram maximum	0–50	50–100	100–150	150–200	200–250
Threshold	20–17	17–15	10–13	8–10	6–7

3.2   Integration of Frames to Measure *Sunshine Hours*

Sunny side changes throughout a day and we integrate day-long video frames to measure *sunshine hours*. Each pixel in the previously obtained shadow (black)-sunny (white) images are analyzed. A matrix of the same size with the input

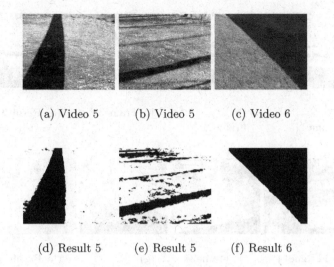

(a) Video 5   (b) Video 5   (c) Video 6

(d) Result 5   (e) Result 5   (f) Result 6

Fig. 2. Frames and its shadow detected by using LAB color space algorithm

images is created. If a pixel in an image is marked in a sunny region, then a value in the matrix corresponding to that pixel is incremented by one. After integrating all the images of the video the matrix is normalized to be from 0 to 1 using an equation (1). The normalized matrix is converted to be a gray-scale final image which shows the *sunshine hours*. This image is further color mapped to distinguish shadow and sunlit regions. Figure 3 represents colored forms of resultant images.

$$Normalised\,Value = \frac{pix\,Value - min\,Value}{max\,Value - min\,Value} \tag{1}$$

3.3   Detecting Sun Direction Using Tracking Algorithm

The videos depict movement of shadow which is opposite to the Sun's direction. In order to detect the Sun's motion we detect the direction in which shadow moves. The direction in which shadow moves is estimated by contouring it. Firstly, a centroid of contour of shadow is found in each frame. Then the movement of the centroid of the shadow contour is tracked by analyzing how it shifts in consecutive frames. This gives the direction in which shadow moves. A mean value of the first 5 frame Centroids and another mean of the last 5 (C2) frames? are calculated. A Difference between two mean values is the direction in which sun moves. This analysis gives a sun vector represented Fig. 4.

4   Evaluation for Shadow Detection and Discussion

To test shadow detection performance of this method, we used two matrices which are shadow Detection Rate (DR) and False Alarm Rate (FAR) [10]. The

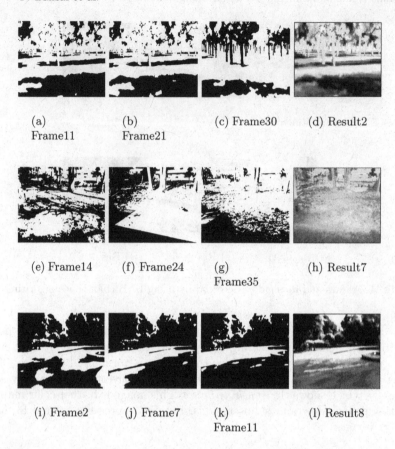

(a)   (b)   (c) Frame30   (d) Result2
Frame11   Frame21

(e) Frame14   (f) Frame24   (g)   (h) Result7
                              Frame35

(i) Frame2   (j) Frame7   (k)   (l) Result8
                              Frame11

Fig. 3. Left three images are examples of shadow detected frames and the right most images are integrated resultant images obtained by merging these frames, which depict the *sunshine hours*. Brighter areas mean more sunny hours than darker areas

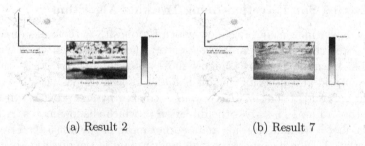

(a) Result 2   (b) Result 7

Fig. 4. Resultant images and sun vector

shadow detection matrix evaluates the number of cast shadow pixels detected as shadow pixels correctly. False Alarm rate gives the probability of non-shadow points classified as shadow pixels. The formulas for calculating these are as follows.

$$Detection\, Rate = \frac{TP}{FN + TP} \tag{2}$$

$$False\, Alarm\, Rate = \frac{FP}{FN + FP} \tag{3}$$

Here TP and FN stand for true positive and false negative pixels with respect to shadow or foreground. To compute the evaluation matrices for images, ground truth images are produced by manually segmenting the area on images into shadow and foreground areas. We compare the proposed method with the original LAB space algorithm [9] in the Table 3. Table 4 represents detection rates by the proposed algorithm and other algorithms. Other algorithms were tested on videos which were taken in just seconds? Length under consistent illumination at a time of a day whereas our experimental videos were captured for a longer duration covering morning to afternoon with variable brightness, which is supposed to be more difficult. Nevertheless, we have achieved higher detection rates with adaptive histogram thresholds method. There were few frames where the purposed algorithm failed. These cases did not have any outstanding maximum histogram value. A further improvement is required to address these cases.

Table 3. Comparing adaptive thresholding method with the LAB color space algorithm [9] for the detection rate (DR) and the False alarm Rate (FAR) on 8 video inputs.

Proposed algorithm	LAB color	Proposed algorithm	LAB color
DR 93.04	FAR 13.65	DR 85.34	FAR 15.58

Table 4. Comparing the algorithm, detection rate (DR) with that of patch-based shadow edge detection (PB) [11] technique, Bayesian Shadow Detection [12] and Statistical Parametric Approach [15]. The values are taken from their papers respectively.

Proposed algorithm	Patch	Bayesian	Parametric
93.04	87.9	93	64

5   Conclusion and Future Work

We have presented that amount of sunlight in an area can be detected by image processing. It does not require any location information of that area or 3D information of surrounding environment to calculate sunlight hours. This is done by detecting shadow in frames and merging them into a single image showing a sunshine hour per pixel. The proposed adaptive thresholding method based on histogram values greatly improved shadow detection rate in images containing leaves, trees and sunset scenes. We also calculate the direction of the sun shown in the video. This work focuses on the out-door scene with only one source of light and a scene without many moving objects. The future work to this will explore a possibility of dynamic scenes.

References

1. Light Trac. http://www.lighttracapp.com/
2. Sun Surveyor. http://www.sunsurveyor.com/
3. Sun Seeker. http://www.ozpda.com
4. SunCalc. http://suncalc.net/
5. Jacques Jr., J.C.S., Jung, C.R., Musse, S.R.: Background subtraction and shadow detection in grayscale video sequences. In: Proceedings of SIBGRAPI, Natal, Brazil, pp. 189–196. IEEE Press (2005)
6. Tian, Y.L., Lu, M., Hampapur, A.: Robust and efficient foreground analysis for real-time video surveillance. IEEE Comput. Vis. Pattern Recognit. **1**, 1182–1187 (2005)
7. Zhang, W., Fang, X.Z., Yang, X.: Moving cast shadows detection based on ratio edge. In: IEEE International Conference on Pattern Recognition, pp. 763–766, November 2006
8. Xu, D., Li, X., Liu, Z., Yuan, Y.: Cast shadow detection in video segmentation. Pattern Recognit. Lett. **26**(1), 5–26 (2005)
9. Murali, S., Govindan, V.K.: Shadow detection and removal from a single image using LAB color space. Cybern. Inf. Technol. **13**(1) (2013). ISSN: 1314-4081
10. Prati, A., Mikic, I., Trivedi, M.M., Cucchiara, R.: Detecting moving shadows: algorithms and evaluation. IEEE Trans. Pattern Anal. Mach. Intell. **25**(7), 918–923 (2003). doi:10.1109/TPAMI.2003.1206520
11. Wu, Q., Zhang, W., Vijaya Kumar, B.V.K.: Strong shadow removal via patch-based shadow edge detection. In: 2012 IEEE International Conference on Robotics and Automation, Saint Paul, MN, pp. 2177–2182 (2012)
12. Benedek, C., Sziranyi, T.: Bayesian foreground and shadow detection in uncertain frame rate surveillance videos. IEEE Trans. Image Process. **17**(4), 608–621 (2008)
13. Cucchiara, R., Grana, C., Piccardi, M., Prati, A.: Detecting objects, sahdows and ghosts in video streams by exploiting color and motion information. In: Proceedings of the IEEE International Conference on Image Analysis and Processing (2001, to appear)
14. Salvador, E., Cavallaro, A., Ebrahimi, T.: Cast shadow segmentation using invariant color features. Comput. Vis. Image Underst. **95**(2), 238–259 (2004)
15. Mikic, I., Cosman, P., Kogut, G., Trivedi, M.M.: Moving Shadow and Object Detection in Traffic Scenes. In: Proceedings of the International Conference on Pattern Recognition, vol. 1, pp. 321–324, September 2000

People-Flow Counting Using Depth Images for Embedded Processing

Guilherme S. Soares$^{(\boxtimes)}$, Rubens C. Machado, and Roberto A. Lotufo

School of Electrical and Computer Engineering, University of Campinas,
Campinas, SP, Brazil
soares.g@gmail.com, rubens.campos.machado@gmail.com,
robertoalotufo@gmail.com
http://www.feec.unicamp.br

Abstract. This paper proposes a people-flow counting algorithm using top-view depth images for implementation on low-power, embedded processors. In the people detection stage the algorithm uses morphological connected filters to find head candidates, and in the tracking stage it uses Kalman filtering in order to obtain good predictions in frames where detection fails. A fast interpolation algorithm is also proposed, which estimates the values of pixels affected by noise and generates an image with a continuous domain. The experiments were done using a Kinect sensor and the processing was performed in real time on a Raspberry Pi 3. The dataset consisted of 4025 short video sequences of people entering and exiting indoor environments, obtained from three different installations. The algorithm proved to be adequate for an embedded application, reaching an accuracy of 98% for frame rates as low as 5.5 FPS.

Keywords: People counting · Depth images · Embedded processing

1 Introduction

The increasing availability of different types of sensors at progressively lower costs is allowing the improvement of many solutions involving automatic people detection, and it is also creating newer demands. Applications include surveillance, security, person-count estimation in public spaces and retail analysis. The types of sensors often used range from simple infrared beam cut detectors to more sophisticated sensors such as RGB and depth cameras.

Many techniques have been proposed for solving problems related to automatic people counting. Most of these techniques rely on conventional PCs to perform the processing. Examples include [9,10] for single RGB cameras, [1] for stereo RGB cameras, and [2,7] for depth cameras.

A few other studies included the particular requirement of performing low-power embedded processing. In [8], a depth camera was used to perform static counting of the number of people in a room, with the processing being made on a Raspberry Pi 2. In [4] a network of depth sensors was used, with sensor

© Springer International Publishing AG 2017
F. Karray et al. (Eds.): ICIAR 2017, LNCS 10317, pp. 239–246, 2017.
DOI: 10.1007/978-3-319-59876-5_27

connected to a Raspberry Pi 2, in order to detect and track people around a scene; the accuracy thus obtained was around 95%.

In this paper we propose an algorithm for automatic people-flow counting using top-view depth images. We also implement and test this algorithm on a Raspberry Pi 3, performing real time processing, using a Kinect depth sensor to capture the images.

2   Algorithm Overview

The proposed algorithm is composed of four stages: preprocessing, in which low-level image processing is performed in order to prepare the image for further analysis; people detection, which attempts to identify the coordinates of all the people present in a single frame; tracking, which attempts to establish a correspondence between the coordinates of people in consecutive frames and determine their trajectory; and counting, which analyzes each trajectory and determines whether the movement is an entrance, an exit, or neither. These steps are executed in sequence for each frame of the video. They are described in more detail further ahead.

2.1   Preprocessing

The preprocessing stage comprises four steps. The first step is a subsampling operation, resulting in an image of 320×240 pixels, in order to achieve faster processing; in the second step the 10-bit raw disparity values provided by the Kinect are mapped into an 8-bit linear image, so that the resulting values are proportional to the distance, in meters, from the sensor to the corresponding points in space; in the third step, the image is negated; and in the fourth step, pixels with unknown values are estimated using an interpolation algorithm.

Interpolation Algorithm. The depth image retrieved by the Kinect sensor typically presents some inherent noise, which leads to pixels with unknown depth values. It is mostly caused by interference of light sources, extremely dark surfaces, translucent objects, specular reflections and occlusions that lead to shadows in the infrared projection. Even though the depth values of pixels with noise cannot be determined, these pixels can be correctly identified by the sensor, which allows us to use methods that attempt to fill in those gaps.

The following method performs an interpolation of valid pixels by setting all invalid pixels to 0 and applying successive conditional dilations. The conditioning image is the mask of all invalid pixels at each iteration. The dilation is repeated until all the initially invalid pixels have a value assigned to them, as shown in Fig. 1. This process is different from the morphological reconstruction in that each pixel has its value changed only once, due to the conditioning image being updated at each iteration. Further details of conditional dilations and other morphological operations can be found in [5].

Morphological operations can have very time-consuming implementations. In this application, we achieve a fast execution by first calculating the distance transform, as in [3], on the invalid pixels, thus obtaining the distance between each invalid pixel and its closest valid one. Next, we sort all the invalid pixels by their distances, in ascending order, and then we sequentially apply the dilation to these pixels simply by assigning to each one the maximum value of its valid neighbors.

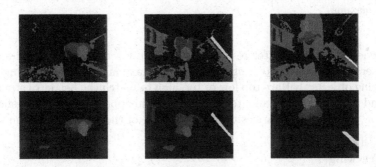

Fig. 1. Examples of the interpolation algorithm for images with different levels of noise. The areas in red represent the pixels with unknown values, as provided by the Kinect sensor. (Color figure online)

2.2 People Detection

This stage of the algorithm consists in attempting to identify the coordinates of all people present in a single image. This is achieved by a two-step process: first, all the regional maxima in the image are retrieved; next, these regional maxima are filtered out according to some parameters, so that only those that are more likely to correspond to people will remain.

According to [5], a regional maximum is a flat zone not adjacent to a flat zone with higher gray value. It is different from a local maximum in that regional maxima are a property of a region, and not of a single pixel. Due to the fine variations in the level of gray throughout the image, some filtering is necessary to retrieve relevant regional maxima, otherwise they would be too many and would correspond to very small regions. By using the operator h-maxima, described in [5], we obtain much more relevant regional maxima, that actually correspond to meaningful objects in the scene.

After obtaining all the regional maxima filtered with the h-maxima operator, the next step consists in selecting the ones that are more likely to correspond to the heads of the people in the scene. This is done by analyzing two criteria of each regional maximum: it must have a minimum area, which eliminates objects that are too small; and it must have a minimum aspect ratio, which eliminates objects that are too elongated. The centroids of the selected regions are then calculated as inputs in the tracking stage.

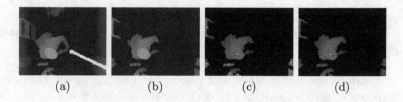

     (a)   (b)   (c)   (d)

Fig. 2. Steps of the people detection stage: (a) input image; (b) thresholding and area open operations; (c) regional maxima after h-maxima operation; and (d) selected regional maxima.

In order to achieve better results, we also apply a thresholding and an area open operations before detecting the regional maxima. The thresholding helps eliminating objects that are too low or too high to match the height of a person's head, and the area open eliminates noise that could appear after the thresholding operation. Figure 2 summarizes the final pipeline of the people detection stage.

2.3 Tracking

The tracking stage is responsible for determining the motion of each person in the scene by establishing the correspondence between the detected centroids in two consecutive video frames. The approach used here only takes into account the spatial coordinates of the centroids; further improvements could also incorporate in the matching process other information about the associated regions, such as shape, area, etc.

Prediction and Matching. In order to achieve a better accuracy in matching the centroids, a Kalman filter [6] is used for each centroid to predict its position in the next frame. It also has the effect of correcting the measured position, thus reducing the measurement noise and producing smoother trajectories. Furthermore, for sequences where a person fails to be detected in some frames, the Kalman filter allows their position to be estimated in those frames, which is essential to keep their tracking.

The first step in the tracking stage is to perform the prediction, using the Kalman filter, of all the centroids that were present in the scene in the previous frame. After that, each predicted centroid is compared to all currently detected centroids, and their Euclidean distances are calculated. The result is an $m \times n$ matrix D with all the pairwise distances, where m is the number of predicted centroids from the previous frame and n is the number of detected centroids in the current frame. The elements of D are then analyzed one at time, in ascending order of distance; for each element $d_{i,j}$, a match between centroids i and j is established if the following criteria are verified:

- distance $d_{i,j}$ is within a maximum distance $d_{\max}$;
- centroids i and j have not matched any centroids yet.

Once all the matches have been determined, the centroids from the current frame that have a correspondence in the previous one have their position corrected by the Kalman filter. Those that do not have a match are assumed to have entered the scene in the current frame, and a new Kalman filter is initialized for each one. Finally, the predicted centroids from the previous frame that do not have a match are propagated to the current frame, except for those having already reached a maximum number of consecutive propagations; those are assumed to have left the scene and are therefore discarded. Figure 3 illustrates the main steps of the tracking stage.

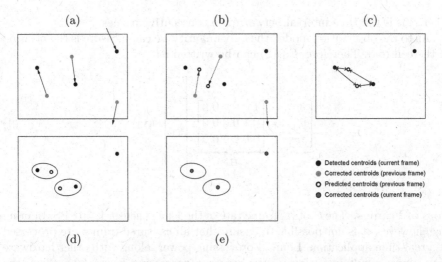

Fig. 3. Steps of the tracking stage: (a) true motion of centroids between two consecutive frames; (b) predictions made by the Kalman filter; (c) distances below the threshold $d_{\max}$; (d) matches between centroids; (e) corrections made by the Kalman filter.

Kalman Filtering. In the Kalman filter theory, the generic model for the evolution of the true state $\mathbf{x}$ of a system, at time k, is

$$\mathbf{x}_k = \mathbf{F}_k\mathbf{x}_{k-1} + \mathbf{G}_k\mathbf{u}_k + \mathbf{w}_k \tag{1}$$

where $\mathbf{F}_k$ is the transition model, $\mathbf{G}_k$ is the control-input model and $\mathbf{w}_k$ is the process noise. The observation $\mathbf{z}_k$ of the true state is given by

$$\mathbf{z}_k = \mathbf{H}_k\mathbf{x}_k + \mathbf{v}_k \tag{2}$$

where $\mathbf{H}_k$ is the observation model and $\mathbf{v}_k$ is the observation noise.

In this particular application, the dynamic system model is quite simple: it assumes that the predicted position of a person depends on their current position and their current velocity, and that in the absence of an input (acceleration) their predicted velocity will remain constant. The actual acceleration, which is unknown, is entirely modeled by the process noise $\mathbf{w}_k$. Thus, Eq. (1) becomes

$$\underbrace{\begin{bmatrix} x_k \\ y_k \\ z_k \\ \dot{x}_k \\ \dot{y}_k \\ \dot{z}_k \end{bmatrix}}_{\mathbf{x}_k} = \underbrace{\begin{bmatrix} 1 & 0 & 0 & \Delta t & 0 & 0 \\ 0 & 1 & 0 & 0 & \Delta t & 0 \\ 0 & 0 & 1 & 0 & 0 & \Delta t \\ 0 & 0 & 0 & 1 & 0 & 0 \\ 0 & 0 & 0 & 0 & 1 & 0 \\ 0 & 0 & 0 & 0 & 0 & 1 \end{bmatrix}}_{\mathbf{F}_k} \underbrace{\begin{bmatrix} x_{k-1} \\ y_{k-1} \\ z_{k-1} \\ \dot{x}_{k-1} \\ \dot{y}_{k-1} \\ \dot{z}_{k-1} \end{bmatrix}}_{\mathbf{x}_{k-1}} + \mathbf{w}_k \tag{3}$$

where Δt is the time interval between two consecutive frames.

As to the observation model, the only quantity we can measure is the position of the centroid. Therefore, Eq. (2) can be written as

$$\underbrace{\begin{bmatrix} x_k \\ y_k \\ z_k \end{bmatrix}}_{\mathbf{z}_k} = \underbrace{\begin{bmatrix} 1 & 0 & 0 & 0 & 0 & 0 \\ 0 & 1 & 0 & 0 & 0 & 0 \\ 0 & 0 & 1 & 0 & 0 & 0 \end{bmatrix}}_{\mathbf{H}_k} \underbrace{\begin{bmatrix} x_k \\ y_k \\ z_k \\ \dot{x}_k \\ \dot{y}_k \\ \dot{z}_k \end{bmatrix}}_{\mathbf{x}_k} + \mathbf{v}_k \tag{4}$$

Loss of Frames. The capture frame rate of the Kinect sensor is 30 FPS. In practice, however, it is not possible to ensure that all captured frames are processed in a real-time application. Limited processing power, along with other hardware limitations, result in a slower, non-constant processing frame rate. Since the tracking is a time-dependent stage, it must take into account the actual time ΔT elapsed between two processed frames. In particular, the following aspects should be observed:

– The prediction phase of the Kalman filter should be repeated n times; n represents the *frame step* between two iterations of the algorithm, i.e., the number of frames captured since the last processed frame, and is given by

$$n = \frac{\Delta T}{\Delta t} \tag{5}$$

– As uncertainty increases with the number of consecutive predictions, the thresholding distance $d_{\max}$ should give more tolerance as n increases, and can be set as

$$d_{\max} = v_{\max} \Delta T \tag{6}$$

where $v_{\max}$ is the maximum speed assumed for a person.

3 Experiments

3.1 Dataset

Three different installations were used in the experiments, with different characteristics, as shown in Table 1 and Fig. 4. As the images were captured, they were processed and recorded, along with the outputs of the algorithm. In a posterior analysis the data were split into 4025 short video sequences, and each sequence was manually labeled with their corresponding number of entrances and exits.

Table 1. Characteristics of the three installations.

	Installation A	Installation B	Installation C
Sensor height (H)	2.90 m	3.80 m	4.00 m
Zenith angle (θ)	5°	15°	15°
Door type	Hinged	Hinged	Sliding
Door material	Glass	Glass	Glass
Room type	Hallway	Open room	Open room
Direct sunlight	Yes	Yes	No
Sunlight source	Skylight above	Front door	-

Fig. 4. Sample images of each installation.

3.2 Results

The processing was performed in real time by a Raspberry Pi 3, which has a 1.2 GHz quad-core ARM processor and 1 GB of RAM. The results of the experiments are shown in Table 2. The average frame rate, for the real-time execution a

Table 2. Results of the experiments for the three installations.

	Valid pixels	TP	FN	FP	Recall	Precision	Accuracy
Installation A	83.0%	964	25	14	0.97	0.99	0.96
Installation B	66.8%	807	147	10	0.85	0.99	0.84
Installation C	86.8%	1462	14	16	0.99	0.99	0.98

the Raspberry Pi 3, was around 5.5 FPS. As can be seen, the maximum accuracy of the system in a particular installation is closely related to the quality of the image acquisition, which is evidenced by the average percentage of valid pixels per frame.

4   Conclusions

Our experiments, conducted in real case scenarios, demonstrate the efficacy of using a depth sensor to perform high-accuracy people-flow counting with real-time, embedded processing. With an accuracy of up to 98%, our algorithm brings results comparable to other experiments of people-flow counting found in the literature, while being able to perform the entire computation on a low-power single-board computer.

References

1. Bernini, N., Bombini, L., Buzzoni, M., Cerri, P., Grisleri, P.: An embedded system for counting passengers in public transportation vehicles. In: 2014 IEEE/ASME 10th International Conference on Mechatronic and Embedded Systems and Applications (MESA), pp. 1–6, September 2014
2. Bondi, E., Seidenari, L., Bagdanov, A.D., Del Bimbo, A.: Real-time people counting from depth imagery of crowded environments. In: 2014 11th IEEE International Conference on Advanced Video and Signal Based Surveillance (AVSS), pp. 337–342. IEEE (2014)
3. Borgefors, G.: Distance transformations in digital images. Comput. Vis. Graph. Image Process. **34**(3), 344–371 (1986)
4. Burbano, A., Bouaziz, S., Vasiliu, M.: 3D-sensing distributed embedded system for people tracking and counting. In: 2015 International Conference on Computational Science and Computational Intelligence (CSCI), pp. 470–475, December 2015
5. Dougherty, E.R., Lotufo, R.A.: Hands-on Morphological Image Processing. SPIE Tutorial Texts in Optical Engineering, vol. TT59. SPIE Publications, Bellingham (2003)
6. Kalman, R.E.: A new approach to linear filtering and prediction problems. J. Basic Eng. **82**(1), 35–45 (1960)
7. Malawski, F.: Top-view people counting in public transportation using Kinect. Chall. Mod. Technol. **5**, 17–20 (2014)
8. Sgouropoulos, D., Spyrou, E., Siantikos, G., Giannakopoulos, T.: Counting and tracking people in a smart room: an IoT approach. In: 2015 10th International Workshop on Semantic and Social Media Adaptation and Personalization (SMAP), pp. 1–5, November 2015
9. Yang, D.B., Gonzlez-Baos, H.H., Guibas, L.J.: Counting people in crowds with a real-time network of simple image sensors. In: 2003 Proceedings of the Ninth IEEE International Conference on Computer Vision, pp. 122–129. IEEE (2003)
10. Zhang, E., Chen, F.: A fast and robust people counting method in video surveillance, pp. 339–343. IEEE, December 2007

Salient Object Detection in Images
by Combining Objectness Clues
in the RGBD Space

François Audet, Mohand Said Allili[✉], and Ana-Maria Cretu

Université du Québec en Outaouais, Gatineau, QC, Canada
{francois.audet02,mohandsaid.allili,ana-maria.cretu}@uqo.ca

Abstract. We propose a multi-stage approach for salient object detection in natural images which incorporates color and depth information. In
the first stage, color and depth channels are explored separately through
objectness-based measures to detect potential regions containing salient
objects. This procedure produces a list of bounding boxes which are
further filtered and refined using statistical distributions. The retained
candidates from both color and depth channels are then combined using a
voting system. The final stage consists of combining the extracted candidates from color and depth channels using a voting system that produces
a final map narrowing the location of the salient object. Experimental
results on real-world images have proved the performance of the proposed method in comparison with the case where only color information
is used.

1 Introduction

Salient object detection is one of the important problems for object recognition and image understanding [3]. It consists of localizing the most informative
objects or regions in images [5,12]. Saliency detection methods rely either on
local or global contrast estimation [3]. Local contrast based methods [11] assume
that regions which stand out from their neighborhoods have high saliency values.
These methods are more suitable to highlight salient object boundaries instead
of entire objects. Global contrast based methods [1,14] express rarity of regions
compared to the overall image in terms of global statistics [5]. They are better
at highlighting entire salient regions. However, they are less accurate to detect
large-sized objects due to the fact the object statistics dominate the global statistics of the image [1].

Most of existing methods for salient object detection show good performance
in general when applied to simple scenarios of images containing single objects
situated against uniform and non-cluttered backgrounds [3,5]. However, when
images contain several objects and/or cluttered backgrounds, the performance of
these methods drastically decreases [6,9]. This stems mainly from the assumption
that salient objects stand out most of the time from the background, where color
contrast is sufficient for their differentiation from the rest of the image. However,

© Springer International Publishing AG 2017
F. Karray et al. (Eds.): ICIAR 2017, LNCS 10317, pp. 247–255, 2017.
DOI: 10.1007/978-3-319-59876-5_28

this assumption is no longer valid when images contain multiple objects and/or cluttered backgrounds where objects can hardly be discriminated using color information.

Recently, another trend of methods aiming at detecting directly objects in the image is gaining popularity [4,8,15]. These methods apply objectness measures in the color space with the aim of detecting bounding boxes surrounding objects [16]. In a nutshell, objectness quantifies how likely it is for an image window to contain an object of any class (e.g., car, dogs, etc.) as opposed to backgrounds, such as grass and water [2]. Objectness-based methods, as their saliency based counterparts, can locate efficiently objects standing out from their immediate surrounding. However, these methods usually produce a huge number of false positives consisting of parts of objects or details of the background. Therefore, an appropriate filtering is required for these methods for better object detection using more discriminative information, such as object appearance, depth information and a priori knowledge such object location, size and shape.

To exploit color and depth information together, Peng et al. [10] have proposed some techniques for locating salient objects in images in the RGBD space. This approach employs local and global contrast measures from color and depth to discriminate the salient objects from the image background. The approach supposes the salient object is centered in the image and relies on depth maps for locating the boundary of salient objects. However, images often times contain multiple objects in the scene, while depth information can miss some parts of objects or merge them with the background, which may cause awed saliency maps.

In this paper, we propose a method combining color and depth information to extract objectness clues for salient object detection and localisation in natural images. Using the algorithm in [16], we first generate separate object proposals for Lab channels of the image and its smoothed version. In addition, given that highest ranked object proposals do not correlate sometimes with salient objects, a filtering step is applied to prune unlikely candidates based on various spatial characteristics. The depth channel is preprocessed using vertical and horizontal scans for removing false object candidates caused by projection effects and noise. Finally, to combine the bounding boxes obtained from all the channels, we propose a voting system taking into account the image layout, object location and size in the image. Experimental results on a set of real-world images have shown that the proposed method yields better localisation of salient objects compared to recent methods in the literature.

This paper is organized as follows: Sect. 2 presents our algorithm for regions of interest proposal. Section 3 presents some experimental results validating our approach. We then end the paper with a conclusion.

2   The Proposed Approach

An outline of the algorithm for salient object detection in the RGBD space is shown in Fig. 1. The algorithm input consists of the color image I and a depth map of the image D. The RGB color image is first transformed into the Lab space which is similar to the human vision perception. Object proposals are then generated from each channel and their combination is performed using a weighted voting and saliency priors. On the other hand, the depth map D is pre-processed to filter projection and noise effects and generate object proposals. Finally, a procedure is proposed to combine color and depth object candidates to form the final bounding box where the salient object is the most likely located. Details of the different steps are presented in the following sections.

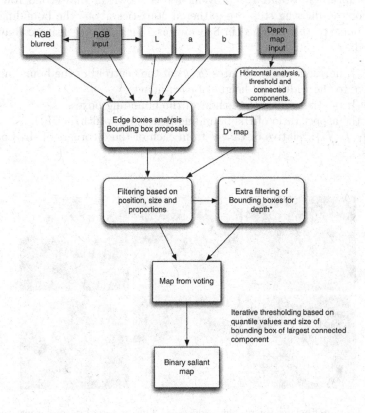

Fig. 1. Outline of the proposed salient object detection algorithm.

2.1   Object Proposal in the RGB Space

As can be seen in Fig. 1, from the RGB image we obtain a blurred version using a 2D Gaussian smoothing kernel with standard deviation $\sigma = 5$. The smoothed version generally allows to reduce the number of returned object proposals corresponding to local object details. The RGB image is also decomposed into normalized Lab channels. The two color maps (RGB and its smoothed version) as well as the three grayscale maps (L, a, b) are passed as inputs to the object proposal algorithm [16], which returns for each map an array of bounding boxes ranked by decreasing objectness score.

Figure 2 first row illustrates some cases where the algorithm puts too much emphasis on minute details, for example around the digits of the clock in the first image. The bounding boxes are represented in decreasing order of objectness, with red showing the highest objectness score, followed by green, blue, yellow and finally magenta, representing the lowest objectness score. The ground truth bounding box is shown in white. By performing a filtering step, as described next, we obtain substantial improvements as shown in the second row of the figure. For the filtering step, we gathered statistics about the bounding boxes of the ground truth contained in 848 images. The following features have been computed:

- $\{f_1, f_2\}$: normalized coordinates (x, y) of the centroid of the bounding boxes relative to the width and height of each image.
- $\{f_3, f_4\}$: relative width and height of the bounding boxes.
- $\{f_5\}$: the aspect ratio of the bounding boxes (i.e. width/height).
- $\{f_6, f_7, f_8, f_9\}$: relative distances from each of the 4 corners of the bounding boxes.

Fig. 2. Effect of filtering on object proposals. The first and second rows represent object proposals before and after filtering, respectively. (Color figure online)

Each of these features has been analyzed for all images grouped together, as well as separately for landscape (width/height ≥ 1) and portrait images (width/height < 1). The marked distinction observed between the two orientations concerns the aspect ratio (width/height) of their bounding boxes, which was found to be on average 0.66 and 1.30 for portrait and landscape, respectively, which incidentally is close to the aspect ratio of the whole images (0.75 and 1.33, respectively). Figure 3 shows the histograms of the first five features. Most of these histograms can be approximated by normal distributions. Thus, using the features computed from the ground truth bounding boxes, we obtain two lists L_p (for portrait) and L_l (for landscape) of 9 normal distributions.

The calculated distributions will serve as a prior knowledge to filter future bounding boxes that will returned by the algorithm in the different channels. Given a bounding box candidate b, the 9 features are first calculated. Let ϕ_i be the normal distribution of feature f_i, and p_i the cumulative probability of its value measured on the bounding box b. The bounding box is rejected if one of the following conditions is satisfied:

1. $\exists f_i : p_i(1 - p_i) < \delta$.
2. $\sum_{i=1}^{9} p_i(1 - p_i) < \eta$

where δ and η are experimental thresholds. Note that the first condition is equivalent to having an observation which exceeds the statistical *p-value* at risk $\alpha\%$ of the feature f_i. A typical value for the risk is $\alpha = 0.02\%$ in which case $\delta = 0.0196$. The second condition aims to enforce global conformity of the bounding box to the prior knowledge. Based on the sum of the p-values for the ground truth (see Fig. 3), a value of 0.9 for η will rejects roughly the 1% least probable candidates.

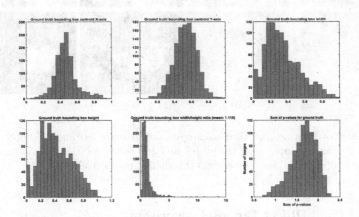

Fig. 3. The histograms of the five first features in the dataset, as well as the p-values. Top row, from left to right, we show histograms of centroid positions x and y and relative width. The second row shows the relative height, aspect ratio and sum of p-values.

2.2   Object Proposal in the Depth Space

In contrast to color information, depth carries information about the relative distance of objects to the camera. Therefore, objects and converging surfaces to the camera can often have similar depth values and be confounded if depth thrsholding is used. To exploit depth information for more precise salient object detection, we should first perform a transformation to remove non-desirable projection effects and noise caused by errors in depth estimation. Given a depth map

D of size $h \times w$, we perform a horizontal scanning using dynamic thresholding to flat out the depth of converging surfaces and noise.

Let σ_l by the standard deviation computed for a given line l in D and let D^* be the transformed map. If $\sigma_l < \epsilon$ (i.e., absence of contrast), the line is not processed any further, and its pixels are set to 0 in D^*. Otherwise, for each column index j of line l, we test if the depth value at the column belongs to an object or the background. For this purpose, we use linear regression to approximate the map of the l-th line. Let $y(j) = a \times j + D(l, 1)$ be the resulted regression line with slope a. Then, if $D(l, j) > y(j)$ and $D(l, j)$ constitutes a local maximum in the line, then $D^*(l, j) \leftarrow D(l, j)$, otherwise $D^*(l, j) \leftarrow 0$. Figure 4 shows some examples of depth filtering, where the salient object has been clearly isolated from the background.

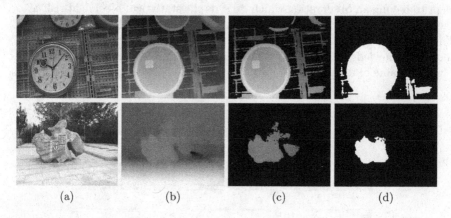

(a)   (b)   (c)   (d)

Fig. 4. Effect of filtering and thresholding on depth map: (a) and (b) represent the RGB and its depth map, respectively, (c) and (d) represent the transformed depth maps after applying the first and second thresholds, respectively.

2.3 The Final Object Proposal Algorithm

From the filtering process as described above, we retain the five most highly ranked bounding boxes for each channel, and therefore a total of at most 30 bounding boxes (in some instances there is less than 5 bounding boxes that are proposed for D^*). Each bounding box votes with 1 for the pixels it contains and the final objectness score is normalized for each pixel. Then, an initial threshold of 25% is used to segment the objectness to obtain a map BC. The size of BC is compared to the ground truth statistics and if its size is above the 95-th percentile, a new map is generated by increasing the threshold by 5%, and the process is repeated until BC area is at a 95-th percentile or below, or that the threshold of 50% has been reached.

The first image of Fig. 5 shows a typical map obtained after the votes of each bounding box. The second image shows in white the area thresholded; the ground truth is represented by the red rectangle.

(a)   (b)   (c)

Fig. 5. Voting and thresholding for the image of the clocks. (Color figure online)

3 Experimental Results

The performance of the proposed model is evaluated quantitatively on the dataset provided by Peng et al. [10]. It is a diversified data set consisting of 1000 images with depth map and ground-truth annotations. It has more than 400 types of common objects under various illumination conditions. Some images contain several salient objects and were not taken into consideration for the tests. Moreover, color and depth areas corresponding to salient objects have wide distributions, and their area varies from 16% and 80% of the image, with a majority under 50%. We evaluated the performance of our method using the *precision*, *recall* and F_α metrics (with $\alpha = 1$). These are calculated by comparing the bounding box of the ground truth in each image with the calculated bounding box obtained by our algorithm.

In Fig. 6, we show ROC curve of the obtained object maps using compared methods, as well as best values for precision, recall and F_α measures. We can note that our ROC curve is better than the other methods. We can see also the improvements in the recall and the F_α values that have been obtained. This gives our method the advantage of better localizing all the parts of salient objects. In

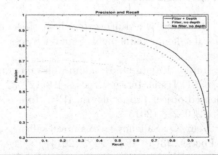

Measure	objectness using [16]	Our method
Precision	0.7166	0.7620
Recall	0.7899	0.8014
F1-measure	0.7024	0.7371

Fig. 6. Comparative results of precision, recall and F_α.

fact, as shown in the results, using only color information as in [16] yields more precision and less recall, which can happen when objects are partially detected. Therefore, our method is better positioned when it is aimed to detect objects for high-level applications such as object recognition and segmentation.

4   Conclusion

We have proposed a multi-stage approach for salient object detection in natural images by combining color and depth information. Color and depth channels are explored through objectness-based measures to detect potential regions containing salient objects. These regions are then filtered and combined using a voting system to produce a final map narrowing the location of the salient object. Experimental results on real-world images have proved the performance of the proposed method in comparison with the case where only color information is used.

References

1. Achanta, R., Hemami, S., Estrada, F., Susstrunk, S.: Frequency-tuned salient region detection. In: IEEE Conference on Computer Vision and Pattern Recognition, pp. 1597–1604 (2009)
2. Alexe, B., Deselaers, T., Ferrari, V.: Measuring the objectness of image windows. IEEE Trans. Pattern Anal. Mach. Intell. **34**(11), 2189–2202 (2012)
3. Borji, A., Cheng, M.-M., Jiang, H., Li, J.: Salient object detection: a benchmark. IEEE Trans. Image Process. **24**(12), 5706–5722 (2015)
4. Cheng, M-M., Zhang, Z., Lin, W-Y., Torr, P.H.S.: BING: binarized normed gradients for objectness estimation at 300 fps. In: IEEE Conference on Computer Vision and Pattern Recognition, pp. 3286–3293 (2014)
5. Cheng, M.-M., Mitra, N.J., Huang, X., Torr, P.H.S., Hu, S.-M.: Global contrast based salient region detection. IEEE Trans. Pattern Anal. Mach. Intell. **37**(3), 569–582 (2015)
6. Filali, I., Allili, M.S., Benblidia, N.: Multi-scale salient object detection using graph ranking and global-local saliency refinement. Signal Process.: Image Commun. **47**, 380–401 (2016)
7. Gopalakrishnan, V., Hu, Y., Rajan, D.: Random walks on graphs for salient object detection in images. IEEE Trans. Image Process. **19**(12), 3232–3242 (2010)
8. He, S., Lau, R.: Oriented object proposals. In: IEEE International Conference on Computer Vision, pp. 280–288 (2015)
9. Liu, Z., Zou, W., Le Meur, O.: Saliency tree: a novel saliency detection framework. IEEE Trans. Image Pocess. **23**(5), 1937–1952 (2014)
10. Peng, H., Li, B., Xiong, W., Hu, W., Ji, R.: RGBD salient object detection: a benchmark and algorithms. In: Fleet, D., Pajdla, T., Schiele, B., Tuytelaars, T. (eds.) ECCV 2014. LNCS, vol. 8691, pp. 92–109. Springer, Cham (2014). doi:10.1007/978-3-319-10578-9_7
11. Seo, H.J., Milanfar, P.: Static and space-time visual saliency detection by self-resemblance. In: IEEE Conference on Computer Vision and Pattern Recognition Workshops, vol. 9, no. 12, pp. 45–52 (2009)

12. Wang, J., Jiang, H., Yuan, Z., Cheng, M.-M., Hu, X., Zheng, N., Detection, S.O.: A discriminative regional feature integration approach. IEEE Int. J. Comput. Vis. 1–18 (2017). doi:10.1007/s11263-016-0977-3

13. Yang, C., Zhang, L., Lu, H., Ruan, X.: Saliency detection via graph-based manifold ranking. In: IEEE Conference on Computer Vision and Pattern Recognition, pp. 3166–3173 (2013)

14. Zhang, L., Gu, Z., Li, H.: SDSP: a novel saliency detection method by combining simple priors. In: IEEE International Conference on Image Processing, pp. 171–175 (2013)

15. Zhang, Z., Torr, P.H.S.: Object proposal generation using two-stage cascade SVMs. IEEE Trans. Pattern Anal. Mach. Intell. **38**(1), 102–115 (2016)

16. Zitnick, C.L., Dollár, P.: Edge boxes: locating object proposals from edges. In: Fleet, D., Pajdla, T., Schiele, B., Tuytelaars, T. (eds.) ECCV 2014. LNCS, vol. 8693, pp. 391–405. Springer, Cham (2014). doi:10.1007/978-3-319-10602-1_26

Feature Extraction

Development of an Active Shape Model Using the Discrete Cosine Transform

Kotaro Yasuda[⊠] and M. Omair Ahmad

Department of Electrical and Computer Engineering,
Concordia University, Montreal, Canada
yasuda.kotaro@gmail.com, omair@ece.concordia.ca

Abstract. In a feature-based face recognition system using a set of features extracted from each of the prominent facial components, automatic and accurate localization of facial features is an essential pre-processing step. The active shape model (ASM) is a flexible shape model that was originally proposed to automatically locate a set of landmarks representing the facial features. This paper is concerned with developing a low-complexity ASM by incorporating the energy compaction property of the discrete cosine transform (DCT). The proposed ASM employs a 2-D profile based on the DCT of the local grey-level gradient pattern around a landmark, and is utilized in a scheme of facial landmark annotation for locating facial features of the face in an input image. The proposed model provides two distinct advantages: (i) the use of a smaller number of DCT coefficients in building a compressed DCT profile significantly reduces the computational complexity, and (ii) the process of choosing the low-frequency DCT coefficients filters out the noise contained in the image. The experimental results show that the use of the proposed model in the application of facial landmark annotation significantly reduces the execution time without affecting the accuracy of the facial shape fitting.

1 Introduction

In recent advances in biometrics, facial recognition has been one of the promising research areas. Facial recognition systems can be categorized into classes of systems employing holistic approaches and feature-based approaches. The feature-based approach is more robust to positional variations of the face in the image compared to the holistic approach [1]. This approach, however, relies heavily on an accurate localization of facial features. Thus, automatic and accurate localization of facial features is an essential pre-processing step in a feature-based facial recognition system.

Active shape model (ASM), which was first introduced in 1994 in [2], is a flexible model that has widely been utilized in order to automatically locate a set of landmarks representing a target object in an image [3–7]. In an application of ASM in facial landmark localization, the facial features are represented by a set of landmarks, and

This work was supported in part by the Natural Sciences and Engineering Research Council (NSERC) of Canada and in part by the Regroupement Strategique en Microelectronique du Quebec (ReSMiQ).

© Springer International Publishing AG 2017
F. Karray et al. (Eds.): ICIAR 2017, LNCS 10317, pp. 259–267, 2017.
DOI: 10.1007/978-3-319-59876-5_29

ASM automatically locates these landmarks by fitting its model shape to a facial shape in an image. The classical ASM performs well if the landmarks of the initial model shape are placed close to their targets. However, the initial model shape could be placed only roughly whereby the landmarks often get placed away from their targets. Thus resulting in long search lines and consequently making the target search computationally expensive. It could also distract the landmarks by local structures in the image. In order to overcome the above drawbacks of the classical ASM, multi-resolution ASM (MRASM) is proposed in [5]. In this scheme, an ASM is first applied to a coarse image to roughly place the model shape near the target object, and then applied to finer images to refine the shape fitting. This approach reduces the risk of distraction of the landmarks by local structures in the image, and decreases the computational complexity [5]. However, the 1-D profile used in the classical ASM and MRASM does not sufficiently represent the grey-level intensity gradient pattern around each landmark to distinguish the landmarks from one another [8]. Consequently, the search can converge to a local minimum and produces poor fitting results. Stacked ASM (STASM), introduced in [6], proposes the use of two ASM searches sequentially, a search using the classical ASM with 1-D profiles to roughly place the model shape, and then a search using an ASM with 2-D profiles to refine the shape fitting. This approach yields relatively more accurate shape fitting results. However, the search can still converge to a local minimum [8]. Furthermore, the use of 2-D profiles significantly increases the computational complexity of the search. In order to further improve the shape fitting accuracy, an ASM with a PCA-based LGGM, is proposed in [7]. It employs the principal component analysis (PCA) in the local grey-level gradient model (LGGM) to model the variations of grey-level intensity gradients in a square region around each landmark. The use of the PCA contributes to improving the shape fitting accuracy and to reducing the risk of the search converging to a local minimum [8]. However, the computational complexity of ASM with a PCA-based LGGM is much larger than that of the techniques using other versions of ASM, since the computation of the PCA is expensive.

This paper is aimed at developing a low-complexity ASM by incorporating the energy compaction property of the discrete cosine transform (DCT). The proposed ASM utilizes a novel 2-D profile of a landmark, which is based on the DCT, in order to reduce the computational complexity without affecting the facial shape fitting accuracy. The development of the proposed model puts emphasis on reducing the execution time of a facial shape search while providing a shape fitting accuracy that is better or the same as that provided by the stacked ASM or ASM with a PCA-based LGGM. This paper is organized as follows: Sect. 2 presents a low-complexity ASM which utilizes a novel 2-D profile. In Sect. 3, a facial shape search scheme using the proposed ASM is presented. Experiments are then performed to examine the effectiveness of the proposed ASM in an automatic facial landmark annotation of frontal faces in Sect. 4. Section 5 concludes this paper by summarizing the main contributions made in this paper.

2 Proposed ASM Using DCT

In order to reduce the size of the 2-D profiles, and subsequently decrease the computational complexity of ASM itself, we propose a low-complexity ASM, which is developed by using a 2-D profile of a landmark of a facial shape based on the DCT of the local intensity gradients of the pixels around it. This 2-D profile captures more information around each landmark of the model shape than the 1-D profile, and the size of the 2-D profile is much smaller than that of the conventional spatial-domain 2-D profile. The process of building such a profile consists of four steps described below.

The building process of the 2D profile for a landmark starts with sampling grey-level intensities from an m × m square region around a particular landmark. Using the grey-level intensities sampled from an m × m region around a landmark, the profile matrix $\mathbf{G}$ of the landmark is computed by calculating the response of a t × q linear spatial filter w(i, j) at each pixel of the square region [6]. The filter response G(x, y) at the pixel position (x, y) is given by the sum of products of the filter coefficients and the corresponding grey-level intensities in the region spanned by the filter mask [9], that is,

$$G(x,y) = \sum_{i=-a}^{a} \sum_{j=-b}^{b} w(i,j)\, i(x+i, y+j) \tag{1}$$

where a = (t − 1)/2 and b = (q − 1)/2.

The profile matrix of the landmark in the spatial domain is transformed to the frequency domain by applying the 2-D DCT to the m × m profile matrix $\mathbf{G}$ of the landmark to obtain an m × m array of the 2-D DCT coefficients. The 2-D DCT of the m × m profile matrix $\mathbf{G}$ is obtained as

$$C_G(k_1, k_2) = \alpha(k_1)\,\alpha(k_2) \sum_{x=0}^{m-1} \sum_{y=0}^{m-1} G(x,y) \cos\left(\frac{(2x+1)k_1\pi}{2m}\right) \cos\left(\frac{(2y+1)k_2\pi}{2m}\right) \tag{2}$$

where

$$\alpha(k_1) = \begin{cases} \sqrt{\frac{1}{m}} & \text{for } k_1 = 0 \\ \sqrt{\frac{2}{m}} & \text{otherwise} \end{cases}$$

and likewise for $\alpha(k_2)$. The resulting m × m array $\mathbf{C_G}$ of the DCT coefficients $C_G(k_1, k_2)$ consists of a DC coefficient representing the zero frequency component and the AC coefficients representing the low- and high-frequency components of the profile matrix.

Making use of the energy compaction property of the DCT, a set of n_c significant coefficients is selected from the m × m array of the DCT coefficients by zig-zag scanning the DCT coefficients in order to select the low-frequency coefficients before

selecting the high-frequency coefficients [11]. The resulting zig-zag sequence of the
$m \times m$ DCT coefficients can be represented by a vector given by

$$\mathbf{Y_{C_G}} = [C_G(0,0),\, C_G(0,1),\, C_G(1,0),\, C_G(2,0),\, C_G(1,1),\, C_G(0,2),\ldots,$$
$$C_G(m-2,m-1),\, C_G(m-1,m-2),\, C_G(m-1,m-1)]^T \qquad (3)$$

The first n_c DCT coefficients representing the low-frequency components of the
profile matrix are selected from the vector $\mathbf{Y_{C_G}}$, giving

$$\mathbf{Y}_{n_c} = [y_0, y_1, y_2, \ldots, y_{n_c-1}]^T \qquad (4)$$

The n_c DCT coefficients are normalized by dividing each coefficient by the mean of
the absolute values of the DCT coefficients as

$$y_i' = \frac{y_i}{\frac{1}{n_c} \sum_{j=0}^{n_c-1} |y_j|} \qquad (5)$$

where y_i is the ith element in $\mathbf{Y}_{n_c}$ and stored in

$$\mathbf{Y}_{n_c}' = [y_0', y_1', y_2', \ldots, y_{n_c-2}', y_{n_c-1}']^T \qquad (6)$$

As a result of the operation, the normalized DCT coefficients are more robust to
variations in brightness and contrast over the images in the training set [6]. In order to
reduce the effect of outliers in the normalized 2-D DCT coefficients, each coefficient is
then equalized by applying a sigmoid transform as

$$y_i'' = \frac{y_i'}{|y_i'| + c} \qquad (7)$$

where c is a shape constant [6]. The resulting n_c DCT coefficients are finally used as a
compressed DCT profile of the landmark, denoted by

$$\mathbf{Y}_{n_c}'' = [y_0'', y_1'', y_2'', \ldots, y_{n_c-2}'', y_{n_c-1}'']^T \qquad (8)$$

3   Annotation Using the Proposed ASM

The shape search method employing the proposed ASM consists of four steps
described below.

The search method begins with localizing a rectangular region in a given image
containing a face using the Viola Jones face detector [10]. The rectangular region,
specified by its centre point (x_c, y_c), width w, and height h, is used for determining the
model parameters $\mathbf{b}$ and the transformation parameters (t_x, t_y, s, θ) to generate an initial

model shape. The mean shape $\bar{x}$ of the training set is mapped onto the image using the centre point as the reference. The resulting shape is then scaled and rotated by assigning a value to the scaling parameter s and the rotation parameter θ to approximately fit it in the rectangular region. The initial model shape represented by the model parameter $b_0 = 0$ and the transformation parameters $(t_x, t_y, s, \theta)_0$ is utilized as an initial shape in LGGM for generating a refined shape.

Each landmark of the shape, represented by the model parameters b and the transformation parameters (t_x, t_y, s, θ), is moved to a suitable position determined by using the LGGM for the landmark to generate a new shape X_S. In the proposed ASM, the LGGM is built by using the compressed DCT profile described in Sect. 2. In order to determine a suitable position for the ith landmark, the cost of fit of the compressed DCT profile Y_i'' of the ith landmark to the mean profile $\bar{Y}_i''$ of the LGGM is evaluated at each candidate position in a 5×5 region around the landmark using the Mahalanobis distance given by

$$f(Y_i'') = (Y_i'' - \bar{Y}_i'')^T S_i^{-1} (Y_i'' - \bar{Y}_i'') \tag{9}$$

where S_i is the covariance matrix for the ith landmark. The position d_{best} at which the profile Y_i'' yields the smallest Mahalanobis distance (the minimum cost of fit) is then selected as the new position of the landmark. The process is repeated for each landmark of the model shape in order to obtain a new shape X_S.

The model parameters b_{best} and the transformation parameters $(t_{xbest}, t_{ybest}, s_{best}, \theta_{best})$ that best approximate the new shape X_S obtained in the previous sub-section are determined by using the similarity transform T and the point distribution model.

The model parameters b and the transformation parameters (t_x, t_y, s, θ) are then updated to the previously determined model and transformation parameters b_{best} and $(t_{xbest}, t_{ybest}, s_{best}, \theta_{best})$ which closely approximate the new shape X_S. The updated model shape is then used as the starting shape in the second iteration of the shape search method. The last three steps are repeated until the iterative process satisfies a pre-specified terminating condition.

4 Experimental Results

In order to study the effectiveness of the proposed ASM in the application of an automatic facial landmark annotation of frontal faces, the facial annotation using the scheme described in Sect. 3 is performed on various facial images. We present and compare the experimental results of an automatic facial landmark annotation of frontal faces using the proposed and various other ASMs.

In order to examine the performance of the proposed ASM experimentally, two groups of training and test sets of images and landmarks of the facial shapes are selected from the three databases, the MUCT database [12], the BioID database [13] and the IMM database [14], containing both images and the landmarks of the corresponding facial shapes. Table 1 summarizes the two groups of training and test sets of images and landmarks of the facial shapes belonging to the three databases.

Table 1. Summary of the number of training and test sets in the two groups of the images chosen from the three databases and the number of landmarks in each facial shape.

	Group 1		Group 2	
	Training set	Test set	Training set	Test set
Database	MUCT	BioID	IMM	
Number of samples	3000	1471	240	240
Number of landmarks	76	20	58	

17 landmarks which are common in each of samples in the training and test sets of samples are used in the experiment using the first group of samples, whereas all the 58 landmarks are utilized in the experiment using the second group of samples.

In order to measure the effectiveness of the proposed ASM in fitting the landmarks of the model shape onto a target facial shape in an image, the fitting accuracy is computed by using an evaluation method known as the average normalized fitting error. The normalized fitting error is obtained by taking the average of the Euclidean distance d between the manually annotated landmarks from a target facial shape and the corresponding landmarks from the model shape, and normalizing the result by a factor that ensures that s, the distance between two landmarks representing the extreme points on the left and right eyes is equal to 50 pixels [7]. In order to compute the overall fitting accuracy of the proposed ASM, the average normalized fitting error is computed by taking the mean of the normalized fitting error over all the samples from the test set as

$$E_{average} = \frac{1}{N_{test}} \sum_{i=1}^{N_{test}} \left\{ \frac{50}{ns} \sum_{j=1}^{n} d_j \right\} \tag{10}$$

where N_{test} is the number of samples from the test set, and n is the number of landmarks in a sample. The execution time of a facial landmark annotation of the frontal face in an image is given by the sum of the execution time for locating the face in the image and the execution time for conducting the facial shape search in the image. In order to conduct the experiments under the same condition for all the ASMs, the execution time for locating the face in the image is a constant time for loading the pre-computed location of the region containing the face. The average execution time is computed by taking the average of the execution time over all the test samples after the first 50, since the initialization of the ASM may have an effect on the computation time for some samples at the beginning of the execution. The experiments are performed on a 2.6 GHz Intel Core i7 CPU with 6-GB RAM and Windows 7 operating system.

The number of DCT coefficients n_c and the filter mask used in building the profile are important choices for the proposed ASM, since they have an effect on the fitting accuracy and the computational complexity. An empirical study is conducted in [15] in order to select the optimal number of DCT coefficients and to choose a suitable filter mask, which are used for building the compressed DCT profile of the ASM. From the results of the experiments, it has been concluded that the use of the gradient filter mask with 25 DCT coefficients used for building the compressed DCT profile provides good performance as well as computational efficiency.

In order to examine the comparative effectiveness of the proposed ASM, we apply this and other ASMs to the problem of facial annotation and compare the performance in terms of the average normalized fitting errors and the average execution times. The parameters utilized for a facial landmark annotation using the three ASMs are summarized in [15]. The two groups specified in Table 1 of training and test sets of images and landmarks of the facial shapes are utilized for conducting experiments for the various ASMs. The facial landmark annotation is conducted 10 times for each ASM using the Group 1 samples and then using the Group 2 samples, and the mean values of the average normalized fitting errors and the average execution times are obtained and compared. Tables 2 and 3 present the mean values of the average normalized fitting error and the average execution time obtained from the experiments using the training and test sets of Group 1 and Group 2, respectively. It is observed from these tables that the lowest average normalized fitting error is obtained by using the proposed ASM. It is also to be noted that the average execution time obtained by using the proposed ASM is much smaller than that obtained by using the stacked ASM and ASM with a PCA-based LGGM. The proposed compressed 2-D DCT profile has provided two distinct advantages: (i) The use of a smaller number of DCT coefficients significantly reduces the computational complexity without sacrificing the fitting accuracy. (ii) The process of choosing the low-frequency DCT coefficients filters out the noise contained in the image. Hence, the use of compressed DCT profile has, in fact, reduced the fitting error as well. Therefore, it can be concluded that the proposed ASM is an attractive and viable alternative to the existing ASMs.

Table 2. Mean values of average normalized fitting error and average execution time obtained by using the proposed and two other ASMs from the experiment using the training and test sets of Group 1.

Method	Average normalized fitting error (pixels)	Average execution time per facial image (seconds)
[6]	1.5807	0.1251
[7]	1.8033	2.8076
Proposed ASM	**1.5220**	**0.0675**

Table 3. Mean values of average normalized fitting error and average execution time obtained by using the proposed and two other ASMs from the experiment using the training and test sets of Group 2.

Method	Average normalized fitting error (pixels)	Average execution time per facial image (seconds)
[6]	3.4666	0.1406
[7]	3.2164	2.1565
Proposed ASM	**3.0652**	**0.0784**

5 Conclusion

In this paper, a low-complexity ASM has been developed by incorporating the energy compaction property of the DCT. The proposed ASM utilizes a novel 2-D profile of a landmark, which is based on the DCT, in order to reduce the execution time of a facial shape search while keeping the shape fitting accuracy of the ASM comparable to that of other improved versions of ASM. The proposed ASM has been employed in a scheme of facial landmark annotation for locating facial features of the face in an input image. It has been noted from the experimental results that the use of a smaller number of DCT coefficients in building the compressed DCT profile significantly reduces the computational complexity, and the process of choosing the low-frequency DCT coefficients filters out the noise contained in the image, which is associated with the high-frequency coefficients. In conclusion, this study has shown that the use of the proposed ASM in an automatic facial landmark annotation of frontal faces has significantly reduced the computational complexity without losing the shape fitting accuracy.

References

1. Hafed, Z.M., Levine, M.D.: Face recognition using the discrete cosine transform. Int. J. Comput. Vis. **43**(3), 167–188 (2001)
2. Cootes, T.F., Hill, A., Taylor, C.J., et al.: The use of active shape models for locating structures in medical images. Image Vis. Comput. **12**(6), 355–366 (1994)
3. Cootes, T.F., Taylor, C.J.: Statistical Models of Appearance for Computer Vision. University of Manchester, England (2004)
4. Cootes, T.F., Taylor, C.J., Cooper, D.H., et al.: Active shape models - their training and application. Comput. Vis. Image Underst. **61**(1), 38–59 (1995)
5. Cootes, T.F., Taylor, C.J., Lanitis, A.: Active shape models: evaluation of a multi-resolution method for improving image search. In: Proceedings of the 5th British Machine Vision Conference, York, pp. 327–336 (1994)
6. Milborrow, S.: Locating facial features with active shape models. MS thesis, University of Cape Town (2007)
7. Seshadri, K., Savvides, M.: Robust modified active shape model for automatic facial landmark annotation of frontal faces. In: Proceedings of the 3rd IEEE International Conference on Biometrics: Theory, Applications Systems, Washington, DC, pp. 1–8 (2009)
8. Seshadri, K., Savvides, M.: An analysis of the sensitivity of active shape models to initialization when applied to automatic facial landmarking. IEEE Trans. Inf. F. Secur. **7**(4), 1255–1269 (2012)
9. Gonzalez, R.C., Woods, R.E.: Digital Image Processing, 3rd edn. Prentice Hall, Upper Saddle River (2008)
10. Viola, P., Jones, M.: Rapid object detection using a boosted cascade of simple features. In: Proceedings of the 2001 IEEE Computer Society Conference, Kauai, HI, vol. 1, pp. I-511–I-518 (2001)
11. Wallace, G.K.: The JPEG still picture compression standard. Commun. ACM 30–44 (1991)
12. Milborrow, S., Morkel, J., Nicolls, F.: The MUCT landmarked face database. In: Proceedings of the Pattern Recognition Association of South Africa, Stellenbosch, South Africa, pp. 179–184 (2010)

13. Jesorsky, O., Kirchberg, K., Frischholz, R.: Robust face detection using the hausdorff distance. In: Proceedings of the 3rd International Conference on Audio- and Video-based Biometric Person Authentication, Halmstad, pp. 90–95 (2001)
14. Stegmann, M.B., Ersbøll, B.K., Larsen, R.: FAME – a flexible appearance modelling environment. IEEE Trans. Med. Imaging **22**(10), 1319–1331 (2003)
15. Yasuda, K.: Development of an active shape model using the discrete cosine transform. MS thesis, Concordia University (2014)

Ground Plane Segmentation Using Artificial Neural Network for Pedestrian Detection

Jorge Candido[✉] and Mauricio Marengoni

Universidade Presbiteriana Mackenzie, São Paulo, SP, Brazil
{jorge.candido,mauricio.marengoni}@mackenzie.br

Abstract. This paper presents a method of ground plane segmentation for urban outdoor scenes using a feedforward artificial neural network (ANN). The main motivation of this project is to obtain some contextual information from the scene for use in pedestrian detection algorithms and to provide an accuracy improvement for this algorithms. The ANN input is fed with features extracted from a patch window of the image scene. The ANN output classifies the patch as belonging or not belonging to the ground plane. After that, the classified patches are joined into a full image with the ground plane area outlined. The images used for training, test and evaluation were obtained from the widely known Caltech-USA database. The accuracy of ground plane segmentation was above 96% in the experiments which improved the precision of the pedestrian detector in 38,5%.

Keywords: Feature extraction · Ground plane segmentation · Pedestrian detection

1 Introduction

Floor and ground plane segmentation plays a very important role in computer vision due to its application in vision-aided navigation systems (INS) for robots and automobiles [3,13]. It is a challenging problem considering that some applications encompass great variations of illumination and structure to handle [14].

This work deals exclusively with outdoor scenes, where variation in illumination and surface textures are bigger than at indoor environments. The focus on outdoor scenes is explained by the possibility of using ground plane area information in pedestrian detection algorithms. According to [8], the use of contextual information can improve detection rate in those tasks, thus reducing the false positives rate.

The proposed framework handles the ground plane segmentation problem as a texture classification problem and uses a ANN as classifier. The image scene is divided into a number of patches and classify them into two classes: ground or non-ground. Ground plane segmentation is achieved with a procedure that joins patches and outlines the scene floor area.

The classifier was trained with a large number of example patches selected from images from the Caltech-USA database [7]. The Caltech-USA database is

© Springer International Publishing AG 2017
F. Karray et al. (Eds.): ICIAR 2017, LNCS 10317, pp. 268–277, 2017.
DOI: 10.1007/978-3-319-59876-5_30

widely used as a benchmark in pedestrian detection algorithms. It consists of approximately 10 h of 30 fps video filmed from inside an automobile driving through urban regular traffic. Positive examples were cropped from regions where the pedestrians can walk: asphalt, sidewalk and grass. Negative examples are patches of buildings, cars, trees etc. The ANN was trained, tested and evaluated using those examples. Figure 1 shows an image annotated with positive and negative examples.

The final experiments were made in a different group of outdoor images. The effectiveness of the method was evaluated as a ground plane segmentation algorithm applied to any scenario.

In order to evaluate the contribution of the contextual information in the task of pedestrian detection, an experiment with a group of images from the Caltech-USA database was conducted and the performance gain of the detector was measured.

Fig. 1. Example of patch extraction. Red boxes are non-ground examples. Light blue boxes are ground plane examples. (Color figure online)

1.1   Related Work

The issue of image segmentation has been studied since the seventies [12,17]. Algorithms that deal with this problem attempt to create separate and non-overlapping regions on the image according to some criteria. In Felzenszwalb and Huttenlocher [9] an algorithm for image segmentation is presented in which boundary regions are represented by a graph-based schema. The method uses two different types of features to describe a neighborhood when constructing the graph. In [1], Arbalaez et al. claims to have reached the state of the art for segmentation by using spectral clustering applied to contour detection. The algorithm uses multiple local cues combined at a global framework. In the image segmentation algorithm presented in [16], Shirakawa and Nagao used an optimization technique called multi-objective where two measures inside regions were made: overall deviation and edge straightness. During the optimization process the image segmentation changes following an objective function. The objective function aims to minimize overall deviation and maximize edge straightness.

There are some works that focus on the specific issue of ground plane segmentation. Recent works have shown some improvements in obstacle avoidance and ground plane detection algorithms. However, these algorithms work with indoor images. The work presented by Kin and Kin [10] estimates the ground normal plane using a homograph-based approach. The normal planes are computed upon a video sequence and the floor area is detected. In Barcel et al. [2] the authors presented a method to segment the floor region of grey scale images that does not require the vanishing points calculation. Thus, the algorithm rapidly adapts itself to changes in the scene caused by the camera movement. In Wang et al. [18] a region based method was presented, in which each region was classified as being an obstacle or not being an obstacle. The classification relies on the region average differences. At the end, a map showing the scene obstacles was created.

Cui et al. [4] presented an algorithm that extracts from the image some geometric features. Thus, the ground normal plane was estimated. In [19] Wang and Wu presented a method to segment floor using a spiking neural network. The network was used to extract from the image information about useful and effective edges. The project purpose was to be an autonomous guide to help visually impaired people to circulate on unknown sites.

2   Proposition

In the present work a framework for ground plane segmentation on outdoor images is proposed. Figure 2 shows the entire process described in a block diagram. The framework receives an image and divides it into a non-overlapping set of patches. The patches are classified as ground or non-ground by subsequent blocks. After classification, the patches belonging to the ground class are linked to form the floor area of the original image. A simple heuristic is used to eliminate some misclassified patches that seem to be isolated and surrounded by patches of the opposite class. In the following subsection the main components of the system are outlined.

2.1   Artificial Neural Network

A typical application of ANN applied on computer vision, directly uses the pixel values in the raw image as input for the ANN [15], but for texture classification the pixel values present a high level of redundancy. To avoid that effect the image features from the patches are used rather than the raw pixel value.

A feed-forward network is used due to its good capacity of generalization when trained with a resilient back-propagation algorithm [5]. The ANN input layer has the number of neurons N defined by:

$$N = \frac{whc}{WH} \tag{1}$$

where w and h represent the patch size, c is the number of image channels and W and H represent the pooling region size.

The activation function used for the neurons in the hidden layer was the sigmoid. The output layer has only one neuron with linear-type activation function which produces an output between 0 and 1. Therefore, responses around 0 represent the non-ground class, while responses around 1 represent the ground class.

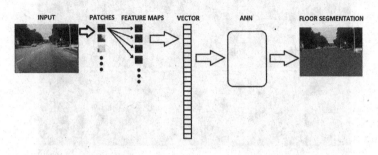

Fig. 2. Proposed framework diagram.

2.2   Feature Extraction

In the proposed framework the ANN input layer receives data from an intermediate step that transforms the original patch into a set of feature channels. Each channel represents the original patch in terms of a particular feature.

The calculation of the feature channels used in this work was inspired by [6]. In that system, feature channels are used for pedestrian detection. We used the same feature combination resulting in 10 feature channels extracted from each patch. The feature channels are normalized gradient magnitude (1 channel), histogram of oriented gradient (6 channels) and LUV color channels (3 channels).

The normalized gradient magnitude (MAG) channel is obtained by convoluting the image patch with a mask M:

$$M = [-1, 0, 1] \tag{2}$$

in both horizontal and vertical directions, resulting in 2 maps that show the intensity differences at pixel level. Those maps are combined and the axis information is presented on the channel.

In the histogram of oriented gradient (HOG), the bins represent the orientation angles proposed for this application. The bins accumulate magnitude-weighted votes for gradients at the respective orientation. In our case we use 6 different orientations, thus resulting in 6 different maps.

The 3 LUV color channels complete our set of features. Among other color space definitions, the LUV space delivered the best results in our experiments. Figure 3 shows an example of channel feature extraction.

Although this feature combination has led to better results on pedestrian detection algorithm, it was tested also if they have significantly contributed for the patch classification task. The groups of features (LUV, MAG and HOG)

Fig. 3. Feature maps proposed. The first line shows the LUV color channels. The second and third lines show the HOG channels in 6 different orientations. The fourth line on the left shows the normalized gradient magnitude channel. The fourth line on the right shows the original image.

were tried individually and in combination. Table 1 shows the best results for each combination.

A pooling process is applied for each channel with the purpose of lowering data dimensionality and improve robustness of the system. The max pooling was calculated over each 4×4 window of the patch.

Table 1. Feature combinations tested in the ANN.

Feature combination	Accuracy
MAG	82.8%
LUV	84.3%
LUV+MAG	85.8%
HOG	89.0%
MAG+HOG	90.2%
LUV+HOG	93.0%
LUV+MAG+HOG	96.5%

2.3   Ground Plane Segmentation

The resulting image showing the ground plane segmentation is constructed based on the patches classified by the ANN. Each patch was analysed according to its class and neighborhood to make the final decision on whether it belongs to the ground plane area.

The ANN output layer has one neuron with a linear transfer function and gives a score in the range [0 1] for each patch submitted to its input. During the evaluation of one full image, we have one response for each patch. Responses in the range between 0,4 and 0,6 do not have enough confidence to define a class and were classified as undefined. To deal with the undefined responses a method that checks the neighborhood classification was used and the undefined patch was classified using the majority operation. However, if the number of ground and non-ground class patches is the same, it was assumed that the patch is in the boundary between ground and non-ground region and classify it as ground if the output neurons is greater than 0.5.

2.4   Pedestrian Detection

The main objective of this work is to improve the performance of pedestrian detection algorithms using the ground plane segmentation. This task was accomplished by combining the pedestrian detection output with the segmentation previously done.

At the detector output a list of pedestrians found is presented with annotation of size and position. The area right below the detection is analysed counting the number os patches of ground in the segmented image. Finally, detection examples that do not have at least 70% of ground patches inside the area were eliminated.

3   Experiments

All experiments conducted in this research were based on the Caltech-USA database.

During the experiments performed in this research, the effectiveness of the framework for ground plane regions detection in a 2D image was tested. The approach consists of working with patch samples, cropped from the images, and training an ANN to classify the patches as ground or non-ground.

The MATLAB Development Environment was used to test the algorithms and to do the image manipulations during all the experiments performed. The remainder of this section is used to describe the experiments done during the research.

3.1   Database

The Caltech-USA image database is one of the most used benchmarks for pedestrian detection. It is made of approximately 10 h of 30 Hz videos taken from inside a car driving in an urban scenario. Extracting every single image from the videos recorded results in 250,000 images. In the present work, 160 images from the Caltech database were selected to construct the dataset for use in the experiments. The patches from those images were annotated and cropped using the toolbox described in [7].

3.2   Procedure

Initially, 1632 patches of ground examples and 4235 patches of non-ground examples were cropped from the images from the dataset. The size of the each patch is 20×20. This size was chosen through experiments with different patch sizes. Patches larger than 20×20 do not bring any improvement and demand much longer training time. However, patches smaller than 20×20 significantly degrade the result.

The next step transforms each patch into a vector containing all the extracted features. Finally, extra data is added to feature vectors defining the target label for use in the ANN training procedure. Value 0 is the label for non-ground examples and 1 is the label for ground examples.

The patch dataset was randomly divided into 3 subgroups. Training subgroup, with 70% of the patches and targets, validation subgroup, with 15% of the patches and targets, and test subgroup, with 15% of the patches and targets.

Training and validation subgroups were used during the ANN training and test subgroup was used after the training to give an independent performance score for the ANN.

Extensive tests were conducted to determinate the number of neurons at the ANN hidden layers. Networks with 1 and 2 hidden layers and number of neurons in each layer between 5 and 20 neurons, were tested. The best results (above 96%) was reached with an ANN having two hidden layers, with 10 neurons at the first hidden layer and 5 neurons at the second hidden layer. The use of more neurons has not shown any substantial improvement in the results; at same time, it increases training time considerably. The use of only 1 hidden layer leads to over 92%, which is a reasonable result.

A min-max pre-processing step was applied on the training feature data in order to have them fit the range $[-1, 1]$. This is necessary for normalizing the characteristics of the feature maps.

The training was executed by the MATLAB's Resilient Backpropagation algorithm.

Since the ANN result is strongly influenced by weights and initialization of biases, the process of initialization, training and test was repeated several times and select the best ANN. Finally, the best performing ANN was selected reaching an accuracy of 96.5%. All numbers showing the ANN performance were obtained at the test round.

The ground plane segmentation process starts with the image being divided into non-overlapping patches. For each patch, we extract the features and calculate the ANN response. Then, we use the results from the patch classification on a whole image to identify the floor region of the image. In this process, we

Fig. 4. Example of floor segmentation performed by the proposed framework.

assume that a group of patches close together in the image is likely to belong to the same class, unless they are in a boundary region between two classes. We use that assumption to classify the patches with uncertainty responses from the ANN (between 0.4 and 0.6) by analyzing the neighborhood of the uncertain patch and give it the same classification as that of the majority of the neighborhood patches. Figure 4 show the result obtained with this approach.

3.3 Pedestrian Detection

The experiments with pedestrian detection was conducted using the state of the art algorithm ACF with local decorrelation (LDCF) [11], running on a reduced group of images from the Caltech database. The approach used here is described in the previous section. The number of false-positives detections has been considerably reduced. While the original algorithm LDCF reached 31,15% of precision, the use of ground plane information raised this number to 43,18%. A 31,5% gain in precision.

4 Conclusions

In this paper, a framework capable of classifying patches of an image and identify the floor region by using an ANN was presented. The effectiveness of the use of an ANN in the patch classifier task was demonstrated. The whole system works as an efficient ground plane segmentation for outdoor images. Even a simple ANN with only 1 hidden layer is capable of efficiently performing the proposed patch classification, reaching a miss rate below 8%. This was possible through the use of features rather than the raw pixels as input to the ANN.

An important contribution of this work is to show that the previous knowledge of the location of the ground plane can improve the performance of the pedestrian detection algorithm.

5 Future Work

The next steps in this research this framework will be tested in a more extensive dataset composed of samples of other elements of the scene, such as trees and cars. These elements are very frequent at urban outdoor images and can give more precise contextual information to use on a pedestrian detector system.

References

1. Arbelaez, P., Maire, M., Fowlkes, C., Malik, J.: Contour detection and hierarchical image segmentation. IEEE Trans. Pattern Anal. Mach. Intell. **33**(5), 898–916 (2011)
2. Barceló, G.C., Panahandeh, G., Jansson, M.: Image-based floor segmentation in visual inertial navigation. In: 2013 IEEE International Instrumentation and Measurement Technology Conference (I2MTC), pp. 1402–1407. IEEE (2013)
3. Conrad, D., DeSouza, G.N.: Homography-based ground plane detection for mobile robot navigation using a modified em algorithm. In: 2010 IEEE International Conference on Robotics and Automation (ICRA), pp. 910–915. IEEE (2010)
4. Cui, X.-N., Kim, Y.-G., Kim, H.: Floor segmentation by computing plane normals from image motion fields for visual navigation. Int. J. Control Autom. Syst. **7**(5), 788–798 (2009)
5. Demuth, H., Beale, M.: Neural network toolbox for use with Matlab (1993)
6. Dollár, P., Belongie, S., Perona, P.: The fastest pedestrian detector in the west. In: BMVC, vol. 2, p. 7. Citeseer (2010)
7. Dollár, P., Wojek, C., Schiele, B., Perona, P.: Pedestrian detection: a benchmark. In: CVPR 2009 IEEE Conference on Computer Vision and Pattern Recognition, pp. 304–311. IEEE (2009)
8. Dollar, P., Wojek, C., Schiele, B., Perona, P.: Pedestrian detection: an evaluation of the state of the art. IEEE Trans. Pattern Anal. Mach. Intell. **34**(4), 743–761 (2012)
9. Felzenszwalb, P.F., Huttenlocher, D.P.: Efficient graph-based image segmentation. Int. J. Comput. Vis. **59**(2), 167–181 (2004)
10. Kim, Y.G., Kim, H.: Layered ground floor detection for vision-based mobile robot navigation. In: Proceedings of the 2004 IEEE International Conference on Robotics and Automation, ICRA 2004, vol. 1, pp. 13–18. IEEE (2004)
11. Nam, W., Dollár, P., Han, J.H.: Local decorrelation for improved pedestrian detection. In: Advances in Neural Information Processing Systems, pp. 424–432 (2014)
12. Pal, N.R., Pal, S.K.: A review on image segmentation techniques. Pattern Recognit. **26**(9), 1277–1294 (1993)
13. Panahandeh, G., Mohammadiha, N., Jansson, M.: Ground plane feature detection in mobile vision-aided inertial navigation. In: 2012 IEEE/RSJ International Conference on Intelligent Robots and Systems, pp. 3607–3611. IEEE (2012)
14. Pears, N., Liang, B.: Ground plane segmentation for mobile robot visual navigation. In: Proceedings of the 2001 IEEE/RSJ International Conference on Intelligent Robots and Systems 2001, vol. 3, pp. 1513–1518. IEEE (2001)
15. Rowley, H.A., Baluja, S., Kanade, T.: Rotation invariant neural network-based face detection. In: Proceedings of the 1998 IEEE Computer Society Conference on Computer Vision and Pattern Recognition 1998, pp. 38–44. IEEE (1998)

16. Shirakawa, S., Nagao, T.: Evolutionary image segmentation based on multiobjective clustering. In: 2009 IEEE Congress on Evolutionary Computation, pp. 2466–2473. IEEE (2009)
17. Szeliski, R.: Computer Vision: Algorithms and Applications. Springer Science & Business Media, London (2010)
18. Wang, H., Yuan, K., Zou, W., Peng, Y.: Real-time obstacle detection with a single camera. In: 2005 IEEE International Conference on Industrial Technology, pp. 92–96. IEEE (2005)
19. Wang, X.W., Wu, Q.X., Zhang, Z., Zhuo, Z., Huang, L.: Segmentation based on spiking neural network using color edge gradient for extraction of corridor floor. In: Farag, A.A., Yang, J., Jiao, F. (eds.) Proceedings of the 3rd International Conference on Multimedia Technology (ICMT 2013). LNEE, vol. 278, pp. 275–285. Springer, Heidelberg (2014). doi:10.1007/978-3-642-41407-7_27

An Improved Directional Convexity Measure
for Binary Images

Péter Bodnár[✉] and Péter Balázs

Department of Image Processing and Computer Graphics,
University of Szeged, Árpád tér 2, Szeged H6720, Hungary
{bodnaar,pbalazs}@inf.u-szeged.hu

Abstract. Balázs et al. (Fundamenta Informaticae 141:151–167, 2015) proposed a measure of directional convexity of binary images based on the geometric definition of shape convexity. The measure is useful for various applications of digital image processing and pattern recognition, especially in binary tomography. Here we provide an improvement of this measure making it to follow better the intuitive concept of geometric convexity and to be more suitable to distinguish between thick and thin objects.

Keywords: Digital geometry · Convexity measure · hv-convexity · Shape descriptor

1 Introduction

Convexity is a widely used shape descriptor for image analysis and classification. There are several ways to measure the convexity of a continuous or digital shape. Area based measures [4,13,14] and boundary-based ones [16] are the most popular, but there are also other methods to determine the grade of convexity [10–12]. In case of digital images, a straightforward alternative of the continuous convexity concept is the horizontal and vertical convexity (or shortly, hv-convexity), arising inherently from the pixel-based representation of the digital image. The property of hv-convexity is widely studied in Binary Tomography [9], where one problem in focus is to reconstruct digital images (binary matrices) from their row and column sums according to several geometrical constraints. Several reconstruction methods utilize the preliminary information of hv-convexity about the binary image to be reconstructed [3,5,6]. Motivated by this problem, in [15] the authors introduced a measure of directional convexity and proved it to be useful in binary tomographic reconstruction. However, they also showed that a 2D extension of this measure is not straightforward [2]. Later, an immediate 2D convexity measure was also proposed in [1]. Independently, the authors of [7,8] came to the same idea as of [15] to use convexity as a shape prior in image segmentation. The aim of this paper is to improve the one-dimensional measure of [15] making it more appropriate for distinguishing between thick and thin obejcts.

The research was supported by the NKFIH OTKA [grant number K112998].

F. Karray et al. (Eds.): ICIAR 2017, LNCS 10317, pp. 278–285, 2017.
DOI: 10.1007/978-3-319-59876-5_31

2   Notation and Definitions

A *binary image* is a digital image containing just black (also called as object) and white (background) pixels. We will number the rows (columns) of a binary image of size $m \times n$ (where $m, n \in \mathbb{Z}$) from top to bottom (left to right), respectively, starting with 1. If M is a binary image then M^T is the binary image we get by interchanging the rows and columns of M.

A *run* of object (background) points within a row is a sequence of consecutive pixels, all of them being object (background) points, respectively, such that it cannot be expanded by further neighboring pixels of the same color. Obviously, each row of the image can be expressed by an alternating sequence of object and background runs. The length of an arbitrary run r will be denoted by $|r|$.

3   Directional Convexity

In the following we shortly recall the directional convexity measure given in [2]. Let M be a digital image of size $m \times n$ and R be an arbitrary row. To compute the *non-convexity* of a R, we split it into a list of object and background runs. Leading and trailing background runs are omitted, thus the rest of the row can be encoded as $R = b_1 w_1 b_2 w_2 \ldots w_{n-1} b_n$, where each b_i is an object run ($i = 1, \ldots, n$) and each w_i ($i = 1, \ldots, n-1$) is a background run. Trivially, taking two object points from the same run, the sequence between them will not contain any background points, thus will not contribute to the non-convexity. Now, let us take two arbitrary object points from different object runs, say the i-th and j-th, such that $i < j$. The contribution to the non-convexity of the background points in between is given by the sum of the lengths of the background runs between the i-th and j-th object runs:

$$\sum_{t=i}^{j-1} |w_t|. \tag{1}$$

For two different object runs (i-th and j-th), we can form $k_i k_j$ possible pairs of object points, by picking one from each. The contribution of these object run pairs is

$$k_i k_j \sum_{t=i}^{j-1} |w_t|. \tag{2}$$

Finally, to get the contributions for the entire row R sum up (2) for all possible combinations of object run pairs:

$$\varphi(R) = \sum_{1 \leq i < j \leq n} k_i k_j \sum_{t=i}^{j-1} |w_t|. \tag{3}$$

The value $\varphi(R)$ actually indicates the *non-convexity* of R, the higher $\varphi(R)$ is, the less convex R is.

In [2] the authors also show that the non-convexity of an arbitrary row R can be normalized (mapped to the interval $[0, 1]$) by $(n/3)^3$, as

$$\hat{\varphi}(R) = \frac{\varphi(R)}{(n/3)^3}, \tag{4}$$

and for all the rows of the image M, it is normalized by m as

$$\Phi(M) = \frac{\sum_{i=1}^{m} \hat{\varphi}(R_i)}{m}. \tag{5}$$

To map non-convexity into convexity, the authors of [2] simply adopt $\Psi(M) = 1 - \Phi(M)$. Then, the whole process can be repeated for M^T, and finally the measure of the (horizontal and vertical) convexity of the image M is defined as

$$\Psi_{hv}(M) = \frac{\Psi(M) + \Psi(M^T)}{2}. \tag{6}$$

4   The Proposed Convexity Measure

We start with the following observation. From the deduction above it follows that the non-convexity measure $\hat{\varphi}$ reaches it maximum when a row is of the form $b_1 w_1 b_2$ such that $|b_1| = |w_1| = |b_2|$ (Fig. 1) (for the sake of presentation simplicity, assume $|b_1| + |w_1| + |b_2| \equiv 0 \pmod{3}$).

Fig. 1. A row in the form of $|b_1| = |w_1| = |b_2|$.

Now, consider a row $R_1 = b_1 w_1 b_2$ of length n with $|b_1| = |b_2| = 1$, $|w_1| = n - 2$, and another one $R_2 = b_1'$ with $|b_1'| = n$ (Fig. 2).

Fig. 2. $R_1 = b_1 w_1 b_2$ with $|b_1| = |b_2| = 1$, $|w_1| = n - 2$ and $R_2 = b_1'$ with $|b_1'| = n$.

Calculating the non-convexity using (4) we get

$$\hat{\varphi}(R_1) = \frac{n-2}{(n/3)^3} \quad \text{and} \quad \hat{\varphi}(R_2) = \frac{0}{(n/3)^3}. \tag{7}$$

Now,

$$\lim_{n\to\infty} \frac{n-2}{(n/3)^3} = 0 \tag{8}$$

yields a full object row having the same convexity measure (equal to 1) as a row with only two object points (infinitely) far from each other. In our opinion, this contradicts the intuitive definition of convexity. Therefore, we follow another way to measure directional convexity.

Let M be a binary image of size $m \times n$ and $R = b_1 w_1 b_2 w_2 \ldots w_{n-1} b_n$ be an arbitrary row (recall that leading and trailing background runs can be omitted). Let B_R be the set of object runs in row R, i.e., $B_R = \{b_1\} \cup \{b_2\} \cup \cdots \cup \{b_n\}$. The sum of object pixels in R is $N_R = |b_1| + |b_2| + \cdots + |b_n|$.

Now, let $b_i, b_j \in B_R$ such that $i < j$. We select one random point from both, say, the k-th from left in b_i denoted by b_{i_k} and the l-th from left in b_j denoted by b_{j_l}. The section connecting these two points is characterized by the non-convexity measure, which value depends on the number of background pixels between b_i and b_j. Let $w = \sum_{l=i}^{j-1} |w_l|$ and d denote the distance of the two chosen points. This distance is made up of the points of b_i to the right of b_{i_k}, the points of b_j to the left of b_{j_l}. There are, $|b_i| - k + 1$ and l such points (including the chosen points, too), respectively. Additionally, the section contains the w background points, and further object point runs between b_i and b_j, if $j > i + 1$. Figure 3 illustrates the calculation.

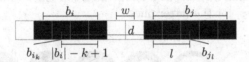

Fig. 3. Calculation of the non-convexity between two object points from different object runs.

Based on the above observations, the normalized non-convexity measure for this section is

$$\varphi(b_{i_k}, b_{j_l}) = \frac{w}{d} = \frac{w}{|b_i| - k + 1 + w + l + \sum_{l=i+1}^{j-1} |b_l|} \tag{9}$$

and the cumulated non-convexity for row R is

$$\varphi_R = \sum_{b_i, b_j \in B_R, i < j} \sum_{k=1}^{|b_i|} \sum_{l=1}^{|b_j|} \varphi(b_{i_k}, b_{j_l}). \tag{10}$$

The number of combinations to select the two object points from different object point runs can be calculated by

$$C_R = \binom{N_R}{2} - \sum_{b \in B_R} \binom{|b|}{2}, \tag{11}$$

thus the normalized non-convexity measure for row R is defined as

$$\hat{\varphi}_R = \frac{\varphi_R}{C_R}. \tag{12}$$

Then, the proposed normalized non-convexity measure for all rows is given by

$$\Phi_p(M) = \frac{\sum_{r=1}^{m} \hat{\varphi}_R}{m}. \tag{13}$$

Finally, the proposed convexity measure for the image M is

$$\Psi_p(M) = 1 - \frac{\Phi_p(M) + \Phi_p(M^T)}{2}. \tag{14}$$

5   Evaluation and Experiments

We conducted several experiments on artificial images to compare the proposed convexity measure to that of Balázs et al. [2]. First, we present a qualitative comparison using a circle and a square as objects (see Fig. 4). One can notice that the latter measure shows high convexity values when the object is thin. Thus, for example, in an image classification task it might provide a misleading result, grouping a thin circle and a full disk in the same class. Our measure can separate thick and thin objects, as it shows high convexity values for the former ones, and low values, for the latter ones.

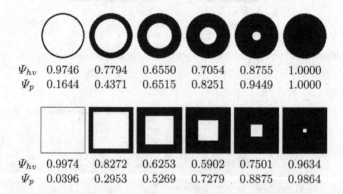

Ψ_{hv}	0.9746	0.7794	0.6550	0.7054	0.8755	1.0000
Ψ_p	0.1644	0.4371	0.6515	0.8251	0.9449	1.0000

Ψ_{hv}	0.9974	0.8272	0.6253	0.5902	0.7501	0.9634
Ψ_p	0.0396	0.2953	0.5269	0.7279	0.8875	0.9864

Fig. 4. The improved convexity measure, demonstrated with an image of a circle (top half) and a square (bottom half). Ψ_{hv} is calculated according to [2], and Ψ_p is the proposed measure.

Figure 5 summarizes the behavior of our convexity measure on an image sequence having a circle with gradually thicker outline (see Fig. 4, again, for some examples from this sequence). From the graphs, it is evident that the

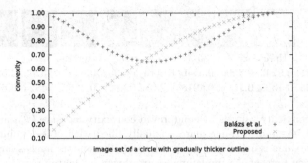

image set of a circle with gradually thicker outline

Fig. 5. The improved convexity measure compared to that of [2], using the image of a circle having 100×100 px resolution. The horizontal axis starts with an image containing a circle object having a thin outline, while the last indicated point corresponds to a full disk.

proposed measure is gradually increasing, according to our expectations, whereas the measure of [2] behaves against our intuitions.

In the next experiment, we created images with randomly placed object points covering the image from 0% to 100% (Fig. 8(a)). Example images with their convexity values are shown in Fig. 6. It shall be noted that for a small percent of object points the image receives high convexity value, according to our proposed measure. The reason is that we only measure the horizontal and vertical convexity value. In case of very few object pixels, these may appear in distinct rows and columns, by which convexity is not violated. We also produced an image sequence having a thin empty square, and gradually filled with object points at random positions. In Fig. 7 we present some examples, while Fig. 8(b) shows the graph of the detailed results. We can observe that the above phenomenon disappeared. From these results we deduce that the proposed measure is more robust against noise, and behaves, again, more intuitively than the measure of [2].

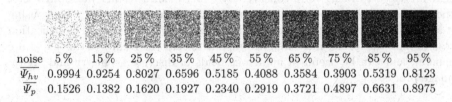

noise	5 %	15 %	25 %	35 %	45 %	55 %	65 %	75 %	85 %	95 %
$\overline{\Psi_{hv}}$	0.9994	0.9254	0.8027	0.6596	0.5185	0.4088	0.3584	0.3903	0.5319	0.8123
$\overline{\Psi_p}$	0.1526	0.1382	0.1620	0.1927	0.2340	0.2919	0.3721	0.4897	0.6631	0.8975

Fig. 6. Mean convexity values of the proposed convexity measure ($\overline{\Psi_p}$) and that of [2] ($\overline{\Psi_{hv}}$) on images with noise only. Each value is calculated as the mean of 100 random images having uniform noise of the given percent. Example images are shown in the top row.

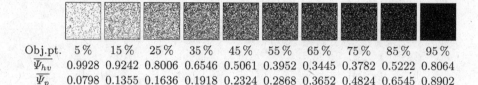

Obj.pt.	5 %	15 %	25 %	35 %	45 %	55 %	65 %	75 %	85 %	95 %
$\overline{\Psi_{hv}}$	0.9928	0.9242	0.8006	0.6546	0.5061	0.3952	0.3445	0.3782	0.5222	0.8064
$\overline{\Psi_p}$	0.0798	0.1355	0.1636	0.1918	0.2324	0.2868	0.3652	0.4824	0.6545	0.8902

Fig. 7. Mean convexity values of the improved convexity measure $(\overline{\Psi_p})$ and that of [2] $(\overline{\Psi_{hv}})$ using an image of a thin square gradually filled with object points in random positions. Each value is calculated as the mean of 100 random images having uniform noise of the given percent. Example images are shown in the top row.

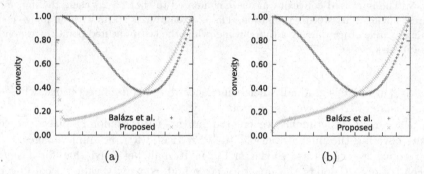

(a)   (b)

Fig. 8. (a) Convexity of images having randomly placed object points from 0–100%. If object points fall into distinct rows and columns, horizontal and vertical convexity is not violated. (b) Having a 1 pixel thin bounding square around the images, the values of the proposed measure become gradually increasing.

6   Conclusion

In this paper we proposed an improved measure of [2] for one-directional convexity of binary images. This measure gives more appropriate values for thin objects that are close to being fully concave. By experimental results we showed that the proposed measure behaves more intuitively than the one of [2]. Thus, using it as an image feature, it may provide more precise recognition and classification results than the former measure. We emphasize that the results of this paper are general, and can be useful in numerous fields of discrete geometry and digital image processing.

References

1. Balázs, P., Brunetti, S.: A measure of Q-convexity. In: Normand, N., Guédon, J., Autrusseau, F. (eds.) DGCI 2016. LNCS, vol. 9647, pp. 219–230. Springer, Cham (2016). doi:10.1007/978-3-319-32360-2_17
2. Balázs, P., Ozsvár, Z., Tasi, T., Nyúl, L.: A measure of directional convexity inspired by binary tomography. Fundam. Informaticae **141**(2–3), 151–167 (2015)

3. Barcucci, E., Del Lungo, A., Nivat, M., Pinzani, R.: Medians of polyominoes: a property for reconstruction. Int. J. Imaging Syst. Technol. 9(2–3), 69–77 (1998)
4. Boxer, L.: Computing deviations from convexity in polygons. Pattern Recognit. Lett. 14(3), 163–167 (1993)
5. Brunetti, S., Lungo, A.D., Ristoro, F.D., Kuba, A., Nivat, M.: Reconstruction of 4- and 8-connected convex discrete sets from row and column projections. Linear Algebra Appl. 339(1), 37–57 (2001)
6. Chrobak, M., Durr, C.: Reconstructing hv-convex polyominoes from orthogonal projections. Inf. Process. Lett. 69(6), 283–289 (1999)
7. Gorelick, L., Veksler, O., Boykov, Y., Nieuwenhuis, C.: Convexity shape prior for segmentation. In: Fleet, D., Pajdla, T., Schiele, B., Tuytelaars, T. (eds.) ECCV 2014. LNCS, vol. 8693, pp. 675–690. Springer, Cham (2014). doi:10.1007/978-3-319-10602-1_44
8. Gorelick, L., Veksler, O., Boykov, Y., Nieuwenhuis, C.: Convexity shape prior for binary segmentation. IEEE Trans. Pattern Anal. Mach. Intell. 39, 258–271 (2017)
9. Herman, G.T., Kuba, A.: Advances in Discrete Tomography and Its Applications. Applied and Numerical Harmonic Analysis. Birkhauser, Baserl (2007)
10. Latecki, L.J., Lakamper, R.: Convexity rule for shape decomposition based on discrete contour evolution. Comput. Vis. Image Underst. 73(3), 441–454 (1999)
11. Rahtu, E., Salo, M., Heikkila, J.: A new convexity measure based on a probabilistic interpretation of images. IEEE Trans. Pattern Anal. Mach. Intell. 28(9), 1501–1512 (2006)
12. Rosin, P.L., Žunić, J.: Probabilistic convexity measure. IET Image Process. 1(2), 182–188 (2007)
13. Sonka, M., Hlavac, V., Boyle, R.: Image Processing, Analysis, and Machine Vision. Cengage Learning, Boston (2014)
14. Stern, H.I.: Polygonal entropy: a convexity measure. Pattern Recognit. Lett. 10(4), 229–235 (1989)
15. Tasi, T.S., Nyúl, L.G., Balázs, P.: Directional convexity measure for binary tomography. In: Ruiz-Shulcloper, J., Sanniti di Baja, G. (eds.) CIARP 2013. LNCS, vol. 8259, pp. 9–16. Springer, Heidelberg (2013). doi:10.1007/978-3-642-41827-3_2
16. Zunic, J., Rosin, P.L.: A new convexity measure for polygons. IEEE Trans. Pattern Anal. Mach. Intell. 26(7), 923–934 (2004)

Learning Salient Structures for the Analysis of Symmetric Patterns

Jaime Lomeli-R.[(⊠)] and Mark S. Nixon

University of Southampton, Southampton, UK
jlr2g12@soton.ac.uk

Abstract. Feature-based symmetry detection algorithms have become popular amongst researchers due to their dominance in performance, nevertheless, these approaches are computationally demanding. Also they are reliant on the presence of matched features, therefore they benefit from the abundance of detected keypoints; this implies that a trade-off between performance and computation time must be found. In this paper both issues are addressed, the detection of large sets of keypoints and the computation time for feature-based symmetry detection algorithms. We present an innovative process to learn rotation-invariant salient structures by clustering self-similarities. Keypoints are detected as local maxima in feature-maps computed using the learnt structures. Keypoints are described using BRISK. We consider an axis of symmetry to be a dense cloud of points in a parameter-space, a density-based clustering algorithm is used to find such clouds. Computing times are drastically shortened taking an average of 0.619 s to process an image. Detection results for single and multiple, straight and curved, reflection and glide-reflection symmetries are similar to the current state of the art.

Keywords: Feature detection · Moments · Matching · Symmetry

1 Introduction

Symmetry is a pervasive cue in both natural and artificial structures. The automated detection of symmetry finds applications in areas such as interest-point detection [7,20], image segmentation [8,9], saliency models [18,21] and medical image analysis [5,6] amongst others.

In 2D there are four recognisable *primitive symmetries* namely, reflection, rotation, translation and glide-reflection [19]. Computer vision tackles the detection of these symmetries in a variety of ways including correlation-based [2], feature-based [3,16] and machine-learning aided methods [4]. In [10], results of reflection, rotation and translation symmetry detection algorithms are shown. Reflection symmetry detection in the literature concentrates predominantly in straight and skewed reflections, few aim at the detection of curved axes; arguably the feature-based approach is the most popular due to its dominance in performance. Yet, the study of the glide-reflection group has not received much attention despite its abundant existence in real-world scenarios.

© Springer International Publishing AG 2017
F. Karray et al. (Eds.): ICIAR 2017, LNCS 10317, pp. 286–295, 2017.
DOI: 10.1007/978-3-319-59876-5_32

Fig. 1. Reflection and glide-reflection example results of the presented method. Yellow indicates the detected axes. Our method finds dense sets of keypoints even in low-textured images (image on the right). (Color figure online)

Feature-based approaches begin by extracting keypoints in the image. These points are then described, mirrored and compared to find local similarities, each successful matching pair of keypoints votes for an axis of symmetry. This voting process benefits from large amounts of matches thus, large sets of detected keypoints are crucial. Lee and Liu highlight this problem, their method detects keypoints in several transformed versions of the image to generate a denser sampling [12]. Manual selection of image transformations requires the implementation of different operators and deep knowledge of various fields.

We address this problem with a novel keypoint detection method that clusters self-similarities. The central idea is to let the computer learn repetitive visual salient structures in a rotation-invariant space. The learning scheme can be depicted as a three step process; randomly extracting image patches from unlabelled data, normalisation, and the use of an unsupervised learning algorithm to learn *keypoint models*. Each keypoint model is used to create a feature-map in which non-maxima suppression can be used to detect keypoints. The detected sets of keypoints are plentiful even in low textured images (Fig. 1).

Extracted keypoints are described with a modified BRISK [13] pattern that can be efficiently mirrored with a look-up table. Successful keypoint matches are mapped into a parameter-space, where axes of reflection and glide-reflection symmetry are recognised as dense clouds of points that are rapidly detected using a density-based clustering algorithm.

Larger sets of keypoints require more computation time, therefore a trade-off between performance and time must be found. The set of techniques we use (Fig. 2), dramatically reduce the computing time in each step of the symmetry detection process. Our approach takes an average of 0.619 s to detect one or many axes of symmetry in an image (6% of the time reported in [12], calculated over a set of 64 images).

We show our results on a meaningful set of images widely used in the symmetry detection community. This set contains real-world and synthetic images with single and multiple, curved and straight and; reflection and glide-reflection symmetries. Our results are comparable with the state-of-the-art, while maintaining small computation times.

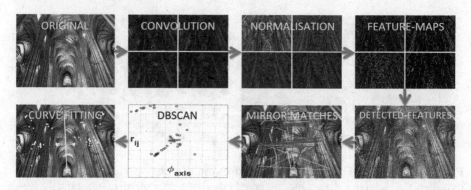

Fig. 2. The image is convolved with M Bessel-Fourier filters, structural-normalisation is computed on the complex modulus of the response of such convolution (as explained in Sect. 2). The probability of each pixel to belong to each of the pre-learnt keypoint models is calculated using Eq. 5, K sets of keypoints are extracted using non-maxima suppression in the resulting feature-maps. We compute and mirror the BRISK descriptors and compare all possible pairs of keypoints that belong to the same keypoint class (same k). Each successful match is transformed into a point in the parameter-space, where a density-based clustering algorithm is used to find dense point-clouds. Polynomial regression is used on each of the clusters found in the previous step to calculate the optimal axes of symmetry.

2   Learning Salient Structures

Keypoints are commonly recognised as local-maxima of a similarity measure on an image. For instance, convolving an image with a *Laplacian of Gaussian* kernel one can find blob-like structures; another example is the task of edge detection, in which a gradient operator is used. Let us analyse such filters in terms of the Fourier transform in cylindrical coordinates.

$$H_m(\rho) = \int_0^{2\pi} \int_0^{\infty} J_m(2\pi r\rho)e^{-im\theta}f(r,\theta)r\delta r\delta\theta, \tag{1}$$

$$F(\rho,\gamma) = \sum_{m=-\infty}^{\infty} (-i)^m H_m(\rho)e^{im\gamma}. \tag{2}$$

Equation 2 shows the Fourier transform of $f(r,\theta)$ in cylindrical coordinates. The function $H_m(\rho)$ is known as the *Hankel transform* of integer order m and, $J_m(2\pi r\rho)$ is the *Bessel function* of the first kind of integer order m (Eq. 1).

The family of functions $H_m(\rho)$ are the *Fourier coefficients* of the function $f(r,\theta)$ in the continuous domain of ρ. This family of functions is discretized by setting the variable $\rho = \lambda_{m,n}/2\pi$, where $\lambda_{m,n}$ are the roots of the Bessel function (corresponding to order m and scaling n). The integration range of r is reduced to obtain the set $B_{m,n}$ in Eq. 3.

$$B_{m,n} = \int_0^{2\pi} \int_0^1 J_m(\lambda_{m,n}r)e^{-im\theta}f(r,\theta)r\delta r\delta\theta. \tag{3}$$

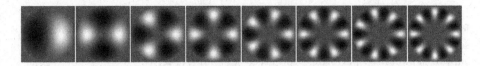

Fig. 3. The real part of the BF moment-generating functions for $n = 0$ and $m = 1, 2, \ldots, 8$.

$B_{m,n}$ is known as the (complex) *Bessel-Fourier* (BF) moment-generating function [14] (orthogonal in the range $0 \leq r \leq 1$). Blob detection operators are filters that have non-zero magnitude in $B_{0,n}$ and, commonly used gradient operators respond to the set $B_{1,n}$. The variable m represents the rotational frequency, and n represents the radially *expanding* waves around the origin (or a given point if the BF functions are used as filters).

We empirically set $n = 0$ and $m = 1, 2, \ldots, M$ where $M = 8$ (Fig. 3). An image $I(x, y)$ is convolved with such M BF-filters at a single scale with a kernel size $kernSize = 0.04 \times max(I.height, I.width)$, then we obtain the magnitude of the complex responses. There are M functions $I_m(x, y)$ thus, each pixel is represented by an M-dimensional vector. This vector is known as the *rotational power spectrum*, we treat it as function of the image space *i.e.* $RPS(x, y)$. The RPS of each pixel is normalised so the sum of all M dimensions sums up to one (Eq. 4, $|x|_1$ denotes the L^1-norm). We call this process *local structural normalisation* as it is independent from the location of the pixel.

$$RPS_{norm}(x_i, y_j) = \frac{RPS(x_i, y_j)}{|RPS(x_i, y_j)|_1}. \tag{4}$$

In practice, all elements of $RPS_{norm}(x_i, y_j)$ are set to 0 when a threshold is not met *i.e.* $\{|RPS(x_i, y_j)|_1 < RPS_{thresh}\}$, this is to avoid detecting features in regions of low energy or prevent division by 0 in its case. Low values of RPS_{thresh} highlight weaker responses, while large values discard detections with low contrast. Any visual cue can be recognised as a keypoint. After the local structural normalisation process, each pixel is a point $RPS_{norm}(x_i, y_j)$ in an M-dimensional simplex and pixels that are visually similar are close to each other in this space.

The probability density function (PDF) of such space describes repeatability, clustering algorithms find local-maxima on the PDF of a discrete random variable. To find repeatable structures, we randomly build set of 10^5 RPS_{norm}'s from various images and run K-Means clustering to find K centres (we set $K = 10$ for the results shown here). During the training process we replace the sample $RPS_{norm}(x_i, y_j)$ for a different one if the condition $\{|RPS(x_i, y_j)|_1 < RPS_{thresh}\}$ is met. The clustering process can be performed off-line, the location of the pixels and the image they came from is not accounted for. The found centres are known as *keypoint models* (denoted as C_k). Centres lie in regions of high-density, this means that visual self-similarities get clustered together. The training process is rapid, usually converging after only eight iterations.

We calculate $RPS_{norm}(x, y)$, we then use Eq. 5 to calculate the probability for each pixel to belong to each of the pre-learnt keypoint models. This creates a set of K rotation-invariant feature-maps $0 \leq M_k(x, y) \leq 1$ in which non-maxima suppression can be performed to detect keypoints (Fig. 4).

$$M_k(x_i, y_j) = \frac{|RPS_{norm}(x_i, y_j) - C_k|_2^{-4}}{\sum_{k'=1}^{K} |RPS_{norm}(x_i, y_j) - C_{k'}|_2^{-4}}. \tag{5}$$

The operator $|\boldsymbol{x}|_2$ denotes the L^2-norm. Keypoints will be denoted henceforth as P_i^k where i is the index of the keypoint and k is the feature-map in which it was found. Keypoints found on the same feature-map are said to be of the same class.

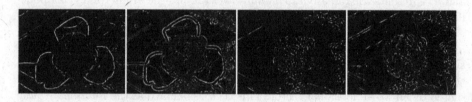

Fig. 4. Six (of the ten) feature-maps created from the image on the left in Fig. 1. Learnt keypoint models represent different structures, for instance the leftmost feature-map has a good response to step-edges and the two rightmost feature-maps respond to highly textured regions on the image.

3   Matching Mirrored Keypoints

We use a BRISK descriptor [13] to compare keypoints across the image and find local similarities, this descriptor is much faster to calculate than SIFT or SURF [1]. BRISK samples the image around a keypoint with a circular pattern. Samples are compared to each other, if a sample a is brighter than sample b the comparison result is a logic 1 or 0 otherwise. The orientation of the keypoint is calculated using the shorter comparisons, the longer ones form the descriptor.

We use a descriptor of 512 comparisons $\{d_p \mid p = 0, 1, \dots, 511\}$ which can be stored in char arrays of size $512/8 = 64$ bytes. The traditional BRISK pattern is slightly modified, we make the pattern symmetric across the orientation of the keypoint and index it in such a way that d_p is the mirror pairing across the keypoint orientation of d_{511-p}. In this manner, we can bit-swap each byte with a look-up table and then swap the order of the array.

The Hamming distance between two binary descriptors can be calculated with an XOR operation and a bit-count. We calculate the distance between all possible pairs of features of the same class, and use only the closest match. A successful matching pair $P_{ij}^k = (P_i^k, P_j^k)$ exists if the Hamming distance between the descriptors of P_i^k and P_j^k is less than a threshold, we empirically choose $BRISK_{thresh} = 120$.

4   Parameter-Space Analysis

A parameter-space is created from successful matches. We use a two-dimensional parameter-space with coordinates (ϕ_{axis}, r_{ij}) calculated with Eqs. 6, 7 and 8 (Fig. 5). In this 2D parameter-space, pure reflection symmetry axes are single points whilst curved reflection axes are smooth contours. As we use a single scale, a keypoint is represented by a location and an orientation *i.e.* $P_i^k = (x_i, y_i, \phi_i)$.

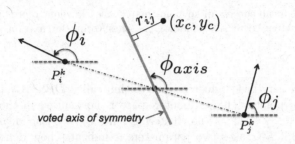

Fig. 5. Every matched pair of features $P_{ij}^k = (P_i^k, P_j^k)$, is mapped to a point in the parameter-space with coordinates (ϕ_{axis}, r_{ij}).

The angle in Eq. 6 is defined in the range $\{-\pi \leq \phi'_{axis} < \pi\}$. As we only care for the angle of the axis and disregard the direction in which it points, we use Eq. 7 to re-bound this parameter in $\{0 \leq \phi_{axis} < \pi\}$. The second coordinate of the parameter-space is the perpendicular distance between the axis and the centre of the image with coordinates (x_c, y_c).

$$\phi'_{axis} = \frac{\phi_i + \phi_j}{2}. \tag{6}$$

$$\phi_{axis} = \begin{cases} \phi'_{axis} & \text{if } 0 \leq \phi'_{axis} \\ \phi'_{axis} + \pi & \text{if } \phi'_{axis} < 0 \end{cases}. \tag{7}$$

$$r_{ij} = \left(\frac{x_i + x_j}{2} - x_c\right) sin(\phi_{axis}) - \left(\frac{y_i + y_j}{2} - y_c\right) cos(\phi_{axis}). \tag{8}$$

In other feature-based symmetry detection approaches, each sample casts a vote in an accumulator matrix. This accumulator matrix is a discrete approximation of a density map of the parameter-space where points of local-maxima represent axes of symmetry. To be able to detect such points, the accumulator matrix has to be smoothed and then a non-maxima suppression algorithm must be used. The set of successful matches that generated a point of high-density represent an axis of symmetry, this procedure is appropriate for straight axes as these are single points in the parameter-space. On the other hand, curved axes of symmetry are contours in the parameter-space; therefore the single-point detection approach is no longer suitable. We consider an axis of symmetry to be a dense cloud of points in the parameter-space.

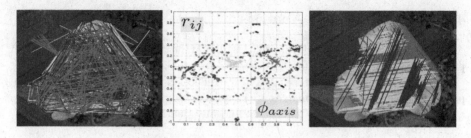

Fig. 6. On the left, all successful keypoint matches; each colour represents a different keypoint class. After DBSCAN (middle), matches voting for axes of symmetry are easily identified (right).

We use a density-based clustering algorithm called *DBSCAN* [15] to detect dense clouds of points. The parameter-space is normalised in the range $\{0 \leq \phi_{axis} \leq 1, -1 \leq r_{ij} \leq 1\}$ as the clustering relies on the Euclidean distance of the points. DBSCAN takes two parameters, ϵ indicates how densely clustered are the points and $MinPts$ indicates the minimum number of density-reachable points for a cluster to exist. We set these parameters empirically as $\epsilon = 0.2^2$ and $MinPts = 50$, although these values may change depending on the application.

DBSCAN does not need to know the number of clusters to be found, this property is useful as the number of axes is usually unknown. It also discards outliers, allowing larger values of $BRISK_{thresh}$ to increase the number of matches; this improves the sampling of the parameter space even further. The middle image in Fig. 6, shows clustering results for the multiple axes detection task on the original image in Fig. 1. The other two images in Fig. 6, show the matched features before and after clustering.

5   Optimal Axes

Each of the L detected clusters is a subset AX_l of the set of all successful matches. As in [12], to find each optimal axis of symmetry we perform twenty polynomial regressions over the middle points of the matching pairs $P_{ij}^k \in AX_l$

Table 1. Results comparing our approach to those showed by Lee and Liu in [12].

Method	Symmetry true positive rate (# False positives)				Timings
	Straight	Straight glide	Curved	Curved glide	Mean(std) secs
Proposed (SIFT)	81%(6)	40%(7)	62%(5)	40%(6)	0.332(0.06)
Proposed	91%(4)	80%(5)	78%(4)	60%(6)	0.619(0.18)
Lee [12]	86%(3)	80%(3)	83%(3)	60%(4)	9.7(10.3)
Loy [11]	91%(9)	7%(46)	0%(38)	0%(26)	6.1(9.4)

Fig. 7. Results of our approach (bottom), compared against ground-truth marked images (top).

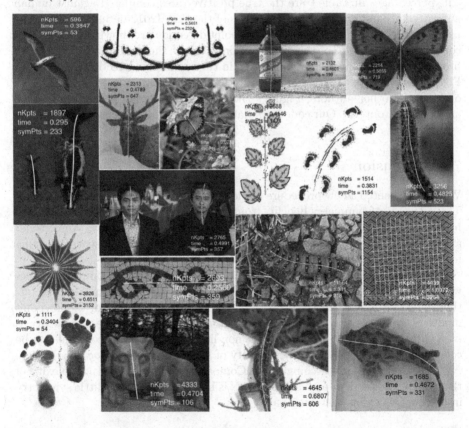

Fig. 8. Single and multiple axes detected on images containing one or more of the reflection, curved reflection and glide reflection symmetry groups. Each image shows the total number of keypoints detected (*nKpts*), the runtime of the algorithm (*time*) and the number of matches supporting the axes of symmetry (*symPts*). For wallpaper images DBSCAN uses *minPts* = 15, the default parameters are used in all other cases. Images were obtained from [2,10–12,17].

(calculated with Eq. 9). Each polynomial regression is performed with degree $d = 1, 2, 3, 4, 5$ and rotation $r = 0°, 45°, 90°, 135°$. To calculate the regression in different angles we rotate the middle points using the rotation matrix in Eq. 10. Out of the twenty possible axes, we consider as optimal the one with the smallest root-mean-square error.

$$\begin{bmatrix} P_{ij,x}^k \\ P_{ij,y}^k \end{bmatrix} = \begin{bmatrix} \frac{x_i + x_j}{2} \\ \frac{y_i + y_j}{2} \end{bmatrix}. \tag{9}$$

$$\begin{bmatrix} P_{ij,x,r}^k \\ P_{ij,y,r}^k \end{bmatrix} = \begin{bmatrix} cos(r) & -sin(r) \\ sin(r) & cos(r) \end{bmatrix} \begin{bmatrix} P_{ij,x}^k \\ P_{ij,y}^k \end{bmatrix}. \tag{10}$$

6 Results

The percentages in Table 1 are the true positive rates compared against human annotated ground truth on the 64 images of the PAMI2012 dataset [12]. Results using both SIFT and our method for keypoint detection are shown. Our approach compares to the state-of-the-art using only 6% of the computation time. We provide result in images commonly used in the symmetry detection literature in Fig. 8, this is to give the reader a means to visually compare our results with those obtained by other authors. Figure 7 shows our results alongside ground truth annotations from [10]. Our code is available at https://github.com/jimmylomro/symmetry.

7 Conclusion

We have presented a new procedure for detecting symmetric structures in images. The detection of symmetry benefits from the plentiful sets of keypoints provided by our method, as the parameter-space is more densely sampled. We consider axes of symmetry as dense clouds of points in the parameter space, allowing more flexibility when detecting curved axes.

It was also shown that it is possible to drastically shorten computing times while maintaining similar detection results. The computation times can be further improved. For example, Eq. 5 is an evaluation of the PDF of the keypoint model C_k generated by all other keypoint models and its time complexity is $O(K^2)$; it is possible to use a different non-linear distance function that can be calculated in $O(K)$. The time complexity of DBSCAN is $O(N_m log(N_m))$ where N_m is the number of matching pairs. Compared to the constant complexity of the Hough transform approach, DBSCAN is better when there is little symmetry in the image *i.e.* few matching pairs.

References

1. Heinly, J., Dunn, E., Frahm, J.-M.: Comparative evaluation of binary features. In: Fitzgibbon, A., Lazebnik, S., Perona, P., Sato, Y., Schmid, C. (eds.) ECCV 2012. LNCS, pp. 759–773. Springer, Heidelberg (2012). doi:10.1007/978-3-642-33709-3_54

2. Lee, S., Collins, R.T., Liu, Y.: Rotation symmetry group detection via frequency analysis of frieze-expansions. In: CVPR (2008)
3. Wang, Z., Tang, Z., Zhang, X.: Reflection symmetry detection using locally affine invariant edge correspondence. IEEE Trans. Image Process. **24**, 1297–1301 (2015)
4. Brachmann, A., Redies, C.: Using convolutional neural network filters to measure left-right mirror symmetry in images. Symmetry **8**, 144 (2016)
5. Yu, C.P., Ruppert, G.C.S., Nguyen, D.T.D., Falcao, A.X., Liu, Y.: Statistical asymmetry-based brain tumor segmentation from 3D MR images. In: Biosignals (2012)
6. Mancas, M., Gosselin, B., Macq, B., et al.: Fast and automatic tumoral area localisation using symmetry. In: ICASSP (2005)
7. Salti, S., Lanza, A., Di Stefano, L.: Keypoints from symmetries by wave propagation. In: CVPR (2013)
8. Levinshtein, A., Dickinson, S.J., Sminchisescu, C.: Multiscale symmetric part detection and grouping. In: ICCV (2009)
9. Teo, C.L., Fermuller, C., Aloimonos, Y.: Detection and segmentation of 2D curved reflection symmetric structures. In: ICCV (2015)
10. Liu, J., Slota, G., Zheng, G., Wu, Z., Park, M., Lee, S., Rauschert, I., Liu, Y.: Symmetry detection from real-world images competition 2013: summary and results. In: CVPR Workshops (2013)
11. Loy, G., Eklundh, J.-O.: Detecting symmetry and symmetric constellations of features. In: Leonardis, A., Bischof, H., Pinz, A. (eds.) ECCV 2006. LNCS, vol. 3952, pp. 508–521. Springer, Heidelberg (2006). doi:10.1007/11744047_39
12. Lee, S., Liu, Y.: Curved glide-reflection symmetry detection. In: TPAMI (2012)
13. Leutenegger, S., Chli, M., Siegwart, R.Y.: BRISK: binary robust invariant scalable keypoints. In: ICCV (2011)
14. Xiao, B., Ma, J.F., Wang, X.: Image analysis by Bessel-Fourier moments. Pattern Recogn. **43**, 2620–2629 (2010)
15. Ester, M., Kriegel, H.P., Sander, J., Xu, X., et al.: A density-based algorithm for discovering clusters in large spatial databases with noise. In: KDD (1996)
16. Patraucean, V., Grompone von Gioi, R., Ovsjanikov, M.: Detection of mirror-symmetric image patches. In: CVPR Workshops (2013)
17. Kovesi, P., et al.: Symmetry and asymmetry from local phase. In: Tenth Australian Joint Conference on Artificial Intelligence (1997)
18. Kootstra, G., Nederveen, A., De Boer, B.: Paying attention to symmetry. In: BMVC (2008)
19. Liu, Y., Hel-Or, H., Kaplan, C.S.: Computational Symmetry in Computer Vision and Computer Graphics. Now Publishers Inc., Breda (2010)
20. Hauagge, D.C., Snavely, N.: Image matching using local symmetry features. In: CVPR (2012)
21. Kootstra, G., Schomaker, L.R.B.: Prediction of human eye fixations using symmetry. In: CogSci (2009)

Triplet Networks Feature Masking
for Sketch-Based Image Retrieval

Omar Seddati[✉], Stéphane Dupont, and Saïd Mahmoudi

Computer Science - TCTS Lab, Université de Mons, Mons, Belgium
{omar.seddati,stephane.dupont,said.mahmoudi}@umons.ac.be

Abstract. Freehand sketches are an intuitive tool for communication and suitable for various applications. In this paper, we present an effective approach that combines triplet networks and an attention mechanism for sketch-based image retrieval (SBIR). The study conducted in this work is based on features extracted using deep convolutional neural networks (ConvNets). In order to overcome the SBIR cross-domain challenge (i.e. searching for photographs from sketch queries), we use triplet loss to train ConvNets to compute shared embedding for both sketches and images. Our main novel contribution is to combine such triplet networks with an attention mechanism. Our approach outperform previous state-of-the-art on challenging SBIR benchmarks. We achieved a recall of 41.66% (at $k = 1$) for the sketchy database (more than 4% improvement), a Kendal score of $42.9\mathcal{T}_b$ on the TU-Berlin SBIR benchmark (close to $5.5\mathcal{T}_b$ improvement) and a mean average precision (MAP) of 31% on Flickr15k (a category level SBIR benchmark).

Keywords: Sketch-based image retrieval · Triplet networks · Feature extraction

1 Introduction

During the last decades, the quantity of available digital media grew exponentially, which raised the interest in the field of large scale multimedia databases retrieval. Despite the significant progress achieved in the fields of text-based image retrieval and content-based image retrieval (CBIR, where we retrieve images similar to a query image), SBIR is still a promising approach. The goal of SBIR is to enable the use of sketches as query to retrieve similar images. A good SBIR system is able to give results that match the query semantics as well as details. SBIR indeed provides a great deal of flexibility when looking for results that go beyond semantic similarities. This is especially true when one lacks image examples to use as a query in CBIR systems and when the database lacks detailed text descriptions of the contained images.

In this work, we use ConvNets (state-of-the-art in the field of sketch recognition [5,8]) for SBIR. When building a SBIR system, we face several challenges related to cross-domain representations (sketches versus natural images).

© Springer International Publishing AG 2017
F. Karray et al. (Eds.): ICIAR 2017, LNCS 10317, pp. 296–303, 2017.
DOI: 10.1007/978-3-319-59876-5_33

In order to deal with this, we combine triplet networks representations and an attention mechanism. We evaluate the proposed approach on three different benchmarks and show that our approach outperforms previous state-of-the-art.

The paper is organized as follows. We start with a brief review of different approaches used in the field of SBIR. Next we present the main idea of our approach. Then we proceed to the evaluation of our sketch based image retrieval system.

2   Related Work

In the last two decades, many sketch based image retrieval methods have been proposed. One of the main challenges in this area is to find a common feature representation that enables to match sketches and images. In order to deal with this challenges, Eitz et al. [2] proposed new descriptors based on bag-of-features. Niu et al. [20] generate visual words for images and sketches and use them to determine the relationship between two visual vocabularies. Li et al. [18] use local descriptors with product quantization and sparse coding. The similarity between the query and test data is achieved by comparing features histograms. Bozas et al. [14] use a hashing algorithm for patch similarity within an overlapping spatial grid to rank sketch-image similarities. Bhattacharjee et al. [15] combine shape and appearance in a graph based re-ranking scheme. Many SBIR approaches use image pre-processing (e.g. Canny edge detection) to generate edgemaps from natural images and match them against the sketch query. Szanto et al. [10] used edge histogram descriptors (EHD) and the histogram of oriented gradients (HOG). Deore et al. [17] use edge detection to generate sketch-like from images and histogram of line relationship to reduce noise. Wang et al. [11] used edge map descriptors and wavelet transform features to match sketches and images. Jin et al. [12] presented an approach based on a visual region descriptor where sketch-like representations are generated from images by using edge map generation and edge pixel screening. Then, the sketch-like representation is used for sketch-image matching. In [13] Seddati et al. proposed a system for real-time sketch recognition and SBIR using ConvNets features and edge detection. In a few recent works, deep neural networks have been used to find an embedding feature space that eases sketch-image matching. In [19] Qi et al. use a siamese ConvNet for SBIR. Sangkloy et al. [6] evaluated several approaches to SBIR taking into account the cross-domain challenge. In their experiences they end up with a multi-task triplet network that gives their best results. Bui et al. [16] explore triplet networks with several strategies for weight sharing.

In order to evaluate SBIR systems, several benchmarks are available for the purpose of comparisons. Hu et al. [3] introduced a category-level benchmark for SBIR. Eitz et al. [2] introduced a benchmark for SBIR fine-grained performance evaluation. Sangkloy et al. [6] introduced a large scale benchmark for SBIR. They asked crowd workers to sketch particular photographic objects for 125 categories. The database contains 75,471 sketches of 12,500 objects.

3   Methodology

In this section, we present the different steps followed in this work in order to improve over earlier SBIR system proposed in [6,16].

Sketch and Image Classification: In order to build our feature extractor, unlike the approach proposed in [6], we train a **single multi-modal ConvNet** for both sketch and image object recognition. During the training of our ConvNet (with a resnet-18 architecture [7]), we follow a standard ConvNet training approach, where crops with a resolution of 224×224 pixels are presented as input to the ConvNet (from an original image resolution of 256×256). The first eight layers of the resulting ConvNet are used as feature extractor for all of our triplet networks.

Triplet Network for SBIR: In order to improve the features extracted with our ConvNet for the SBIR application, we first train a triplet network. In this kind of architecture, we have three copies of the same network with shared parameters. The input is in the form of anchor, positive and negative samples. The goal is minimizing the distance between an anchor and a positive while maximizing the distance between this anchor and a negative. In our case, the anchor is a sketch (for instance a sitting dog). The positive sample is an image of the sketched object in the same pose (sitting dog), while the negative sample is an image for an object from the same category but in a different pose (standing dog). We add an embedding layer with 128 neurons on the top of our feature extractor and we normalize the outputs (l_2-norm). For each training triplet (anchor, positive, negative), the distances of the positive and negative sample features to the anchor features can be computed and used to compute a triplet loss that can be used for gradient descent training using back-propagation.

The authors of [6,16] found that using multi-task learning improves triplet network features for SBIR. To do so, an auxiliary object category classification task is used. A **linear** classification branch is added to our triplet networks architecture. The multi-task triplet network can hence learn to extract features that can be good for both object classification and similarity search. The auxiliary task enhance the embedding layer features for semantic retrieval.

Combining Triplet Networks with Attention Mechanism for SBIR: Two alternative approaches for including an attention mechanism within the SBIR ConvNet feature extractor are proposed. The first approach acts directly on a high-level representation and does not take into account the spatial configuration, while the second approach acts in a more localized manner. Figure 1 summarizes the first approach proposed in this work. The main idea is to use the outputs of our feature extractor to train a model (attention mechanism) that generates a binary mask. These features are also passed to an embedding layer that generates a first embedding, the product element-by-element of this embedding and the generated mask provides the final embedding vector. This "feature" hard attention mechanism can learn to generate masks depending on the category of the input and enhance the embedding representation

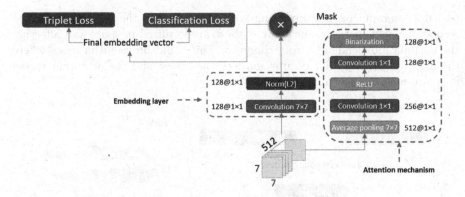

Fig. 1. Our first approach combining triplet networks with an attention mechanism.

category-level performance. (*Note:* in our experiment, we use two independent attention models, one for sketches and the other for images).

In order to understand the novel use of an attention mechanism for SBIR, we briefly introduce its components. First, we have the binarization layer, a cascade of simple operations (sign, addition, division Fig. 2) to produce binary outputs (0 or 1). Second, a repeatTensor module that transforms one 2D feature map (an input feature map with ony one channel $1 \times W \times H$ where W and H are the respectively the width and height) into C stacked feature maps (where C is the number of channels) by stacking C copies of the input. Figure 2 shows the different modules used to build the proposed attention mechanism model.

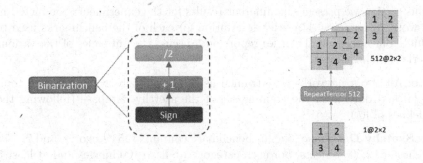

Fig. 2. Two different modules used to build our attention mechanism model.

Figure 3 shows an alternate approach combining triplet networks with an attention mechanism. In this second approach, the attention model generates a mask formed by a set of "tubes" (a tube is a 3D matrix $C \times 1 \times 1$ where C is the number of channels and all the values of this matrix are equal) that covers all the area of the feature maps (the output of the feature extractor). The product element-by-element of this mask with the outputs of our feature extractor is feeded as an input to the embedding layer. Compared to the first approach, the second one acts spatially, since our "tubes" are localized in space and affecting the outputs independently from channels. This could be useful in cases where the main object does not occupy most of the image and the attention model learns to focus on the current object area.

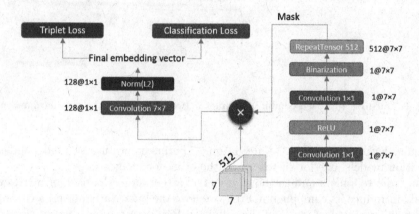

Fig. 3. Our second approach combining triplet networks with an attention mechanism.

4   Experimental Results

In this section we present the different results for the experiments conducted in this work. We start with a brief description for each of the benchmarks used to evaluate our approach. Then we report and discuss the outcomes of the various experiences carried out.

Note: All of our networks were trained using the publicly available Torch toolbox [4] and data from the training-set of the sketchy database (following the guidelines of [6]).

The Sketchy Database [6]: this benchmark contains 125 categories and 75,471 sketches of 12,500 objects. For each category we have 100 images and at least 5 sketches associated to each image. In Fig. 4 we have the average recall at $K = 1$ (K: the number of nearest neighbors that we retrieve) for our different approaches (Triplet: triplet network; Triplet_MT: multi-task triplet network; Emb-Mask: the mask is applied to the outputs of the embedding layer; Tubes_Mask: the mask is applied to spatial feature maps). We can see that training a model that generates masks for the outputs of the embedding layer improves multi-task triplet networks performance for SBIR.

TU-Berlin SBIR Benchmark [1]: this benchmark contains 31 sketches and 40 ranked images for each sketch (1,240 benchmark images). This benchmark

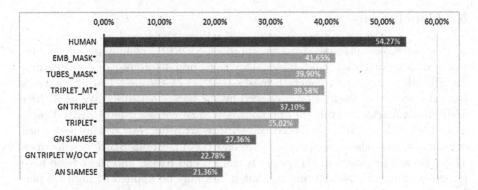

Fig. 4. Comparison of average recall at $K = 1$ on the sketchy database with the results reported in [6] (*: our approaches).

Table 1. SBIR results comparison: Kendal's correlation on TU-Berlin SBIR benchmark (left) and MAP comparison on Flickr15k (right).

Method	T_b
Emb-Mask	42.9
Triplet (final model) [16]	37.4
Triplet (sketch-photo) [16]	33.3
SHoG [1]	27.7
Triplet (edge) [16]	22.3
HoG (global) [1]	22.3
Structure Tensor [2]	22.3
Spark [1]	21.7
HoG (local) [22]	17.5
Shape Context [21]	16.1

Method	MAP (%)
Triplet (final model)* [16]	36.17
Triplet (no-share)* [16]	32.87
Triplet (half-share)* [16]	31.38
Emb-Mask	31
Triplet_MT	27.2
Triplet (edge,half-share)* [16]	24.45
Siamese CNN [19]	19.54
Perceptual Edge [23]	18.37
Extended GF-HoG [24]	18.20
GF-HoG [3]	12.22
SHoG [1]	10.93

contains 100,000 more images collected from Flickr and used as distractor. As suggested by the authors of [1], we use Kendal score for evaluation. As we can see in Table 1 (left) our approach achieves $42.9T_b$ and exceeds previous state-of-the-art with a large margin.

Flickr15k [3]: this is a category level SBIR benchmark, which means that if the retrieved image belongs to the same category as the query, we make the assumption that the result is correct. This benchmark has 33 sketch categories, for each category we have 10 sketches. It also contains 15,024 images with more than 60 categories and different number of images per category. We computed the mean average precision (MAP) as suggested by the authors of [3]. A comparison with previous works is reported in Table 1 (right), we can notice that our system is still showing good performance (even if the authors of [16] used two times more object categories during the training and knowing that this benchmark is category-level).

5   Conclusions

In this paper, we introduced a new approach for SBIR combining triplet networks with an attention mechanism. This new approach improves deep features for SBIR by learning to generate masks that select the best feature elements from the vector representation produced by the triplet networks. We used three SBIR benchmarks to evaluate our approach. The category level SBIR benchmark Flickr15k, where we achieved state-of-the-art result. We also outperformed previous state-of-the-art on the sketchy and the TU-Berlin SBIR benchmarks. The approach proposed in this paper is trained end-to-end and there is no need for extra annotation, which means that it can may be applied to improve performance in other domains where triplet networks are currently used. Another perspective would be to look into soft-attention mechanisms where feature vector element importance are weighted by the computed attention value.

References

1. Eitz, M., Hildebrand, K., Boubekeur, T., et al.: Sketch-based image retrieval: benchmark and bag-of-features descriptors. IEEE Trans. Vis. Comput. Graph. **17**(11), 1624–1636 (2011)
2. Eitz, M., Hildebrand, K., Boubekeur, T., et al.: An evaluation of descriptors for large-scale image retrieval from sketched feature lines. Comput. Graph. **34**(5), 482–498 (2010)
3. Hu, R., Collomosse, J.: A performance evaluation of gradient field hog descriptor for sketch based image retrieval. Comput. Vis. Image Underst. **117**(7), 790–806 (2013)
4. Collobert, R., Kavukcuoglu, K., Farabet, C.: A matlab-like environment for machine learning. In: BigLearn, NIPS Workshop (No.EPFL-CONF-192376)
5. Seddati, O., Dupont, S., Mahmoudi, S.: Deepsketch: deep convolutional neural networks for sketch recognition and similarity search. In: 2015 13th International Workshop on Content-Based Multimedia Indexing (CBMI), pp. 1–6. IEEE (2015)
6. Sangkloy, P., Burnell, N., Ham, C., et al.: The sketchy database: learning to retrieve badly drawn bunnies. ACM Trans. Graph. (TOG) **35**(4), 119 (2016)
7. He, K., Zhang, X., Ren, S., et al.: Deep residual learning for image recognition. arXiv preprint arXiv:1512.03385 (2015)
8. Seddati, O., Dupont, S., Mahmoudi, S.: DeepSketch 2: deep convolutional neural networks for partial sketch recognition. In: 2016 14th International Workshop on. IEEE Content-Based Multimedia Indexing (CBMI), pp. 1–6 (2016)
9. Schroff, F., Kalenichenko, D., Philbin, J.: Facenet: a unified embedding for face recognition and clustering. In: Proceedings of the IEEE Conference on Computer Vision and Pattern Recognition. pp. 815–823 (2015)
10. Sznt, B., Pozsegovics, P., Vmossy, Z., et al.: Sketch4matchContent-based image retrieval system using sketches. In: 2011 IEEE 9th International Symposium on Applied Machine Intelligence and Informatics (SAMI), pp. 183–188. IEEE (2011)
11. Wang, Y., Yu, M., Jia, Q., et al.: Query by sketch: an asymmetric sketch-vs-image retrieval system. In: 2011 4th International Congress on Image and Signal Processing (CISP), pp. 1368–1372. IEEE (2011)

12. Jin, C., Wang, Z., Zhang, T., et al.: A novel visual-region-descriptor-based approach to sketch-based image retrieval. In: Proceedings of the 5th ACM on International Conference on Multimedia Retrieval, pp. 267–274. ACM (2015)
13. Seddati, O., Dupont, S., Mahmoudi, S.: DeepSketch2Image: deep convolutional neural networks for partial sketch recognition and image retrieval. In: Proceedings of the 2016 ACM on Multimedia Conference, pp. 739–741. ACM (2016)
14. Bozas, K., Izquierdo, E.: Large scale sketch based image retrieval using patch hashing. In: Bebis, G., Boyle, R., Parvin, B., Koracin, D., Fowlkes, C., Wang, S., Choi, M.-H., Mantler, S., Schulze, J., Acevedo, D., Mueller, K., Papka, M. (eds.) ISVC 2012. LNCS, vol. 7431, pp. 210–219. Springer, Heidelberg (2012). doi:10. 1007/978-3-642-33179-4_21
15. Bhattacharjee, S.D., Yuan, J., Hong, W., et al.: Query adaptive instance search using object sketches. In: Proceedings of the 2016 ACM on Multimedia Conference, pp. 1306–1315. ACM (2016)
16. Bui, T., Ribeiro, L., Ponti, M., et al.: Generalisation and Sharing in Triplet Convnets for Sketch based Visual Search. arXiv preprint arXiv:1611.05301 (2016)
17. Deore, A., Gunjal, B.L.: Advanced sketch based image retrieval system using object boundary selection algorithm (2016)
18. Li, Q., Han, Y., Dang, J.: Sketch4Image: a novel framework for sketch-based image retrieval based on product quantization with coding residuals. Multimed. Tools Appl. **75**(5), 2419–2434 (2016)
19. Qi, Y., Song, Y-Z., Zhang, H., et al.: Sketch-based image retrieval via Siamese convolutional neural network. In: 2016 IEEE International Conference on Image Processing (ICIP), pp. 2460–2464. IEEE (2016)
20. Niu, J., Ma, J., Lu, J., Liu, X., Zhu, Z.: M-SBIR: an improved sketch-based image retrieval method using visual word mapping. In: Amsaleg, L., Guðmundsson, G., Gurrin, C., Jónsson, B., Satoh, S. (eds.) MMM 2017. LNCS, vol. 10133. Springer, Cham (2017). doi:10.1007/978-3-319-51814-5_22
21. Belongie, S., Malik, J., Puzicha, J.: Shape matching and object recognition using shape contexts. IEEE Trans. Pattern Anal. Mach. Intell. **24**(4), 509522 (2002)
22. Dalal, N.,Triggs, B.: Histograms of oriented gradients for human detection. In: Proceedings of the CVPR, vol. 1, pp. 886–893 (2005)
23. Qi, Y., Song, Y.-Z., Xiang, T., Zhang, H., Hospedales, T., Li, Y., Guo, J.: Making better use of edges via perceptual grouping. In: Proceedings of the CVPR (2015)
24. Bui, T., Collomosse, J.: Scalable sketch-based image retrievalusing color gradient features. In: Proceedings of the IEEE International Conference on Computer Vision Workshops, p. 18 (2015)

Are You Smiling as a Celebrity? Latent Smile and Gender Recognition

M. Dahmane$^{(\boxtimes)}$, S. Foucher, and D. Byrns

Computer Research Institute of Montreal,
405, avenue Ogilvy, bureau 101, Montreal, QC H3N 1M3, Canada
dahmanmo@crim.ca
http://crim.ca

Abstract. Person gender detection is an important feature in many vision-based research fields including surveillance, human computer interaction, Biometrics, stratified behavior understanding, and content-based indexing. Researchers are still facing big challenges to establish automated systems to recognize gender from images where human face represents the most important source of information. In the present study, we elaborated and validated a methodology for gender perception by transfer learning. First, the face is located and the corresponding cropped image is fed to a pre-trained convolutional neural network, the generated deep "latent" features are used to train a linear-SVM classifier. The overall classification performance reached 90.69% on the FotW validation set and 91.52% on the private test set.

In this paper, we investigated also whether these features can deliver a smile recognizer. A similar trained architecture for classification of smiling and non-smiling faces gave a rate of 88.14% on the validation set and 82.12% on the private test set.

Keywords: Face analysis · Smile detection · Gender recognition · Deep learning

1 Introduction

Computerized facial analysis from images presents an important challenge arising mainly from personal appearance (identity) and facial dynamics under different head orientations. Environment changes, lighting conditions are also an additional source of image processing difficulties.

Human can easily differentiate between a male and female. However, it is a challenging task for computer vision. Authors in [9] have surveyed the methods used for human gender recognition in images and videos using computer vision techniques. Most of the reported approaches were evaluated in laboratory conditions. Gender classification have not been as well studied compared to individual recognition or facial expression recognition. Demirkus *et al.* [4] used a Markovian temporal model for gender classification from unconstrained video sequences. In [2] the authors tested both feature level and score level fusion in a multi-expert

© Springer International Publishing AG 2017
F. Karray et al. (Eds.): ICIAR 2017, LNCS 10317, pp. 304–311, 2017.
DOI: 10.1007/978-3-319-59876-5_34

architecture for gender classification. Nian *et al.* [10] proposed a convolutional neural network (CNN)-face representation to handle unconstrained face images. A CNN-based ensemble model was also designed for gender recognition from face images [1].

Authors in [14] found a troublesome aspect in automatic facial analysis as there is a gap between the levels of performance on the standard datasets and the performances in real life conditions. The Chalearn-2016 contest was one of the several challenges that was organized to help bridging this gap.

1.1   The ChaLearn-2016 Challenge Context

Facial expression analysis and gender perception are becoming a growing research area due to the constantly increasing interest in applications for digital humanity and smart systems.

In the literature, the proposed systems generally use the same processing pipelines (face localization, feature extraction and classification). They are still using non-unified sparse data with non-standardized evaluation protocols, suffering therefore of a lack of compatibility and comparability. Over the two last decades, the challenge organizers are trying to bring up the level of compatibility to advance the understandings and findings of the established problems in facial analysis.

In this context, ChaLearn-2016 organized three parallel quantitative challenge tracks on face analysis to help bridging the gap between the research environment on facial analysis and the real-world context. The first track was about *apparent age estimation*, the second track concerned the *accessories classification*, and the third track was about *smile and gender classification*. This paper focuses on this two tasks in which participants were invited to classify images of the FotW dataset according to gender (male, female or other) and basic expression (smiling, neutral or other expression). Along with accurate face detection and alignment, this track will require robust feature selection and extraction in order to identify the subject's gender and expression, which can be difficult to classify even to the human eye in uncontrolled environments such as those present in the FotW dataset (Fig. 1). The annotated dataset is composed of training set (6, 171 images), validation set (3, 086 images) and test set (8, 505 images). The data reparation is given in Table 1 and some training images showing the real life imaging conditions are presented in Fig. 1.

To tackle this problem, our method used a pre-trained CNN that generates a set of robust "latent" descriptors.

2   Method Description

Convolutional neural networks are defined as hierarchical representations of convolution, subsampling and fully connected layers (e.g. Neocognitron82, LeNet89). They appear with their weight sharing and sparse connectivity features as

Fig. 1. Some face images from the FotW training dataset showing the real-life imaging conditions.

Table 1. FotW - smile and gender database characteristics.

Class	Subsets		
	Train	Val	Private test[a]
Male	2946	1691	4614
Female	3318	1361	3799
Not sure	93	34	92
Smile	2234	1969	4411
No smile	3937	1117	3849

[a] Notice that the final evaluation was done on a private test subset that was not available to the participants

a biological-inspired successor of the multilayer perceptron neural network in processing two dimensional data.

CNN-based techniques have revolutionized computer vision providing significant performance gains compared to traditional techniques in various challenging vision problems. The year 2012 was the turning point for the problem of recognition of large-scale objects where a new generation of deep CNN models won both the classification and localization task of ILSVR [12] by introducing a set of novel features such as non-saturating neurons to enhance the nonlinearity of the network, the dropout regularization method to prevent overfitting, and a GPU implementation of the convolution operations to enlarge the learning capacity.

In the literature, there are generally two approaches to use CNN for detection *(a)* as a full system with end-to-end training (e.g. [7]) or *(b)* as a feature extractor from pre-trained networks.

CNN-parameters learning usually requires very large training datasets when using end-to-end or fine tuning. The advantage of using deep features is to exploit already trained networks in order to produce high level features without the burden of having to train the model from scratch.

In this paper, we investigate the transfer learning of deep features from face recognition task to *gender/smile* recognition tasks.

First, the face is located and the corresponding cropped image is fed to the pre-trained CNN that generates, at a specified level of abstraction (here, the fully connected layer 'FC6'), a set of deep features which are used then to train a SVM. The diagram of the method is given in Fig. 2 showing different steps.

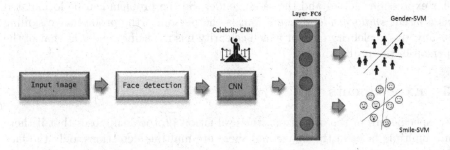

Fig. 2. Descriptive flowchart of the procedures used in the method.

2.1 Face Detection

Faces are detected using the retrained deformable part models [5,8]. In this work, We did not use any face alignment however we cropped each image to the best scoring face over a hierarchical search on its pyramidal representation.

2.2 Gender and Smile Detection

A CNN architecture is defined as a hierarchical representation alternating convolution layers with non-linear transformations and sampling layers. The training which consists of tuning a set of parameters associated with a set of bias and weights has introduced for the new generation of CNN a set of novel aspects to enhance the non-linearity of the network and to prevent overfitting. Thus, a substantial improvement was observed in performance using these models.

For both gender and smile classification tasks we used the same paradigm based on task learning transfer using a convolutional neural net architecture trained on different task but using similar input data (i.e. faces). In this work we used VGG-Face [11] as a generic deep feature extractor and a linear-SVM [3] as a base learner of facial attributes.

VGG-Face is a very deep architecture based model trained from scratch using 2.6 Million images of celebrities and public figures collected from the web. The model has been trained using MatConvNet library [13] to recognize 2622 faces. Our motivation behind using this model is that if in the identity verification, the goal is to tell whether two persons has the same faces; the gender classification task could also be seen as a gender verification. The same principle of verification can be applied to smile detection.

A CNN based gender classification was already introduced by Levi and Hassncer [7]) which, however, used an end-to-end architecture. Our proposed solution instead is based on learning transfer. The problem can be seen as *smile* and *gender* verification since originally VGG-Face was designed to recognize and verify person identity. First, VGG-Face makes specific neurons that correspond to a specific learned person having the most similar face switch 'on'. Second, the SVM compare the two faces to infer information about whether they have a similar expression (smile) and the same gender. So the problem can be formulated as a latent *smile/gender* recognition: Is the person in the probe image smiling as any given celebrity face, or which celebrity has the same gender as that of the person in the probe image?

3   Experiments

As specified from the organizers, the evaluation protocol supposes that if there are multiple faces in the image and there are multiple detections, only the face closer to the center (central face) has to be considered and thus each image should only have one prediction. The face location was previously introduced as a sub-challenge, but it was found that the face location labeling differs too much from one rater to another to be used as ground-truth for a Jaccard index. In this contest, the evaluation metric consists only of smile/gender attributes and not face localization. The measurement of error was given by the mean square error between the predictions and the ground-truth attributes.

Some considerations have been taken into account in the experiments: (a) LFW dataset [6] was used as external data but did not really improved the accuracy, (b) we also used the non-central faces to extend the training set, and (c) when L_2 normalized features were used, we saw an improvement in accuracy.

The feature extractor was associated with different abstraction layers and we found that the fully connected layer "FC6" gave the best performance on both tasks. On the gender classification task, the accuracy reached 90.69% on the validation set of FotW. For the second task, the classification of smiling faces gave a rate of 88.14% of correct predictions. Results were improved by performing L_2 normalization on deep features before they were fed to the SVM (Tables 2 and 3).

Table 2. Recognition performance of different features on *Gender* classification task.

Layers	Linear-SVM			RBF-SVM		
	$\|\cdot\|_2$	$\|\cdot\|_\infty$	None	$\|\cdot\|_2$	$\|\cdot\|_\infty$	None
FC6	**90.69**	87.87	85.75	90.28	87.63	90.18
FC7	90.52	83.80	87.05	54.68	54.01	86.95
FC8	84.01	87.42	83.71	87.01	84.05	85.73

Table 3. Recognition performance of different features on *Smile* classification task.

Layers	Linear-SVM			RBF-SVM		
	$\|\cdot\|_2$	$\|\cdot\|_\infty$	None	$\|\cdot\|_2$	$\|\cdot\|_\infty$	None
FC6	**88.14**	84.94	83.27	85.40	83.01	80.91
FC7	86.84	82.40	81.83	62.79	62.01	83.10
FC8	79.12	78.81	78.50	84.60	83.50	84.55

Using different SVM kernels did not really improve the performance but "FC6" still gave the best validation performances on both tasks (Tables 2 and 3). We guess that the fully connected layer "FC6" generalized well because at this level of abstraction the CNN has not been specialized yet for identity recognition. So it can deliver more generic deep features.

The ROC curves on the validation set for smile and gender detection are shown on Figs. 3 and 4 with AUC values of 0.89 and 0.91 respectively. This explain that the detection results are very good with respect to the degree of difficulty of the face images captured from an uncontrolled environment such as those present in the FotW dataset.

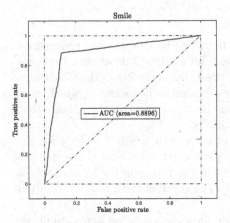

Fig. 3. ROC curve for *Smile* detection on the FotW validation set.

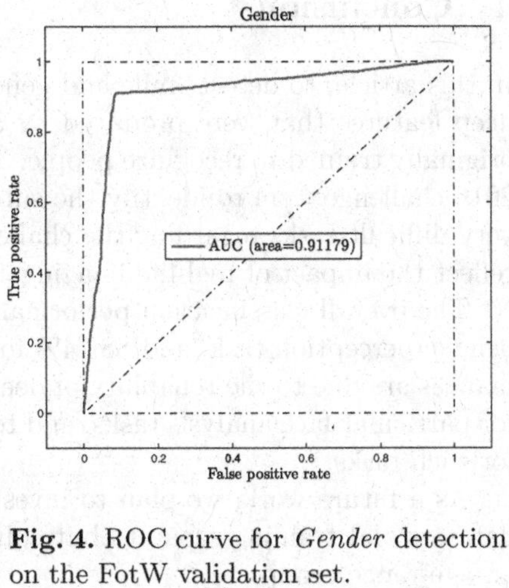

Fig. 4. ROC curve for *Gender* detection on the FotW validation set.

The organizers report that around 60 participants registered for the ChaLearn competition. The 7 teams that finally submitted predictions[1] are listed in Table 4. The accuracy in *Gender* and *Smile* classification of each team is shown.

Our method denoted by "VISI.CRIM" achieved the third place on the private test subset of the FotW dataset (Table 4). Notice that the test subset was not

---

[1] http://chalearnlap.cvc.uab.cat/challenge/13/track/20/result/.

Table 4. ChaLearn accuracy on *Smile* and *Gender* recognition.

Team	Gender	Smile	Mean
SIAT_MMLAB	0.9269	0.8583	0.8926
IVA_NLPR	0.9152	0.8252	0.8702
VISI.CRIM	0.9016	0.8212	0.8614
SMILELAB NEU	0.8999	0.8148	0.8574
Lets Face it!	0.8454	0.8439	0.8446
CMP+ETH	0.7465	0.7189	0.7327
SRI	0.5716	0.5853	0.5784

available to the competitors. The evaluation was performed by the organizers with the code that we have uploaded to them. The winning solution proposed by the "SIAT_MMLAB" team used three cascaded convolutional neural networks which were fine-tuned with external data from Faces of the World[2] database and CelebA[3] which is a large-scale face attributes dataset with more than 200K celebrity face images (32 times larger than FotW subset).

4   Conclusion

In this article, to detect smile and gender from face images we used generalized deep features that were produced by a convolutional neural network that was originally trained to recognize people. The two tasks in track-3 of the ChaLearn-2016 challenge were to identify the subject's gender and expression, which were very difficult tasks regarding the challenging images present in the dataset that reflect the impact of real-life imaging conditions.

The overall classification performance of the system reached 90.69% for the *gender* perception task, and 88.14% for the *smile* detection task. These performances are due to the reliability of deep features to represent the visual content for particular face analysis tasks, and to faithfully extract the latent information for such tasks.

As a future work, we plan to investigate data consolidation across multiple datasets and training separate but collaborative networks (e.g. one network by an age-range group).

Acknowledgements. This work has been made possible the Ministère de l'Économie, des Sciences et de l'Innovation (MESI) of Québec, and the Natural Sciences and Engineering Research Council of Canada (www.nserc-crsng.gc.ca). We are grateful to NVIDIA corporation for the Tesla K40 GPU Hardware Grant to support this work.

---

[2] https://www.zooniverse.org/projects/pszmt1/faces-of-the-world/about/research.
[3] http://mmlab.ie.cuhk.edu.hk/projects/CelebA.html.

References

1. Antipov, G., Berrani, S.-A., Dugelay, J.-L.: Minimalistic CNN-based ensemble model for gender prediction from face images. Pattern Recogn. Lett. **70**, 59–65 (2016)
2. Castrillón-Santana, M., Marsico, M.D., Nappi, M., Riccio, D.: MEG: texture operators for multi-expert gender classification. Comput. Vis. Image Underst. **156**, 4–18 (2017). Image and Video Understanding in Big Data
3. Chih-Chung, C., Chih-Jen, L.: LIBSVM: a library for support vector machines. ACM Trans. Intell. Syst. Technol. **2**(3), 1–27 (2011)
4. Demirkus, M., Toews, M., Clark, J.J., Arbel, T.: Gender classification from unconstrained video sequences. In: 2010 IEEE Computer Society Conference on Computer Vision and Pattern Recognition - Workshops, pp. 55–62, June 2010
5. Felzenszwalb, P.F., Girshick, R.B., McAllester, D.A.: Cascade object detection with deformable part models. In: The Twenty-Third IEEE Conference on Computer Vision and Pattern Recognition, CVPR 2010, San Francisco, CA, USA, pp. 2241–2248 (2010)
6. Huang, G.B., Ramesh, M., Berg, T., Learned-Miller, E.: Labeled faces in the wild: a database for studying face recognition in unconstrained environments. Technical report 07–49. University of Massachusetts, Amherst (2007)
7. Levi, G., Hassncer, T.: Age and gender classification using convolutional neural networks. In: 2015 IEEE Conference on Computer Vision and Pattern Recognition Workshops (CVPRW), pp. 34–42, June 2015
8. Mathias, M., Benenson, R., Pedersoli, M., Van Gool, L.: Face detection without bells and whistles. In: Fleet, D., Pajdla, T., Schiele, B., Tuytelaars, T. (eds.) ECCV 2014. LNCS, vol. 8692, pp. 720–735. Springer, Cham (2014). doi:10.1007/978-3-319-10593-2_47
9. Ng, C.B., Tay, Y.H., Goi, B.M.: Vision-based human gender recognition: a survey (2012). arXiv preprint arXiv:1204.1611
10. Nian, F., Li, L., Li, T., Xu, C.: Robust gender classification on unconstrained face images. In: Proceedings of the 7th International Conference on Internet Multimedia Computing and Service, ICIMCS 2015, pp. 77:1–77:4. ACM, New York (2015)
11. Parkhi, O.M., Vedaldi, A., Zisserman, A.: Deep face recognition. In: Proceedings of the British Machine Vision Conference (BMVC) (2015)
12. Russakovsky, O., Deng, J., Su, H., Krause, J., Satheesh, S., Ma, S., Huang, Z., Karpathy, A., Khosla, A., Bernstein, M., Berg, A.C., Fei-Fei, L.: Imagenet large scale visual recognition challenge. IJCV **115**(3), 211–252 (2015)
13. Vedaldi, A., Lenc, K.: Matconvnet - convolutional neural networks for MATLAB. CoRR, abs/1412.4564 (2014)
14. Whitehill, J., Littlewort, G., Fasel, I., Bartlett, M., Movellan, J.: Toward practical smile detection. IEEE Trans. Pattern Anal. Mach. Intell. **31**(11), 2106–2111 (2009)

An Empirical Analysis of Deep Feature Learning for RGB-D Object Recognition

Ali Caglayan[✉] and Ahmet Burak Can

Department of Computer Engineering, Hacettepe University, Ankara, Turkey
{alicaglayan,abc}@cs.hacettepe.edu.tr

Abstract. Conventional deep feature learning methods use the same model parameters for both RGB and depth domains in RGB-D object recognition. Since the characteristics of RGB and depth data are different, the suitability of such approaches is suspicious. In this paper, we empirically investigate the effects of different model parameters on RGB and depth domains using the Washington RGB-D Object Dataset. We have explored the effects of different filter learning approaches, rectifier functions, pooling methods, and classifiers for RGB and depth data separately. We have found that individual model parameters fit best for RGB and depth data.

Keywords: RGB-D object recognition · Deep feature learning

1 Introduction

Object recognition is a challenging problem and has an important role for autonomous systems. The task has certain difficulties such as high variations in the context of viewpoints, illumination, scaling and deformations, and thus it still remains a challenging problem. The popularity of cheap RGB-D sensors such as the Kinect has extended the problem to the depth domain. Kinect's depth data enables us to overcome lighting variations and enriches object representation with additional information about object shape. The basic requirement of a robust object recognition system is to extract meaningful and complementary features. Until recently, hand-crafted features which require domain expertise have been extracted. With the advance of deep learning techniques, algorithms such as convolutional feature learning have eliminated the need for hand-crafted feature representations and features are learnt from the low-tier of data. These new techniques have led to drastic improvements in the field and provided state-of-the-art object recognition performance.

Deep architectures have different characteristics in terms of having model properties including layer stack-ups and the methods used in these layers as well as "meta-parameters" such as receptive field size and number of filter size. Within this context, numerous feature learning algorithms have been proposed in recent years [1,3,5–7,9,11,16]. Typically, most of these methods have concentrated on the same modularity both for RGB and depth domains.

© Springer International Publishing AG 2017
F. Karray et al. (Eds.): ICIAR 2017, LNCS 10317, pp. 312–320, 2017.
DOI: 10.1007/978-3-319-59876-5_35

Since the characteristics of RGB and depth data are different, the suitability of such approaches is not clear. However, the decision about which model parameters should be used for which domain is still an open problem.

In this work, we empirically investigate the effects of different model parameters for both RGB and depth data. To do this, we use the well-known Convolutional-Recursive Neural Network (CNN-RNN) method [16] as the initial model and perform experiments on various combinations of model parameters. We experimentally evaluate the effects of patch extraction approaches, rectifier units, pooling methods, and various classifiers on the popular Washington RGB-D Object Dataset [13]. We have found that different model parameters for RGB and depth data can produce better results than using same parameters.

2   Related Work

During recent years, many new feature learning methods have been presented for RGB-D object recognition. Bo et al. [5] present a hierarchical matching pursuit (HMP) method using sparse coding to learn features in an unsupervised way. Jhuo et al. [11] propose a feed-forward model that attempts to determine the relationship between the gray-scale and depth images corresponding to an object. Blum et al. [3] propose the convolutional k-means descriptor (CKM), in which they learn features around interest points. In [16], Socher et al. present a convolutional-recursive neural network (CNN-RNN) to learn RGB and depth features separately and combine them for a softmax classifier. Cheng et al. [6] propose a semi-supervised learning model using CNN-RNN framework. Later then, the same authors extend this work by using a spatial pyramid pooling layer to prevent performance degradation of learned features by cropping and warping of images [7]. They utilize grayscale images and surface normals in addition to the RGB and depth images. CNN-RNN [16], CKM [3], HMP [5] and depth kernel descriptors [4] are also evaluated. Bai et al. [1] advocate using rather a subset based approach instead of the common extraction of random patches to learn filters. Guo et al. [9] present an integrated model by combining CNN and Hidden Markov Model (HMM) for recognizing house numbers captured in street view images. They use HMM to model the dynamics of frames which extracted by sliding window and CNN for dealing with the frame appearance. Coates et al. [8] present a similar work to ours in RGB domain. They give a detailed analysis of several unsupervised learning methods and the effect of changes in the meta-parameters including receptive field size, number of filter size, stride and the effect of whitening. In this study, we analyze the effects of different model parameters on RGB and depth images. Since the characteristic of RGB and depth images are different, we come to the conclusion that using the same models for both domains as the above conventional methods is not an effective solution.

3   Learning Method

Deep feature learning methods attempt to learn feature hierarchies automatically from the low-level raw pixels to the high-level object parts. Contrary to hand-designed features, deep features give better results because they are learnt from the low-level tier by considering the spatial arrangement between the raw pixel values. In this work, the effects of different model parameters in deep feature learning are investigated based on the CNN-RNN framework of [16]. The CNN-RNN framework is taken as an initial model to compare the results of our different deep learning approaches. The learning framework has a filter learning module, a single CNN layer including a convolution, a rectified unit and a pooling step, an RNN layer with fixed tree structures and finally a classifier to determine object categories. An overview of the learning method is shown in Fig. 1.

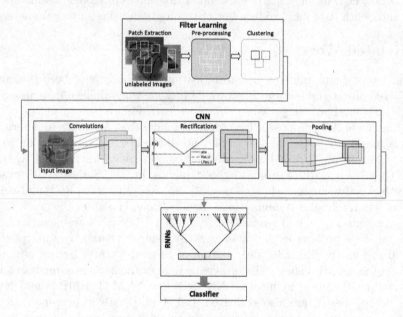

Fig. 1. Schematic overview of the learning method

3.1   Filter Learning Module

The model starts with random patch extraction from the input images. Each patch has a dimension of w-by-w and d channels ($d = 3$ for RGB and $d = 1$ for depth images). Then, these patches are whitened and normalized as preprocessing steps. After preprocessing, k-means clustering algorithm is applied to learn K centroids from these patches. All these steps are applied for both RGB and depth images separately. Random patch extraction leads to the question of how distinctive and meaningful these patches are. Thus, as a first step in this module, we investigated to see if we could learn more distinctive filters by extracting patches around the interest points of algorithms such as SIFT [14] and SURF [2] instead of extracting patches from the random regions. Secondly, when random patches are extracted, it is not guaranteed that patches are obtained from all instances under an object category. This may decrease the success of the system if there are missed instances, especially in the categories with rich intra-class variations. We adapt the initial model to ensure that patches are extracted from all instances.

3.2   CNN Layer

CNN is used to extract features from the input images using the learned K filters in the previous module. Since CNN requires fixed equal-sized input images, the images are resized to n-by-n pixels. Images are convolved with K learned filters of size w-by-w with d channels and $(n - w + 1) \times (n - w + 1) \times K$ convolutional responses are obtained for each image. Then, it is followed by a rectified unit and local contrast normalization. Finally, the standard practice of downsampling is used to obtain a pooled response with $(n - w + 1 - p)/s + 1$ sized square regions. Here p is the size of square pooling window and s is the stride size. At the end of this procedure, low-level translational invariant features are learnt. This procedure is also applied for both RGB and depth images separately.

Rectifier Unit. Rectifier is an activation function that takes numeric input values and performs a specific mathematical operation on it to obtain non-linearity. The initial model [16] uses absolute function in the rectifier unit (Eq. 1). However, ReLU (Eq. 2) or leaky ReLU (Eq. 3) are frequently used in recent years of deep learning studies (e.g. [12,15]). In this study, we investigate the effects of ReLU and leaky ReLU in addition to the absolute rectifier.

$$f(w^T x) = \left| w^T x \right| = \begin{cases} w^T x, & \text{if } w^T x \geq 0 \\ -w^T x, & \text{if } w^T x < 0 \end{cases} \tag{1}$$

$$f(w^T x) = max(0, w^T x) = \begin{cases} w^T x, & \text{if } w^T x > 0 \\ 0, & \text{if } w^T x \leq 0 \end{cases} \tag{2}$$

$$f(w^T x) = max(0.01w^T x, w^T x) = \begin{cases} w^T x, & \text{if } w^T x > 0 \\ 0.01w^T x, & \text{if } w^T x \leq 0 \end{cases} \tag{3}$$

Pooling. Pooling is a commonly used method in deep learning studies that summarizes statistics of a larger input in a lower dimension. It reduces the number of parameters to yield computational efficiency and provides robustness to small distortions in the role as the cells in visual perception [10]. In this study, we investigate the effects of two predominant pooling methods, max pooling and average pooling of the initial model. The input is downsampled by a factor of p window size and s stride along the spatial dimensions with their average values and max values in the average and max pooling respectively.

3.3   RNN Layer

Recursive neural networks are used to learn high-level hierarchical feature representations by using balanced fixed-trees. For a given input image, the downsampled output from the previous layer $X \in \mathbb{R}^{K \times r \times r}$ is transformed to a new representation $y \in \mathbb{R}^K$. Each layer of the tree merges the blocks of adjacent vectors into a parent vector with the same randomly initialized weights. Then, the parent vector is passed through *tanh* nonlinearity. The same procedure is repeated in the tree until a single parent vector remains. This whole process is done for a single RNN. In this work, multiple randomly initialized N RNNs are used. Each RNN has a K-dimensional vector and a total of $(N \times K)$-dimensional final matrix is produced. Then this final representation containing the learned features is given to the classifier.

3.4   Classification

The final step of the model is classification. The softmax classifier is used without the need for back-propagation. It is possible to test the $(N \times K)$-dimensional final representation with different classifiers as well. In this study, we use Nearest Neighbor (NN), linear Support Vector Machines (SVM) and Naive Bayes (NB) classifiers in addition to the Softmax.

4   Experimental Evaluation

In this section, the impacts of various learning parameters are evaluated for both RGB and depth data on the above framework. For the experiments, the Washington RGB-D Object Dataset introduced by Lai et al. [13] is used. The dataset consists of 51 categories with a total of 207.662^1 images under 300 category instances. When evaluating the effects of different model parameters, we use the 10 testing instances provided in [13,16] but we subsample every 10th image of dataset instead of every 5th image. In this way, we reduce the computation time for testing parameters. However, in the final comparison, we subsample every 5th image of the dataset as in the literature to make comparison with the related works. All the input images are resized to 148×148 and a total of 400.000 patches are extracted. The filter size $w = 9$, number of filter $K = 128$ and number of RNNs $N = 64$ are used for the experiments. To evaluate the impact of each of our changes, we modify the initial model with the related changes step by step.

4.1   Contribution of Filter Learning

In unsupervised filter learning module, mainly patches are extracted from the images randomly. What we experimentally investigate here is whether more meaningful patches can be found by extracting around interest points instead of random extraction. Patch size is $w = 9 \times 9 \times 3$ for RGB and $w = 9 \times 9$ for depth images. In our experiments, we use the cropped images of just the object on the turntable instead of the full 640×480 RGB-D frames. Thus, there may not be found enough interest points in a cropped image. In such a case, random

---

[1] In fact, there are 207.920 images in total, but 258 of them do not have object mask.

Table 1. Contribution of filter learning approaches

	RGB	Depth	Both
Random filters	80.05 ± 2.03	76.40 ± 1.50	85.68 ± 1.31
SIFT filters	80.35 ± 1.98	76.67 ± 1.71	85.96 ± 1.40
SURF filters	80.46 ± 1.93	76.74 ± 1.66	86.01 ± 1.29
Filters from all instances	80.51 ± 2.13	76.85 ± 1.54	85.77 ± 1.27

Table 2. Impact of different rectifiers

	RGB	Depth	Both
Absolute	80.05 ± 2.03	76.40 ± 1.50	85.68 ± 1.31
ReLU	77.88 ± 2.79	77.66 ± 1.64	85.66 ± 1.55
Leaky ReLU	77.82 ± 2.71	77.55 ± 1.65	85.67 ± 1.61

patches are used. As in Table 1, there is a small difference between randomly patching or patching around SIFT/SURF interest points. Since cropped images are used, this may be due to insufficient interest points. Finally, even patches are extracted from all the categories, it is not guaranteed to extract them from all the instances. In the last row of the table, we can see that the results are improved slightly if patches are extracted from all instances.

4.2 Effect of Rectifiers

Rectifier functions are expected to provide nonlinearity, while preventing vanishing gradient problem and preserving zero-centered data. All the three functions we use are the same on the positive side, and they behave differently on the negative side. As shown in Table 2, the best result for RGB data is obtained with absolute function, while the best result for depth data is obtained with ReLU. The use of different rectifiers gives a difference up to 2.2% for RGB and 1.3% for depth data. It should be noted that since absolute function does not saturate on either side of zero and preserves strong gradients on the negative side as well, it is an interesting possibility for deeper models with back-propagation.

4.3 Effect of Pooling Methods

The pooling window size is 10 and the stride size is 5 in our experiments. Thus, $(148 - 9 + 1 - 10)/5 + 1 = 27$ sized outputs are obtained as a result of pooling operation. Table 3 shows the accuracy results of different pooling methods for RGB and depth data. Max pooling is slightly better for RGB data whereas average pooling gives a better accuracy for depth data with a significant difference. This shows that using different pooling methods for different data can improve performance.

Table 3. Effect of pooling methods

	RGB	Depth	Both
Average	80.05 ± 2.03	76.40 ± 1.50	85.68 ± 1.31
Max	80.52 ± 1.88	68.18 ± 1.09	83.82 ± 1.54

4.4 Classifier Comparisons

Figure 2 indicates the accuracy results of all classifiers we used in this work for RGB, depth, and combined (RGB+depth). In this experiment, we use the model parameters that give the best results for RGB and depth images separately. Absolute rectifier and max pooling for RGB images, ReLU and average pooling for depth images are used. Patches are extracted from all instances around the SURF interest points. As shown in the figure, SVM and Softmax give the best results close to each other. We use Softmax for the final comparisons with the related works because it works faster than SVM.

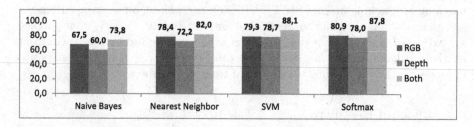

Fig. 2. Accuracy comparison between different classifiers

4.5 Discussion

We have shown that separate model parameters fit best for RGB and depth data. Random patches can actually work as well as patches around interest points like SIFT, SURF. Nevertheless, extracting patches around interest points for large-sized input images may be slightly better. Also extracting patches from all instances gives better results. For rectifiers, the absolute function performs best on RGB data and ReLU produces the best result for depth data. We have found interesting results for pooling methods. While max pooling produces more successful results for RGB images, average pooling performs significantly better than max pooling on depth images. For classifiers, we have seen that Softmax and SVM can be used alternatively to each other. Finally, we test all the best model parameters for RGB and depth images separately and compare the results to the initial model of [16] as shown in Table 4. These results show that by choosing suitable model parameters for RGB and depth data, it is possible to improve the results up to 2%.

Table 4. Effect of all the changes together

	RGB	Depth	Both
CNN-RNN [16]	80.05 ± 2.03	76.40 ± 1.50	85.68 ± 1.31
All-together	80.85 ± 1.76	78.02 ± 1.66	87.80 ± 1.16

Table 5. Comparison of our learning pipeline containing best model parameters with the related works

	RGB	Depth	Both
EMK-SIFT [13]	74.5 ± 3.1	64.7 ± 2.2	83.8 ± 3.5
KDES [4]	77.7 ± 1.9	78.8 ± 2.7	86.2 ± 2.1
CKM [3]	-	-	86.4 ± 2.3
CNN-RNN [16]	80.8 ± 4.2	78.9 ± 3.8	86.8 ± 3.3
HMP [5]	82.4 ± 2.1	81.2 ± 2.3	87.5 ± 2.9
Subset-RNN [1]	82.8 ± 3.4	81.8 ± 2.6	88.5 ± 3.1
CNN-SPM-RNN [7]	85.2 ± 1.2	83.6 ± 2.3	90.7 ± 1.1
This work	81.8 ± 1.7	79.7 ± 1.9	88.6 ± 1.4

When exploring the best model parameters in the previous experiments, we subsample every 10th image for each instance and set the number of patches and RNNs to 400.000 and 64 respectively. Finally, to compare our pipeline with the literature, we conduct experiments by subsampling every 5th image and setting the number of patches and RNNs to 500.000 and 128 respectively as in [16]. The results of these experiments and other related works on RGB-D dataset are presented in Table 5. As can be seen in the table, the performance of the system can be increased up to 2% by using the appropriate model parameters without extra requirement. We have found that it is possible to improve the performance of a system by using individual model parameters for RGB and depth images.

5   Conclusion

In this paper, we have applied different model parameters for RGB and depth data to explore best fitting model parameters on the Washington RGB-D Object Dataset. In contrast to the conventional methods that are using the same parameters for RGB and depth data, our experiments show that different model parameters produce the best results for RGB and depth images separately. While absolute rectifier and max pooling have produced the best results for RGB, ReLU and average pooling have performed best on depth images. We also have investigated the role of patches extracted around interest points instead of random patch extraction and have concluded that the difference is insignificant. Finally, we have confirmed that selecting right model parameters can improve learning pipeline's performance significantly.

References

1. Bai, J., Wu, Y., Zhang, J., Chen, F.: Subset based deep learning for RGB-D object recognition. Neurocomputing **165**, 280–292 (2015)
2. Bay, H., Ess, A., Tuytelaars, T., Van Gool, L.: Speeded-up robust features (SURF). Comput. Vis. Image Underst. **110**(3), 346–359 (2008)
3. Blum, M., Springenberg, J.T., Wülfing, J., Riedmiller, M.: A learned feature descriptor for object recognition in RGB-D data. In: 2012 IEEE International Conference on Robotics and Automation (ICRA), pp. 1298–1303. IEEE (2012)
4. Bo, L., Ren, X., Fox, D.: Depth kernel descriptors for object recognition. In: 2011 IEEE/RSJ International Conference on Intelligent Robots and Systems (IROS), pp. 821–826. IEEE (2011)
5. Bo, L., Ren, X., Fox, D.: Unsupervised feature learning for RGB-D based object recognition. In: Desai, J.P., Dudek, G., Khatib, O., Kumar, V. (eds.) Experimental Robotics, pp. 387–402. Springer, Cham (2013)
6. Cheng, Y., Zhao, X., Huang, K., Tan, T.: Semi-supervised learning for RGB-D object recognition. In: 2014 22nd International Conference on Pattern Recognition (ICPR), pp. 2377–2382. IEEE (2014)
7. Cheng, Y., Zhao, X., Huang, K., Tan, T.: Semi-supervised learning and feature evaluation for RGB-D object recognition. Comput. Vis. Image Underst. **139**, 149–160 (2015)
8. Coates, A., Lee, H., Ng, A.Y.: An analysis of single-layer networks in unsupervised feature learning. Ann Arbor **1001**(48109), 2 (2010)
9. Guo, Q., Wang, F., Lei, J., Tu, D., Li, G.: Convolutional feature learning and Hybrid CNN-HMM for scene number recognition. Neurocomputing **184**, 78–90 (2016)
10. Jarrett, K., Kavukcuoglu, K., Lecun, Y., et al.: What is the best multi-stage architecture for object recognition? In: 2009 IEEE 12th International Conference on Computer Vision, pp. 2146–2153. IEEE (2009)
11. Jhuo, I.-H., Gao, S., Zhuang, L., Lee, D.T., Ma, Y.: Unsupervised feature learning for RGB-D image classification. In: Cremers, D., Reid, I., Saito, H., Yang, M.-H. (eds.) ACCV 2014. LNCS, vol. 9003, pp. 276–289. Springer, Cham (2015). doi:10.1007/978-3-319-16865-4_18
12. Krizhevsky, A., Sutskever, I., Hinton, G.E.: Imagenet classification with deep convolutional neural networks. In: Advances in Neural Information Processing Systems, pp. 1097–1105 (2012)
13. Lai, K., Bo, L., Ren, X., Fox, D.: A large-scale hierarchical multi-view RGB-D object dataset. In: 2011 IEEE International Conference on Robotics and Automation (ICRA), pp. 1817–1824. IEEE (2011)
14. Lowe, D.G.: Distinctive image features from scale-invariant keypoints. Int. J. Comput. Vis. **60**(2), 91–110 (2004)
15. Maas, A.L., Hannun, A.Y., Ng, A.Y.: Rectifier nonlinearities improve neural network acoustic models. In: Proceedings of the ICML, vol. 30 (2013)
16. Socher, R., Huval, B., Bath, B., Manning, C.D., Ng, A.Y.: Convolutional-recursive deep learning for 3D object classification. In: Advances in Neural Information Processing Systems, pp. 665–673 (2012)

Image Registration Based on a Minimized Cost Function and SURF Algorithm

Mohannad Abuzneid[✉] and Ausif Mahmood

Department of Computer Science and Engineering,
University of Bridgeport, Connecticut, USA
mohannad@my.bridgeport.edu, mahmood@bridgeport.edu

Abstract. Computer vision and image recognition became one of the interesting research areas. Image registration has been widely used in fields such as computer vision, MRI images, and face recognition. Image registration is a process of aligning multiple images of the same scene which are taken from a different angle or at a different time to the same coordinate system. Image registration transforms the target image to the source image based on the affine transformation such as translation, scaling, reflection, rotation, shearing etc. It is a challenging task to find enough matching points between the source and the target images. In the proposed method, we used Speeded-Up Robust Features (SURF) and Random sample consensus (RANSAC) to find the best matching points between the pair images in addition to the minimized cost function which enhances the image registration with a few matching points. We took in our concentration some of the affine transformation which is translation, rotation, and scaling. We achieved a higher accuracy in the image registration with few matching points as low as two matching points. Experimental results show the efficiency and effectiveness of the proposed method.

Keywords: Image registration · Speeded-Up Robust Features (SURF) · Key-points · Random sample consensus (RANSAC) · Cost function · Affine transformation

1 Introduction

Image registration is an essential method used in the image processing systems such as face and object recognition [1, 5], object detection, motion estimation [2], and medical application [3]. The information inherited from two related images for the same scene are different. Therefore, they need a suitable registration to make the two images uniform and transfer them to the same coordinate system. The classical steps of the image registration are divided into 4 steps as shown in Fig. 1 [4]. The first step is to extract the most significant features from the source image and the target image such as edges, corners, and intersections, etc. The key-points can be found using methods such as Gaussians difference algorithm [6], segmentation methods [7], representations of general line segments or elongated anatomic structures [8], virtual circles [9], and local curvature discontinuities detected using the Gabor wavelets [10]. These algorithms are recommended if the image contains detectable objects. On the other hand, the medical

© Springer International Publishing AG 2017
F. Karray et al. (Eds.): ICIAR 2017, LNCS 10317, pp. 321–329, 2017.
DOI: 10.1007/978-3-319-59876-5_36

images usually have one object and considered as a lake of details. Fast Fourier transform (FFT) used to extract the features in the frequency domain and obtain the parameters based on cross correlation [11]. The discrete wavelet transform is another method used for feature extraction with root mean square error (RMSE) method [12]. The second step is to find the matching features between the two images from the features which we extracted in step one. Obtaining the corresponding points between the images has been a motivation of many invariant algorithms such as the Scale-Invariant Feature Transform (SIFT) [6], Speeded-Up Robust Features (SURF) [13–15], and Binary Robust Independent Elementary Feature (BRIEF) [16]. Even with these methods, it is still a challenge to obtain the appropriate matches. To Handel the mismatching points, usually by using RANSAC method will eliminate all the mismatching points by finding the best fitting on random subsets of the matches then selecting the best fitting subset. RANSAC [17] is robust to mismatches but finds a sub-optimal estimation, where LMedS [18] is a more accurate estimation, however, requires at least 50% correct matches. The third step is to find the affine transformation parameters such as translation, scaling, reflection, rotation, and shearing using some of the methods such as the minimized cost function. The last step is transforming the target image to the source image coordinate system using the affine transformation parameters which obtained from the third step.

We proposed in this paper, an enhanced image registration method based on SURF algorithm for the features extraction and minimized cost function to find the affine transformation parameters. The effectiveness of the proposed method is validated by the experiment results. This paper is organized as follows. In Sect. 2, Overview of SURF, minimized cost function, and RANSAC. The details of the proposed method will be discussed in Sect. 3. Experimental results will be discussed in Sect. 4. Section 5 describes the conclusion of this paper and the future work.

Fig. 1. Classical image registration.

2 SURF, Minimized Cost Function, and RANSAC

2.1 Speeded-Up Robust Features (SURF)

The Speeded-Up Robust Features (SURF) is a method to obtain the local features which will be used to align two related images which are taken at a different time or a different position. SURF process includes feature detection, feature description, and feature matching. The SURF steps are:

(1) Find the integral image $(I\Sigma)$ based on the input image I to achieve fast computation of the convolution filters. The value for each point $P = (x, y)$ is represented by the sum of all pixels in the input image I.

Then we calculate the sum of the intensities within a rectangular region formed by the origin and P using Eq. (1).

$$\sum = I\Sigma(C) - \Sigma(B) - \Sigma(D) + \Sigma(A) \tag{1}$$

(2) Find the interesting points using hessian matrix by finding the maximum hessian matrix corresponding to a point $P = (x, y)$.
(3) Subtract the adjacent Gaussian images to obtain the difference of Gaussian images by repeatedly convolved with Gaussians.
(4) Optimize the key-points after obtaining image gradients using three methods to obtain the descriptor.

   (a) Local extrema detection
   (b) Accurate key-point Localization
   (c) Eliminating edge responses

(5) Finding the orientation Assignment: using the pixel differences, we compute orientation $\theta(x, y)$ and gradient magnitude $M(x, y)$ for each image sample $L(x, y)$.

$$M(x, y) = \sqrt{(A)^2 + (B)^2} \tag{2}$$

(6) Obtaining Key-point descriptors: 64 element vector obtain by combining all the orientation histogram entries.
(7) Perform the k-nearest-neighbor (KNN) on the feature values to find the distance between the features using four steps:

- Compute the Euclidean distance on the obtained features.
- Descending sort of the labeled example.
- Based on root mean square deviation (RMSE), find the optimal K of the KNN.
- Represent the image based on these KNN.

2.2 Minimized Cost Function

Matrix Eq. (3) is used to minimize the cost of the image registration (without shear) when the related points in two images X and Y are identified. Summation indicates the sum over all points in an image. We used in our implementation the sigmoidal function as the neuron activation functions.

$$Cost = \sum (I1 - T(I2))^2 = \sum \left(\begin{pmatrix} X1 \\ Y1 \end{pmatrix} - \left(\begin{pmatrix} a & b \\ -a & a \end{pmatrix} \begin{pmatrix} X2 \\ Y2 \end{pmatrix} + \begin{pmatrix} t1 \\ t2 \end{pmatrix} \right) \right)^2 \tag{3}$$

To find the optimal transformation that will align image 2 to image 1, take the partial derivatives of the above cost with respect to *a, b, t1 and t2* and set these to 0 ($\partial C/\partial a = 0$, $\partial C/\partial b = 0$, $\partial C/\partial t1 = 0$, $\partial C/\partial t2 = 0$). We express the four resulting equations in matrix form as is shown in Eq. (4).

In matrix form, it can be written as:

$$\sum \begin{pmatrix} 2x_2^2 + 2y_2^2 & 0 & 2x_2 & 2y_2 \\ 0 & 2y_2^2 + 2x_2^2 & 2y_2 & -2x_2 \\ 2x_2 & 2y_2 & 2 & 0 \\ 2y_2 & -2x_2 & 0 & 2 \end{pmatrix} \begin{pmatrix} a \\ b \\ t_1 \\ t_2 \end{pmatrix}$$

$$= \sum \begin{pmatrix} 2x_1x_2 + 2y_1y_2 \\ 2x_1y_2 - 2x_2y_1 \\ 2x_1 \\ 2y_1 \end{pmatrix} \tag{4}$$

After calculating the sum over all points for the 4 × 4 matrix and the right-hand 4 × 1 vector in Eq. (4), we can compute the required transformation by:

```
Matrix Ainv = A.Inverse;
Matrix Res = Ainv * B;
```

2.3 Random Sample Consensus (RANSAC)

Fischer et al. introduced RANSAC algorithm in 1981. RANSAC is one of the most applicable algorithms to eliminate the false matched points in the source and target images in the presence of noise [17]. Some of the disadvantages of RANSAC are computing time, correct matches count, and the dependency of mismatches removal on the amount of threshold value. RANSAC algorithm is divided into four steps. The first step is to select a suitable model based on the transformation model. Equation (5) is used to calculate the number of related points which are required to calculate the transformation parameters. q is the minimum number of the related points and p is the number of the parameters need to calculate.

$$q = \frac{p}{2} \tag{5}$$

The second step is selecting the best model in a specified iterations. In each iteration, the minimum number of the related points randomly selected to estimate the transformation parameters based on Eq. (6) for 6 parameters a, b, c, d, e, and f.

$$\begin{bmatrix} x_1 \\ y_1 \\ 1 \end{bmatrix} = \begin{bmatrix} x_1' \\ y_1' \\ 1 \end{bmatrix} \Rightarrow \begin{cases} x_1' = ax_1 + by_1 + c \\ y_1' = ax_1 + ey_1 + f \end{cases} \tag{6}$$

The third step is to calculate the distance between the source image and the transformed image. We consider a point is a right match if the distance is less than the threshold, otherwise we eliminate the point since it is not a true match. In the last step, for each iteration, we count the numbers of the true match and if they are more than the desired value, or reaches the predetermined maximum number of iterations, then the algorithm stops. Transformation model selected which has the highest matching count.

3 Proposed Method

Image registration objective to align two images of the same object to the same coordinator. Suppose we obtain two images for the same parson $I1$ and $I2$ which are acquired at a different angle or distance. In image $I1$ (x, y) is a point and (x', y') is a related point in the image $I2$ and the transformation model between the two images include scale, rotation, and translation which can be expressed in Eq. (7).

$$\begin{bmatrix} x_1' \\ y_1' \\ 1 \end{bmatrix} = s \begin{bmatrix} cos(\theta) & -sin(\theta) & tx \\ sin(\theta) & cos(\theta) & ty \\ 0 & 0 & 01 \end{bmatrix} \begin{bmatrix} x_1 \\ y_1 \\ 1 \end{bmatrix} \tag{7}$$

Where tx and ty are horizontal, vertical translation parameters, s is the scale parameter, θ is the rotation angle parameter. Our purpose is to find these key-points then the matching points, calculate the affine transformation parameters, then align two images as is shown in Fig. 2.

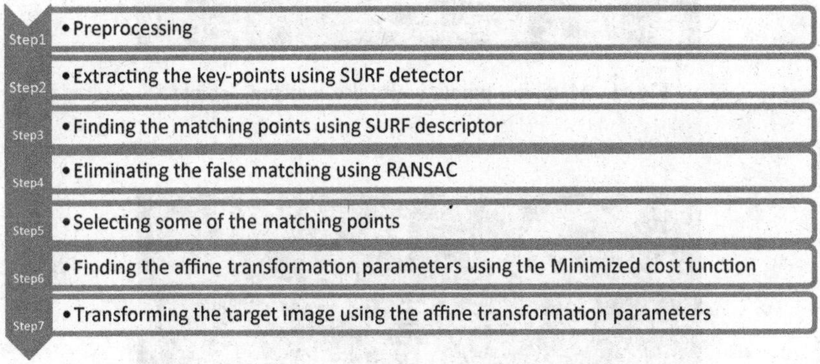

Step1	• Preprocessing
Step2	• Extracting the key-points using SURF detector
Step3	• Finding the matching points using SURF descriptor
Step4	• Eliminating the false matching using RANSAC
Step5	• Selecting some of the matching points
Step6	• Finding the affine transformation parameters using the Minimized cost function
Step7	• Transforming the target image using the affine transformation parameters

Fig. 2. The proposed image registration method.

We used SURF algorithm in our proposed method to find the key-points for both source and target images. We choose a high threshold to reduce the number of the key-points since we are looking for few of the matching points to register the image using the minimized cost function. Figure 3 shows the key-points for an image with different threshold (100, 50 and 20).

Fig. 3. Key-points for Lena with variant threshold. (a) Threshold = 100. (b) Threshold = 50. (3)Threshold = 20.

After we obtain the key-points, we used the SURF descriptor to find the matching points. With a huge number of key-points, we found some of the false matching points which we have to eliminate to achieve higher registration accuracy. Figure 4 shows the matching points between two images with some of the false matching points. Then, we used RANSAC algorithm to eliminate all the false matching points as shown in Fig. 5.

Fig. 4. Matching points with false matching points.

Fig. 5. Matching points after elemintaing the fasle points.

The next step is to find the affine transformation parameters using the minimized cost function with few matching points. We have been able to achieve the lowest error with only two matching points. Figure 6 shows the registration example with only two matching points (Fig. 7).

Fig. 6. Example of image registration. (a) Source and target images with two matching points. (b) Registered image.

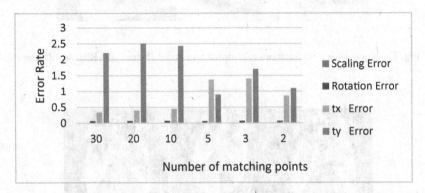

Fig. 7. Registration error rate for variant matching point number on Lena image.

4   Experimental Result

In this experiment, we applied our proposed method on 256×256 pixels Lena image. After converting Lena image to gray scale image and adding some noise, we applied some transformation to Lena image such as 0.2 scaling, 0.2 rotation, -15 translation tx, and -5 translation ty. We achieved a high registration accuracy with the lowest error by using few matching points. We tried variant matching point scenarios and we achieved a steady error rate. The main contribution in this paper is registering the images with only two sufficient matching points. Table 1 shows the error rate for affine parameters

Table 1. Registration error rate for variant matching point number

Number of matching points	Scaling error	Rotation error	tx error	ty error
30	0.007	0.066	0.34	2.21
20	0.01	0.067	0.4	2.5
10	0.012	0.07	0.45	2.43
5	0.015	0.064	1.37	0.9
3	0.017	0.07	1.4	1.7
2	0.019	0.065	0.86	1.09

with variant matching point number. This table shows the proposed algorithm achieved the lowest passable error rate with few matching points. The proposed methods can be applied on the human face images as shown in Fig. 6. We can use it for all type of objects as shown in Fig. 8, and medical images as shown in Fig. 9.

(a)   (b)

Fig. 8. Object image registration. (a) Source and target images. (b) Registered image.

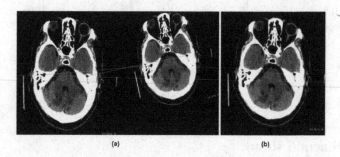

(a)   (b)

Fig. 9. Medical image registration. (a) Source and target images. (b) Registered image.

5 Conclusion and the Future Work

In this paper, we proposed a method for the image registration using the minimized cost function and Speeded-Up Robust Features (SURF). SURF algorithm usually produces some of the false matching points which are effecting the image registration accuracy. Therefore, we applied RANSAC algorithm to eliminate all the false matching points. This experiment showed that the error rate is steady with a variant number of the matching points. Thus, the main contribution of this proposed method is registering any type of images using only two matching points. We achieved a high registration accuracy with the lowest error rate.

Future work will be based on this algorithm, using a deep neural network and the minimized cost function since the deep neural network is performing well as feature extraction.

References

1. Yang, Z., Cohen, F.S.: Image registration and object recognition using affine invariants and convex hulls. IEEE Trans. Image Process. **8**(7), 934–946 (1999)
2. Dufaux, F., Konrad, J.: Efficient, robust, and fast global motion estimation for video coding. IEEE Trans. Image Process. **9**(3), 497–501 (2000)
3. Maintz, J.B.A., Viergever, M.A.: A survey of medical image registration. Med. Image Anal. **2**(1), 1–36 (1998)
4. Dou, H., Yao, L.: Medical image registration based on edge inflection point. Life Sci. Instrum. **10**(4), 15–18 (2006)
5. Hartley, R., Zisserman, A.: Multiple View Geometry in Computer Vision. Cambridge University Press, Cambridge (2003)
6. Lowe, D.G.: Distinctive image features from scaleinvariant keypoints. Int. J. Comput. Vis. **60**(2), 91–110 (2004)
7. Zitova, B., Flusser, J.: Image registration methods: a survey. Image Vis. Comput. **21**, 977–1000 (2003)
8. Vujovic, N., Brzakovic, D.: Establishing the correspondence between control points in pairs of mammographic images. IEEE Trans. Image Process. **6**, 1388–1399 (1997)
9. Alhichri, H., Kamel, M.: Virtual circles: a new set of features for fast image registration. Pattern Recogn. Lett. **24**, 1181–1190 (2003)
10. Manjunath, B., Shekhar, C., Chellapa, R.: A new approach to image feature detection with applications. Pattern Recogn. **29**, 627–640 (1996)
11. Wisetphanichkij, S., Dejhan, K.: Fast fourier transform technique and affine transform estimation-based high precision image registration method. GESTS Int. Trans. Comput. Sci. Eng. **20**(1), 179–191 (2005)
12. Karani, R., Sarode, T.: Image registration using discrete cosine transform and normalized cross correlation. In: IJCA Proceedings on International Conference and Workshop on Emerging Trends in Technology, pp. 28–34 (2012)
13. Paul, E., Ajeena Beegom, A.S.: Mining images for image annotation using SURF detection technique. In: 2015 International Conference on Control Communication and Computing India (ICCC), Trivandrum, pp. 724–728 (2015)
14. Chen, W., et al.: FPGA-based parallel implementation of SURF algorithm. In: 2016 IEEE 22nd International Conference on Parallel and Distributed Systems (ICPADS), Wuhan, China, pp. 308–315 (2016)
15. Bay, H., Ess, A., Tuytelaars, T., Van Gool, L.: Speeded-up robust features (SURF). IEEE Trans. **14**(1), 346–359 (2008)
16. Calonder, M., Lepetit, V., Strecha, C., Fua, P.: BRIEF: binary robust independent elementary features. In: Daniilidis, K., Maragos, P., Paragios, N. (eds.) ECCV 2010. LNCS, vol. 6314, pp. 778–792. Springer, Heidelberg (2010). doi:10.1007/978-3-642-15561-1_56
17. Fischler, M.A., Bolles, R.C.: Random sample consensus: a paradigm for model fitting with applications to image analysis and automated cartography. Commun. ACM **24**(6), 381–395 (1981)
18. Rousseeuw, P.J.: Least median of squares regression. J. Am. Stat. Assoc. **79**(388), 871–880 (1984)

A Better Trajectory Shape Descriptor for Human Activity Recognition

Pejman Habashi$^{(\boxtimes)}$, Boubakeur Boufama, and Imran Shafiq Ahmad

School of Computer Science, University of Windsor, Windsor, Canada
{habashi,boufama,imran}@uwindsor.ca

Abstract. Sparse representation is one of the most popular methods for human activity recognition. Sparse representation describes a video by a set of independent descriptors. Each of these descriptors usually captures the local information of the video. These features are then mapped to another space, using Fisher Vectors, and an SVM is used for clustering them. One of the sparse representation methods proposed in the literature uses trajectories as features. Trajectories have been shown to be discriminative in many previous works on human activity recognition. In this paper, a more formal definition is given to trajectories and a new more effective trajectory shape descriptor is proposed. We tested the proposed method against our challenging dataset and demonstrated through experiments that our new trajectory descriptor outperforms the previously existing main shape descriptor with a good margin. For example, in one case the obtained results had a 5.58% improvement, compared to the existing trajectory shape descriptor. We run our tests over sparse feature sets, and we are able to reach comparable results to a dense sampling method, with fewer computations.

Keywords: Human activity recognition · Trajectory descriptor · Trajectory encoding · Shape descriptor · Shape encoding

1 Introduction

Human activity recognition (HAR) has been a hot topic of research. During the past decade, sparse representation gained more attention [1–5,17]. Sparse representation methods describe a given video by a set of sparsely sampled features, regardless of their location (spatial and temporal). Then, this feature set is mapped to a fixed-sized vector, to be used by an existing machine learning method. Many sparse representation methods proposed in the literature rely on local appearance information. There are few other methods that use the motion of feature points for this matter (Sect. 2).

One of the methods based on motion information for HAR uses *trajectories*, as they have been shown to be discriminative for human activity recognition [1]. Trajectories are defined as the trails of interest point over time. Trajectories are also useful for aligning small neighborhoods in consecutive frames for extraction of traditional local features, namely Histogram of Oriented Gradient (HOG),

© Springer International Publishing AG 2017
F. Karray et al. (Eds.): ICIAR 2017, LNCS 10317, pp. 330–337, 2017.
DOI: 10.1007/978-3-319-59876-5_37

Histogram of Optical Flow (HOF), Motion Boundary Histogram (MBH), and even deep features [4]. In fact, trajectory aligned feature points produce better results compared to trajectory shape descriptors in the experiments of Wang [1,5,6].

Although the trajectory aligned descriptors produce slightly better results, they are more demanding in term of computations. Once the trajectories are obtained, one has to align the frames based on trajectories and then calculate traditional descriptors. There are many such trajectories in each video, which means a lot of processing time is needed. In this paper, we aim at overcoming this issue by finding a better way of using trajectory shape information directly for classification.

To be useful for any machine learning algorithm, the information of the phenomenon of interest should be encoded appropriately. Wang proposed a naive method to encode the shape of a trajectory that despite its simplicity is effective. In this paper, we examine the trajectories more in-depth and suggest a better alternative for trajectory shape descriptor (Sect. 3).

2 Background

Sparse representation is a group of methods which represent each video by a set of independent features. Each one of these features usually captures the characteristics of a small neighborhood in the spatiotemporal space of the video. These features are often represented by a vector of the same size, but the number of these features might be different for each video. Hence, in sparse representation, each video will be represented by a set of independent features. To be used by the existing learning methods, these features should be mapped into a vector of fixed-size. Initially, the bag of words (BOW) have been used for this purpose, but recently, the Fisher Vector (FV) encoding is being used as it outperforms the BOW by a good margin [4]. Finally, an SVM [7] or a similar learning method can be used for classification.

Traditionally, sparse feature points are extracted in two steps. The first step, called feature extraction, is used to find the locations (spatial and temporal) in the video that are interesting. Having these interest points, the second step, called feature descriptor extraction, aims at encoding the information in video around these feature points. These descriptors should be as discriminant as possible.

2.1 Feature Extraction Methods

In 2D still images, feature points are defined as points that carry more texture information. These are points with severe changes in the intensity. From the visual point of view, this could be a boundary line. Corners have even more information as they have intensive changes in both directions. This simple idea leads to several corner detectors algorithms to find interest points (e.g. Harris

Interest Point Detector). In the video paradigm, the same idea inspired the invention of many algorithms.

Laptev et al. [8,9] used the idea behind standard Harris interest point detector and extended it to the 3D spatiotemporal space. Laptev et al. looked for significant changes in both spatial and temporal domain. In other words, they looked for spatial corners with meaningful motion (non-constant motion) [8].

The spatiotemporal corners are very robust feature points, but they are rare in videos. Furthermore, these interest points do not occur in all kinds of human activity videos. Dollar et al. [10] mentioned these limitations and introduced another local feature point extractor named cuboid. The cuboid feature points are extracted by using a response function. The authors mentioned that cuboid locations and types should be sufficient for many recognition tasks.

Another approach that interestingly produced a comparable result is the random selection of interest points. Instead of searching videos for interest points, Scovanner et al. proposed to use random points as interest points [11].

Trajectory features are slightly different from traditional appearance-based feature points, as the only critical component for them is the frame and the spatial location in which they are starting. Only a few works have been done on trajectory descriptors. One of the significant works on trajectories was proposed by Wang and Schmid [6].

In particular, they have shown that dense sampling could outperform other methods. The dense sampling is easier to implement and is fast to calculate. These methods usually put a grid on top of each frame; then the grid corners are examined to find out how much visual information exist in each area. If these grid points happen to be on a smooth area of an image (no texture), they are not suitable for tracking, and they are deleted. This method usually produces lots of feature points, which increases the computation time of next steps.

Some methods might provide more interest points compared to others. As a rule of thumb, more interest points means more accurate results, but also more CPU time for further processing. In the current paper, we are working with trajectories and, we have used the OpenCV implementation of FAST corner detector [12,13] for finding the initial locations of trajectories. FAST corner detector is a simple corner detector that based on a machine learning method, selects points that are brighter or darker from the majority of their neighborhoods.

2.2   Feature Description Methods

Different sparse feature descriptor methods have been proposed in the literature. We only mention the most important ones in this paper. HOG and HOF, proposed by Laptev [9,14], Motion Boundary Histograms, proposed by Dalal et al. [15] and Trajectory shape descriptors, proposed by Wang et al. [16], are the ones discussed in this section.

Histogram of Oriented Gradient (HOG) and Histogram of Optical Flow (HOF), are two popular methods of sparse feature extraction [14], where HOG captures the visual appearance information and HOF captures the dynamics of an action in consecutive frames. To calculate HOG, a dense grid is put around each interest point first. Then, a local histogram of orientation for each cell is calculated separately. By combining these histograms, the ultimate HOG feature descriptor is obtained. To make a more robust descriptor, a contrast normalization is usually done on overlapping blocks (spatial regions larger than individual cells). HOF is very similar to HOG but, instead of appearance information, optical flow information is used to build the histogram.

Dalal et al. [15] proposed to calculate the optical flow derivatives for horizontal and vertical directions separately. This leads to MBHx and MBHy components, which will be combined later. Wang et al. [16] argued that the motion captured by HOF may come from different sources (object motion v.s. camera motion), so the camera movement can degrade the accuracy of HOF. They proposed and showed that since MBH uses derivatives, the constant camera motion will be removed.

Trajectory shape descriptors proposed by Wang were compared to HOG, HOF and MBH. Also in their tests, dense sampling outperformed other sampling methods. They showed that the trajectory aligned HOF, HOG and MBH could beat other descriptors, including the trajectory shape descriptor.

3   Proposed Method

The shape descriptor proposed by Wang is a simple but effective. To better formulate the trajectory shape, let us define trajectory more formally.

Trajectories are trails of 2D spatial feature points in time (Fig. 1). Any feature detector algorithm could be used to extract the 2D feature points. Formally, a trajectory T_k is an ordered list of spatial locations, in $l + 1$ consecutive frames, where l is called the length of the trajectory.

$$T_k = (p_0, p_1, p_2, \ldots, p_l), p_i \in \mathbb{R}^2, i = 0 \ldots l \tag{1}$$

In other words, each trajectory is simply the trails of a spatial point (feature point) in time. Each trajectory can be denoted by a function in a small domain. Let us define $T_k : D \to \mathbb{R}^2$ where $D = \{n \in \mathbb{N}_0 | n \leq l\}$ and $T_k(i) = p_i$. In fact,

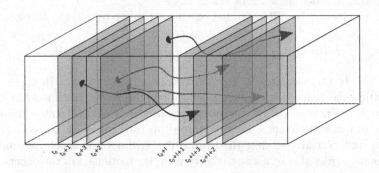

Fig. 1. Tracking of feature points for l consecutive frames in spatiotemporal space

this function pinpoints the feature point location in relative time, so it is a 2D position. If we differentiate it with respect to time, we get the instance velocity in that particular moment of that particular trajectory. Since the measurements have been done in discrete times (frames), the differentiation should be defined in discrete space. We have used the forward difference for this matter.

$$V_k(t) = \frac{dT_k}{dt} = \lim_{t \to 0} \frac{T_k(t + dt) - T_k(t)}{dt} \tag{2}$$

The smallest value for t is the difference between two consecutive frames (usually between 30 ms and 40 ms). Since the videos usually have a fixed frame rate, we can measure the time in frame unit instead of seconds to make our calculations easier. Hence, the smallest value for t would be one, making $dt = 1$ and we can write:

$$V_k(t) = T_k(t + 1) - T_k(t) \tag{3}$$

Alternatively, we may write all the values of the velocity function as:

$$V_k = (v_0, v_1, \ldots, v_{l-1}), v_i = p_{i+1} - p_i, p_i = T_k(i), i = 0 \ldots l - 1 \tag{4}$$

Let V_k be the velocity representation of the trajectory T_k. The following equation shows the shape descriptor proposed by Wang as defined in their paper [1].

$$W_k = \frac{(w_0, w_1, \ldots, w_{l-1})}{\Sigma_i \|w_i\|}, w_i = p_{i+1} - p_i, p_i = T_k(i), i = 0 \ldots (l - 1) \tag{5}$$

where $\|.\|$ denotes the L^2 norm. As it can be seen, the w_i's are simply defined the same way as instant velocities, i.e., $v_i = w_i$. So from (4) and (5), it is easy to show that:

$$W_k = \frac{V_k}{\Sigma_i \|v_i\|} \tag{6}$$

This means that the trajectory shape descriptor proposed by Wang et al. [1], as shown in (5), can be interpreted as the normalized version of velocity that is proposed in this paper and given in (4). From the physical point of view, Wang's descriptors still have the sense of velocity.

The velocity can be differentiated one more time to obtain the acceleration.

$$A_k = \frac{dV_k}{dt} = (a_0, a_1, \ldots, a_{l-1}), a_i = v_{i+1} - v_i, v_i = V_k(i), i = 0 \ldots (l - 2) \tag{7}$$

We call A_k the acceleration representation of trajectory T_k. In theory, one may continue to proceed with higher order differentiations. Each level of differentiation introduces possibly new information, but it also increases the effect of noise. In our experiments, we have tested up to the seventh derivative. It is expected that derivatives, after the third order, will not add much information, as human activities do not consist of very complex motions. On the other hand, after the third derivation, the effect of noise will affect the results and reduces the accuracy of the system.

Having the derivatives, we have defined different descriptors by concatenating them. For example, the first descriptor D_1 is simply the first derivative. The second descriptor D_2 is the concatenation of first and second descriptors together, and the third one D_3 is the concatenation of D_1, D_2 and D_3 derivatives together, and so on. We have also defined the zero order derivative to be the same as trajectory minus its starting location. The zero order descriptor D_0 only contains zero order derivatives.

4   Experiments

We have tested our new method against our challenging dataset. Our dataset contains 27 different activities, performed by 11 different actors and each actor repeated each activity for four times. The dataset recorded with two off the shelf cameras, originally intended for stereo vision, but for this paper, we have only used one camera output only. The cameras are fixed in the everyday office environment. The dataset is challenging since it has many activities and some activities are similar from motion point of view.

The tests ran over various trajectory lengths. Fisher Vector is used to prepare data for learning and a multi-class C-SVM used for classification. The reported results obtained by N-Fold cross-validation, where N is the number of actors. In each run, we left all instances of one actor out and trained the SVM with the other actors; then the tests have been done on the left out actor.

Table 1 summarizes the results. Each row represents one encoding method, while each column is dedicated to a particular trajectory length.

Table 1. A summary of the obtained results

Method	11	13	15	17
D_0	83.24%	81.21%	83.32%	84.08%
D_1	84.50%	85.72%	84.37%	84.54%
D_2	**85.80%**	87.46%	86.48%	86.94%
D_3	85.30%	87.95%	87.07%	**89.09%**
D_4	84.84%	**88.42%**	**88.55%**	88.42%
D_5	84.08%	85.68%	88.21%	87.11%
D_6	84.79%	86.18%	86.44%	86.20%
D_7	82.10%	83.15%	84.96%	85.43%
NormDiff (Wang)	83.78%	82.84%	84.58%	85.89%

The baseline for comparison is the last row of the table, *NormDiff*. This row represents the results is obtained by the Wang [1,16] trajectory encoding algorithm (5). The proposed algorithm has outperformed Wang's in almost all cases. As expected, D_0 results are below the baseline. D_1 which is the non-normalized version of *NormDiff* performs better in some cases. This means, normalization of trajectory might remove some of the discriminant information. This makes sense as the magnitude of trajectories can be discriminative for some types of activities. In all cases, D_2 outperforms D_1 and in most cases, D_3 outperforms D_2.

As it can be seen, D_3 and D_4 outperforms other trajectory descriptors. For example, D_3 produces best results for length 17, with more than 89% accuracy. It is a 3.2% improvement, compared to the baseline. For trajectory length of 13, D_4 reaches more than 88.42% accuracy, which is a 5.58% improvement over the baseline. After D_4, there is no significant improvement. This is because human activities do not have, or have a little, higher order complexities. On the other hand, differentiation increases the effect of noise which results in losing the accuracy.

Table 2. Results of dense sampling versus sparse sampling

Method/encoding	11	13	15	17
Dense trajectory/NormDiff encoding	89.54%	88.90%	88.74%	87.80%
Sparse trajectory/proposed encoding	85.80%	88.42%	88.55%	89.09%

We also compared our method with dense sampling results as well. Dense sampling uses much more sampling points, compared to our sparse sampling technique. The best results we obtained by our method compared with dense sampling results are presented in Table 2. As can be seen, with new trajectory encoding algorithm, we can obtain competitive results with the ones using dense trajectories. Note that this is not a fair comparison as the number of dense trajectories are higher than the sparse trajectories and as a result, they need more CPU time, while sparse trajectories are lighter and faster.

5   Conclusion

In this paper, we have reviewed different encoding methods used in the literature for human action recognition. We have examined the trajectory shape encoding method more intensively and proposed a new encoding method, which is a superset of the existing trajectory shape encoding. We have defined the trajectories as functions and have formally defined our descriptors. We have also provided rational reasons for why this new encoding method is a superset of the existing method, and why it should provide more accurate results.

Finally, our experiments have shown that the new encoding method outperforms the existing encoding method with a good margin. We have also demonstrated that the new encoding method applied to sparse sampling can achieve competitive results to dense sampling method with fewer computations.

References

1. Wang, H., Klaser, A., Schmid, C., Liu, C.-L.: Action recognition by dense trajectories. In: 2011 IEEE Conference on Computer Vision and Pattern Recognition (CVPR), pp. 3169–3176. IEEE (2011)
2. Habashi, P., Boufama, B., Ahmad, I.S.: The bag of micro-movements for human activity recognition. In: Kamel, M., Campilho, A. (eds.) ICIAR 2015. LNCS, vol. 9164, pp. 269–276. Springer, Cham (2015). doi:10.1007/978-3-319-20801-5_29
3. Mohammadi, E., Wu, Q.J., Saif, M.: Human action recognition by fusing the outputs of individual classifiers. In: 2016 13th Conference on Computer and Robot Vision (CRV), pp. 335–341. IEEE (2016)
4. Wang, Y., Tran, V., Hoai, M.: Evolution-preserving dense trajectory descriptors. arXiv preprint arXiv:1702.04037 (2017)
5. Wang, H., Oneata, D., Verbeek, J., Schmid, C.: A robust and efficient video representation for action recognition. Int. J. Comput. Vis. **119**(3), 219–238 (2016)
6. Wang, H., Schmid, C.: Action recognition with improved trajectories. In: 2013 IEEE International Conference on Computer Vision (ICCV), pp. 3551–3558. IEEE (2013)
7. Chang, C.-C. Lin, C.-J.: LIBSVM: a library for support vector machines. ACM Trans. Intell. Syst. Technol. **2**, 27:1–27:27 (2011). Software available at http://www.csie.ntu.edu.tw/cjlin/libsvm
8. Laptev, I., Lindeberg, T.: Interest point detection and scale selection in spacetime. In: Griffin, L.D., Lillholm, M. (eds.) Scale-Space 2003. LNCS, vol. 2695, pp. 372–387. Springer, Heidelberg (2003). doi:10.1007/3-540-44935-3_26
9. Laptev, I.: On space-time interest points. Int. J. Comput. Vis. **64**(2–3), 107–123 (2005)
10. Dollár, P., Rabaud, V., Cottrell, G., Belongie, S.: Behavior recognition via sparse spatio-temporal features. In: 2nd Joint IEEE International Workshop on Visual Surveillance and Performance Evaluation of Tracking and Surveillance, pp. 65–72. IEEE (2005)
11. Scovanner, P., Ali, S., Shah, M.: A 3-dimensional sift descriptor and its application to action recognition. In: Proceedings of the 15th international conference on Multimedia, pp. 357–360. ACM (2007)
12. Rosten, E., Drummond, T.: Machine learning for high-speed corner detection. In: Leonardis, A., Bischof, H., Pinz, A. (eds.) ECCV 2006. LNCS, vol. 3951, pp. 430–443. Springer, Heidelberg (2006). doi:10.1007/11744023_34
13. Rosten, E., Porter, R., Drummond, T.: Faster and better: a machine learning approach to corner detection. IEEE Trans. Pattern Anal. Mach. Intell. **32**(1), 105–119 (2010)
14. Laptev, I., Marszalek, M., Schmid, C., Rozenfeld, B.: Learning realistic human actions from movies. In: IEEE Conference on Computer Vision and Pattern Recognition, CVPR 2008, pp. 1–8. IEEE (2008)
15. Dalal, N., Triggs, B., Schmid, C.: Human detection using oriented histograms of flow and appearance. In: Leonardis, A., Bischof, H., Pinz, A. (eds.) ECCV 2006. LNCS, vol. 3952, pp. 428–441. Springer, Heidelberg (2006). doi:10.1007/11744047_33
16. Wang, H., Kläser, A., Schmid, C., Liu, C.-L.: Dense trajectories and motion boundary descriptors for action recognition. Int. J. Comput. Vis. **103**(1), 60–79 (2013)
17. Mohammadi, E., Wu, Q.J., Saif, M.: Human activity recognition using an ensemble of support vector machines. In: 2016 International Conference on High Performance Computing and Simulation (HPCS), pp. 549–554. IEEE (2016)

Detection and Classification

Gaussian Mixture Trees for One Class Classification in Automated Visual Inspection

Matthias Richter[1,2]([✉]), Thomas Längle[2], and Jürgen Beyerer[1,2]

[1] Karlsruhe Institute of Technology, Adenauerring 4, 76131 Karlsruhe, Germany
matthias.richter@kit.edu
[2] Fraunhofer IOSB, Fraunhoferstr. 1, 76131 Karlsruhe, Germany
{thomas.laengle,juergen.beyerer}@iosb.fraunhofer.de

Abstract. We present Gaussian mixture trees for density estimation and one class classification. A Gaussian mixture tree is a tree, where each node is associated with a Gaussian component. Each level of the tree provides a refinement of the data description of the level above. We show how this approach is applied to one class classification and how the hierarchical structure is exploited to significantly reduce computation time to make the approach suitable for real time systems. Experiments with synthetic data and data from a visual inspection task show that our approach compares favorably to flat Gaussian mixture models as well as one class support vector machines regarding both predictive performance and computation time.

Keywords: One class classification · Anomaly detection · Density estimation · Automated visual inspection

1 Introduction

During the last few years, computer vision applications such as object recognition and scene understanding have seen significant advances, most of which are backed by machine learning methods. Yet automated visual inspection for industrial manufacturing processes still largely relies on rule based expert systems [1]. We identify two main reasons for this.

First, while positive training samples are plentiful, defective examples are hard to obtain and therefore vastly underrepresented in the training set. Many classifiers allow to weight the training samples to deal with imbalanced data, and for those that do not, the training set can be over- or under-sampled. However, the training set will still not be representative. As a result, classification performance in production often lacks behind the performance on a validation set.

Second, classification takes place in an open world setting: only a small subset of the *classes* encountered in operation are represented in the training set. Objects of an unknown class must be recognized as such. There are plug-in rules to introduce a reject option to any classifier, e.g. [2,3], as well as classifier-specific methods such as [4], but these methods generally only reject objects

© Springer International Publishing AG 2017
F. Karray et al. (Eds.): ICIAR 2017, LNCS 10317, pp. 341–351, 2017.
DOI: 10.1007/978-3-319-59876-5_38

who fall close to the decision boundary. Unknown objects that are far from the decision boundary will not be rejected, even if they are distant from the known samples.

Both issues—unrepresentative training sets and open world classification—can be solved with one class classification (OCC). Here, the goal is to separate the target (foreground) class from all other classes (jointly denoted background class), when only samples from the foreground class are available for training.

1.1 Related Work

One class classification falls into the broader scope of anomaly detection, but here we will focus on OCC only. Readers interested in anomaly detection are referred to the extensive survey by Chandola et al. [5]. A common approach to OCC is to estimate the target class' feature density $p(\mathbf{x})$ (parametric or non-parametric, e.g., Gaussians or Parzen windows [6]) and reject a sample if $p(\mathbf{x}) < \theta$. The underlying assumption is that the background class follows a uniform distribution. While straightforward, this approach breaks down when the model assumption does not hold or when the training set is small in comparison to the number of features.

Many non-density-based approaches utilize support vector machines to work around this issue. Prominent examples are SVDD [7], where a minimal enclosing hypersphere is constructed around the training samples, and the one class SVM [8], where the target class is linearly separated from the origin with maximum margin. In both, the kernel trick enables complex nonlinear decision boundaries. Furthermore, both approaches have been shown to be equivalent if radial basis functions (RBF) kernels are used [8]. Many refinements of this approach are available (see, e.g., [9,10], or [11] for an overview), but here we will focus on non-SVM methods utilizing tree structures.

Desir et al. convert OCC into a binary classification task by generating artificial outlier samples that complement the target class, and solve the task using a Random Forest classifier [12]. Sampling from the outlier distribution becomes feasible using random subspace projection, bagging and exploiting the distribution of the known class. Targeting clustering instead of OCC, Vasconcelos and Lippman propose an expectation maximization method to learn hierarchical mixture models [13]. Similar to agglomerative clustering, the model is built bottom up, where they show that in the case of Gaussian components, the parameters of a given level can be directly computed from the parameters of the level below, without needing to consult the training set. In a related work, Williams presents a Bayesian approach to hierarchical clustering [14]. He proposes a generative hierarchical model of Gaussian, where the tree is represented by a structure vector and a connectivity matrix. Both are estimated from data using MCMC. Williams posits that the method can be used in arbitrary feature spaces, but only shows application in the univariate case. Since both approaches also produce density estimates, they can be used in OCC as described above. A density estimate can also be obtained using Ram and Gray's Density estimation trees [15], where the feature density is approximated as a piecewise constant function. This, however, unnecessarily restricts the shape of the density estimate, which leads to relatively poor prediction performance in a classification setting [15].

Perhaps closest to our approach are Criminisi et al.'s Density Forests [16]. Similar to decision tree learning [17], the trees in the forest are constructed by recursively finding splits that maximize information gain on the training set. They show that under a Gaussian assumption, this is equivalent to minimizing the volume of ellipsoids around the data on either side of the split. Leaf nodes are associated with a multivariate Gaussian density. As the leaves partition the space into non-overlapping regions, the overall density estimate is composed of non-overlapping truncated Gaussians, which in turn results in a complex partition function—the main drawback of this approach.

2   Methods

Gaussian mixture trees draw inspiration from decision trees as well as Gaussian mixture models (GMMs). A GMM is a convex combination of densities,

$$p(\mathbf{x}) = \sum_{k=1}^{K} P(z_k)\, p(\mathbf{x}|z_k) \tag{1}$$

where the components $p(\mathbf{x}|z_k) = \mathcal{N}(\mathbf{x}; \boldsymbol{\mu}_k, \boldsymbol{\Sigma}_k)$ are Gaussians and z_k indicates whether $\mathbf{x}$ was generated by the k-th component. The parameters $P(z_k)$, $\boldsymbol{\mu}_k$ and $\boldsymbol{\Sigma}_k$ can be estimated from training data $\mathcal{D} = \{\mathbf{x}_n | n = 1, \ldots, N\}$ using expectation maximization. OCC with GMMs is, however, wasteful: all K components have to be evaluated to obtain $p(\mathbf{x})$, even if they contribute little to the density at $\mathbf{x}$. Density estimation trees [16] require only one Gaussian to be evaluated, but the resulting density has a very complex partition function with no closed form solution. Axis aligned splits simplify the partition function [16], but at the same time severely restrict the shape of the density estimate.

A Gaussian mixture tree has no such limitations. It is fast to evaluate, straightforward to implement and offers multiple opportunities to speed up computation.

2.1   Gaussian Mixture Trees

A Gaussian mixture tree (GMT) is a tree structure, where each node is associated with a Gaussian density $p(\mathbf{x}|z_{\mathbf{k}})$ and a prior $P(z_{\mathbf{k}})$. Each level of the tree represents a density estimate, but with varying level of detail (see Fig. 1). To estimate the density $p(\mathbf{x})$ at a given point $\mathbf{x}$, one must consider the contribution of every path through the tree. Let $\mathbf{k} = (k_1 k_2 \ldots k_d)$ denote the path of length d that is obtained by first branching to k_1, then to the k_2-th child of k_1, etc. Let further denote $z_{\mathbf{k}}$ the node reached by following $\mathbf{k}$.

If the tree is one level deep ($d = 1$, Fig. 1b), then the GMT degenerates to a GMM and $p(\mathbf{x})$ is computed according to Eq. (1). If $d = 2$, then

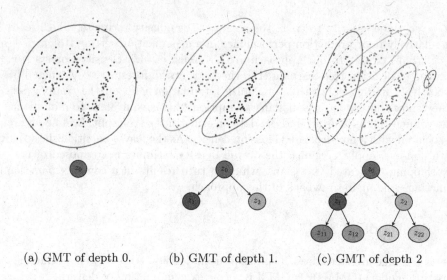

(a) GMT of depth 0.   (b) GMT of depth 1.   (c) GMT of depth 2

Fig. 1. GMTs of increasing depth describe the underlying data with an increasing level of detail. (a) and (b) are degenerate cases: a single Gaussian and a two-component GMM, respectively.

$p(\mathbf{x}) = \sum_{\mathbf{k}} p(\mathbf{x}, z_{\mathbf{k}})$ marginalizes over all possible paths $\mathbf{k} = (k_1 k_2)$ through the tree, where

$$p(\mathbf{x}, z_{\mathbf{k}}) = p(\mathbf{x}, z_{(k_1 k_2)}) = P(z_{(k_1)}|\mathbf{x})P(z_{(k_1 k_2)})p(\mathbf{x}|z_{(k_1 k_2)}). \qquad (2)$$

That is, the joint density estimate a leave is weighted by the *posterior* probability that $\mathbf{x}$ was generated its parent node. In general, the overall density marginalizes over all paths through the tree and the joint density of $\mathbf{x}$ and $z_{\mathbf{k}}$ is computed by weighting the estimate in the leaf node by the posteriors of all intermediate nodes, i.e., $p(\mathbf{x}) = \sum_{\mathbf{k}} p(\mathbf{x}, z_{\mathbf{k}})$, where

$$p(\mathbf{x}, z_{\mathbf{k}}) = P(z_{(k_1)}|\mathbf{x}) \cdots P(z_{(k_1 \dots k_{d-1})}|\mathbf{x})P(z_{(k_1 \dots k_d)})p(\mathbf{x}|z_{(k_1 \dots k_d)}). \qquad (3)$$

2.2   Gaussian Mixture Tree Learning

Learning a GMT from data is similar to decision tree training [17]. At the root node, a single Gaussian is fit to the data $\mathcal{D}$ using maximum likelihood and the prior is set to $P(z_0) = 1$. In the first level, the priors and Gaussian components are obtained by fitting a GMM to $\mathcal{D}$. Then, a reduced training set $\mathcal{D}_k = \{\mathbf{x} \in \mathcal{D}|P(z_k|\mathbf{x}) \geq \phi\}$ is constructed for each node such that $\mathcal{D}_k$ contains the samples that were likely generated by the corresponding GMM component z_k. Note that depending on ϕ and the number of child nodes, the same training sample may appear in more than one $\mathcal{D}_k$. Using $\mathcal{D}_k$ as training set for the corresponding branch, the procedure is applied recursively until a maximal depth is reached, or until $\mathcal{D}_k$ does not contain enough samples to reliably learn a GMM. Algorithm 1 summarizes this procedure.

```
function fit(node, 𝒟, d):
    if d ≤ d_max and |𝒟| ≥ N_min:
        for (μ_k, Σ_k, P(z_k)) ∈ GMM(𝒟, K):
            node.child_k ← fit(Node(μ_k, Σ_k, P(z_k)), {x ∈ 𝒟|P(z_k|x) ≥ φ}, d + 1)
    return node
```

Algorithm 1. Gaussian mixture tree learning.

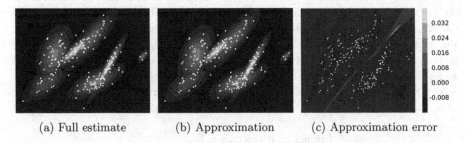

(a) Full estimate   (b) Approximation   (c) Approximation error

Fig. 2. Full and approximate density estimation with a binary GMT of depth 3. The approximation is erroneous in regions where two components of the same level overlap.

2.3 Fast Density Estimation

Density estimation according to Eq. (3) requires to evaluate all components in the tree. In consequence, the runtime is worse than estimation with a GMM with as many components as leaves on the GMT. However, the density can be approximated by considering only the path most likely to have generated x. Formally, let $c = \arg\max_{c \in \text{children}(k)} P(z_c|x)$ denote the path to z_k's child node responsible for x. Starting from the root node, the density is approximated recursively,

$$p_a(x, z_k) = \begin{cases} P(z_k)\, p(x|z_k) & \text{if } k \text{ is leaf node,} \\ p_a(x, z_c) & \text{otherwise.} \end{cases} \qquad (4)$$

Note that the approximation is erroneous in regions where components of the same level overlap (see Fig. 2). Still, for the purpose of one class classification, these errors can usually be tolerated. Density estimation can further be accelerated by computing log probabilities, in which case the greedy path search simplifies to

$$c = \arg\max_{c \in \text{children}(k)} (x - \mu_c)^\top \Sigma_c^{-1} (x - \mu_c) + \log|\Sigma_c| - \log P(z_c)^2. \qquad (5)$$

2.4 One Class Classification

Since GMTs are density estimators, a one class classifier can be constructed by thresholding, $h(x) = \mathbf{1}[p(x) \geq \theta]$. However, the choice of a suitable threshold θ is not straightforward, as the range of $p(x)$ depends on the dimensionality of the feature space as well as the distribution of the data. Here, we opt to determine θ using a proxy parameter $\rho \in [0, 1]$, so that $(N\rho)$ training samples are accepted,

$$\mathbb{E}_{\mathcal{D}}\left[\hat{p}(\mathbf{x}) \geq \theta\right] \approx \rho. \tag{6}$$

While straightforward, this scheme is still somewhat wasteful: Because the parent node already describes the data that reaches the child nodes, and because the covariance of a child node is smaller than or equal to the covariance of the parent, a sample can be rejected early if it falls sufficiently far from the Gaussian stored in the parent. For this, a rejection parameter $\theta_{\mathbf{k}}$ is associated with every node $\mathbf{k}$ and a sample $\mathbf{x}$ is rejected if $p(\mathbf{x}|z_{\mathbf{k}}) < \theta_{\mathbf{k}}$. Otherwise, if $\mathbf{k}$ denotes a leaf node, the sample is accepted. If $\mathbf{k}$ is not a leaf and $\mathbf{x}$ is not rejected, the procedure is recursed with the child node with largest posterior probability. Again, starting from the root node and with $\mathbf{c}$ computed as in Eq. (5),

$$h(\mathbf{x}, \mathbf{k}) = \begin{cases} \text{reject} & \text{if } p(\mathbf{x}|z_{\mathbf{k}}) < \theta_{\mathbf{k}}, \\ \text{accept} & \text{if } p(\mathbf{x}|z_{\mathbf{k}}) \geq \theta_{\mathbf{k}} \text{ and } \mathbf{k} \text{ is leaf node}, \\ h(\mathbf{x}, \mathbf{c}) & \text{otherwise}, \end{cases} \tag{7}$$

where $\theta_{\mathbf{k}}$ is determined using Eq. (6) considering only the samples in $\mathcal{D}_{\mathbf{k}}$. Note that unlike density-based rejection, the decision boundary will be composed of superimposed ellipsoids. As a result, their classification performance will differ.

3   Experiments

We compared our approach to OCC with Gaussian mixture models as well as one class SVMs with RBF kernel using both synthetic and real data. The GMM and SVM implementation were provided by the well tested scikit-learn library [18]. The GMTs were implemented using the Julia language [19]. All experiments were run on a consumer grade PC with 8 GB of RAM and an Intel Core i7 CPU clocked at 2.4 GHz.

We used binary GMTs with $\phi = 0.5$, a maximum depth of $d_{\max} = 4$ and an acceptance ratio of $\rho = 0.99$ in all experiments. The GMM consisted of $K = 5$ components with full covariance matrices and the rejection parameter θ was fixed so that $\rho = 0.99$. The SVM parameter ν, which can be interpreted as upper bound on the fraction of training errors, was set to $\nu = 0.1$, and the kernel width was set to the dimensionality of the feature space D.

3.1   Synthetic Data

Our first set of experiments was conducted on synthetic data with varying dimensionality $D \in \{2, 5, 10, 20, 50\}$. The foreground samples were generated by a GMM with $K = 5$ components, where the means were normally distributed, $\boldsymbol{\mu}_k \sim \mathcal{N}(\mathbf{0}, 2\sqrt{D}\mathbf{I})$, the covariances followed a Wishart distribution,

Table 1. True and false positive rates for the evaluated one class classifiers using synthetic data of increasing dimensionality.

Method	$\mathbf{x} \in \mathbb{R}^2$		$\mathbf{x} \in \mathbb{R}^5$		$\mathbf{x} \in \mathbb{R}^{10}$		$\mathbf{x} \in \mathbb{R}^{20}$		$\mathbf{x} \in \mathbb{R}^{50}$	
	TPR	FPR	TPR	FPR	TPR	FPR	TPR	FPR	TPR	FPR
GMM	0.98	0.20	0.94	0.01	0.84	-	0.50	-	-	-
oc SVM	0.94	0.17	-	-	-	-	-	-	-	-
full GMT	0.97	0.20	0.89	-	0.74	-	0.98	-	0.76	-
approx. GMT	0.97	0.20	0.88	-	0.72	-	0.98	-	0.76	-
oc GMT	0.99	0.30	0.99	0.01	0.97	-	0.79	-	0.51	-
optimal classifier	0.99	0.20	0.99	0.01	0.99	-	1.00	-	0.99	-

$\Sigma_k \sim \mathcal{W}_D(\mathbf{I}, D^2)$, and the priors $P(z_k)$ were drawn from a uniform distribution and normalized to sum to one. The background class was sampled from a uniform distribution over a region as large as the range of values of the foreground samples. For every D, we generated 250 training samples and 200 000 testing samples, where half of the samples were drawn from the foreground, and the other half were drawn from the background distribution. The experiments were repeated 20 times and the results were averaged.

Table 1 shows the average true and false positive rates (TPR and FPR) obtained by the different classifiers. The error rates of the Bayes optimal classifier were computed using the ground truth distributions. Surprisingly, one class SVM (oc SVM) was not able to derive reliable classifiers for $D > 2$. This result was observed for other values of ν as well. OCC with a GMM works well in low dimensional spaces, but performance deteriorates for $D > 10$—even though the number of components is the same as in the ground truth distribution. Density based rejection with GMT estimates (full and approximate) generally work well, but results are unreliable with $D = 10$. One class GMT (oc GMT) according to Eq. (7) gives the most stable results, but is biased toward accepting samples, resulting in a larger FPR than the other approaches. Note that this bias can be controlled using the parameter ρ.

The average classification time per sample is shown in Table 2. Note that the timings for GMM and SVM are not directly comparable to our approach because of the different implementation backends. Still, it can be seen that computation time required by oc SVM grows much quicker with the number of features than the other methods. This can be explained by the larger number of support vectors required to define the decision boundary in higher dimensional spaces. Computation time for both GMM and GMT grows much slower. Indeed, the complexity of GMT seems to be near constant expect for $D = 50$. The likely cause is that here the CPU's cache is too small to contain the whole dataset. Recalling that the trees are of the same depth, this result is explained by tree traversal dominating computation of the Gaussian densities. Finally, it can be

Table 2. Average classification time per sample in μs. Note that the timings are not directly comparable due to different implementation languages.

Method	$\mathbf{x} \in \mathbb{R}^2$		$\mathbf{x} \in \mathbb{R}^5$		$\mathbf{x} \in \mathbb{R}^{10}$		$\mathbf{x} \in \mathbb{R}^{20}$		$\mathbf{x} \in \mathbb{R}^{50}$	
	Accept	Reject	Accept	Reject	Accept	Reject	Accept	Reject	Accept	Reject
GMM	0.79	0.71	0.97	0.97	1.32	1.36	2.20	2.25	4.99	4.86
oc SVM	3.48	3.41	8.20	14.44	9.47	22.23	23.10	23.83	33.19	33.02
full GMT	9.61	9.45	9.69	9.62	10.31	10.26	8.68	6.93	17.35	15.79
approx. GMT	2.59	2.67	2.76	2.68	3.06	3.08	2.74	2.46	10.20	10.16
oc GMT	2.54	1.03	2.51	0.52	2.97	0.53	2.51	0.70	4.04	1.66

(a) Lego bricks: 2×2, 2×4 and connector   (b) healthy/infected berries and twigs

Fig. 3. Example images from the Lego and wine datasets (images not to scale).

seen that on average, rejection is more than three times as fast as acceptance with oc GMT.

3.2 Real Data

In the second set of experiments, we used two datasets from two automated visual inspection tasks: sorting of Lego bricks and sorting of wine berries. The images for both datasets were recorded using a bulk material sorting machine. Example images are shown in Fig. 3. In the first dataset, nine canonical geometry features[1] were extracted from images the images. In three experiments, the target classes were bricks of size 2×2 (377 samples), 2×4 (433 samples), and a special connector brick (415 samples). More information about this dataset can be found in [20]. For the wine dataset, ten dimensional color and texture descriptors were extracted from the images as described in [21]. Here, the goal was to discriminate healthy wine berries (145 samples, target class) from infected fruit and other parts of the plant (633 samples, unknown background class). Stratified five fold cross validation was used in all experiments.

The true positive rates are shown in Fig. 4, where we omitted GMT with full density estimates. As with the synthetic data, one class SVM completely fails to classify the Lego dataset, while both GMM and GMT produce reliable classifiers. On the wine dataset, density based OCC with both GMM and GMT results in true positive rates below 60%, but one class SVM as well as one class

---

[1] Diameter, density, area, convex area, compactness, extent, roundness, perimeter, and convex hull perimeter.

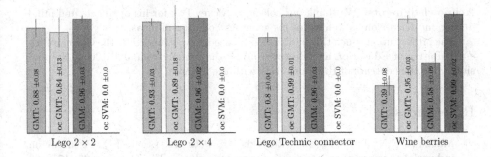

Fig. 4. True positive rates obtained on data from a real world sorting application. Error bars mark one standard deviation over the 5-fold cross validation. False positive rates were below 0.03 in all experiments.

GMT give true positive rates above 90%. The false positive rate was below 3% in all experiments. All in all, one class GMT provides the most consistent results over all datasets.

4 Conclusion

In this paper we have presented Gaussian mixture trees for density estimation and one class classification. Each node in the tree is associated with a Gaussian distribution. Different levels of the tree represent different levels of detail in the density approximation. This structure is exploited to significantly speed up the rejection of unknown samples. Both the training and classification algorithm are simple and straightforward to implement. Evaluation with synthetic and recorded data show that GMTs produce reliable classificators that outperform both OCC with Gaussian mixture models and one class SVMs. At the same time, classification requires very little computation time, which makes this approach suitable for real time applications like automated visual inspection.

In the future, we plan to improve the training algorithm by investigating different stopping criteria (e.g., based on information gain) and regularizing the Gaussian components to deal with small datasets (e.g., using covariance shrinkage). Multi-class classification with reject option could also be handled using GMTs by additionally associating a classifier with each leaf node. Preliminary investigations show that a GMT distributes the training set so that the leaf nodes are almost class pure, which simplifies classification. The levels of detail in a GMT may also find use in approximate computing, where an inaccurate, but timely result may be favored over one that is exact, but delivered too late. Finally, large parts of Sect. 2 are valid for *any* type of base density $p(\mathbf{x}|z_{\mathbf{k}})$. Swapping Gaussians for a suitable model could make the approach applicable in other domains such as natural language processing.

Acknowledgments. We thank our colleague Alexey Pak for his open ears and mind during our discussion as well as for his generous donation of ideas.

The project upon which this publication is based on was funded by the German Federal Ministry of Education and Research under the funding code 01IS12051. Responsibility for the contents of this work rests with the authors.

References

1. Malamas, E.N., Petrakis, E.G., Zervakis, M., Petit, L., Legat, J.D.: A survey on industrial vision systems, applications and tools. Image Vis. Comput. **21**, 171–188 (2003)
2. Herbei, R., Wegkamp, M.H.: Classification with reject option. Can. J. Stat. **34**, 709–721 (2006)
3. Yuan, M., Wegkamp, M.: Classification methods with reject option based on convex risk minimization. J. Mach. Learn. Res. **11**, 111–130 (2010)
4. Dubuisson, B., Masson, M.: A statistical decision rule with incomplete knowledge about classes. Pattern Recogn. **26**, 155–165 (1993)
5. Chandola, V., Banerjee, A., Kumar, V.: Anomaly detection: a survey. ACM Comput. Surv. (CSUR) **41**, 15 (2009)
6. Parzen, E.: On estimation of a probability density function and mode. Ann. Math. Stat. **33**, 1065–1076 (1962)
7. Tax, D.M., Duin, R.P.: Support vector domain description. Pattern Recogn. Lett. **20**, 1191–1199 (1999)
8. Schölkopf, B., Platt, J.C., Shawe-Taylor, J., Smola, A.J., Williamson, R.C.: Estimating the support of a high-dimensional distribution. Neural Comput. **13**, 1443–1471 (2001)
9. Huang, G., Yang, Z., Chen, X., Ji, G.: An innovative one-class least squares support vector machine model based on continuous cognition. Knowl.-Based Syst. **123**, 217–228 (2017)
10. Utkin, L.V., Zhuk, Y.A.: An one-class classification support vector machine model by interval-valued training data. Knowl.-Based Syst. **120**, 43–56 (2017)
11. Khan, S.S., Madden, M.G.: One-class classification: taxonomy of study and review of techniques. Knowl. Eng. Rev. **29**, 345–374 (2014)
12. Désir, C., Bernard, S., Petitjean, C., Heutte, L.: One class random forests. Pattern Recogn. **46**, 3490–3506 (2013)
13. Vasconcelos, N., Lippman, A.: Learning mixture hierarchies. In: NIPS, pp. 606–612 (1998)
14. Williams, C.K.: A MCMC approach to hierarchical mixture modelling. In: NIPS, pp. 680–686 (1999)
15. Ram, P., Gray, A.G.: Density estimation trees. In: 17th ACM SIGKDD International Conference on Knowledge Discovery and Data Mining, pp. 627–635. ACM (2011)
16. Criminisi, A., Shotton, J., Konukoglu, E.: Decision forests: a unified framework for classification, regression, density estimation, manifold learning and semi-supervised learning. Found. Trends® Comput. Graph. Vis. **7**, 81–227 (2012)
17. Breiman, L., Friedman, J., Stone, C.J., Olshen, R.A.: Classification and Regression Trees. CRC Press, Boca Raton (1984)

18. Pedregosa, F., Varoquaux, G., Gramfort, A., Michel, V., Thirion, B., Grisel, O., Blondel, M., Prettenhofer, P., Weiss, R., Dubourg, V., Vanderplas, J., Passos, A., Cournapeau, D., Brucher, M., Perrot, M., Duchesnay, E.: Scikit-learn: machine learning in python. J. Mach. Learn. Res. **12**, 2825–2830 (2011)
19. Bezanson, J., Karpinski, S., Shah, V.B., Edelman, A.: Julia: a fast dynamic language for technical computing (2012)
20. Richter, M., Maier, G., Gruna, R., Längle, T., Beyerer, J.: Feature selection with a budget. In: EECSS 2016, Budapest, Hungary, pp. 104.1–104.8. Avestia Publishing (2016)
21. Richter, M., Längle, T., Beyerer, J.: Visual words for automated visual inspection of bulk materials. In: Machine Vision Applications, Tokyo, Japan, pp. 210–213 (2015)

Shadow Detection for Vehicle Classification in Urban Environments

Muhammad Hanif[1], Fawad Hussain[1(✉)],
Muhammad Haroon Yousaf[1], Sergio A. Velastin[2],
and Zezhi Chen[3]

[1] Department of Computer Engineering,
University of Engineering and Technology, Taxila, Taxila, Pakistan
engr.muhammadhanif@yahoo.com,
{fawad.hussain,haroon.yousaf}@uettaxila.edu.pk
[2] Department of Computer Science,
University Carlos III de Madrid, Getafe, Spain
sergio.velastin@ieee.org
[3] School of Computing and Information Systems,
Kingston University, London, UK
Zezhi.chen@gmail.com

Abstract. Finding an accurate and computationally efficient vehicle detection and classification algorithm for urban environment is challenging due to large video datasets and complexity of the task. Many algorithms have been proposed but there is no efficient algorithm due to various real-time issues. This paper proposes an algorithm which addresses shadow detection (which causes vehicles misdetection and misclassification) and incorporates solution of other challenges such as camera vibration, blurred image, illumination and weather changing effects. For accurate vehicles detection and classification, a combination of self-adaptive GMM and multi-dimensional Gaussian density transform has been used for modeling the distribution of color image data. RGB and HSV color space based shadow detection is proposed. Measurement-based feature and intensity based pyramid histogram of orientation gradient are used for classification into four main vehicle categories. The proposed method achieved 96.39% accuracy, while tested on Chile (MTT) dataset recorded at different times and weather conditions and hence suitable for urban traffic environment.

Keywords: Gaussian mixture model · Histogram of oriented gradient · ITS · RGB · HSV · Inter-channel intensity deviation

1 Introduction

Technologies to count, track and classify vehicles have already been used widely in many fields of research and monitoring. These applications are being extended to intelligent transportation systems, surveillance, security, e-tagging toll plazas, avoiding traffic jams and advanced driver assistance systems [1]. The tendency to use surveillance cameras in order to prevent and handle traffic congestion and intelligently

© Springer International Publishing AG 2017
F. Karray et al. (Eds.): ICIAR 2017, LNCS 10317, pp. 352–362, 2017.
DOI: 10.1007/978-3-319-59876-5_39

monitoring is increasing day by day [2]. Transportation monitoring is categorized mainly in three categories. Manual monitoring by humans requires much effort. Physical devices based monitoring [3] is costly and unreliable system while computer vision based systems offer much promise. Many algorithms proposed for highway traffic have achieved sufficient efficiency and reliability because the environment on highways is not so complex [4]. However, for urban environments this is much more difficult [5]. Shadow detection and removal is one of the main challenge faced in vision based traffic monitoring because shadows cause misdetection and misclassification of vehicles [6]. Still various challenges exist for the automated and reliable surveillance realization [7], especially in urban environments [8]. In this paper we propose an RGB and HSV inter-channel intensity deviation based shadow detection technique which provides efficient results when tested on complex dataset. Improving shadow detection also increases the accuracy of vehicles detection and classification.

The proposed technique, dataset details, experimentation results and conclusion is explained in detail in subsequent Sects. 2, 3, 4 and 5 respectively.

2   Proposed Technique for Vehicle Detection and Classification

The proposed scheme is divided into three main steps as shown in Fig. 1 as explained below.

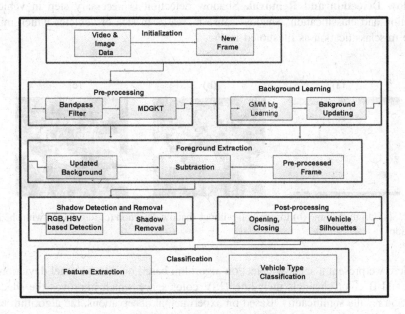

Fig. 1. Proposed frame work

2.1   Initialization and Pre-processing

The process requires a video or a sequence of images as input. As the data is captured in various environments and under different conditions which contains noise. In the pre-processing step a band-pass filter in the frequency domain is performed. The filtered output is then back transformed into the spatial domain [12]. An image may contain low frequencies noise due to illumination condition which varies with data captured time. The average brightness determines the DC component of the information captured while illumination contains only low frequencies or high frequencies noise which causes due to the sudden change occur for a short interval of time or might be repetitive such as thermal. Such noises are removed with the band-pass filter. Thereafter, as a preprocessing step, Multi-dimensional Gaussian Kernel density (MDGKT) is used. To control the size of kernel, it only requires pair of bandwidth parameters (h_s, h_t) which in turns determine the time interval and resolution of GMM. Experiments show improvements in terms of stability and accuracy of GMM when using the MDGKT as demonstrated in [7, 10].

2.2   Foreground Extraction

Foreground is extracted by subtracting learned background (carried out on-line using MDGKT and self-adaptive Gaussian mixture model [1]) from the current frame. The subtraction result gives the moving foreground part in the current frame [11].

Shadow Detection and Removal. Shadow detection is necessary step in vehicles detection and classification. Shadow cause increases in size of vehicles which might cause misclassification as illustrated in Fig. 2.

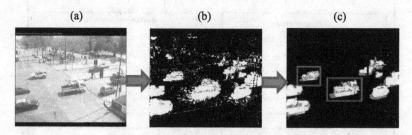

Fig. 2. (a) Original frame (b) foreground without shadow detection (c) foreground with shadow detection

Here we present a shadow detection algorithm based on inter-channel deviation of R, G and B color channels including HSV color space which improves the shadow detection results significantly. Based on experimental observations, the algorithm uses the following criteria (a block diagram is shown in Fig. 3):

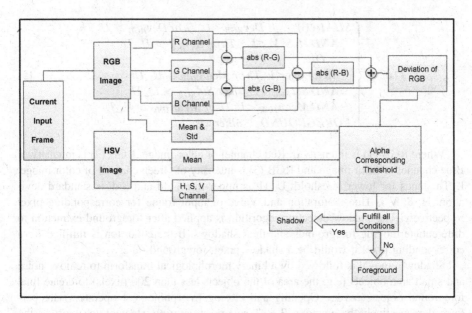

Fig. 3. Flow chart of proposed technique for shadow detection

i. RGB inter-channel color deviation of vehicles with blue/bluish color and red/reddish color will be greater than the threshold as the corresponding color channels value will be much greater than the other color channel values.

ii. For black/blackish color vehicles mean value of RGB color channel will be smaller than lower the threshold which will be even lower than the mean value of shadow this assumption helps to detect dark black color vehicles accurately.

iii. For shadow, RGB inter-channel color deviation will be smaller than threshold and the V-channel value from HSV color model is used for shadow verification as V-channel value for shadow will be smaller than the threshold.

iv. For vehicles, whose color properties matches with shadow such as dark gray or similar to shadow whose RGB inter-channel color deviation and V-channel value is same as shadow for those vehicles another threshold is applied on S-channel from HSV color space which differentiate vehicles from shadow based on saturation of foreground. The mathematical form of algorithm is as follow,

$$D1 = abs(RGB(1) - RGB(2)) \tag{1}$$

$$D2 = abs(RGB(2) - RGB(3)) \tag{2}$$

$$D3 = abs(RGB(1) - RGB(3)) \tag{3}$$

$$Dev_{RGB} = D1 + D2 + D3 \tag{4}$$

$$f(x) = \begin{cases} SHADOW, & \text{if } Dev_{RGB} < L\_Th \text{ or } Dev_{RGB} > U\_Th \\ \quad AND \, H, S, V < L\_\_Th \text{ or } H, S, V > U\_Th \\ \quad OR \\ Mean_{RGB} < L\_Th \text{ or } Mean_{RGB} > U\_Th \\ \quad AND \, Std_{RGB} < L\_Th \text{ or } Std_{RGB} > U\_Th \\ \quad AND \, Mean_{HSV} < L\_Th \text{ or } Mean_{HSV} > U\_Th \\ FOREGROUND, & \text{other wise} \end{cases} \qquad (5)$$

Where RGB (1) is intensity of Red channel of color image, RGB (2) is intensity of Blue channel of color image and RGB (3) is intensity of Green channel of color image. L_Th stands for lower threshold, U_Th is upper threshold and Std. is standard deviation. H, S, V is Hue, Saturation and Value of input frame for corresponding pixel respectively. The shadow detection algorithm is applied after foreground extraction to differentiate moving foregrounds and their shadows. If the condition is fulfilled for a corresponding pixel it would be a shadow pixel, foreground otherwise.

Shadow removal is followed by a binary morphological transform to remove noise and small area objects (e.g. the area of the objects less than 200 pixels) to create final foreground objects masks. Opening and closing (morphological operators) are performed respectively by using a 3×3 square structuring element to improve the detection results.

2.3   Vehicles Classification

A set of measurement-based features (MBF) is used to classify vehicles. MBF is employed due to less computational and memory complexity for vehicle feature database generation [1]. Pyramid histogram of gradient (PHOG) based feature is used from [9] to improve the classification accuracy. The used descriptor consists of HOG over all sub-regions on an image with equivalent resolution level of each bounding box used for detection. Combining MBF with IPHOG gives motivation to utilize pros of both features for better vehicle classification accuracy, which is discussed in Sect. 4.

For classification, blocks of detected vehicles are selected as 50×50 cells size, producing a 1×1485 length feature vector. This feature vector consists of the parameters shown in Table 1:

Table 1. Feature vector's length

Feature	Length
Perimeter	1×1
Minor axis length of best-fitting ellipse	1×6
Width of bounding box	1×1
Area	1×1
3-level IPHOG	1×1476
Total length of feature vector (for one vehicle)	1×1485

3 Dataset

The Chile (MTT) dataset (available upon request from the authors) is used in this paper which is contains many of the main challenges faced in the detection and classification. The MTT dataset consists of 5 h of video recoded under different weather conditions and different times. Image size is 704 × 576 pixels, recorded at 25 frames per second. The videos are compressed using Xvid MPG4 codec. The dataset is captured under following different weather conditions (Fig. 4) and at different times of day (Fig. 5).

Fig. 4. Samples of Chile (MTT) dataset (a) cloudy (b) raining (c) strong sunlight

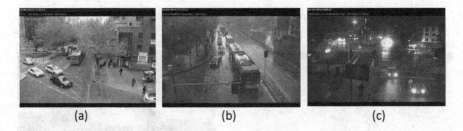

Fig. 5. Samples of Chile (MTT) dataset showing (a) day (b) evening (c) night time

4 Experimental Results

Vehicles detection and classification approach is presented which is combination of MDGKT and self-adaptive GMM for detection along with RGB and HSV based shadow detection technique. And IPHOG and MBF are used for classification. The efficiency of algorithm is tested on MATLAB R2013a with Intel core i3, 1.70 GHz, 8 GB RAM, 2 GB graphic RAM and 64-bit operating system. In the following section experiments based on two different classifiers, SVM and RF for GMM are done.

Figures 6 illustrate the detection of vehicles under different weather conditions and times. Shadow detection algorithm is evaluated by comparing with annotated ground truths of the dataset because there is no standard performance evaluation method in intelligent transportation system particularly for Chile (MTT) dataset.

Shadow detection results of our proposed method are more accurate than [1], particularly for white color vehicles as explained earlier in Sect. 2.2. Figure 7 illustrate the improvement. Model-based classification depicts the poor results as presented in Table 2. The True Positive Rate (TPR) of different vehicle types (van, car, motorcycle and bus) based on model-based classification are (0.9613, 0.7929, 0.7759, 1.0) respectively, resulting in an un-weightage TPR average of 0.8895.

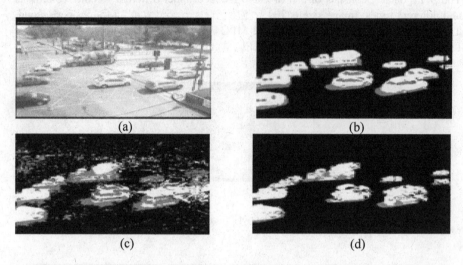

(a)   (b)

(c)   (d)

Fig. 6. (a) Input frame (b) ground truth (c) detected foreground (d) cleaned foreground

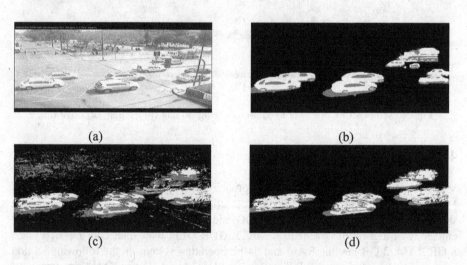

(a)   (b)

(c)   (d)

Fig. 7. (a) Original image (b) ground truth (c) results of [1] and (d) our result

The data set is classified by using the support vectors of best performing SVM classifier which is used for classification of dataset [1]. The resulting TPR for car, van, bus, and motorcycle/bicycle are 0.9661, 0.8913, 0.9931 and 1.0, respectively. The average TPR is 0.9509. Associated confusion matrix is given in Table 3.

Table 2. Confusion matrix for MBC

Vehicles types	Car	Van	Bus	Motorcycle
Car	993	40	0	0
Van	106	467	16	0
Bus	0	65	225	0
Motorcycle	0	0	0	143

Table 3. Confusion matrix for SVM

Vehicles types	Car	Van	Bus	Motorcycle
Car	998	35	0	0
Van	63	525	1	0
Bus	0	2	288	0
Motorcycle	0	0	0	143

Comparison of TPR results between SVM and RF is presented in Table 4 using MBF, IPHOG and MBF+IPHOG features. TPR is the median of the 10-fold cross validation results.

Table 4. The median TPR of SVM vs. RF (four different features used)

Features vs. classifiers	MBF	IPHOG	MBF+IPHOG
SVM	0.954	0.985	0.991
RF	0.951	0.973	0.981

For classification of vehicle types, SVM outperforms RF, using MBF or IPHOG feature sets individually and jointly. In SVM, IPHOG performs better than MBF. Table 4 clearly shows that SVM using a combination of MBF+IPHOG is the best for classification of vehicle type. Shadow removal improves the classification results significantly. If shadow is not detected and removed from foreground it causes mis-classification particularly when shadow of a vehicle extended up to the other nearby vehicle, then whole region is detected as foreground and is misclassified or it increases the foreground region of vehicle casting shadow as shown in Fig. 8.

Classification results without shadow detection and removal shows lower accuracy than the classification after shadow removal. Accuracy difference increases as the difference between the sizes of types of vehicles lowers as shown in Fig. 9 due to the increase of size, motorcycles tend to be categorized as cars, cars as vans and vans as buses.

This type of misclassification is reduced by shadow removal. Classification accuracy after shadow removal is improved from 78.5% to 94.3% for cars, 76.5% to 89.1% for vans, 92.6% to 98.2% for buses and 84.7% to 99% for motorcycles, which shows an average accuracy increases from 83.07% to 95.15%. Post-processing also improves the accuracy with a countable figure because post-processing clarifies the vehicles occupied region and remove noise connected with vehicles. Post-processing improves classification accuracy from 94.3% to 96.6% for cars, 87.3% to 89.1% for vans, 98.2% to 99.3% for buses and 99.1% to 100% for motorcycles, which is an increase of average accuracy from 95.15% to 96.39%. This is summarized in Fig. 9.

It is not possible to address all challenges simultaneously. Our proposed system is robust to shadow detection more accurately, illumination changes over long period of times and blurring problem. However, there are certain limitations in the proposed system for scenarios like strong light reflection, road surface reflection after rain and sudden illumination changes due to switching ON/OFF the street lights during night time.

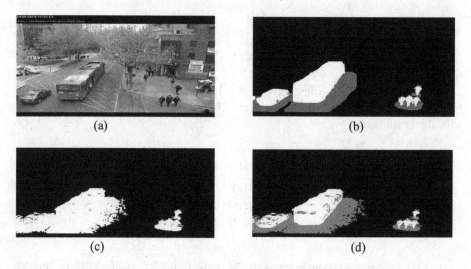

(a)   (b)

(c)   (d)

Fig. 8. (a) Original frame (b) ground truth (c) moving foreground without shadow detection (d) moving foreground with shadow detection

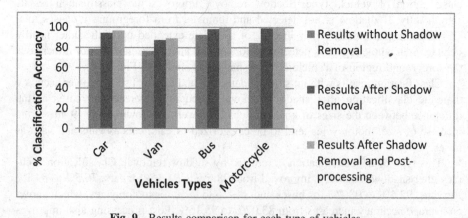

Fig. 9. Results comparison for each type of vehicles

5 Conclusion

The proposed algorithm of shadow detection based on RGB and HSV color space provides good accuracy for vehicles detection an urban environment. Comparisons with methods reported in [1] show that the proposed algorithm provides better results of shadow detection especially for vehicles of white colors and vehicles whose properties matches with shadow e.g. gray color vehicles. The results demonstrate a vehicle detection rate of 96.39% and classification accuracy of 94.69% under varying illumination and weather conditions (from cloud to rain). There is no perfect system. Background modeling and subtraction are naturally application oriented due to different semantic interpretation of the moving foreground and background, thus leaving few unsolved problems. In future, approach can be extended to incorporate the view independence and to select the better classifiers dealing with automatically segmented binary silhouettes of vehicles detected from video sequences in multi-camera tracking network.

The authors are grateful to the Chilean Ministry of Transport and Telecommunications (MTT) for providing video material of urban traffic captured in Santiago de Chile.

References

1. Chen, Z., Ellis, T., Velastin, S.: Vehicle detection, tracking and classification in urban traffic. In:15th IEEE Annual Conference on Intelligent Transportation Systems, Anchorage, Alaska, USA, 16–19 September 2012, pp. 951–956 (2012)
2. Ozkurt, C., Camci, F.: Automatic traffic density estimation and vehicle classification for traffic surveillance systems using neural networks. Math. Comput. Appl. **14**(3), 187–196 (2009)
3. Day, C., Premachandra, H., Brennan, T., Sturdevant, J., Bullock, D.: Operational evaluation of wireless magnetometer vehicle detectors at a signalized intersection (2009). http://trrjournalonline.trb.org/doi/abs/10.3141/2192-02
4. Yousaf, K., Iftikhar, A., Javed, A.: Comparative analysis of automatic vehicle classification techniques: a survey (2012). https://www.researchgate.net/publication/266288728_Comparative_Analysis_of_Automatic_Vehicle_Classification_Techniques_A_Survey
5. Chen, Z., Ellis, T., Velastin, S.: Confidence based active learning for vehicle classification in urban traffic. In: The Fourth Chilean Workshop on Pattern Recognition (CWPR 2012), Valparaiso, Chile, 12–16th November 2012 (2012)
6. Chen, Z., Ellis, T., Velastin, S.: Vehicle type categorization: a comparison of classification schemes. In: 14th IEEE Annual Conference on Intelligent Transportation Systems, The George Washington University, Washington, DC, USA, 5–7 October 2011, pp. 74–79 (2011)
7. Chen, Z., Ellis, T.: Self-adaptive Gaussian mixture model for urban traffic monitoring system. In: 11th IEEE International Workshop on Visual Surveillance (ICCV 2011 Workshop), Barcelona, Spain, 13 November, pp. 1769–1776 (2011)
8. Cucchiara, R., Piccardi, M., Prati, A.: Detecting moving objects, ghosts and shadows in video streams. IEEE Trans. Pattern Anal. Mach. Intell. **25**(10), 1337–1342 (2003)

9. Chen, Z., Pears, N., Freeman, M., Austin, J.: Background subtraction in video using recursive mixture models, spatio-temporal filtering and shadow removal. In: Bebis, G., et al. (eds.) ISVC 2009. LNCS, vol. 5876, pp. 1141–1150. Springer, Heidelberg (2009). doi:10.1007/978-3-642-10520-3_109

10. Kaew TraKul Pong, P., Bowden, R.: An improved adaptive background mixture model for real-time tracking with shadow detection. In: Remagnino, P., Jones, G.A., Paragios, N., Regazzoni, C.S. (eds.) Video-Based Surveillance Systems, pp. 135–144. Springer, New York (2001)

11. Izadi, M., Saeedi, P.: Robust region-based background subtraction and shadow removing using color and gradient information. In: 19th International Conference on Pattern Recognition (ICPR), pp. 1–5 (2008)

12. Buch, N., Orwell, J., Velastin, S.A.: 3D Extended Histogram of Oriented Gradients (3DHOG) for Classification of Road Users in Urban Scenes (2009)

Input Fast-Forwarding for Better Deep Learning

Ahmed Ibrahim[1,4(✉)], A. Lynn Abbott[1], and Mohamed E. Hussein[2,3]

[1] Virginia Polytechnic Institute and State University, Blacksburg, USA
{nady,abbott}@vt.edu
[2] Egypt-Japan University of Science and Technology, New Borg El Arab, Egypt
mohamed.e.hussein@ejust.edu.eg
[3] Alexandria University, Alexandria, Egypt
[4] Benha University, Banha, Egypt

Abstract. This paper introduces a new architectural framework, known as input fast-forwarding, that can enhance the performance of deep networks. The main idea is to incorporate a parallel path that sends representations of input values forward to deeper network layers. This scheme is substantially different from "deep supervision", in which the loss layer is re-introduced to earlier layers. The parallel path provided by fast-forwarding enhances the training process in two ways. First, it enables the individual layers to combine higher-level information (from the standard processing path) with lower-level information (from the fast-forward path). Second, this new architecture reduces the problem of vanishing gradients substantially because the fast-forwarding path provides a shorter route for gradient backpropagation. In order to evaluate the utility of the proposed technique, a Fast-Forward Network (FFNet), with 20 convolutional layers along with parallel fast-forward paths, has been created and tested. The paper presents empirical results that demonstrate improved learning capacity of FFNet due to fast-forwarding, as compared to GoogLeNet (with deep supervision) and CaffeNet, which are 4× and 18× larger in size, respectively. All of the source code and deep learning models described in this paper will be made available to the entire research community (https://github.com/aicentral/FFNet).

1 Introduction

Developments in deep learning have led to networks that have grown from 5 layers in LeNet [10], introduced in 1998, to 152 layers in the latest version of ResNet [5]. One consequence of deeper and deeper networks is the problem of vanishing gradients during training. This problem occurs as error values, which depend on the computed gradient values, are propagated backward through the network to update the weights at each layer. With each additional layer, a smaller fraction of the error gradient is available to guide the adjustment of network weights. As a result, the weights in early layers are updated very slowly; hence, the performance of the entire training process is degraded.

© Springer International Publishing AG 2017
F. Karray et al. (Eds.): ICIAR 2017, LNCS 10317, pp. 363–370, 2017.
DOI: 10.1007/978-3-319-59876-5_40

Many models have been proposed to overcome the vanishing-gradient problem. One approach is to provide alternative paths for signals to travel, as compared to traditional layer-to-layer pathways. An example of this approach is the Deeply-Supervised Network (DSN) [11], where a companion objective function is added to each hidden layer in the network, providing gradient values directly to the hidden layers. DSN uses Support Vector Machines (SVM) [3] in its companion objective function, which means that end-to-end training of the network is not supported. Another example is relaxed deep supervision [12], where an improvement over a holistic edge detection model [19] is made by providing relaxed versions of the target edge map to the earlier layers of the network. This approach provides a version of the gradient directly to the early layers. However, relaxed deep supervision is suitable only for problems where relaxed versions of the labels can be created, such as for maps of intensity edges. GoogLeNet [13] is another model that uses a mechanism to address the problem of vanishing gradients. More relevant details about GoogLeNet will be given in Sect. 2 because it serves as a baseline for comparison with our proposed model.

The novel approach that is proposed here provides parallel signal paths that carry simple representations of the input to deeper layers through what we call a fast-forwarding branch. This approach allows for a novel integration of "shallower information" with "deeper information" by the network. During training the fast-forwarding branch provides an effective means for back-propagating errors so that the vanishing-gradient problem is reduced.

To demonstrate the efficacy of the model, we created a 20 layer network with fast-forwarding branches, which we call FFNet. To study the effect of the fast-forwarding concept, the network layers are made of simple convolutional layers followed by fully connected layers with no additional complexities. The results that we have obtained using the relatively small and simple FFNet model have been surprisingly good, especially when compared with the performance of bigger and more complex models.

The rest of this paper is organized as follows. Section 2 presents a brief survey of related work, including a discussion of the models that will be used as a baseline to be compared with FFNet. Section 3 provides details concerning the proposed model. In order to gauge the performance of this approach, experimental results from FFNet were compared with results from several well-known deep models. These experiments are described in Sect. 4. Finally, concluding remarks are given in Sect. 5.

2   Related Work

2.1   Deep Learning

Deep learning is a machine-learning technique that has become increasingly popular in computer vision research. The main difference between classical machine learning (ML) and deep learning is the way that features are extracted. For classical ML techniques such as support vector machines (SVM) [3], feature extraction is performed in advance using techniques crafted by the researchers. Then, the training procedure develops weights or rules that map any given feature vector to an output class label. In contrast, the typical deep-learning procedure is to directly feed signal values as inputs to the training procedure, without any preliminary efforts at feature extraction. The network takes the input signal (pixel values, in our case), and assigns a class label based on those signal values directly. Because the deep-learning approach implicitly must derive its own features, many more training samples are required than for traditional ML approaches.

Several deep-learning packages are available for researchers. The popular package that we have used to evaluate the proposed model is Caffe [7], which was created with computer vision tasks in mind. Caffe is relatively easy to use, flexible, and powerful. It was developed in C++ using GPU optimization libraries, such as CuDNN [2], BLAS [18], and ATLAS [17]. In the next sections, we will discuss briefly two well-known deep models, AlexNet and GoogLeNet. These two models will be used as a baseline for comparison with the proposed FFNet model.

2.2 AlexNet and CaffeNet

AlexNet [9] was the first deep model to win the ILSVRC [4] challenge. For the ILSVRC-2012 competition, AlexNet won with a top-5 test error rate of 15.3%, compared to 26.2% achieved by the second-best entry. This model consists of five convolutional layers followed by three fully-connected layers. The creators of Caffe [7] introduced a slightly modified version of AlexNet by switching the order of pooling and normalization layers. They named the modified version CaffeNet [1]. As the only modification done to the network is switching the order of pooling and normalization layers, the size of the network is exactly the same as AlexNet.

AlexNet and CaffeNet will be used to provide baseline cases of simple architectures that rely on huge numbers of parameters. The number of filters in the convolutional layers range from 96 to 384 in AlexNet, while the proposed FFNet model uses only 64 filters in each convolutional layer. AlexNet uses a 4069-node fully-connected layer followed by another layer of the same size, whereas FFNet uses only a 400-node fully connected layer followed by a 100-node layer. The total size of AlexNet is therefore approximately 18 times bigger than FFNet.

2.3 GoogleNet

GoogLeNet [13] is another winner of the ILSVRC challenge. This model won the ILSVRC-2014 competition with a top-5 test error rate of 6.6%. The network consists of 22 layers with a relatively complex design called "inception". The inception module, which is used to implement the layers of GoogLeNet, consists of parallel paths of convolutional layers of different sizes concatenated together. The number of filters in the convolutional layers inside the inception modules ranges from 16 to 384. (By comparison, in FFNet the number of filters in each convolutional layer is fixed.) In addition to using the inception design, GoogLeNet uses three auxiliary classifiers connected to the intermediate layers during training. GoogLeNet is of interest to us as a baseline for comparison because of its depth, because of its complex architecture, and especially because of the auxiliary classifiers. GoogLeNet is 4 times bigger and far more complex than the proposed FFNet model.

2.4 Benchmarking Datasets

Many datasets have been created to aid in machine learning for computer vision. To evaluate the proposed FFNet model, we selected two publicly available datasets, COCO-Text-Patch and CIFAR-10.

COCO-Text-Patch [6], contains approximately $354,000$ images of size 32×32 that are each labeled as "text" or "non-text". This dataset was created to address the problem of text verification, which is an essential stage in the end-to-end text detection and recognition pipeline. The dataset is derived from COCO-Text [15], which contains $63,686$ images of real-world scenes with $173,589$ instances of text.

CIFAR-10 [8] is a labeled subset of the "80 million tiny images" dataset [14]. They were collected by the creator of AlexNet. The CIFAR-10 dataset consists of $60,000$ color images of size 32×32 in 10 classes, with $6,000$ images per class.

3   Proposed Model: FFNet

The new FFNet model consists of convolutional units that are organized into a sequence of stages. Within each stage, as illustrated in Fig. 1, computations are performed in 2 parallel paths. The left branch in the figure represents a standard convolutional path, whereas the right branch represents an extra parallel data path. It is this parallel, "fast-forwarding", path that delivers the improved performance of the network.

The input to the stage, $S1$, arrives from the previous layer, and the output to the next layer is shown as $S2$. The standard (deep) branch consists of three consecutive $3 \times 3 \times 64$ convolutional layers. Each layer is followed by an in-place Rectified Linear Unit (ReLU). The last layer of the deep branch is padded with zeros, for reasons that are described below.

Let the input $S1$ be of size $N \times N \times C$. The value of C is the number of channels, which is typically 128 except for the first stage where $C = 3$ to match the input data. Refer to a stage's deep convolutional layers as $S2C1$, $S2C2$, and $S2C3$, as shown in the figure. The deep branch's output $S2C3$ can be represented as follows, where $CONV$ is the convolutional operation, s is the stride, and p is the padding:

$$S2C3 = CONV_{3\times3,s=1,p=1}(CONV_{3\times3,s=1,p=0}(CONV_{3\times3,s=1,p=0}(S1))) \quad (1)$$

The size of $S2C3$ will be $(N - 2) \times (N - 2)$.

The fast-forwarding branch consists of a single $5 \times 5 \times 64$ convolutional layer followed by a ReLU. This branch takes $S1$ as input, and generates the output $B2C1$ that can be represented as follows:

$$B2C1 = CONV_{5 \times 5, s=1, p=0}(S1) \qquad (2)$$

No padding is used for the fast-forwarding branch, so that the resulting output size is also $(N-2) \times (N-2)$. This branch will provide a "shallower" representation of the input $S1$ to the next stage.

The outputs of the deep branch and of the fast-forwarding branch are concatenated to create the single stage output $S2$. The size of $S2$ will be $(N-2) \times (N-2) \times 128$. Because the last layer of the deep branch is padded with zeros, both branches provide data of the same size to the output.

To evaluate the fast-forwarding concept, we built a Fast-Forwarding Network (FFNet) that consists of 6 consecutive fast-forwarding stages followed by two fully connected layers plus an output layer, as shown in Fig. 2. The 6 fast-forwarding stages consist of a total of 18 convolutional layers, each of size $3 \times 3 \times 64$. The first layer of the two fully-connected layers consists of 400 nodes, while the second layer consists of 100 nodes.

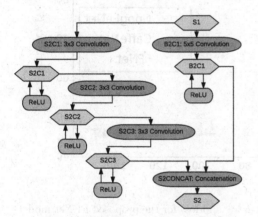

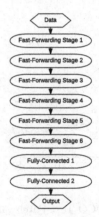

Fig. 1. A single fast-forwarding stage. Node $S1$ represents the input, and $S2$ is the output. The left pathway contains common convolutational blocks. At the right is the fast-forward path.

Fig. 2. Proposed FFNet model. Because of fast-forwarding, this relatively small network has yielded empirical results that are better than much larger deep networks.

4 Evaluation

To evaluate the performance of the proposed model, a number of experiments were conducted that compare FFNet to AlexNet, CaffeNet, and GoogLeNet. The publicly available datasets CIFAR-10 [8] and COCO-Text-Patch [6] were

Table 1. Performance comparison of the proposed FFNet model with several common alternatives. Although FFNet is much smaller than the other models, its error rate was lower than the others (with one exception), using publicly available test sets.

Model				Error rate (%)	
Description	Layers	Size (MB)	Time*(ms)	CIFAR-10	CTP**
AlexNet with dropout [9]	8	181.3	-	15.6	-
AlexNet with stoch. pooling [20]	8	181.3	-	15.3	-
AlexNet with channel-out [16]	8	181.3	-	**13.2**	-
GoogLeNet [6]	22	41.2	9.4	-	9.9
AlexNet [9], CaffeNet [6]	8	181.3	5	18.0	9.1
FFNet (the proposed model)	20	**10.8**	**2.8**	13.6	**9.0**

*Average forward path time per image on a K80 GPU
**CTP: COCO-Text-Patch dataset [6]

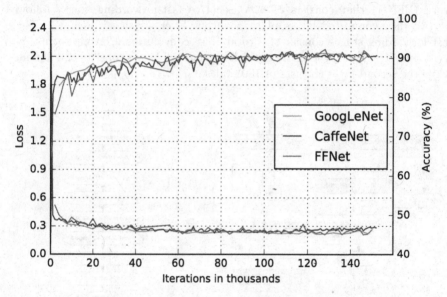

Fig. 3. COCO-Text-Patch validation accuracy and loss for the proposed FFNet model (red), CaffeNet (blue), and GoogLeNet (green). (Color figure online)

used in the evaluation, as described previously. FFNet was implemented using Caffe [7]. Standard 10-crop augmentation was applied to the datasets. All the training and testing were performed on a GPU with batch size 32. The training was stopped after 150,000 iterations as the validation accuracy and loss started to plateau.

A summary of results is provided in Table 1. Despite its relatively small size, the performance of the proposed FFNet model exceeded the performance of CaffeNet and GoogleNet in these experiments. The accuracy and validation loss graphs shown in Fig. 3 demonstrate how the proposed model converges with the same speed as CaffeNet and GoogLeNet. These trends provide evidence of the effectiveness of the fast-forwarding approach in fighting the vanishing-gradient problem.

5   Conclusion

This paper has presented a new concept, called *input fast-forwarding*, which results in improved performance for deep-learning systems. The approach utilizes parallel data paths that provide two advantages over previous approaches. One advantage is the explicit merging of higher-level representations of data with lower-level representations. A second advantage is a substantial reduction to the effects of the vanishing gradients problem.

To evaluate the model, we built a 20-layer network (FFNet) that implements the fast-forwarding concept. The network consists of simple convolutional layers, with no added complexities, to prove that the outstanding performance of the model is primarily the result of the fast-forwarding approach. Empirical results also showed convergence during training at virtually the same rate as the bigger and more complex models. FFNet achieved an error rate of 13.6% on the CIFAR-10 dataset, which is on par with one variation of AlexNet. When tested on COCO-Text-Patch, FFNet's performance surpassed that of CaffeNet and GoogLeNet, which are all significantly larger in size.

These results suggest that similar advantages may be obtained through the application of fast-forwarding to other models, and with different benchmark datasets.

References

1. BVLC reference CaffeNet model. https://github.com/BVLC/caffe/tree/master/models/bvlc_reference_caffenet. Accessed June 2016
2. Chetlur, S., Woolley, C., Vandermersch, P., Cohen, J., Tran, J., Catanzaro, B., Shelhamer, E.: cuDNN: efficient primitives for deep learning. arXiv preprint arXiv:1410.0759 (2014)
3. Cristianini, N., Shawe-Taylor, J.: An Introduction to Support Vector Machines and Other Kernel-Based Learning Methods. Cambridge University Press, Cambridge (2000)
4. Deng, J., Dong, W., Socher, R., Li, L.J., Li, K., Fei-Fei, L.: ImageNet: a large-scale hierarchical image database. In: Proceedings of the IEEE Conference on Computer Vision and Pattern Recognition (CVPR), pp. 248–255 (2009)
5. He, K., Zhang, X., Ren, S., Sun, J.: Deep residual learning for image recognition. In: Proceedings of the IEEE Conference on Computer Vision and Pattern Recognition (CVPR), pp. 770–778 (2016)
6. Ibrahim, A., Abbott, A.L., Hussein, M.E.: An image dataset of text patches in everyday scenes. In: Bebis, G., et al. (eds.) ISVC 2016. LNCS, vol. 10073, pp. 291–300. Springer, Cham (2016). doi:10.1007/978-3-319-50832-0_28
7. Jia, Y., Shelhamer, E., Donahue, J., Karayev, S., Long, J., Girshick, R., Guadarrama, S., Darrell, T.: Caffe: convolutional architecture for fast feature embedding. In: 22nd ACM International Conference on Multimedia, pp. 675–678 (2014)

8. Krizhevsky, A., Hinton, G.: Learning multiple layers of features from tiny images. Master's thesis, Department of Computer Science, University of Toronto (2009)
9. Krizhevsky, A., Sutskever, I., Hinton, G.E.: ImageNet classification with deep convolutional neural networks. In: Proceedings of Advances in Neural Information Processing Systems (NIPS), pp. 1097–1105 (2012)
10. LeCun, Y., Bottou, L., Bengio, Y., Haffner, P.: Gradient-based learning applied to document recognition. Proc. IEEE **86**, 2278–2324 (1998)
11. Lee, C.Y., Xie, S., Gallagher, P.W., Zhang, Z., Tu, Z.: Deeply-supervised nets. In: Proceedings of the 18th International Conference on Artificial Intelligence and Statistics (AISTATS), vol. 2, pp. 562–570 (2015)
12. Liu, Y., Lew, M.S.: Learning relaxed deep supervision for better edge detection. In: Proceedings of the IEEE Conference on Computer Vision and Pattern Recognition (CVPR), pp. 231–240 (2016)
13. Szegedy, C., Liu, W., Jia, Y., Sermanet, P., Reed, S., Anguelov, D., Erhan, D., Vanhoucke, V., Rabinovich, A.: Going deeper with convolutions. In: Proceedings of the IEEE Conference on Computer Vision and Pattern Recognition (CVPR), pp. 1–9 (2015)
14. Torralba, A., Fergus, R., Freeman, W.T.: 80 million tiny images: a large data set for nonparametric object and scene recognition. IEEE Trans. Pattern Anal. Mach. Intell. (PAMI) **30**(11), 1958–1970 (2008)
15. Veit, A., Matera, T., Neumann, L., Matas, J., Belongie, S.: COCO-Text: dataset and benchmark for text detection and recognition in natural images. arXiv preprint arXiv:1601.07140 (2016). http://vision.cornell.edu/se3/wp-content/uploads/2016/01/1601.07140v1.pdf
16. Wang, Q., JaJa, J.: From maxout to channel-out: encoding information on sparse pathways. In: Wermter, S., et al. (eds.) ICANN 2014. LNCS, vol. 8681, pp. 273–280. Springer, Cham (2014). doi:10.1007/978-3-319-11179-7_35
17. Whaley, R.C., Dongarra, J.J.: Automatically tuned linear algebra software. In: Proceedings of the ACM/IEEE Conference on Supercomputing, pp. 1–27 (1998)
18. Xianyi, Z., Qian, W., Chothia, Z.: OpenBLAS (2014). http://xianyi.github.io/OpenBLAS
19. Xie, S., Tu, Z.: Holistically-nested edge detection. In: Proceedings of the IEEE International Conference on Computer Vision (ICCV), pp. 1395–1403 (2015)
20. Zeiler, M., Fergus, R.: Stochastic pooling for regularization of deep convolutional neural networks. In: Proceedings of the International Conference on Learning Representation (ICLR) (2013)

Improving Convolutional Neural Network Design via Variable Neighborhood Search

Teresa Araújo[1,2]([✉]), Guilherme Aresta[1,2], Bernardo Almada-Lobo[1,2],
Ana Maria Mendonça[1,2], and Aurélio Campilho[1,2]

[1] INESC TEC - Institute for Systems and Computer Engineering,
Technology and Science, Porto, Portugal
{tfaraujo,gmaresta}@inesctec.pt
[2] Faculdade de Engenharia da Universidade do Porto, Porto, Portugal

Abstract. An unsupervised method for convolutional neural network (CNN) architecture design is proposed. The method relies on a variable neighborhood search-based approach for finding CNN architectures and hyperparameter values that improve classification performance. For this purpose, t-Distributed Stochastic Neighbor Embedding (t-SNE) is applied to effectively represent the solution space in 2D. Then, k-Means clustering divides this representation space having in account the relative distance between neighbors. The algorithm is tested in the CIFAR-10 image dataset. The obtained solution improves the CNN validation loss by over 15% and the respective accuracy by 5%. Moreover, the network shows higher predictive power and robustness, validating our method for the optimization of CNN design.

Keywords: Machine learning · Convolutional neural network · Parameter optimization · Variable neighborhood search

1 Introduction

Convolutional Neural Networks (CNNs) [8] are a machine learning technique capable of automatically learn relevant features from multi-dimensional data, such as images. These methods have shown to achieve state-of-the-art performance in several complex computer-vision tasks including object recognition [6] and biomedical image analysis [1] since they have in account spatial information.

Machine Learning algorithms often require careful tuning of their hyperparameters. This is specially true for deep learning methods, in which CNNs are included [11]. In fact, the classification accuracy of a CNN is influenced by multiple factors, such as the number of neurons and organization of layers of each type, the regularization strength and the dropout percentage. Furthermore, exhaustive parameter search is particularly difficult for these networks since the high training time hinders the repetitive evaluation of the goal function.

CNNs' architecture and hyperparameters are usually manually tuned for a particular dataset [5]. This approach hardly retrieves a close-to-optimal model

© Springer International Publishing AG 2017
F. Karray et al. (Eds.): ICIAR 2017, LNCS 10317, pp. 371–379, 2017.
DOI: 10.1007/978-3-319-59876-5_41

and can be time-consuming. Thus, automatic methods that perform the optimization of the CNN design are desirable. Some works in the literature aim at the development of automatic methods for optimizing the CNN architecture and/or hyperparameters. Jin *et al.* [5] proposed a submodularity and supermodularity method for optimizing neural network architectures, while other hyperparameters were optimized using grid-search or, alternatively, set to recommended values. Snoek *et al.* [11] presented methods for Bayesian optimization of machine learning algorithms' hyperparameters. The method was tested in the CIFAR-10 dataset [7], using the network described in [6]. A total of nine hyperparameters were optimized, among which are the number of epochs, the learning rate and the pooling size. The authors report an improvement of 3% in the validation accuracy relative to the initial parameter setting.

Heuristic and metaheuristic techniques are commonly used for solving optimization problems. However, their implementation for CNN improvement is still little explored. Further, most optimization methods do not focus on the architecture design but rather on the optimization of hyperparameters such as learning rate, weight initialization and number of epochs. We propose a Variable Neighborhood Search-like (VNS) approach for CNN architecture and hyperparameter optimization. The VNS algorithm is composed of two steps. The first one is a diversification of the search in which a new neighborhood is generated. Then, an intensification step allows to achieve a new local optima. This process is repeated for increasing neighborhoods, allowing to obtain different local optima and thus, most likely, improve the incumbent solution [4].

2   Materials and Methods

2.1   Solution Representation and Neighborhood Definition

Let S be a $m \times n$ matrix, where m is the number of layers and n the number of parameters of a network. S is characterized by a set of parameter indices s:

$$S_{m,n} = \begin{pmatrix} s_{1,1} & s_{1,2} & \cdots & s_{1,n} \\ s_{2,1} & s_{2,2} & \cdots & s_{2,n} \\ \vdots & \vdots & \ddots & \vdots \\ s_{m,1} & s_{m,2} & \cdots & s_{m,n} \end{pmatrix} \tag{1}$$

where s has correspondence to a given layer parameter as function of its column coordinate in S. For instance, the type of layer is mapped by:

$$s_{i,1} \in \{1, 2, 3, 4\} \mapsto \{\mathbf{C}, \mathbf{P}, \mathbf{D}, \mathbf{FC}\} \tag{2}$$

where $\mathbf{C}$ stands for convolutional, $\mathbf{P}$ for pooling, $\mathbf{D}$ for dropout, and $\mathbf{FC}$ for fully-connected layers. Both $\mathbf{C}$ and $\mathbf{FC}$ are followed by rectified linear units (ReLU), except for the last $\mathbf{FC}$ layer, which is activated with a softmax function to perform the final classification. The remaining columns of the matrix contain the values of the parameters needed for the network design. The fifth column

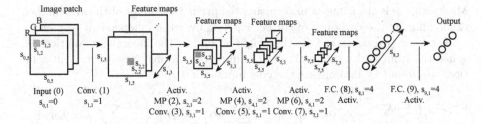

Fig. 1. Example of a convolutional neural network architecture according to the proposed solution representation. ▓ convolutional kernel; ▓ max-pooling kernel. (Color figure online)

stores the output size of each layer, allowing to assess the validity of the network architecture. The second column of S defines the size of the filters of **C** (3, 5, 7 or 9 in our experiments) and **MP** layers (2, 3, 4), the dropout percentage of **D** (0, 0.25, 0.50) layers and the number of neurons of **FC** (4, 16, 32, 64, 128). Like-wise, the third column indicates the number of maps of **C** layers (8, 16, 32, 64) and the fourth the L2 regularization (0.01) of **C**, **M** and **FC** layers. A CNN architecture illustrating the proposed solution representation is shown in Fig. 1.

Neighbors of the current solution S are networks that result from movements of *insert*, *remove* and *swap* applied to S. A neighbor is considered valid if it respects the following rules: (1) it does not have 2 consecutive **MP** layers or **D** layers; (2) the output size of the layer before **MP** must be divisible by the **MP** filter size; (3) **MP** cannot come after **D** layers, following the traditional CNN design; (4) **FC** must appear only at the end of the CNN; (5) the first layer must be **C**; *Remove* movements tend to be easier to perform. To avoid shallow architectures, these movements are penalized relatively to the other two. Similarly, *swap* movements are penalized relative to *insert* movements. For this purpose, an exponential function is used for biasing the operation decision.

2.2 Network Performance Evaluation

In this study, the networks' performance is evaluated on the independent test set in terms of (i) accuracy (acc), corresponding to the ratio of correctly predicted image classes (Eq. 3) and (ii) loss (L), which is related to how confident the CNN is in the predicted label (Eq. 4):

$$acc(u,g) = \frac{\sum_{i=1}^{N} \sum_{j=1}^{M} y_{ij}\hat{y}_{ij}}{N} \quad (3) \qquad L(u,g) = -\frac{1}{N}\sum_{i=1}^{N}\sum_{j=1}^{M} y_{ij}log(p_{ij}) \quad (4)$$

where u is the model in study, g the current training epoch, N is the number of test samples, M is the number of classes, $y_{ij} \in \{0,1\}$ is a binary variable with value 1 if observation i belongs to class j, $\hat{y}_{ij} \in \{0,1\}$ a binary variable with value 1 if the predicted class is j and p_{ij} the respective prediction probability.

Minimizing L is the same as maximizing the predictive power of the system, i.e., of maximizing the certainty of correct labeling over the independent test set.

Let $U = \{u_1, u_2, \ldots u_p\}$ be the neighborhood of the incumbent solution u_0, generated from u_0 by randomly performing r operations. New incumbent solutions are found by minimizing the cost function $C(u)$:

$$
C(u) = \left(\frac{1}{v} \sum_{g=V-v}^{V} acc_{tr}(u, g) \right)^{-1} \left(\frac{1}{v} \sum_{g=V-v}^{V} acc_{tt}(u, g) \right)^{2} \left(\frac{1}{v} \sum_{g=V-v}^{V} L_{tt}(u, g) \right)^{-1}
$$
(5)

where V is the number of training epochs, v the number of epochs to analyze and u the current network design. The first two terms of Eq. 5 state that the model should have similar accuracy performance for both training (tr) and test samples (tt), while prioritizing test accuracy. The third term minimizes the test loss, thus increasing the generalization capability of the network. A new incumbent solution u_0 is considered every time that $C(u)^{-1} > C(u_0)^{-1}$. This strategy is known as first neighbor search.

The training process has a near-constant duration that depends on the complexity of the network. Since predicting the behavior of the network based on few training epochs is a complex task [2], it is important to select V such that the obtained performance results are representative without substantially increasing the tunning time. In this work, V is manually selected by analyzing the behavior of the initial solutions.

2.3 Solution Space Exploration

The incumbent solution's neighborhood is defined as the set of neighbors obtained from u_0 by performing r randomly selected viable operations. These neighbors are characterized by a set of features related to the architecture of the network which are used for describing the solution space: (i) number of **C**, **MP** and **FC** layers; (ii) average number of maps of **C** layers; (iii) average filter size of **C** layers and (iv) average pooling size of **MP**. Note that although these features do not fully describe the solution, their network characterization is more complete than a measurement of similarity such as the edit distance.

The dimension of the feature space is then reduced to 2D by applying a t-Distributed Stochastic Neighbor Embedding (t-SNE) [9]. Based on the natural aggregation of the neighbors, a K-means [10] approach is used for dividing the solution space. The discussed pre-processing steps allow to obtain a well defined solution space. This solution space is now searched through a variable neighborhood search (VNS) approach, as detailed in Algorithm 1.

The proposed method performs a search inside each of the k groups of the solution space. Each time a network with higher performance is found, as defined by Eq. 5, the incumbent solution u_0 is updated. For that purpose, the algorithm starts by evaluating randomly selected neighbors in the same cluster of u_0. A preliminary accuracy evaluation after 3 training epochs guarantees that the current solution is viable. Otherwise, a different neighbor is explored. Similarly, redundant neighbors are skipped. When a better solution is found, or alternatively if

Algorithm 1. Variable Neighborhood Search for Convolutional Neural Network design improvement. t_r: overall elapsed time; t_c: time elapsed in each cluster; $t_{r,max}$: maximum running time; $t_{c,max}$: cluster-search maximum running time; u_0: initial solution; U: set of neighbors of u_0; n_{op}: number of operations (insertion, deletion, swap) to perform; k: number of clusters Clu to form from U.

Require: u_0, $t_{c,max}$, $t_{r,max}$, n_{op}, k
  $t_r \leftarrow 0$, $loss(u_0) \leftarrow \text{train}(u_0)$
  while $t_r < t_{r,max}$ **do**
    $U \leftarrow \text{generateNeighbors}(u_0, n_{op})$, $C \leftarrow \text{cluster}(U, k)$
    for c in Clu **do**
      $t_c \leftarrow 0$
      while $t_c < t_{c,max}$ **do**
        $u \leftarrow \text{randomSelection}(c)$, $loss(u) \leftarrow \text{train}(u)$
        if $loss(u)$ **better than** $loss(u_0)$ **then**
          $u_0 \leftarrow u$, **break**
        update(t_c)
    update(t_r), $n_{op} \leftarrow n_{op} + 1$

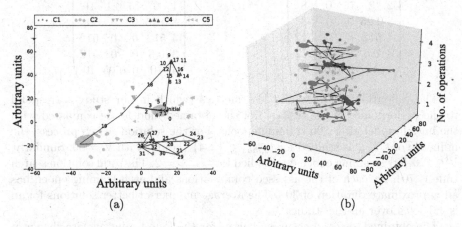

Fig. 2. 2D and 3D schemes of the proposed variable neighborhood search method for convolutional neural network design improvement. The solution space is reduced via t-SNE and clustering is performed using k-means. 2a: example of visited solutions of a neighborhood generated from the incumbent solution by performing one operation; 2b: visited neighbors with one to five operations.

the pre-established cluster running time is exceeded, the algorithm moves to a different cluster. After visiting all clusters, the number of operations r is incremented, a new solution space is generated and the process is repeated until the overall running time is exceeded. Figure 2 illustrates the proposed scheme.

3   Results

The proposed VNS-based approach for CNN architecture optimization is evaluated on the CIFAR-10 dataset [7]. This dataset is composed of 50 000 training images and 10 000 test images (RGB) of 10 different classes. The small size of the images lowers the training time by reducing the number of parameters to learn and memory requirements, thus allowing a simpler assessment of the proposed methodology. The incumbent solution is the network from [3] (Eq. 6). The achieved architecture is shown in Eq. 7. The influence of the total number of training epochs, V, and the maximum search time per cluster, $t_{c,max}$ is assessed. Experiments were performed using a CPU Intel i7-5960X workstation, 32 GB RAM and GPU Nvidia GTX1080. Python 3.5 and Keras framework were used for experiment design and evaluation. Experiments were performed in GPU.

$$
S_{init} = \begin{pmatrix}
0 & 0 & 0 & 0 & 32 \\
1 & 3 & 32 & 0 & 32 \\
1 & 3 & 32 & 0 & 32 \\
2 & 2 & 0 & 0 & 16 \\
3 & 0.25 & 0 & 0 & 16 \\
1 & 3 & 64 & 0 & 16 \\
1 & 3 & 64 & 0 & 16 \\
2 & 2 & 0 & 0 & 8 \\
3 & 0.25 & 0 & 0 & 8 \\
4 & 512 & 0 & 0 & 512 \\
3 & 0.5 & 0 & 0 & 512 \\
4 & 10 & 0 & 0.05 & 10
\end{pmatrix}
\quad (6)
\qquad
S_f = \begin{pmatrix}
0 & 0 & 0 & 0 & 32 \\
1 & 3 & 32 & 0 & 32 \\
1 & 3 & 32 & 0 & 32 \\
2 & 2 & 0 & 0 & 16 \\
3 & 0.25 & 0 & 0 & 16 \\
1 & 3 & 64 & 0 & 16 \\
1 & 3 & 64 & 0 & 16 \\
1 & 9 & 16 & 0.1 & 16 \\
3 & 0.25 & 0 & 0 & 16 \\
4 & 512 & 0 & 0 & 512 \\
3 & 0.5 & 0 & 0 & 512 \\
4 & 10 & 0 & 0.05 & 10
\end{pmatrix}
\quad (7)
$$

The validation accuracy and loss metrics (which in our study correspond to the performance in the test set) of the obtained models are compared with the initial model after 200 training epochs. For the solution search process, the influence of $t_{c,max}$ is studied for $t_{c,max} \in \{2, 4\}$ (hours) with $V = 50$. Similarly, $V \in \{20, 50\}$ (training epochs) is studied for $t_{c,max} = 2$. The total solution search time is 70 h for each of the assessed combinations and each training epoch has an approximate duration of 30 s. The average number of better solutions found is 4.3 ± 0.4 over all the studies.

The obtained loss and accuracy curves for the found solutions are shown in Fig. 3a and b, respectively, for 120 training epochs. Figure 3a shows that the found solutions yield a lower and more constant loss than the initial model. The best architectures provide a higher degree of confidence in their predictions and are less prone to over-fit to the training data. Accuracy-wise, the new models show to either perform similarly or slightly better than the initial model. Overall, the found solutions are more robust than the initial model without compromising the accuracy. This shows that the proposed VNS-based heuristic is successfully capable of improving the design of the CNN.

The key results achieved for the studied V and $t_{c,max}$ values are summarized in Tables 1a and b. During the 70 h of search, in the **S1** and **S3** approaches (refer to Table 1a) approximately 240 different solutions are visited while in **S2** the number decreases to 100. The minimum achieved loss is 0.52, corresponding to an improvement of 17% relatively to the original solution loss. For 30 training epochs, number for which the lowest loss value of the original solution is obtained,

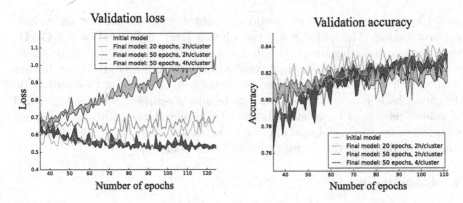

Fig. 3. Validation loss and accuracy curves of the obtained solutions. Two repetitions are performed for the initial model and for the 50 epochs with 4 h per cluster study. 3a: validation loss of the best found solutions; 3b: validation accuracy of the best found solutions.

Table 1. Validation loss and accuracy of the obtained solutions for key training epochs. **S0** - initial model; **S1** - 2 h/cluster with 20 training epochs; **S2** - 2 h/cluster with 50 training epochs; **S3** - 4 h/cluster with 50 training epochs.

(a) Loss

Epoch	S0	S1	S2	S3
30	0.63	**0.60**	0.73	0.67
65	0.81	**0.53**	0.63	0.54
70	0.81	0.56	0.61	**0.52**
111	0.98	0.61	0.68	**0.52**

(b) Accuracy

Epoch	S0	S1	S2	S3
30	**0.80**	0.79	0.78	0.77
65	0.82	**0.84**	0.82	0.82
70	0.82	**0.84**	0.83	0.83
111	0.82	**0.84**	0.83	0.83

the improvement is 5%. Accuracy-wise, the proposed system allows a maximum 5% accuracy improvement. For 30 epochs the obtained accuracy is lower only because the increased complexity of the obtained solution tends to slow down the training process in terms of epochs. Furthermore, although not directly comparable because of the difference in the initial network and implementation, this value is greater than the 3% achieved by [11] for the same running time. Note that, despite the success of the proposed methodology, further improvements could be achieved by increasing the solution space to include different network optimizers, regularization values, etc. For instance, it would be easy to vary the stride of **C** layers, which has shown to be an alternative to adding **MP** layers [12]. Furthermore, the selection of the V and $t_{c,max}$ parameters seems not to be crucial since in the performed study, all the obtained models achieve high performance given enough number of training epochs.

The comparison between the studied approaches indicates that **S1** shows the best compromise between loss and accuracy improvement as function of the number of training epochs. For this model, shown in Eq. 7, a significant performance boost is achieved by doubling the network training time in relation to the initial solution. Note that in terms of CNN architecture the obtained model is uncommon and thus would not probably be studied during manual network tuning. The added 9×9 convolution layer in the end of the **C-MP** sequence is aggregating the previously learned filters into a smaller feature space, which may contribute to the improvement of the performance. However, further studies need to be done to confirm this hypothesis. Because of this, automatic design methods such as this may prove to be advantageous. The lower $t_{c,max}$ and V values, reinforced by cluster-wise division achieved via t-SNE and k-means methods, allow the VNS-based approach to visit neighbors with lower similarity, which contributes to the identification of better solutions.

4   Conclusions

An unsupervised method for CNN architecture design is proposed. A VNS-based approach is used for finding solutions that improve a problem-oriented cost function. For that purpose, t-SNE allows to effectively represent the solution space in a 2D space, which is easier to interpret and k-Means clustering divides the solution space having in consideration the distance between neighbors. The obtained solution improves the CNN loss by over 15% and the respective accuracy by 5%, allowing higher predictive power and robustness.

Based on the achieved results, further improvements to the system can be performed. Generically, the initial solution could be randomly generated instead of being user provided. This could diversify the solution space and thus allow for further performance improvement. Ultimately, the entire network design process could be automated. It would be of interest to study classification tasks of higher complexity, such as medical image analysis. For that purpose, other relevant parameters, such as the learning rate, weight initialization, loss function, among others, could be optimized. A more complete set of features to describe the solution space could be used. Similarly, the number of epochs needed to evaluate to network could be adapted to the complexity of the network i.e., reducing the number of epochs to a minimum should allow to increase the number of neighbors visited for the same period of time.

Acknowledgements. Teresa Araújo and Guilherme Aresta equally contributed to this work. Project "NanoSTIMA: Macro-to-Nano Human Sensing: Towards Integrated Multimodal Health Monitoring and Analytics/NORTE-01-0145-FEDER-000016" is financed by the North Portugal Regional Operational Programme (NORTE 2020), under the PORTUGAL 2020 Partnership Agreement, and through the European Regional Development Fund (ERDF). Teresa Araújo is funded by the FCT grant contract SFRH/BD/122365/2016. Guilherme Aresta is funded by the FCT grant contract SFRH/BD/120435/2016.

References

1. Cireşan, D.C., Giusti, A., Gambardella, L.M., Schmidhuber, J.: Mitosis detection in breast cancer histology images with deep neural networks. In: Mori, K., Sakuma, I., Sato, Y., Barillot, C., Navab, N. (eds.) MICCAI 2013. LNCS, vol. 8150, pp. 411–418. Springer, Heidelberg (2013). doi:10.1007/978-3-642-40763-5_51

2. Domhan, T., Springenberg, J.T., Hutter, F.: Speeding up automatic hyperparameter optimization of deep neural networks by extrapolation of learning curves. In: IJCAI International Joint Conference on Artificial Intelligence 2015, pp. 3460–3468, January 2015
3. Github: Cifar 10 CNN. https://github.com/fchollet/keras/blob/master/examples/cifar10_cnn.py
4. Hansen, P., Mladenovi, N.: Variable neighborhood search: principles and applications. Eur. J. Oper. Res. **130**, 449–467 (2001)
5. Jin, J., Yan, Z., Fu, K., Jiang, N., Zhang, C.: Neural Network Architecture Optimization through Submodularity and Supermodularity, pp. 1–10 (2016)
6. Krizhevsky, A., Sutskever, I., Hinton, G.E.: ImageNet classification with deep convolutional neural networks. Adv. Neural Inf. Process. Syst. **25**, 11061114 (2012)
7. Krizhevsky, A.: Learning multiple layers of features from tiny images. Master's thesis, Department of Computer Science, University of Toronto (2009)
8. LeCun, Y., Bottou, L., Bengio, Y., Haffner, P.: Gradient based learning applied to document recognition. Proc. IEEE **86**(11), 2278–2324 (1998)
9. Maaten, L., Hinton, G.E.: Visualizing high-dimensional data using t-SNE. J. Mach. Learn. Res. **9**, 2579–2605 (2008)
10. Macqueen, J.: Some methods for classification and analysis of multivariate observations. In: Proceedings of the Fifth Berkeley Symposium on Mathematical Statistics and Probability, vol. 1, no. 233, pp. 281–297 (1967)
11. Snoek, J., Larochelle, H., Adams, R.: Practical Bayesian optimization of machine learning algorithms. In: Advances in Neural Information Processing Systems, pp. 1–9 (2012)
12. Springenberg, J.T., Dosovitskiy, A., Brox, T., Riedmiller, M.: Striving for simplicity: the all convolutional net. In: ICLR 2015, pp. 1–14 (2015)

Fast Spectral Clustering Using Autoencoders and Landmarks

Ershad Banijamali[1(✉)] and Ali Ghodsi[2]

[1] School of Computer Science, University of Waterloo, Waterloo, Canada
ershad.banijamali@gmail.com
[2] Department of Statistics and Actuarial Science,
University of Waterloo, Waterloo, Canada

Abstract. In this paper, we introduce an algorithm for performing spectral clustering efficiently. Spectral clustering is a powerful clustering algorithm that suffers from high computational complexity, due to eigen decomposition. In this work, we first build the adjacency matrix of the corresponding graph of the dataset. To build this matrix, we only consider a limited number of points, called landmarks, and compute the similarity of all data points with the landmarks. Then, we present a definition of the Laplacian matrix of the graph that enable us to perform eigen decomposition efficiently, using a deep autoencoder. The overall complexity of the algorithm for eigen decomposition is $O(np)$, where n is the number of data points and p is the number of landmarks. At last, we evaluate the performance of the algorithm in different experiments.

1 Introduction

Clustering is a long-standing problem in statistical machine learning and data mining. Many different approaches have been introduced in the past decades to tackle this problem. Spectral clustering is one of the most powerful tools for clustering. The idea of spectral clustering originally comes from the min-cut problem in graph theory. In fact, if we represent a dataset by a graph where the vertices are the data points and edges are the similarity between them, then the final clusters of spectral clustering are the cliques of the graph that are formed by cutting minimum number of the edges (or minimum total weight in weighted graphs).

Spectral clustering is capable of producing very good clustering results for small datasets. It has application in different fields of data analysis and machine learning [4,10,11,14]. However, the main drawback of this algorithm comes from the eigen decomposition step, which is computationally expensive ($O(n^3)$, n being the number of data points). To solve this problem, many algorithms have been designed. These algorithms are mainly based on sampling from the data (or the affinity matrix), solving the problem for the samples, and reconstruction of the solution for the whole dataset based on the solution for the samples. In [6,9,15], Nystrom method has been used to sample columns from affinity

© Springer International Publishing AG 2017
F. Karray et al. (Eds.): ICIAR 2017, LNCS 10317, pp. 380–388, 2017.
DOI: 10.1007/978-3-319-59859-5_42

matrix and the full matrix is approximated using correlation between the selected columns and the remaining columns. In [2], a performance guarantee for these approaches has been derived and a set of conditions have been discussed under which this approximation performs comparable to the exact solution. [7] suggests an iterative process for approximating the eigenvector of the Laplacian matrix. In [16], k-means or random projection is used to find centroids of the partitions of the data. Then, they perform spectral clustering on the centroid, and finally, assign each data point to the cluster of their corresponding centroids.

The most relevant work to ours, however, is by Chen and Cai [1]. The authors proposed a method for accelerating the spectral clustering based on choosing p landmarks in the dataset and computing the similarities between all points and these landmarks to form a $p \times n$ matrix. Then the eigenvectors of the full matrix is approximated by the eigendecomposition of this $p \times n$ matrix. The overall complexity of this method is $O(np^2 + p^3)$. The results of the method are close to the actual spectral clustering on n data points, but with much less computation time.

Multi-layer structures have been used for spectral clustering in some recent works [12,13]. Training deep architectures is done much faster than eigendecomposition, since it can be easily parallelized over multiple cores. However, in the mentioned works, the size of input layer of the network is equal to the number of data points, i.e. n, and consequently the whole network is drastically enlarged as n grows. Therefore, it will be infeasible to use these structures for large datasets. In this paper, we combine the idea of landmark selection and deep structures to achieve a fast and yet accurate clustering. The overall computational complexity of the algorithm, given the parallelization of the network training, is $O(np)$.

2   Background

2.1   Spectral Clustering

Mathematically speaking, suppose we have a dataset $\mathbf{X}$ with n data points, $\{\mathbf{x}_1, \mathbf{x}_2, \ldots, \mathbf{x}_n\}$. We want to partition this set to k clusters. To do spectral clustering, we first form the corresponding graph of the dataset, where the vertices are the points, and then obtain the adjacency matrix of the graph, denoted by W. Each entry of W shows the similarity between each pair of points. So W is a symmetric matrix. The degree matrix of the graph, denoted by D, is a diagonal matrix and its nonzero elements are summation over all the elements of rows (or columns) of W, $d_{ii} = \sum_j w_{ij}$.

Based on D and W, the Laplacian matrix of the graph, denoted by L, is obtained. There are different ways for defining the Laplacian matrix. The unnormalized Laplacian matrix is defined as: $L = D - W$. To do spectral clustering, we can get the final clusters of the dataset, by running k-means on the k eigenvectors of L, corresponding to the k smallest eigenvalues (the smallest eigenvalue of L is 0 and its corresponding eigenvector, which is constant, is discarded). The are some normalized versions of L, which usually yield a better clustering results. One of them is defined as $L_{norm} = D^{-1/2}LD^{-1/2}$ [3]. We can also use the k eigenvectors corresponding to the k smallest eigenvalues of L_{norm} (or equivalently k largest eigenvalues of $L_{norm} = D^{-1/2}WD^{-1/2}$, according to [10]).

2.2 Autoencoders and Eigendecomposition

The relation between autoencoder and eigendecomposition was first revealed in Hinton and Salakhutdinov's paper [8]. Consider the Principal Component Analysis (PCA) problem. We would like to find a low-dimensional representation of data, denoted by Z, which preserves the variation of the original data, denoted by X, as much as possible. Suppose U is the linear transformation for PCA problem, which projects the data points in a low-dimensional space, $Z = U^\top X$. To keep the maximum variation of the data, it is known that, the basis of the low-dimensional space (or columns of matrix U) are the eigenvectors of the covariance matrix, $\text{Cov}_X = XX^\top$, corresponding to the largest eigenvalues. PCA can also interpreted as finding a linear transformation that minimizes the reconstruction loss, i.e.

$$\min_U \| X - UZ \|^2 \tag{1}$$

The above objective function is exactly used in conventional autoencoders. In fact, a single layer autoencoder with no nonlinearity spans the same low-dimensional space as its latent space. However, deep autoencoders are capable of finding better low-dimensional representations than PCA.

Training an autoencoder using backpropagation is much faster than solving the eigen decomposition problem. In [13], the authors used autoencoders for graph clustering. Instead of using an actual data point as the input of the autoencoder, they use vector of similarity of that point with other points. Their results show benefit of the model compared to some rival model, in terms of Normalized Mutual Information (NMI) criterion. However, since the length of the similarity vector is equal to the number of data points, n, extending the idea of this work for large datasets, with hundred of thousands or even million data points, is not feasible.

Inspired by these works, we introduce a simple, but fast and efficient algorithm for spectral clustering using autoencoders. In the next section we describe the model.

3   Model Description

As described in the previous section, spectral clustering can be done by decomposing the eigenvalues and eigenvectors of $L_{norm} = D^{-1/2}WD^{-1/2}$. In our work, we do this decomposition using an autoencoder. Instead of original feature vectors, we represent each data point by its similarity to other data points. However, instead of calculating the similarity of a given data point with all other data points and forming a vector of length n, we only consider some landmarks and compute the similarity of the points with these landmarks. Lets denote the p selected landmarks ($p \ll n$) by $\{\boldsymbol{\ell}_1, \boldsymbol{\ell}_2, \ldots, \boldsymbol{\ell}_n\}$. Then we compute the similarity of all data points with those landmarks and form a $p \times n$ matrix W. In this work, we used Gaussian kernel as the similarity measure between the points, i.e.:

$$w_{ij} = \exp(-\frac{\|\boldsymbol{\ell}_i - \mathbf{x}_j\|^2}{\sigma}) \tag{2}$$

where σ is the parameter of the model. To make the model more robust, we always set σ to be the median of the distance between data points and landmarks.

$$\sigma = \text{median}\{\|\boldsymbol{\ell}_i - \mathbf{x}_j\|^2\}_{i,j=1}^{p,n} \tag{3}$$

This way, we also guarantee that the value of similarities are well spread in $[0, 1]$ interval. Each column of matrix W, denoted by $\mathbf{w}_i$'s, represents a data point in the original set X based on its similarity to the landmarks. Constructing matrix W takes $O(npd)$ (d being the number of features), which is inevitable in all similar algorithms. However, our main contribution is in decreasing computational cost in decomposition step.

Next, we have to form the Laplacian matrix. We should notice that $L_{norm} = D^{-1/2}WD^{-1/2}$ is no longer a valid matrix, since W is $p \times n$. Now, we are looking for a Laplacian matrix that can be written in the form of $L_{norm} = SS^{\top}$, so that we can use S as the input to our autoencoder in order to eigen decomposition. To do so, we define another matrix $M = W^{\top}W$. Based on this definition, M is also a similarity matrix over the data points. However, since $m_{ij} = \mathbf{w}_i^{\top}\mathbf{w}_j$, M is a more local measure than W, which is a good property for spectral clustering.

The diagonal matrix D can be obtained by summing over elements in columns of M. However, since our goal in this work is to minimize the computational cost of the algorithm, we would like to avoid computing M, directly, which has a computation cost of $O(n^2p)$. Instead, we compute D another way. In fact, we know $d_{ii} = \sum_i m_{ij} = \sum_i \mathbf{w}_i^{\top}\mathbf{w}_j$. We can write this as follows:

$$d_{ii} = \sum_{j=1}^{n} m_{ij} = \sum_{j=1}^{n} \mathbf{w}_i^{\top}\mathbf{w}_j = \mathbf{w}_i^{\top} \sum_{j=1}^{n} \mathbf{w}_j = \mathbf{w}_i^{\top}\mathbf{w}^s \tag{4}$$

where $\mathbf{w}^s$ is a $p \times 1$ vector and its kth elements is a sum over elements in the kth row of W. Therefore, D can be written as:

$$D = diag(W^{\top}\mathbf{w}^s) \tag{5}$$

Note that calculation of $\mathbf{w}^s$ and D this way has complexity of $O(np) + O(np) = O(np)$, which is a significant improvement compared to $O(n^2p)$.

We can then obtain the Laplacian matrix:

$$\mathrm{L}_{norm} = D^{-1/2}MD^{-1/2} = D^{-1/2}W^{\top}WD^{-1/2} \tag{6}$$

By putting $S = WD^{-1/2}$ as the input of our autoencoder, we can start training the network. The objective function for training the autoencoder is minimizing the error of reconstructing S. After training the network, we obtain the representation of all data points in the latent space and run k-means on the latent space. Again, instead of computing $WD^{-1/2}$, we can simply multiply each diagonal element of $D^{-1/2}$ by the corresponding column of W, i.e. $\mathbf{s}_i = d_{ii}^{-1/2}\mathbf{w}_i$. This operation also has computational complextiy $O(np)$.

Figure 1 shows the proposed model. The objective function for training the autoencoder, as described above, is to minimize the euclidean distance between the input and the output of the network, i.e. S and $\tilde{S}$. Training a network can be done very efficiently using backpropagation and mini-batch gradient descent, if the number of hidden units in each layer be in order of p, which is the case. Furthermore, in contrast to eigen decomposition problem, the training phase can be easily distributed over several machines (or cores). These two facts together helps us to keep the computational complexity of decomposition step in $O(np)$.

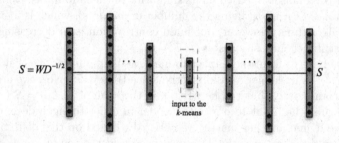

Fig. 1. Input to the network is $WD^{-1/2}$ and k-means is performed in the latent space

Algorithm 1 describes the steps of the proposed method. Note that the p landmarks can be obtained in different ways. They can be randomly sampled from the original dataset or be the centroids of p clusters of the dataset by running k-means or be picked using column subset selection methods, e.g. [5].

Algorithm 1. Spectral clustering using autoencoders and landmarks

---

Input: Dataset $\mathbf{X}$ with n samples $\{\mathbf{x}_1, \mathbf{x}_2, ..., \mathbf{x}_n\}$
Output: k clusters of the dataset
1: Select p landmarks
2: Compute the similarities between data points and landmarks and store
    them in matrix W
3: Compute the degree matrix: $D = diag(W^\top \mathbf{w}^s)$
4: Compute S, the input to the autoencoder: $\mathbf{s}_i = d_{ii}^{-1/2}\mathbf{w}_i$
5: Train an autoencoder using S as its input
6: Run k-means on the latent space of the trained autoencoder

---

4   Experiment Results

In the following two subsections, we present the results of applying our clustering algorithm on different sets of data. In all of these experiments, the autoencoder has 5 hidden layers between input and output layer. Only number of units in the layers changes for different datasets. The activation function for all hidden layer is ReLU, except the middle layer that has linear activation. The activation for output layer is sigmoid.

4.1   Toy Datasets

To demonstrate the performance of the proposed method, we first show the results for some small 2-dimensional datasets. Figure 3 shows the performance of the algorithm on four different datasets. As we can see in this figure, the natural clusters of the data have been detected by a high accuracy. Number of landmarks for all of these experiments is set to 200, and they are drawn randomly from the datasets. For all of these experiments, number of units in the hidden layers is: 64, 32, 2, 32, and 64, respectively (Fig. 2).

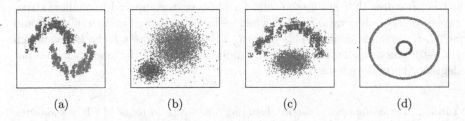

(a)   (b)   (c)   (d)

Fig. 2. Clustering using the proposed method: (a) two-moon dataset, $n = 4000$ (b) two-circle dataset $n = 4500$ (c) moon-circle dataset $n = 4000$ (d) concentric rings $n = 3000$

4.2   Real-World Datasets

In this section we evaluate the proposed algorithm with more challenging and larger datasets. To measure the performance here, we use Clustering Purity (CP) criterion. CP is defined for a labeled dataset as a measure of matching between classes and clusters. If $\{C^1, C^2, \ldots, C^L\}$ are L classes of a dataset $\mathbf{X}$ of size n, then a clustering algorithm, $\mathcal{A}$, which divides $\mathbf{X}$ into K clusters $\{X^1, X^2, \ldots, X^K\}$ has $CP(\mathcal{A}, X)$ as:

$$CP(\mathcal{A}, X) = \frac{1}{n} \sum_{j=1}^{K} \max_i |C^i \cap X^j|. \tag{7}$$

Table 1 contains a short description about each of the datasets.

Table 1. Specification of the datasets

Dataset	Size (n)	# of classes	Description
MNIST	60000	10	28×28 grayscale images of digits
Seismic	98528	3	Types of moving vehicle in a wireless sensor network
CIFAR-10	50000	10	32×32 colored images of 10 different objects
LetterRec	20000	26	Capital letters in English alphabet

In Table 2, we can compare the performance of our proposed algorithm, SCAL (Spectral Clustering with Autoencdoer and Landmarks), with some other clustering algorithms. SCAL has two variants: (1) SCAL-R where landmarks are selected randomly, (2) SCAL-K where landmarks are centroids of k-means. LSC-R and LSC-K are methods from [1], which also uses landmarks for spectral clustering. Based on this table, SCAL-K outperforms SCAL-R in almost all cases. As we increase the number of landmarks, the performance improves, and in some cases we get better result than original spectral clustering, which is an interesting observation. This may suggest that deep autoencoders are able to extract features that are more useful for clustering, compared to shallow structures. Another observation is that when we increase p, gap between SCAL-R and SCAL-K becomes smaller. This suggests that if we choose p to be large enough (but still much smaller than n), even random selection of the landmark does not degrade the performance too much.

Table 2. Performance of different clustering algorithms in terms of CP. p shows the number of landmarks. For all of these result we used 10 epochs of data.

Algorithm	MNIST	Seismic	CIFAR-10	LetterRec
Spectral clustering	71.54	66.68	**60.13**	33.19
k-means	57.31	62.82	40.12	30.01
LSC-R $(p = 500)$	62.94	66.19	47.16	29.44
LSC-K $(p = 500)$	68.10	67.71	50.40	31.59
SCAL-R $(p = 500)$	64.13	64.41	49.41	29.52
SCAL-K $(p = 500)$	69.14	68.43	54.64	32.88
SCAL-R $(p = 1000)$	70.61	67.55	56.19	33.94
SCAL-K $(p = 1000)$	**72.98**	**68.61**	58.02	**34.70**

Figures below show the performance and runtime of the algorithm versus LSC-R and LSC-K methods, as a function of number of landmarks.

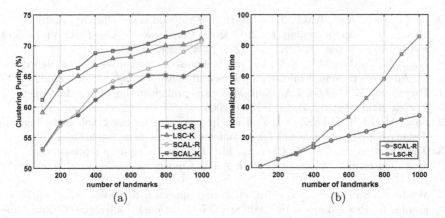

Fig. 3. Performance of different methods versus number of landmarks (a) clustering purity (b) normalized run time; SCAL-K and LSC-K have similar behavior, since they just have an additional overhead

5 Conclusion

We introduced a novel algorithm using landmarks and deep autoencoders, to perform spectral clustering efficiently. The complexity of the algorithm is $O(np)$, which is much faster than the original spectral clustering algorithm as well as some other approximation methods. Our experiment shows that, despite the gain in computation speed, there is no or limited loss in clustering performance.

References

1. Chen, X., Cai, D.: Large scale spectral clustering with landmark-based representation. In: Twenty-Fifth AAAI Conference on Artificial Intelligence (2011)
2. Choromanska, A., Jebara, T., Kim, H., Mohan, M., Monteleoni, C.: Fast spectral clustering via the Nyström method. In: Jain, S., Munos, R., Stephan, F., Zeugmann, T. (eds.) ALT 2013. LNCS, vol. 8139, pp. 367–381. Springer, Heidelberg (2013). doi:10.1007/978-3-642-40935-6_26
3. Chung, F.R.: Spectral Graph Theory, vol. 92. American Mathematical Soc., Providence (1997)
4. Dhillon, I.S.: Co-clustering documents and words using bipartite spectral graph partitioning. In: Proceedings of the Seventh ACM SIGKDD International Conference on Knowledge Discovery and Data Mining, pp. 269–274. ACM (2001)
5. Farahat, A.K., Elgohary, A., Ghodsi, A., Kamel, M.S.: Greedy column subset selection for large-scale data sets. Knowl. Inf. Syst. **45**(1), 1–34 (2015)
6. Fowlkes, C., Belongie, S., Chung, F., Malik, J.: Spectral grouping using the Nystrom method. IEEE Trans. Pattern Anal. Mach. Intell. **26**(2), 214–225 (2004)
7. Gittens, A., Kambadur, P., Boutsidis, C.: Approximate spectral clustering via randomized sketching. Ebay/IBM Research Technical Report (2013)
8. Hinton, G.E., Salakhutdinov, R.R.: Reducing the dimensionality of data with neural networks. Science **313**(5786), 504–507 (2006)

9. Li, M., Lian, X.C., Kwok, J.T., Lu, B.L.: Time and space efficient spectral clustering via column sampling. In: 2011 IEEE Conference on Computer Vision and Pattern Recognition (CVPR), pp. 2297–2304. IEEE (2011)

10. Ng, A.Y., Jordan, M.I., et al.: On spectral clustering: analysis and an algorithm. In: Advances in Neural Information Processing Systems, pp. 849–856 (2002)

11. Paccanaro, A., Casbon, J.A., Saqi, M.A.: Spectral clustering of protein sequences. Nucleic Acids Res. **34**(5), 1571–1580 (2006)

12. Shao, M., Li, S., Ding, Z., Fu, Y.: Deep linear coding for fast graph clustering. In: Twenty-Fourth International Joint Conference on Artificial Intelligence (2015)

13. Tian, F., Gao, B., Cui, Q., Chen, E., Liu, T.Y.: Learning deep representations for graph clustering. In: Twenty-Eighth AAAI Conference on Artificial Intelligence (2014)

14. White, S., Smyth, P.: A spectral clustering approach to finding communities in graphs. In: Proceedings of the 2005 SIAM International Conference on Data Mining, pp. 274–285. SIAM (2005)

15. Williams, C.K., Seeger, M.: Using the Nyström method to speed up kernel machines. In: Proceedings of the 13th International Conference on Neural Information Processing Systems, pp. 661–667. MIT press (2000)

16. Yan, D., Huang, L., Jordan, M.I.: Fast approximate spectral clustering. In: Proceedings of the 15th ACM SIGKDD International Conference on Knowledge Discovery and Data Mining, pp. 907–916. ACM (2009)

Improved Face and Head Detection Based on Traditional Middle Eastern Clothing

Abdulaziz Alorf$^{(\boxtimes)}$ and A. Lynn Abbott

Bradley Department of Electrical and Computer Engineering,
Virginia Polytechnic Institute and State University (Virginia Tech),
Blacksburg, VA 24061, USA
{aaalorf,abbott}@vt.edu

Abstract. This paper is concerned with the detection of individuals in images who wear traditional Middle Eastern clothing. Traditional headwear for men includes a scarf known as the *shemagh* that often occludes the face or causes significant shadows. State-of-the-art face-detection systems do not perform well for these cases. To address this problem, we have developed a novel approach that detects a distinctive part of traditional headwear known as the *igal*. This is a band or cord, typically black, that rests on the shemagh to hold it in place. Our approach starts by applying multiscale SVM classification with a HoG descriptor to perform tentative detection. The proposed detections are then refined using a bag of visual words categorization system. Experimental results have shown significantly better performance for our technique over several face-detection systems. Our technique yielded an F1 score of 80% with a low false-positive rate, showing an improvement of 15% over the best face detector.

Keywords: Face detection · Head detection · Clothing detection · Headwear · Surveillance

1 Introduction

A population of approximately 366 million people live in the Middle East, with 25 Arab countries lying between the Atlantic Ocean and the Arabian Gulf [1]. Many individuals wear traditional clothing within this geographic region of the world. Several examples of traditional dress are shown in Fig. 1.

For men, traditional Middle Eastern clothing includes headwear known as the *shemagh* or *keffiyeh*, which is a scarf that covers the head. The shemagh is commonly held in place by a dark band known as the *igal*. For today's automated face-detection systems, occlusion and shadows from the shemagh present significant problems. The shemagh often obscures portions of the face, particularly causing dark shadows near the eyes. As a result many face detectors do not perform well for these subjects as shown in Fig. 1.

This paper introduces a new approach that addresses problematic situations involving the traditional Middle Eastern headwear. The main idea is to shift the

© Springer International Publishing AG 2017
F. Karray et al. (Eds.): ICIAR 2017, LNCS 10317, pp. 389–398, 2017.
DOI: 10.1007/978-3-319-59876-5_43

Fig. 1. These two images show several examples of traditional Middle Eastern dress. In addition, these images show examples of poor performance by 2 face detectors: (left) Face++ [6,7,20], and (right) Microsoft Cognitive Services [14].

detection efforts from the face to the igal. Example igal areas are highlighted in Fig. 2(a). Additional advantages are achieved by exploiting the distinctive appearance of the igal. For example, the igal is nearly always exposed, and its appearance is invariant to in-plane rotations. This means that a single classifier can cope with yaw rotations of 360° by the head. To our knowledge, this research is the first to consider the problem of igal detection. The approach is well suited to such tasks as surveillance, and to other tasks for which face detection would commonly be applied. Another contribution of this work is a new dataset of images, available in [22], containing traditional Middle Eastern dress.

The next section provides an overview of related work. Section 3 introduces our 2-stage approach to igal detection, including the training methodologies. Section 4 presents experimental results which demonstrate performance that is significantly better than several state-of-the-art face detectors. Finally, Sect. 5 presents concluding remarks.

2   Related Work

Our problem falls within the scope of object detection [2,5–7,9,13,17,19,20]. A typical approach starts by computing a multiscale image representation, which is screened using a sliding window approach to find positive candidates that are selected based on similarity with positive examples. Finally, non-maxima suppression is applied to remove false positives.

The problem of igal detection is, in terms of its applications to real life problems, very similar to the problems of face detection [13,19]. One of the best known face detectors was developed by Viola and Jones [19]. They introduced the so called integral image representation, which allows the computation of a fast image representation using an image derivative approximated by box filters. Such features are then used with Adaboost, which is used in the cascade combination of a classifier, to quickly discard background regions and then focus the computation efforts on the most face-like regions.

Despite the fact that face detection has been considered to be essentially a solved problem, several researchers have recently proposed new approaches to the problem, improving over baseline methods [17,19,21].

One such face detector is Face++ [6,7,20], which performed well in the Labeled Faces in the Wild study [11]. The system uses four-level convolutional network cascades to deal with the problem of face detection in a coarse-to-fine fashion. Each level is trained to optimize a subset of face landmarks detected by the previous levels. In addition, geometric constraints are incorporated to improve the accuracy of landmark detection.

Recently, Mathias *et al.* [13] proposed a new evaluation framework for face detection. In addition, they have proposed two novel versions of classical object detectors. The first one is the evolution of a deformable parts model, while the second one is a version of the Viola-Jones algorithm that exploits an idea from [3] to improve face detection. The authors have shown that both detectors deliver similar performance.

To the best of our knowledge, this paper presents the first research on the problem of igal detection. Our idea has multiple advantages over face detection: (1) the igal is independent of facial appearance (i.e., it does not matter if a face is frontal or non-frontal); and (2) the igal is usually visible even if a face is not seen.

3   Igal Detection

3.1   Overview

Igal detection is difficult because the object is typically small, and is present in places with cluttered and noisy background. To the best of our knowledge, this is the first work proposing a pipeline for igal detection as a means to improve face detection performance.

In this paper, we adopt traditional gradient-based description with a classical support vector machine (SVM) formulation instead of deep learning for multiple reasons: (1) SVM reaches the global optimum rather than a local optimum, (2) it is easy to optimally choose the set of SVM parameters with less pruning to avoid overfitting, and (3) the approach requires less data for training.

Initially we planned to use local binary patterns (LBP) and local phase quantisation (LPQ) features, but we decided to use the histogram of oriented gradients (HoG) and SIFT descriptors. The decision was made based on the fact that the igal is commonly worn over different shemagh (scarf) patterns, especially checkered and pure white patterns. We found that texture descriptors were overly sensitive to the visual texture of the different shemagh patterns, whereas gradient-based descriptors were better suited to capture the shape of the igal. This enables the training of a single SVM classifier for each stage to detect the igal for different shemagh types.

As described previously, the proposed method has two main stages. The first stage uses a multiscale SVM classifier with HoG features. The classifier is refined through hard-mining training, by explicitly creating negative examples out of wrong patches, and adding the negatives to our training set again. In the second stage, we have used object classification as a means to reject outliers in the image. Dense object classification is computationally heavy. To speed up the process, we have used the set of the detected windows as an entry point to our one-vs.-all classification problem.

(a) Igal regions.   (b) Positive training set.   (c) Average igal image.

Fig. 2. (a) Examples of igal areas, highlighted in red. (b) Set of images used for both detection and classification. (c) A simple average of all 2545 igal images used for training. (Color figure online)

3.2   Igal Detection and the Hard-Mining Learning Model

For detection purposes, we have used the HoG descriptor proposed by Dalal and Triggs [5] that encodes local object shape by using the distribution of first order derivatives. To compute image gradients, we replaced the traditional difference operators by the ones proposed by Farid and Simoncelli [8], that have proved to be more accurate in slant directions. Afterwards, a set of cell histograms is computed by quantizing the image space. Each pixel within the rectangular cell casts a weighted vote for an orientation-based histogram channel based on the values found in the gradient computation. The histogram channels are evenly spread over 0 to $180°$.

For training the linear SVM for igal detection, we manually collected a set of ≈ 2000 images from the web. The images were manually segmented to obtain 2545 igal examples for training as shown in Fig. 2(b). As a negative set, we have collected unrelated data of around 7000 images for training the first model. In the iterative process of negative hard mining, the linear binary SVM model is trained using 9000 samples. This initial training set is progressively augmented with false positive examples produced while scanning the images with the model learned so far. At the end of this process, we ended up with 10000 negative samples.

Candidate selection is implemented using convolutional blocks [4] where the input feature map $\mathbf{X} \in \mathbb{R}^{W \times H \times D}$ is convolved with the learned model $\mathbf{F} \in \mathbb{R}^{W' \times H' \times D'}$ to obtain the output signal $\mathbf{Y} \in \mathbb{R}^{W'' \times H'' \times D''}$. W, H, and D are the width, height, and depth of the input feature map; W', H', and D' are the width, height, and depth of the learned model (filters); lastly, W'', H'', and D'' are the width, height, and depth of the output signal (stacked activation maps for all filters). Formally, the convolution is implemented as:

$$Y_{i'',j'',d''} = b_{d''} + \sum_{i'=1}^{H'} \sum_{j'=1}^{W'} \sum_{d'=1}^{D'} F_{i',j',d'} \times X_{i''+i'-1,j''+j'-1,d''+d'-1} \qquad (1)$$

where $b_{d''}$ refers to the usual constant bias. The convolution is applied in a multiscale framework for scale invariant detection. I.e., we start by processing

(a) Permissive detection (b) Outlier removal stage.
stage.

Fig. 3. Comparison of the output from the two stages of our algorithm. It can be seen that the pruning (outlier removal) stage discards all false positive detections in this example.

the image at full resolution, and we subsequently down-sample the image until a valid HoG descriptor can be extracted. Usually the final step on object detection is non-maxima suppression to reduce the number of candidates. However, we have observed that this approach does not provide satisfactory results, with a very large pool of false positive (and negative) detections being returned as shown in Fig. 3(a).

3.3 Pruning the False Positives

In the previous section, we observed that the permissive detection policy results in a large set of false positives but also provides good detection recall. To deal with the false positives, we propose a robust filtering approach based on dense region description. Applying this on the full image would be computationally intractable; therefore, we choose to apply it on the full set of detections.

In this outlier removal step, we used the standard SIFT descriptor [12], which computes a 128-D gradient histogram as a description vector. After normalizing all HoG detections to a canonical patch, the system extracts SIFT descriptors on a densely grid sampled at every 3 pixels [10]. The descriptors are extracted around each point of the grid at 3 local square patches with side sizes of 6, 12 and 16 pixels. This data could be directly fed to a classifier but we would have to deal with a set of features of high dimensionality, which would make training very time consuming.

To reduce the dimensionality of the igal description space, we trained a vocabulary tree using hierarchical k-means clustering, where k denotes the branch factor of the tree. Each branch is recursively split into k new groups along L levels of the tree, which sums up to L visual words. This vocabulary was trained using the same sample set as was used for the HoG detection step. At processing time, this vocabulary is used to obtain a histogram **h** of visual words.

It is well known that a non-linear SVM tends to deliver better performance than a linear SVM, but at the cost of computational efficiency in the learning step [18]. In this work, we have exploited the outcomes of [18] and employed the kernel trick. The kernel trick is based on the observation that the non-linear SVM can be approximated by mapping the features with a non-linear kernel and using the linear SVM for learning [18]. Therefore, we have used a linear SVM model with a Gaussian kernel (also called a Radial Basis Function kernel):

$$K(\mathbf{h}, \mathbf{h}') = \exp(\gamma \|\mathbf{h} - \mathbf{h}'\|) \tag{2}$$

The coefficient γ is inversely proportional to the variance of the Gaussian kernel. For learning the SVM model, we used a 10-fold cross validation [18] approach, which yielded to $\gamma = 1$ as an optimal parameter.

4 Experiments and Results

This section shows the experiments that have been carried out, and discusses the experimental results of our algorithm.

4.1 Methods Under Evaluation

To the best of our knowledge, this paper is the first to propose a mechanism to detect the Arabian igal. For comparison purposes, and given that the application scenarios of our igal detector are very similar to face detection, we have compared our algorithm against four face detectors: (1) the standard Viola-Jones algorithm as included in the MATLAB Computer Vision toolbox, (2) Face++ Matlab API [15], (3) Microsoft Cognitec Services [14], and (4) HPE Haven OnDemand API by HP [16].

Please keep in mind that these face detectors are not direct competitors with igal detection, and in fact may be used cooperatively. It is also important to notice that we have not performed modification of hyperparameters for those algorithms and APIs, but we have just used their optimized versions.

4.2 Experiment Details

We collected a database that contains 1169 images for detection evaluation. This database comprises a total number of 7854 people present in the scenes. Since our method is based on using traditional Middle Eastern clothing as a cue to improve face/head detection results, we have split our database into two sets:

1. Subjects who do not wear the shemagh. In our database, there are 1981 people who do not wear the traditional Middle Eastern clothing.
2. Subjects who wear the shemagh and possibly the igal. Our database comprises a total of 5873 people who are wearing the traditional Middle Eastern clothing.

The face-detection algorithms listed in the previous subsection were evaluated using both sets. On the other hand, our algorithm was exclusively evaluated using the second set, where it is applicable.

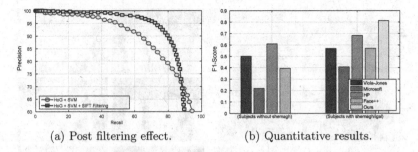

(a) Post filtering effect.   (b) Quantitative results.

Fig. 4. (a) The comparison between the use of HoG + SVM only, and the use of the full algorithm for igal detection. (b) Performance evaluation for face/igal detection: Even without the shemagh, the F1 scores are relatively low for traditional face detectors. This fact illustrates the difficulty of our dataset. Our proposed igal detector outperforms traditional face detectors on the set where subjects wear the shemagh and possibly the igal. Note that, F1 scores for people without shemagh can not be compared to the scores for people with shemagh/igal, because they are different people, and also the number of people in each set (i.e., number of people who wear and do not) is different.

4.3 Results and Discussions

Figure 4(a) depicts the comparison for igal detection between the use of HoG + SVM only, and the use of the full detection algorithm. It can be seen that post filtering using SIFT is very helpful because it reduces the false positive rate with almost no loss on the detection recall.

The quantitative comparison of all methods is performed using the F1-score, which is given by

$$F1 = 2 \cdot \frac{\text{Recall} \cdot \text{Precision}}{\text{Recall} + \text{Precision}} \tag{3}$$

This score is particularly useful when it is not possible to tune the algorithm. In our case, we do not have the ability to tweak the online APIs, where they are working in their optimal settings. Figure 4(b) shows the quantitative results obtained by applying the face-detection algorithms on the dataset that contains subjects who do not wear the shemagh, and then applying them with our algorithm on the set where subjects wear shemagh and possibly the igal. By applying only the face-detection algorithms, Viola-Jones achieved a higher score than Microsoft and Face++. However, it has a false positive rate of 35.9%, against 1.2% and 2.2% for Microsoft and Face++, respectively.

Regarding the detection of people wearing shemagh and possibly the igal, it can be seen in Fig. 4(b) that our algorithm outperforms the second best algorithm (i.e., HP) by a large margin of ≈15% in F1 score. We decided to operate our algorithm at a recall of 70%, and a false positive rate of 3%. At these operating settings, HP has provided a false positive rate of 9.5%. For the same false positive rate (i.e., 9.5%), our algorithm achieved a recall of ≈80%.

Figure 5 shows some qualitative results in which the state-of-art face detectors do not perform well. Most of the problematic cases are associated with

Fig. 5. Comparison of different face detectors and our igal detector. It can be seen that most of the face detectors have difficulties with faces that are rotated or away from the optical axis. Our detector performs well under such circumstances, outperforming the state-of-the-art face detectors.

non-frontal faces, with both roll and yaw orientations of the head. This is expected since most of the face detectors were developed for frontal captures, disregarding non-frontal faces. This is problematic in surveillance scenarios where the subject rarely faces the camera.

Overall, it can be seen that our algorithm has performed significantly better at igal detection than the traditional methods for face detection, for the same human subjects. The main reason for the improvement in performance is the fact that the igal detection is relatively invariant to face orientation, since the igal appearance remains the same for any yaw angle. A head can therefore be detected even if the face is not looking to the camera at all. This is a very important feature of our algorithm that can be explored when merging face detection with our framework.

In general, we have observed that our algorithm fails when the person is leaning back, and the igal is therefore hidden from view. In such cases, we could combine our detector with a face detector for improving the overall detection results. In this paper, we have chosen not to perform this merging because our goal has been to demonstrate that the igal alone is very distinctive.

We have implemented our method in MATLAB, and the detector runs in 2 s on average on a full 1920×1080 HD frame when using a standard laptop (Intel Core i5, 2.5 GHz, 8 GB RAM). From our experience, an optimized C++ version could achieve a processing frame rate of ≈ 10 frames per second on images of this size.

5 Conclusion

This paper has presented a novel technique for detecting heads and faces of men wearing traditional Middle Eastern clothing. The algorithm is based on the standard multiscale HOG detector, trained using an SVM to detect the headwear item known as the igal. However, instead of relying solely on HOG for detection, we relaxed the decision threshold to allow an increase in the number of false positives. Then we implemented an efficient post-classifier filter that utilizes dense SIFT filtering, bag-of-visual-words decomposition, and nonlinear SVM classification to discard false positives.

For subjects wearing the shemagh and igal, experimental results have demonstrated that our technique exhibits much better performance at detecting those subjects as compared to state-of-the-art face-detection methods. The best face detector that we tried exhibited an F1-score that was approximately 15% lower.

We have collected and labelled a dataset, available in [22], to encourage further research on this topic. This new dataset comprises 1169 images with a total of 7854 people, with ground-truth head/face locations indicated. Of that total, 5873 subjects are wearing Middle Eastern clothing. In addition, the evaluation protocol as well as the source code for our method are released in [22], to act as a baseline for further comparisons.

This work represents an important step toward significantly improved head/face detection in images from the Middle East. This paper has demonstrated that state-of-the-art face-detection techniques do not work well when traditional headwear is present. Focusing image analysis on the igal rather than the face has broad implications for the detection of human subjects in images.

References

1. Central Intelligence Agency. The world fact book (2014). https://www.cia.gov/library/publications/the-world-factbook/
2. Angelova, A., Krizhevsky, A., Vanhoucke, V., Ogale, A., Ferguson, D.: Real-time pedestrian detection with deep network cascades. In: British Machine Vision Conference (BMVC) (2015)
3. Benenson, R., Markus, M., Tuytelaars, T., Van Gool, L.: Seeking the strongest rigid detector. In: IEEE Conference on Computer Vision and Pattern Recognition (CVPR) (2013)
4. Browne, M., Ghidary, S.S.: Convolutional neural networks for image processing: an application in robot vision. In: Gedeon, T.T.D., Fung, L.C.C. (eds.) AI 2003. LNCS, vol. 2903, pp. 641–652. Springer, Heidelberg (2003). doi:10.1007/978-3-540-24581-0_55

5. Dalal, N., Triggs, B.: Histograms of oriented gradients for human detection. In: IEEE Conference on Computer Vision and Pattern Recognition (CVPR), pp. 886–893 (2005)

6. Fan, H., Cao, Z., Jiang, Y., Yin, Q., Doudou, C.: Learning deep face representation. arXiv preprint 1403.2802 (2014)

7. Fan, H., Yang, M., Cao, Z., Jiang, Y., Yin, Q.: Learning compact face representation: packing a face into an int32. In: 22nd ACM International Conference on Multimedia, pp. 933–936 (2014)

8. Farid, H., Simoncelli, E.P.: Differentiation of discrete multidimensional signals. IEEE Trans. Image Process. **13**, 496–508 (2004)

9. He, K., Zhang, X., Ren, S., Sun, J.: Deep residual learning for image recognition. In: IEEE Conference on Computer Vision and Pattern Recognition (CVPR) (2016)

10. Lazebnik, S., Schmid, C., Ponce, J.: Beyond bags of features: spatial pyramid matching for recognizing natural scene categories. In: IEEE Conference on Computer Vision and Pattern Recognition (CVPR), pp. 2169–2178 (2006)

11. Learned-Miller, E., Huang, G.B., RoyChowdhury, A., Li, H., Hua, G.: Labeled faces in the wild: a survey. In: Kawulok, M., Celebi, M.E., Smolka, B. (eds.) Advances in Face Detection and Facial Image Analysis, pp. 189–248. Springer, Cham (2016)

12. Lowe, D.G.: Distinctive image features from scale-invariant keypoints. Int. J. Comput. Vis. (IJCV) **60**, 91–110 (2004)

13. Mathias, M., Benenson, R., Pedersoli, M., Van Gool, L.: Face detection without bells and whistles. In: Fleet, D., Pajdla, T., Schiele, B., Tuytelaars, T. (eds.) ECCV 2014. LNCS, vol. 8692, pp. 720–735. Springer, Cham (2014). doi:10.1007/978-3-319-10593-2_47

14. Microsoft. Microsoft cognitive services (face detection) (2017). https://www.microsoft.com/cognitive-services/en-us/face-api

15. Face++. Face++ detection API (2017). https://www.faceplusplus.com/face-detection/

16. HPE Haven OnDemand API. HPE Haven OnDemand API (2017). https://dev.havenondemand.com/apis/detectfaces#try

17. Mikolajczyk, K., Schmid, C., Zisserman, A.: Human detection based on a probabilistic assembly of robust part detectors. In: Pajdla, T., Matas, J. (eds.) ECCV 2004. LNCS, vol. 3021, pp. 69–82. Springer, Heidelberg (2004). doi:10.1007/978-3-540-24670-1_6

18. Vedaldi, A., Zisserman, A.: Efficient additive kernels via explicit feature maps. IEEE Trans. Pattern Anal. Mach. Intell. (TPAMI) **34**(3), 480–492 (2012)

19. Viola, P., Jones, M.J.: Robust real-time face detection. Int. J. Comput. Vis. (IJCV) **57**(2), 137–154 (2004)

20. Zhou, E., Cao, Z., Yin, Q.: Naive-deep face recognition: touching the limit of LFW benchmark or not? arXiv preprint arXiv:1501.04690 (2015)

21. De Marsico, M., Nappi, M., Riccio, D.: Face: face analysis for commercial entities. In: IEEE International Conference on Image Processing (ICIP) (2010)

22. Alorf, A., Abbott, A.L.: A dataset for face detection based on traditional middle eastern clothing (2017). https://drive.google.com/drive/folders/0B1FXrfJPvPmYNXkwQnNZV21nNnc?usp=sharing

Unsupervised Group Activity Detection by Hierarchical Dirichlet Processes

Ali Al-Raziqi[✉] and Joachim Denzler

Computer Vision Group, Friedrich-Schiller-Universität Jena, Jena, Germany
{ali.al-raziqi,Joachim.Denzler}@uni-jena.de

Abstract. Detecting groups plays an important role for group activity detection. In this paper, we propose an automatic group activity detection by segmenting the video sequences automatically into dynamic clips. As the first step, groups are detected by adopting a bottom-up hierarchical clustering, where the number of groups is not provided beforehand. Then, groups are tracked over time to generate consistent trajectories. Furthermore, the Granger causality is used to compute the mutual effect between objects based on motion and appearances features. Finally, the Hierarchical Dirichlet Process is used to cluster the groups. Our approach not only detects the activity among the objects of a particular group (intra-group) but also extracts the activities among multiple groups (inter-group). The experiments on public datasets demonstrate the effectiveness of the proposed method. Although our approach is completely unsupervised, we achieved results with a clustering accuracy of up to 79.35% and up to 81.94% on the Behave and the NUS-HGA datasets.

1 Introduction

Public spaces are characterized by the existence of several activities. Many researchers have contributed in activity recognition. The approaches can be divided into three categories: (I) Action recognition, which is handled by analyzing the action of a single object through extracting features of the whole object or the segmented body parts [1]. (II) Pair activity, which is interpreted by analyzing the relationships of a pair of objects [8,11]. (III) Group activity, which is considered as coherent activities performed by multiple objects. In this paper, we focus on the group activity recognition. The analysis of group activity plays an important role in video analysis. Accordingly, localizing and understanding the group activity is an important topic in many applications such as security and surveillance interaction detection. In addition, it can help in detecting suspicious and illegal group behavior.

Most of the group activity recognition methods are supervised [4,13,14,16,18–20]. In contrast, we focus on detecting the group activity in an unsupervised manner using *Hierarchical Dirichlet Processes* (HDP). Although recent work applied HDP in interaction detection [2], they rely on optical flow features, which is not helpful in case of the fixed objects. The intuition behind their work is to segment activities into spatio-temporal patterns. Also, a video sequence was divided temporally into equally sized clips without overlap. As a consequence, too short clips will split up an activity into sub-activities, and thus too long clips might join non-relevant activities.

© Springer International Publishing AG 2017
F. Karray et al. (Eds.): ICIAR 2017, LNCS 10317, pp. 399–407, 2017.
DOI: 10.1007/978-3-319-59876-5_44

We tackle this problem by dividing a video automatically into clips using an unsupervised clustering approach. As a result, the clips might have overlap and have different lengths. To this end, first, relevant groups of objects are detected using a *bottom-up hierarchical clustering*. The groups are tracked over time to form consistent trajectories, then, each group is treated as one clip. Finally, the HDP is used to cluster the clips.

The main contributions of this paper are as follows: (I) We presented a novel approach for detecting meaningful groups without training. This addresses (1) a varying number of involved objects and (2) an unknown number of groups. (II) The Granger causality is used to measure the mutual effect among objects in a particular group and among groups as well based on motion trajectories and appearances features.

The rest of this paper is organized as follows. Section 2 provides an overview of the existing literature on group activity recognition. The proposed framework of group activity is described in Sect. 3. The experiments and results conducted on the Behave and NUS-HGA datasets are described in Sect. 4 along with results.

2 Related Work

Ni *et al.* in [14], analyzed the self, pair, and inter-group causalities to detect group activities based on trajectories. They assumed that there is only one group activity in the scene. Hence, they cannot handle more complicated environments of simultaneous activities. In contrast, we handle all the activities in the scene. Zhang *et al.* in [19] tried to detect multiple group activities. This was achieved by clustering the objects into subgroups using K-means. But, providing the number of groups in advance is not a robust solution. An interesting work has been presented in [13], Kim *et al.* tried to overcome the fixed number of groups. They recognized the groups by modeling proxemics. This was achieved by defining Interaction Potential Zone (IPZ) around each object (bubbles with 58 pixels). However, using the same IPZ value whether objects are far or near from the camera leads to dispersing the relevant objects in irrelevant groups and vice versa. In contrast to them, we cluster the objects using bottom-up hierarchical clustering based on the velocity and motion direction.

Additionally, deep neural networks have been recently applied for group activity detection [6,7,10]. Thus, they are more robust and effective, but they used supervised learning methods. In work presented in [2], spatio-temporal patterns are analyzed automatically by using HDP to extract the hidden topics. They divided the video into short and equally sized clips without overlap, where the clip size effects on the performance. Another interesting method which tried to tackle this problem using an extended probabilistic Latent Semantic Analysis (pLSA) [21]. Unlike them, our approach extracts the number of activities automatically. Unlike many of the approaches described above, our approach extracts the group activity by dividing a video automatically into clips without further knowledge.

3 Framework

Our framework for group activity detection has several stages as shown in Fig. 1. Given an input video, objects are detected using YOLO [15], which is a unified neural network

Fig. 1. Proposed framework for automatic group activity detection.

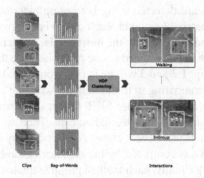

Fig. 2. Process of extracting BoWs and clips clustering.

based approach. All detected objects are tracked by the GMMCP tracker [5]. After generating the groups of objects, each group is tracked over time and treated as one clip. Afterward, these clips are clustered by the HDP as shown in Fig. 2. The last two steps will be described in detail as follows.

3.1 Groups Detection and Tracking

In common situations, multiple objects are involved simultaneously in separate activities, and those objects may further interact with each other. To detect all activities in the scene, the main step is the detection of groups. The key assumption of our approach is to cluster objects that are spatially close and moving in the same direction with the same speed. Given objects trajectories, each object represented by 3 tuples $(\mathcal{P}, \mathcal{V}, \theta)$, where $\mathcal{P}$ is the center of mass coordinate (x, y), $\mathcal{V}$ represents the velocity and θ is the motion direction. However, the pairwise distance $d^t(i, j)$ is computed for the trajectories i, j as

$$d^t(i, j) = \xi_{i,j}^t \cdot \vartheta(i, j)$$

$$\xi_{i,j}^t = |\mathcal{P}_i^t - \mathcal{P}_j^t| \tag{1}$$

$$\vartheta_t(i, j) = \begin{cases} 1 & \text{if } |\mathcal{V}_i^t - \mathcal{V}_j^t| < \mathcal{T}_v \wedge |\theta_i^t - \theta_j^t| < \mathcal{T}_\theta \\ 0 & \text{otherwise} \end{cases} \tag{2}$$

Consequently, we compute d^t by multiplying ξ^t and ϑ^t and normalized, which ensures that the objects are spatially close and moving in the same direction with the same speed. Where $\mathcal{T}_v$ and $\mathcal{T}_\theta$ are predefined thresholds. Then, the adjacency matrix $\mathcal{A}$ is built as a result of $1 - d^t$ for each pair of detections. Equation 3 shows an example of 4 objects. In matrix $\mathcal{A}$, a large value means that the two objects are most close and they are moving with the same speed and in the same direction.

$$\mathcal{A} = \begin{pmatrix} 1-d^t(1,2) & 1-d^t(1,3) & 1-d^t(1,4) \\ & 1-d^t(2,3) & 1-d^t(2,4) \\ & & 1-d^t(3,4) \end{pmatrix} \rightarrow \begin{pmatrix} 0.5 & 0.2 & 0 \\ & 0.3 & 0.4 \\ & & 0.1 \end{pmatrix} \tag{3}$$

After that, the groups are detected using bottom-up hierarchical clustering. In the first step, by taking the matrix $\mathcal{A}$ as input, each object is assigned to a separate cluster (4 clusters in this case) and merged with the most similar clusters in the next iterations. For instance, objects $1, 2$ will be assigned to one group because they have the maximum value (0.5). In the next step, $1,2$ and 3 will be in one group, etc. In our case, the number of groups is not required comparing to traditional methods (e.g., K-means or spectral).

Once the groups are generated, dense SIFT features are extracted for each group. All features are clustered into k clusters using K-means. Then the Bag of Words (BoWs) histograms are extracted. Hence, each group is described by BoWs. The matching between the groups' BoWs is computed by the distance of the histogram intersection, which is bounded by $[0, 1]$. Finally, each tracked group is treated as one clip.

3.2 Activity Clustering

When the clips have been generated, our approach does not only detect the activity among the objects of a particular group (intra-group) but also extracts the activities among multiple groups (inter-group). Suppose that G_j is a group of size n objects, the group center is determined by the average position of all objects $G_j(c^t, y^t) = \frac{1}{n}\left(\sum_{i=1}^{|n|} x_i, \sum_{i=1}^{|n|} y_i\right)$. In order to describe the activities of intra-group and inter-group, trajectories-based features are extracted for every time window $\mathcal{K}$. Some important features are as follows.

Causality: The temporal causality is an usual way to recognize the group activity. In this paper, Granger causality (GC) [9] is used to measure the causal relationships between objects. Generally speaking, given two time series A and B, A is said to be Granger-cause B ($A \rightarrow_G B$) if the past values of A with the past values of B provide significant information of B. Many approaches [13, 14, 19, 21] have focused on measuring GC between objects trajectories in terms of center mass coordinates (x, y). In our approach, we compute the causality of both, the objects coordinates and the appearance features. The appearance SIFT features are extracted for each object, the dictionary is built using K-means, then each object represented by concatenating BoWs over the window $\mathcal{K}$.

To infer the GC, the null hypothesis $A \nrightarrow_G B$ has to be tested first, by evaluating the autoregression as

$$\begin{aligned} B^{(i)} &= \beta_0 + \beta_1 B^{(i-1)} + \cdots + \beta_l B^{(i-l)} \\ \bar{B}^{(i)} &= B^{(i)} + \vartheta_0 + \vartheta_1 A^{(i-1)} + \cdots + \vartheta_l A^{(i-l)}, \end{aligned} \tag{4}$$

where β_l and ϑ_l are the model parameters. Therefore, the residual sum of square errors $RSS_B, RSS_{\bar{B}}$ are used for the evaluation. Finally, the causality calculated by

$$F_{A \to B} = \frac{(RSS_{\bar{B}} - RSS_B)/l}{RSS_{\bar{B}}/(K - 2l - 1)} \tag{5}$$

where K is the number of samples considered for the analysis and l represents the lag.

Shape Similarity: Suppose we have two trajectories $\mathcal{A}$ and $\mathcal{B}$, Dynamic Time Warping (DTW) is used to map one trajectory to another by minimizing the distance between the two. In particular, the sum of the Euclidean distances is used as feature.

Velocity and Distance Features: We extracted some other features like velocity $\mathcal{V}$, absolute change $\hat{\mathcal{V}}$ of the same object and the absolute difference $\mathcal{V}_{[i,j]}$ in velocity of a pair of objects i and j. In addition to that, vorticity $\mathcal{V}$ can be measured as a deviation of the center mass of an object from a line. The line is calculated by fitting a line to the positions of the trajectory in window $\mathcal{K}$. Moreover, computing the distance $d_{[i,j]}$ and the difference in distance $\hat{d}_{[i,j]}$ are useful to distinguish the interaction among objects. Since they are sensitive to tracking errors, the distance $\tilde{d}_{[i,j]}$ is calculated for every point in window $\mathcal{K}$. More information can be found in [3]. Since the number of features varies, encoding those normalized features using K-means is required. The BoWs histograms are extracted, that each activity is represented by BoWs.

Hierarchical Dirichlet Processes. Once the BoWs histograms are computed, the HDP is used to extract the activities as shown in Fig. 2. HDP is a generative clustering technique used to cluster words in documents into K latent *topics* [17]. In HDP, the number of topics is inferred automatically from the data and the hyper-parameters.

In HDP, given a video, the whole activities G_0 are generated using Dirichlet Process (DP). Then, for each clip, DP generates specific activities G_m which are drawn from the global list G_0. The generative HDP formulation is written as follows

$$G_0 \mid \gamma, H \sim DP(\gamma, H)$$
$$G_m \mid \alpha, G_0 \sim DP(\alpha, G_0) \quad \text{for}, m \in [1, M]. \tag{6}$$

where the hyper-parameters α and γ are the concentration parameters and the parameter H is called the base distribution (which is a Dirichlet distribution in our case).

Bayesian Inference. Given the observed BoWs for all clips, to estimate the hidden states $\phi_{m,n}$, we perform Bayesian inference using Gibbs sampler. Therefore the word-topic association conditional distribution is formulated as

$$p(\phi_{m,n} = k, \alpha, \gamma, \eta, \theta, H) \propto (n_{m,k}^{\neg m,n} + \alpha \theta) \cdot \frac{n_{k,t}^{\neg m,n} + \eta}{n_k^{\neg m,n} + V.\eta} \tag{7}$$

where $n_{m,k}$, $n_{k,t}$, n_k and V represent statistics of the word-topic, topic-document and the topic-wise word counts and the dictionary. Assigning the current word $x_{m,n}$ to a topic k_i is proportional to the number of words already assigned to a topic k_i. A more detailed explanation is presented in [17].

4 Experimental Results

We validate our proposed approach on two benchmark datasets, the Behave and NUS-HGA [3,14]. Since the NUS-HGA dataset does not contain the tracking ground truth, the objects are detected and tracked using YOLO detector and GMMCP tracker. For evaluation purposes, the extracted HDP topics are mapped to the ground truth labels by voting among the topics. From the perspective of the HDP theory, each document is a distribution over all extracted topics, so we restricted one topic for one class. Then the evaluation is done as a classification problem. Due to the randomness in the Bayesian inference, each experiment runs ten times, and we report the average performance.

Concerning the parameters analyzing in Sect. 4.2, the parameters of the experiments are chosen for both datasets as follows. For feature extraction, the dictionary size of the SIFT features for computing the Granger causality are 20 and 90 for Behave and NUS-HGA. For the group activity recognition experiments, the dictionary sizes for K-means of the whole features are 50 and 150 for Behave and NUS- HGA. HDP hyperparameters α and η are set as 0.8 and 0.1 for both dataset.

4.1 Behave and NUS-HGA Datasets

In the Behave dataset, multiple objects ranging from two to five are involved in each activity. We achieved clustering accuracy of up to 79.35 ($\pm 5\%$). As can be seen in Fig. 3(a), when the dictionary size is increased, the performance is further decreased. The Behave dataset is represented significantly by dictionary size 50. We compare our approach with [2,3,12–14,16,18], according to Table 1, we outperformed all unsupervised approaches and the one that presented in [13] on the Behave dataset.

Table 1. Comparison with other works on the Behave and NUS-HGA datasets

	Method	Accuracy %	
		Behave dataset	NUS-HGA dataset
Supervised	[4]	42.50	93.50
	[13]	**93.74**	96.02
	[14]	-	74.16
	[16]	-	**98.00**
	[18]	93.65	-
	[3]	93.67	-
Unsupervised	[2]	65.95	-
	[12]	66.25	-
	Our approach	**79.35**	81.94

NUS-HGA dataset has different group activities, *WalkInGroup, Gather, RunInGroup, Fight, StandTalk, and Ignore*. We achieved clustering accuracy of up to 81.94 ($\pm 3.07\%$). As can be seen in Fig. 3(a), as the dictionary size is increased, the performance is further decreased. The performance is compared with [4, 13, 14, 16] as shown in Table 1. Despite the fact that our approach is unsupervised, we outperformed the supervised work that is presented in [14].

4.2 Influence of Parameters

We studied the influence of the GC, the extracted features and HDP parameters.

Causality: As can be seen in Fig. 3(b), we achieved the highest accuracy with the Granger causality of SIFT BoW as 20 and 90 combined with the other features on both datasets respectively. This shows that the Granger causality of SIFT BoWs is robust even with high dimensional data.

Features: As can be seen from Fig. 3(c), the most represented feature is the shape similarity DTW of the trajectories for both datasets. The combination of all features improves the performance significantly.

HDP Hyper-Parameters: In these experiments, η ranges from 0.1 to 2. With increasing the η, the number of extracted topics increases linearly, as can be seen in Fig. 3(d). From the perspective of the HDP theory, an infinite number of topics are extracted. Therefore, a too small number of topics would under-represent the group activities, which causes joining of similar activities into one. On the other hand, a too large number of topics would lead to over-fitting.

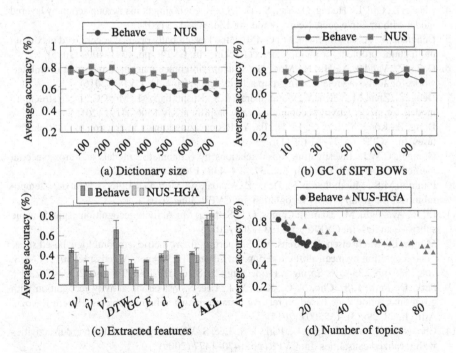

Fig. 3. Illustration of impact of the dictionary size, the GC of SIFT BOWs, the extracted features and HDP η hyperparameter on the Behave and NUS-HGA datasets.

5   Conclusion

The aim of this paper was to address the problem of group activity detection in an unsupervised manner. We introduced a new approach to segment video sequences automatically into clips based on the occurring activities. The main step was the detection of groups using the bottom-up hierarchical clustering. Furthermore, the Granger causality is used to measure the mutual effect among objects in a particular group and among groups as well based on motion trajectories and appearances features. Finally, the activities are extracted by using HDP. We achieved results with a clustering accuracy of up to 79.35% on the Behave dataset and up to 81.94% on the NUS-HGA dataset.

Acknowledgements. The authors are thankful to Mahesh Krishna and Manuel Amthor for useful discussions and suggestions.

References

1. Aggarwal, J.K., Xia, L.: Human activity recognition from 3D data: a review. Pattern Recogn. Lett. **48**, 70–80 (2014)
2. Al-Raziqi, A., Denzler, J.: Unsupervised framework for interactions modeling between multiple objects. In: VISAPP, vol. 4, pp. 509–516 (2016)
3. Blunsden, S., Fisher, R.B.: The BEHAVE video dataset: ground truthed video for multi-person behavior classification. BMVA **4**, 1–12 (2010)
4. Cheng, Z., Qin, L., Huang, Q., Yan, S., Tian, Q.: Recognizing human group action by layered model with multiple cues. Neurocomputing **136**, 124–135 (2014)
5. Dehghan, A., Assari, S., Shah, M.: GMMCP tracker: globally optimal generalized maximum multi clique problem for multiple object tracking. In: CVPR, pp. 4091–4099 (2015)
6. Deng, Z., Vahdat, A., Hu, H., Mori, G.: Structure inference machines: recurrent neural networks for analyzing relations in group activity recognition. In: CVPR (2016)
7. Deng, Z., Zhai, M., Chen, Y., Muralidharan, L.S., Roshtkhari, M., Mori, G.: Deep structured models for group activity recognition. arXiv preprint arXiv:1506.04191 (2015)
8. Dong, Z., Kong, Y., Liu, C., Li, H., Jia, Y.: Recognizing human interaction by multiple features. In: ACPR, pp. 77–81 (2011)
9. Granger, C.W.J.: Investigating causal relations by econometric models and cross-spectral methods. Econom.: J. Econom. Soc. **37**, 424–438 (1969)
10. Ibrahim, M.S., Muralidharan, S., Deng, Z., Vahdat, A., Mori, G.: A hierarchical deep temporal model for group activity recognition. In: CVPR, June 2016
11. Li, B., Ayazoglu, M., Mao, T., Camps, O.I., Sznaier, M.: Activity recognition using dynamic subspace angles. In: CVPR, pp. 3193–3200 (2011)
12. Münch, D., Michaelsen, E., Arens, M.: Supporting fuzzy metric temporal logic based situation recognition by mean shift clustering. In: Glimm, B., Krüger, A. (eds.) KI 2012. LNCS, vol. 7526, pp. 233–236. Springer, Heidelberg (2012). doi:10.1007/978-3-642-33347-7_21
13. Park, U., Park, J.-S., Cho, N.-G., Kim, Y.-J., Lee, S.-W.: Group activity recognition with group interaction zone based on relative distance between human objects. Pattern Recogn. Artif. Intell. **29**(05), 1555007 (2015)
14. Cho, N.-G., Kim, Y.-J., Park, U., Park, J.-S., Lee, S.-W.: Recognizing human group activities with localized causalities. In: CVPR, pp. 1470–1477 (2009)
15. Redmon, J., Divvala, S., Girshick, R., Farhadi, A.: You only look once: unified, real-time object detection. arXiv preprint arXiv:1506.02640 (2015)

16. Stephens, K., Bors, A.G.: Group activity recognition on outdoor scenes. In: Advanced Video and Signal Based Surveillance (AVSS), pp. 59–65 (2016)
17. Teh, Y.W., Jordan, M.I., Beal, M.J., Blei, D.M.: Hierarchical Dirichlet processes. JASA 101(476), 1566–1581 (2006)
18. Yin, Y., Yang, G., Xu, J., Man, H.: Small group human activity recognition. In: Image Processing (ICIP), pp. 2709–2712. IEEE (2012)
19. Zhang, C., Yang, X., Lin, W., Zhu, J.: Recognizing human group behaviors with multi-group causalities. In: WI-IAT, vol. 3, pp. 44–48 (2012)
20. Zhou, Y., Ni, B., Yan, S., Huang, T.S.: Recognizing pair-activities by causality analysis. ACM - TIST 2(1), 5 (2011)
21. Zhu, G., Yan, S., Han, T.X., Xu, C.: Generative group activity analysis with quaternion descriptor. In: Lee, K.-T., Tsai, W.-H., Liao, H.-Y.M., Chen, T., Hsieh, J.-W., Tseng, C.-C. (eds.) MMM 2011. LNCS, vol. 6524, pp. 1–11. Springer, Heidelberg (2011). doi:10.1007/978-3-642-17829-0_1

Classification Boosting by Data Decomposition Using Consensus-Based Combination of Classifiers

Vitaliy Tayanov[✉], Adam Krzyżak, and Ching Suen

Department of Computer Science and Software Engineering,
Concordia University, Montreal, QC H3G 1M8, Canada
{tayanov,krzyzak,suen}@encs.concordia.ca

Abstract. This paper is devoted to data decomposition analysis. We decompose data into functional groups depending on their complexity in terms of classification. We use consensus of classifiers as an effective algorithm for data decomposition. Present research considers data decomposition into two subsets of "easy" and "difficult" or "ambiguous" data. The easiest part of data is classified during decomposition using consensus of classifiers. For other part of data one has to apply other classifiers or classifier combination. One can prove experimentally that afore mentioned data decomposition using optimal consensus of classifiers leads to better performance and generalization ability of the entire classification algorithm.

1   Introduction

Problems investigated in this paper belong to the area of combination learning. Combination learning means that one has to find the optimal combination of classifiers as well as their quality and quantity to achieve better performance than that of any particular classifier involved in the combination. Also combining scheme has to possess be good enough generalization ability. However Kuncheva [6] gave an example involving majority voting algorithm when unweighted majority voting performs worse than some classifiers in ensemble and only weighted scheme of majority voting had better performance than any of classifiers used in the combination. Our objective is the following: having a pool of available classifiers we aim at creation of their combination that has better performance than any of the arbitrary classifier in the pool.

To reach the stated objective one needs to solve the following problems: optimize classifier selection and fusion to optimally fuse their outputs. To generate a pool of classifiers different classifier training techniques can be used: random forests, rotation forests and random space method [2,5,8]. All these algorithms are rather simple and perform well in a number of applications. Unstable classifiers like random forests trained by Adaboost and Bagging [1,4] have low level of overfitting. However very complex classifiers with a very large number of parameters usually suffer from overfitting [7]. So we have to try some simple models as well.

© Springer International Publishing AG 2017
F. Karray et al. (Eds.): ICIAR 2017, LNCS 10317, pp. 408–415, 2017.
DOI: 10.1007/978-3-319-59876-5_45

Present research is mainly devoted to the following problem: we have to test effectiveness of rejection algorithm based on consensus to verify if our algorithm rejects objects with high enough probability of misclassification. To do that we are going to use k-NN classifiers from the subset of the set of generated classifiers. This subset is restricted to have only strong enough classifiers. Thus we can make sure that our diverse classifiers generate different decisions but are close in terms of errors they produce. So we are going to split them only by similarity using some diversity scores [10]. Finally to have a subset of required classifiers we limit the subset of potential "good" classifiers using error rates they produce.

2   Algebraic Approach

Assume that there is a set of strong classifiers F, such that any strong classifier belongs to this set, e.g. $f_i \in F, i = 1, \dots n$, where n is a number of strong classifiers in the set, thus $|F| = n$.

Consider N class classification problem. Let I be the number of objects in the set and P be the number of classification results.

Every classification p $(p = 1, \dots, P)$ associates probe k from the sample with one and only one class. Elementary co-association matrix $\mathbf{A}^k$ contains the information according to which algorithms u and v have a consensus or not with respect to some class label i for probe k:

$$a_{u,v}^k = \begin{cases} 1 \text{ for } u \equiv v \\ 0 \text{ otherwise} \end{cases}, \tag{1}$$

where $\equiv$ denotes consensus between classifiers f_u and f_v. Because $u \equiv v \Rightarrow v \equiv u$, $\mathbf{A}^k$ is a $n \times n$ symmetric binary matrix. The number of different consensuses that could be created is equal to $P = \binom{n}{2} = \frac{n(n-1)}{2}$.

From the entire set of possible consensuses equal to P one needs to take two of them that are maximally dissimilar in terms of Hamming distance between binary results of classification for objects $k \in I$ of the training sample of the set χ. Finding maximal Hamming distance between classifiers spans entire classifier space F. This allows to consider all possible combinations among classifiers and use all possible Hamming distances between classifiers which is important during optimization of consensus. Results of classification are represented as binary vector $\mathbf{v}^p$ that consists of zeros and ones. The elements of vector $\mathbf{v}^p$ are taken from the set of matrices $\mathbf{A}^k$:

$$v_k = a_{u,v}^k. \tag{2}$$

The size of vector $\mathbf{v}^p$ is equal to the number of objects I. Then tensor $\mathbf{A}$ has size $n \times n \times I$. Now the only task left to do is to find the pair of algorithms that have the maximal Hamming distance. Determination of appropriate indexes could be done by the following procedure

$$\left\{ i, j \right\} = \arg\min_{u,v} \sum_k A_{u,v}^k. \tag{3}$$

To estimate the pair of the most dissimilar classifiers one needs to use some training set. That is why solution (3) is approximate.

3   Optimization Techniques for Consensus of Combined Classifiers

General performance can be optimized using a subset of strong classifiers $F^* = \bigcup_i f_i \subset F$ from an initial pool of classifiers F to decompose data into functional groups depending on their complexity. Strong classifiers are selected from the pool F, using a threshold γ for the error rate ϵ, where ϵ is a random variable. New subset of classifiers has lower entropy of the error rate $\mathbb{H}(\epsilon(\gamma))$ due to limited amount of errors caused by threshold γ. Value of γ can be determined experimentally using a training subset of data. However it is also important to evaluate the conditional entropy $\mathbb{H}(D_H|\epsilon(\gamma))$ of Hamming distance or some other similarity measure between classifiers for some fixed error rate. It is possible to use mutual information as well [7]

$$\mathbb{I}(D_H; \epsilon(\gamma)) = \mathbb{H}(D_H(\gamma)) - \mathbb{H}(D_H((\gamma))|\epsilon(\gamma)) = \mathbb{H}(\epsilon(\gamma)) - \mathbb{H}(\epsilon(\gamma)|D_H(\gamma)). \quad (4)$$

So one needs to know the influence of diversity of error rates on diversity of classifiers in terms of Hamming distance or some other diversity score (Q statistics, correlation coefficient, disagreement measure, etc.) [6,9]. To build a reliable consensus it is very important to have the most diverse classifiers. However they can be erroneous at the same time. To have a set of strong classifiers it is important to limit classifier space in terms of error rate preserving large enough value of diversity score. Changes in diversity can be evaluated via entropy rate $\frac{\mathbb{H}(D_H|(\gamma))}{\mathbb{H}(D_H|\gamma=1)}$. For simplicity we use notation $\mathbb{H}(D_H|(\gamma))$ remembering that $\mathbb{H}(D_H) = f(\gamma)$ as well as $\mathbb{H}(\epsilon) = g(\gamma)$.

Optimization of the set of classifiers to perform data decomposition has direct influence on the computational complexity due to limitation of classifier pool and should be carried out during training.

We are going to consider the optimization procedure concerning initial data decomposition into subsets of "easy" and "difficult" ("ambiguous") data. In general all "ambiguous" data can be characterized by very different probabilities of correct classification. Some of them might have an error rate more than 0.5 while others less than 0.5 for some basic classifiers. One has to split the initial data set in such a way that average error probability of all "ambiguous" patterns is high enough. This allows to obtain a very good classification accuracy by training some powerful classifier to reclassify only such "ambiguous" data. In other words we propose to detect "difficult" or "ambiguous" patterns on which some basic classifiers operate as random ones or close to that. So this actually determines how good is our detection of such patterns. The quality of such a detection can also be expressed in terms of information gain:

$$\mathbb{E}(\epsilon(A(x \in \Im_2))) > 0.5, \quad (5)$$

where $A = \otimes(A_1, A_2, \ldots, A_n) = \otimes_{i=1}^{n}\left\{A_i\right\}$ is a consensus consisting of n algorithms, x is a pattern and $\epsilon(\cdot)$ is a consensus error or the probability that x

belongs to a group of misclassified patterns. The maximal average accumulated gain in accuracy will be

$$\mathbb{E}(\epsilon(A(x \in \Im_2))) - 0.5, \qquad (6)$$

when $\epsilon(A(x)) > 0.5, \forall x \in \Im_2$.

As it can be seen all optimization techniques are directed on selecting "ambiguous" data in such a way to maximize the accumulated average gain (6).

4   Experiments

We used 5 data sets from UCI repository [3] as indicated in Table 1. Three of them ('pima', 'bupa' and 'heart') belong to medical domain and two others ('german' and 'sonar') to non-medical domain. Even though they are all binary problems, they are one of the most difficult problems from UCI repository used for different tests of machine learning algorithms.

Table 1. Summary of characteristics of used data sets

Method/task	'pima'	'bupa'	'heart'	'german'	'sonar'
Size	768	345	270	1000	208
Features	8	6	13	20	60
Classes	2	2	2	2	2

Table 2. Accuracy of classification for different combining algorithms: decision trees

Method/task	'pima'	'bupa'	'heart'	'german'	'sonar'
Bagging	75.4	68.7	78.7	75.6	79.3
AdaBoost	69.8	65.8	78.1	73.3	79.8
Random subspace	73.6	65.8	80.5	73.9	80.3
Rotation forest	75.4	69.5	79.4	77.0	81.8
Random oracle	77.2	69.3	78.1	74.4	78.4
Proposed (NB)	78.6	69.6	83.6	71.2	73.7
Proposed (LR)	78.6	60.3	76.9	68.0	82.2
Proposed (AB)	**86.0**	**84.3**	**88.2**	**81.3**	**92.4**

To build a consensus one uses a set of k-NN classifiers with different L_p norms as dissimilarity functions. We varied p in a norm L_p from 0.1 to 5 with a step of 0.1. Value of k was changed from 1 to 30 using only odd values of k nearest neighbors yielding a pool of 800 k-NN classifiers. Classification results were compared to those obtained by the most popular ensembles of classifiers or their modifications, e.g., bagging, boosting, random forests, random subspace and random oracle. In one scenario (Table 2) as a week classifier decision tree

Table 3. Accuracy of classification for different combining algorithms: linear classifiers

Method/task	'pima'	'bupa'	'heart'	'german'	'sonar'
Bagging	76.8	67.8	65.3	77.7	74.5
AdaBoost	77.3	68.6	69.4	76.7	73.6
Random subspace	75.9	59.7	80.1	74.1	78.0
Rotation forest	63.4	67.8	63.6	76.9	73.6
Random oracle	76.0	73.0	67.7	76.4	78.9
Proposed (NB)	78.6	69.6	83.6	71.2	73.7
Proposed (LR)	78.6	60.3	76.9	68.0	82.2
Proposed (AB)	**86.0**	**84.3**	**88.2**	**81.3**	**92.4**

has been taken and in the second scenario (Table 3) we used linear classifier. The accuracy of these five ensembles of classifiers is compared with three versions of developed algorithm for each of two scenarios. The difference between them is only in reclassification of "difficult" or "ambiguous" patterns. So first version uses naive Bayes (NB) classifier to reclassify these objects, the second one uses logistic regression (LR) and the third one uses AdaBoost (AB) with decision stumps as week classifiers. All three versions use 90% of "ambiguous" objects for training and 10%–for testing. In some situations when the portion of difficult data is small in comparison with all available data the ratio between training and testing partitions should be reduced.

Comparing naive Bayes and logistic regression based on Tables 2 and 3 one may conclude that naive Bayes outperforms logistic regression on "difficult" data in most cases. This means that generative model might better fit "difficult" data than the discriminative one which is geared towards training data and completely learns from the data. Nevertheless AdaBoost learns very well on entirely "difficult" patterns and actually does not overfit. This explains that reclassifying "ambiguous" data with boosting algorithms one might obtain very small error rates even on testing examples. Such a behavior of AdaBoost can be explained by the fact that one reduced the hypothesis space by selecting two the most dissimilar classifiers from the pool of generated classifiers. These classifiers determine the set of "ambiguous" data. If one selects other two classifiers from the pool they can detect different subsets of such data. So appropriate selection of classifiers is very important issue. Thus learning classifiers on a partition of "difficult" patterns which are grouped near the separating hyperplane and are characterized by lower variance in comparison to entire set allows us to reduce the set of all potential classification hypotheses. This leads to better learning, less error rate on the test data and hence less overfitting.

Data separation into functional groups is given in Table 4. So any object can either be misclassified, considered as ambiguous or correctly classified by consensus. The difficulty of data set can be estimated by a number of misclassified patterns. This is irreducible error. To have lots of "ambiguous" patterns

Table 4. Distribution of different types of data (%)

Method/task	'pima'	'bupa'	'heart'	'german'	'sonar'
Misclassified	14.0	12.7	10.2	15.0	7.6
Ambiguous	21.6	42.7	32.2	34.0	34.0
Correctly classified	64.4	44.6	57.6	51.0	58.4

Table 5. Reclassification accuracy of ambiguous data %

Method/task	'pima'	'bupa'	'heart'	'german'	'sonar'
naive Bayes	65.8	58.5	80.8	59.5	45.0
Logistic regression	65.7	36.7	60.0	50.0	70.0
AdaBoost	**100**	**93.0**	**95.0**	**89.0**	**100**

might not be very bad since they can be learned effectively using, for example, AdaBoost. Table 5 shows how "good" are some classifiers in reclassifying "ambiguous" data. The best results of such reclassification are achieved by the Adaboost. It outperforms the naive Bayes and logistic regression quite a lot. As can be seen in the last two tables reclassification error of "ambiguous data" using naive Bayes and logistic regression is high enough implying that selecting the most dissimilar classifiers is reasonable. However if we need less dissimilar classifiers we can also find them in the entire pool of generated classifiers. So selecting algorithms to build a consensus should be done in a way that allows us to optimize the total error which is trade-off between misclassification error and error caused by "ambiguous" data.

Figure 1 presents performance results of our algorithm for five data sets mentioned above. First row shows different errors produced by consensus and the second one is about how close the real consensus error might be to the one in case of error independence. All dependencies are built as a function of the normalized Hamming distance. From the results given in the first row we see that total consensus error drops out having more dissimilar classifiers in consensus. The second row shows that in case of more dissimilar classifiers the real consensus error approaches the error obtained under assumption of error independence in consensus.

Figure 2 presents the local and global diversity scores for all five data sets. The first row corresponds to the local diversity scores, i.e., the diversity scores between any of two classifiers as a function of the normalized Hamming distance. To evaluate that one uses coefficients based on correlation and Q-statistics. The second row represents the global diversity score defined as an entropy of generated and preselected classifiers as a function of the normalized Hamming distance. This shows how diverse is our subset of generated classifiers after the limitation of maximal error they can produce.

Figure 3 shows the results of reclassification of ambiguous data using AdaBoost with decision stumps for the same five problems from UCI repository.

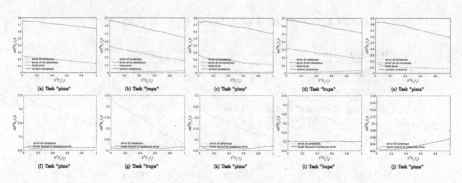

Fig. 1. Performance of consensus: different errors during consensus and lower bound of consensus error

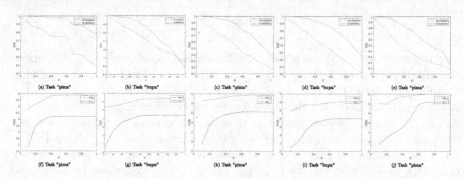

Fig. 2. Local and global diversity scores

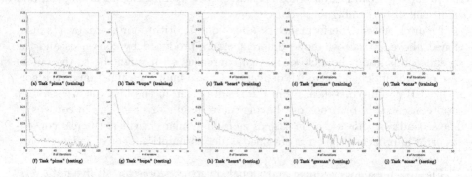

Fig. 3. Reclassification of ambiguous data with decision stumps trained by AdaBoost

The first row represents the training subset of ambiguous data and the second one the testing subset. As can be seen from the presented results AdaBoost performs very similarly on the training and testing subsets of ambiguous data. For some problems error approaches zero both during training and testing. This validates our hypothesis that splitting data into difficult and easy parts with further reclassification of difficult one allows for radical improvement of learning in general. Learning only how to reclassify difficult part of data needs lower size hypothesis space or lower capacity of our classification model which means better learning and less overfitting as observed in a number of results.

5   Conclusions

Experiments show that separation of objects into functional groups depending on classification difficulty allows to obtain better results during training and testing. To perform this separation one used consensus-based combination learning. This means that it is possible to obtain better classification performance and less overfitting. Classification results are rather stable for the correct consensus probability. The easiest part of data is classified by a consensus of classifiers. The hardest part of data should be classified using another classifier or ensemble of classifiers which should be trained on some part of data. Data decomposition into functional groups allows us to use simpler hypotheses leading to lower hypothesis space. One does not need to build a classifier with high classification capacity. Using classifiers with lower capacity means that they overfit less. Generally speaking in such a case the overall learning will be substantially improved.

References

1. Breiman, L.: Bagging predictors. Mach. Learn. **24**(2), 123–140 (1996)
2. Breiman, L.: Random forests. Mach. Learn. **45**(1), 5–32 (2001)
3. Frank, A., Asuncion, A.: UCI repository of machine learning databases. Technical report. University of California, School of Information and Computer Sciences Irvine, CA (2010)
4. Freund, Y., Schapire, R.E.: A desicion-theoretic generalization of on-line learning and an application to boosting. In: Vitányi, P. (ed.) EuroCOLT 1995. LNCS, vol. 904, pp. 23–37. Springer, Heidelberg (1995). doi:10.1007/3-540-59119-2_166
5. Ho, T.K.: The random subspace method for constructing decision forests. IEEE Trans. Pattern Anal. Mach. Intell. **20**(8), 832–844 (1998)
6. Kuncheva, L.: Combining Pattern Classifiers. Wiley, Hoboken (2014)
7. Murphy, K.P.: Machine Learning: A Probabilistic Perspective. MIT press, Cambridge (2012)
8. Rodriguez, J.J., Kuncheva, L.I., Alonso, C.J.: Rotation forest: a new classifier ensemble method. IEEE Trans. Pattern Anal. Mach. Intell. **28**(10), 1619–1630 (2006)
9. Rokach, L.: Pattern Classification Using Ensemble Methods. World Scientific, Hackensack (2009)
10. Vorontsov, K.V.: Splitting and similarity phenomena in the sets of classifiers and their effect on the probability of overfitting. Pattern Recogn. Image Anal. **19**(3), 412–420 (2009)

Classification Using Mixture of Discriminative Learners: The Case of Compositional Data

Elvis Togban[✉] and Djemel Ziou

Département d'Informatique, Faculté des Sciences, 2500 Bl. Université,
Université de Sherbrooke, Sherbrooke, Québec J1K2R1, Canada
{E.Togban,Djemel.Ziou}@usherbrooke.ca

Abstract. Compositional data arise in many fields and their analysis
has to be done with care since these data are bounded and summing up to
a constant. In this paper, we propose a mixture model which combines
several discriminative models through a set of Dirichlet-based weights.
It is worth noticing that the Dirichlet distribution is not used here as a
prior to the mixing coefficients but instead to model the repartition of
the tasks among the classifiers. By doing so, we do not need to transform
the data while keeping interpretable results. Experiments on synthetic
and real-world data sets show the efficiency of our model.

Keywords: Compositional data · Dirichlet mixture model · Discriminative classification · Multi-classifiers

1 Introduction

Many fields such as ecology, chemical, geology, economics, etc., deal with data
which represent parts of a whole (e.g. molarities, percentage of income, rate).
These data are bounded, positive and summing up to constant (usually assumed
to be 1 for convenience). As on any data, statistical analysis such as classification
can be performed on the compositional data. Classification consists in assigning
an object based on its attributes to a predefined class or group and can be done
with either a generative model or a discriminative model. The former relies on
the distribution that generates the attributes to classify an object while the
latter determines the class boundaries without any assumption about how the
attributes are generated. In this work, we will focus on the discriminative models
then, we are limited to fully labeled data.

Sometimes, a discriminative model can fail to classify objects having a complex relationship between the classes and the attributes. By example, one class is
included in another. A natural way to deal with complex tasks is to break them
down into easier tasks. Since the complex task is an ensemble of easier ones,
we can use the word *"mixture"* to refer to it. In the context of discriminative
learning, Mixture of Experts (ME) was introduced by Jacobs et al. [6]. The idea
is to divide a complex problem in a set of sub-problems identified by regions
which are delimited by a set of hyperplanes. These latter are described by an

© Springer International Publishing AG 2017
F. Karray et al. (Eds.): ICIAR 2017, LNCS 10317, pp. 416–425, 2017.
DOI: 10.1007/978-3-319-59876-5_46

input-dependent weight which indicates how a data sample is inside that region and each component distribution is a discriminative model acting as an *expert*. Since its introduction, they have been many variants which essentially deal with the weight, the component distribution as well as the learning algorithm. Readers can refer to the following reviews for more insight [8,15].

The sum constraints of the compositional data make the standard statistical method like ME not suitable for a reasonable classification task. In fact, the standard statistical method assume that the data are normally-distributed which is the case for compositional data. Often, the data can be transformed by using a centered-log-ratio (clr) defined as: $clr(x) = (\log(x_1/\mathcal{G}), \cdots, \log(x_D/\mathcal{G}))$ with $\mathcal{G}$ the geometric mean of the compositional vector $x = (x_1, \cdots, x_D)$ [1]. However, the resulting models are only interpretable in the transformed space and have no straightforward meaning. In addition, popular classification methods such as SVM and deep neuronal networks (DNNs) can also be used to classify compositional data ignoring or not the nature of such data. However, while DNNs models act like a *"black box"*, SVM usually uses a kernel which maps the data into a new feature space having no straightforward meaning. Simple methods like decision trees can be used without any transformation and with better interpretation of the output result but which is not probabilistic.

We partially address these limitations by using a Dirichlet-based weight-function to split the input space and to affect the tasks among the discriminative models. Indeed, the Dirichlet distribution is a natural choice to model the compositional since it is defined on the simplex $S_{D-1} = (x_1, \cdots, x_D), \sum_{d=1}^{D-1} x_d < 1$. It is worth noticing that the Dirichlet distribution are not used here as a prior to the mixing coefficients contrary to [12] but instead to model the repartition of the tasks among the classifiers. The negative covariance structure of the Dirichlet distribution is often evoked to reject its use in the analysis of compositional data [1]. However, since the weight function we used is based on mixture of Dirichlet distribution, this rejection is not justified [5]. Moreover with this weight function we can understand the repartition of the data among the classifiers without any transformations. We show by experiments on both artificial and real-word datasets that our model can achieve at least the same classification score as the standard ME using clr-transformation. The remaining of this paper is organized as follows: in Sect. 2, we introduce the Dirichlet mixture of discriminative learners. In Sect. 3, we present the learning algorithms. Finally, some experimental results are shown in Sect. 4, followed by a conclusion.

2   Dirichlet Mixture of Discriminative Learners

2.1   The Dirichlet Distribution

Let $\mathbf{x} = (x_1, \cdots, x_D)^T \in \mathbb{R}_{>0}^D$ be a vector having a Dirichlet distribution with density function:

$$\mathcal{D}(\mathbf{x} \mid \mu) = \frac{\Gamma(\sum_{d=1}^{D} \mu_d)}{\prod_{d=1}^{D} \Gamma(\mu_d)} \prod_{d=1}^{D} x_d^{\mu_d - 1} \tag{1}$$

where $\Gamma(.)$ is the Gamma function and $\sum_{d=1}^{D} x_d = 1;\ 0 \le x_d \le 1$. When $D = 2$, Eq. (1) is reduced to the Beta distribution. Many forms can be taken by the Dirichlet distribution ranging from an Uniform to a Gaussian-like distribution according to the choice of μ. The Dirichlet distribution is a natural choice to model the compositional since it is defined on the simplex $S_{D-1} = (x_1, \cdots, x_D),\ \sum_{d=1}^{D-1} x_d < 1$. However, this distribution has a negative covariance structure and can not be used to properly model positive correlated data. The Dirichlet distribution and its variants have been widely used for both regression [9] and in supervised discriminative model [7].

2.2 Dirichlet Mixture of Discriminative Learners

Let $\mathbf{x} = (x_1, \cdots, x_D)^T \in \mathbb{R}_{>0}^{D} \mid \sum_{d=1}^{D} x_d = 1$; be a compositional vector and $y \in \{1 \cdots C\}$ denotes its corresponding class label. A discriminative model is defined by the distribution $p(\mathbf{y}|\mathbf{x}, \Theta)$ of the class label y, conditioned on the variable x and a set of parameters θ. The goal of this model is to predict the class label $\hat{y}$ given a new $\hat{\mathbf{x}}$. However, if the relationship between $\mathbf{x}$ and $\mathbf{y}$ is too complex to be understood by the distribution $p(\mathbf{y}|\mathbf{x}, \Theta)$, this prediction task will fail. For example, when one class is included in the other, a mixture of several discriminative models can be used to locally approximate this complex class boundary. Formally, a Dirichlet mixture of discriminative model (DMD) can be defined as:

$$p(\mathbf{y}|\mathbf{x}, \Omega) = \sum_{i=1}^{K} \pi_i(\mathbf{x}) p(\mathbf{y}|\tilde{\mathbf{x}}, \Theta_i) \mid \pi_i(\mathbf{x}) = \frac{\alpha_i \mathcal{D}(\mathbf{x}|\mu_i)}{\sum_{l=1}^{K} \alpha_l \mathcal{D}(\mathbf{x}|\mu_l)};\ i = 1 \cdots K$$

$$S.t. \sum_{l=1}^{K} \pi_l(\mathbf{x}) = 1\ and\ \sum_{l=1}^{K} \alpha_l = 1;\ \alpha_l > 0; \tag{2}$$

where $\mathcal{D}(\mathbf{x}|\mu_l)$ is the Dirichlet distribution with parameter μ_l, Ω is the set of all parameters, K the number of mixture components and Θ_i is the parameters of the ith classifier $p(\mathbf{y}|\tilde{\mathbf{x}}, \Theta_i)$. Intuitively, the input space containing $\mathbf{x}$ is softly split into K regions and the probability that the data instance $\mathbf{x}$ lies in the ith region is given by $\pi_i(\mathbf{x})$. Then, each classifier $p(\mathbf{y}|\tilde{\mathbf{x}}, \Theta_i)$ discriminates among the data belonging to the ith region. The mixture weight $\pi_i(\mathbf{x})$ is an input-dependent parametric function describing the spatial boundary of the ith region. This weight '*distributes*' the tasks among the classifiers.

The model described in (2) is related to the well-known mixture of experts (ME) [6] and localized mixture of Experts (LME) [14]. However, ME uses the linear softmax weight to split the input space when LME used a Gaussian-based weight. These two types of weights are not suitable for compositional data since they assume that the data come from a Gaussian distribution. This problem can be circumvented by using the proposed transformations in [1]. Unfortunately, the resulting models are only interpretable in the transformed space and have no straightforward meaning. Other classification methods such as SVM and DNNs can also be used to classify compositional data ignoring or not (in this case the data need to be transformed to remove the sum-constraint) the nature of such data. However, while DNNs models act like a *"black box"*, SVM usually uses a kernel which maps the data into a new feature space having no straightforward meaning. By contrast, since our model used a Dirichlet-based weight function its interpretation is more straightforward. Indeed, the expression of $\pi_i(\mathbf{x})$ in (2) can be viewed as the posteriori probability that $\mathbf{x}$ is drawn from the ith component of the following mixture of Dirichlet: $\sum_{l=1}^{K} \alpha_l \mathcal{D}(x_n \mid \mu_l)$. The Dirichlet distribution is often rejected in the analysis of compositional due to its negative covariance structure. However, this rejection is not justified in the case of mixture of Dirichlet since the resulting covariance is not necessarily negative [5]. In the remaining of this paper, each component distribution is chosen as a multinomial logistic regression and we have for $i = 1 \cdots K$:

$$
p\left(y = c | x; \Theta_i\right) = \begin{cases} \dfrac{e^{\theta_{ic}^T \tilde{x}}}{1 + \sum_{h=1}^{C-1} e^{\theta_{ih}^T \tilde{x}}} & c \neq C \\[4mm] \dfrac{1}{1 + \sum_{h=1}^{C-1} e^{\theta_{ih}^T \tilde{x}}} & c = C \end{cases} \tag{3}
$$

where C is the number of class labels, $\Theta_i = \{\theta_{ih}\}_{h=1}^{C}$, $\theta_{ih} \in \mathbb{R}^{D^*}$, $\tilde{x} = \mathcal{F}(x)$, $\mathcal{F}$ is a preprocessing function and D^* is the dimension of $\tilde{x}$.

3   Parameters Estimation Using Maximum Likelihood (ML)

Let $\mathbf{Obs} = \{\mathbf{X}, \mathbf{Y}\}$ be a set of observed data where $\mathbf{X} = \{\overrightarrow{x_1}, \cdots, \overrightarrow{x_N}\} \in \mathbb{R}^{D \times N}$ is a set of i.i.d. vectors and $\mathbf{Y} = (y_1, \cdots, y_N)^T \in \mathbb{R}^N$ their equivalent class labels; N is the number of observations. Since our model (Eq. 2) is a mixture model, EM algorithm can be used for the training process. Let $\mathcal{L}$ be the *incomplete-log-likelihood*:

$$
\mathcal{L} = \sum_{n=1}^{N} \ln \left[\sum_{i=1}^{K} \pi_i(x_n) p(y_n | \tilde{x}_n, \Theta_i) \right] \tag{4}
$$

As is usual in the EM literature, we introduce a *'hidden'* binary random variable for the nth data: $z_i^{(n)}$. If x_n belongs to the region i then $z_i^{(n)} = 1$ and 0, otherwise and the *complete-log-likelihood* can be written as follows:

$$
\mathcal{L}^c = \sum_{n=1}^{N} \sum_{i=1}^{K} z_i^{(n)} \ln[\pi_i(x_n) p(y_n | \tilde{x}_n, \Theta_i)] \tag{5}
$$

In the E-step we compute the following expectation:

$$
\Phi\left(\Omega, \Omega^{(t)}\right) = \sum_{n=1}^{N} \sum_{i=1}^{K} h_i^{(n)} \ln\left[\pi_i\left(x_n\right) p\left(y_n | \tilde{x}_n, \Theta_i\right)\right] \tag{6}
$$

where $\Psi^{(t)}$ is the parameters at the tth iteration; $h_i^{(n)}$ is the expected value of $z_i^{(n)}$ and is expressed as:

$$h_i^{(n)} = \frac{\pi_i(x_n)p(y_n|\tilde{x}_n,\theta_i)}{\sum\limits_{l=1}^{K}\pi_l(x_n)p(y_n|\tilde{x}_n,\Theta_l)} \tag{7}$$

Equation (6) can be split in two parts where each one is containing respectively the parameters of the weight functions as well as the experts parameters. The following equations give the maximization operations performed in the M-Step:

$$G^{(t+1)} = argmax_G\ \Phi_1\ |\ \Phi_1 = \sum_{n=1}^{N}\sum_{i=1}^{K}h_i^{(n)}\ln\left[\pi_i(x_n)\right] \tag{8}$$

$$\Theta_i^{(t+1)} = argmax_{\Theta_i}\ \Phi_2\ |\ \Phi_2 = \sum_{n=1}^{N}h_i^{(n)}\ln\left[p(y_n|\tilde{x}_n,\Theta_i)\right] \tag{9}$$

where $G^{(t+1)} = \left\{\alpha_i^{(t+1)}, \mu_i^{(t+1)}\right\}_{i=1}^{K}$ is the set of parameters of the weight functions at the $(t+1)^{th}$ iteration. Substituting the expression of $\pi_i(x_n)$ (Eq. 2) in Eq. (8) will lead to a highly nonlinear optimization problem. To deal with this, the EM algorithm can be performed on the joint probability $p(y_n|x_n,\Psi)p(x_n|\Psi)$ where $P(x_n|\Psi) = \sum\limits_{l=1}^{K}\alpha_l\mathcal{D}(x_n\ |\ \mu_l)$ [14]. This is equivalent to replace the expression of $\pi_i(x_n)$ in Eqs. (4–8) by $\alpha_i\mathcal{D}(x_n,\mu_i)$. Since μ_{id} has to be strictly positive, we reparametrize by setting $\mu_{id} = e^{\xi_{id}}$. We estimate the parameters $G^{(t+1)}$ via the Newton-Raphson method and the partial derivative of Φ_1 (Eq. 8) with respect to ξ_{id} is given by:

$$g_{id} = \frac{\partial}{\partial\xi_{id}}\Phi_1 = \mu_{id}\left(\Psi(\sum_{d=1}^{D}\mu_{id}) - \Psi(\mu_{id})\right)\sum_{n=1}^{N}h_i^{(n)} + \mu_{id}\sum_{n=1}^{N}h_i^{(n)}\log(x_{nd}) \tag{10}$$

where Ψ is the Digamma function. The second and mixed derivatives are given by:

$$\frac{\partial^2}{\partial\xi_{id}^2}\Phi_1 = \frac{\partial}{\partial\xi_{id}}\Phi_1 + \mu_{id}^2\left(\Psi'(\sum_{d=1}^{D}\mu_{id}) - \Psi'(\mu_{id})\right)\sum_{n=1}^{N}h_i^{(n)} +$$
$$\mu_{id}\sum_{n=1}^{N}\frac{\partial}{\partial\xi_{id}}h_i^{(n)}\left[(\Psi(\sum_{d=1}^{D}\mu_{id}) - \Psi(\mu_{id}) + \log(x_{nd}))\right] \tag{11}$$

and

$$\frac{\partial^2}{\partial\xi_{id_1}\xi_{id_2}}\Phi_1 = \mu_{id_1}\sum_{n=1}^{N}\frac{\partial}{\partial\xi_{id_2}}h_i^{(n)}\left[(\Psi(\sum_{d=1}^{D}\mu_{id}) - \Psi(\mu_{id}) + \log(x_{nd}))\right] +$$
$$\mu_{id_1}\mu_{id_2}\Psi'(\sum_{d=1}^{D}\mu_{id})\sum_{n=1}^{N}h_i^{(n)} \tag{12}$$

where Ψ' is the Trigamma function. Experiments have shown that the results are not affected when we assumed $\dfrac{\partial}{\partial \xi_{id_2}} h_i^{(n)} \simeq 0$. A similar approximation is made in [2] and we can now rewrite the Hessian matrix H_i as:

$$H_i = Q_i + a_i b_i b_i^T \mid a_i = \Psi'(\sum_{d=1}^{D} \mu_{id}) \sum_{n=1}^{N} h_i^{(n)}; \ b_i^T = (\mu_{i1}, \cdots, \mu_{iD})$$

$$Q_i = Diag(q_{i1}, \cdots, q_{iD}) \mid q_{id} = \frac{\partial}{\partial \xi_{id}} \Phi_1 - \mu_{id}^2 \Psi'(\mu_{id}) \sum_{n=1}^{N} h_i^{(n)}$$

Giving an initial ξ_{id}^{old} (we used the procedure described in [3]), we have the following update equation:

$$\xi_{id}^{new} = \xi_{id}^{old} - \frac{g_{id} - c}{q_d} \mid c = a_i^* \mu_{id} (\sum_{d=1}^{D} \frac{\mu_{id} g_{id}}{q_d}); \ a_i^* = \frac{a_i}{1 + a_i \left(\sum_{d=1}^{D} \frac{\mu_{id}^2}{q_d} \right)} \qquad (13)$$

Due to the limited size, we do not give a full demonstration of the Eq. (13) here. Deriving Φ_1 (Eq. 8) with respect to α_l and taking into account the constraint $\sum_{l=1}^{N} \alpha_l = 1$, we have the following update equation: $\alpha_i^{(t+1)} = \frac{1}{N} \sum_{n=1}^{N} h_i^{(n)}$. By substituting (3) in (9), we can estimate the parameters θ_{ic} via a standard non-linear optimization algorithm (e.g. quasi-newton methods) and the gradients are expressed as:

$$\nabla_{\theta_{ic}} \Phi \left(\Omega, \Omega^{(t)} \right) = \sum_{n=1}^{N} h_i^{(n)} \left[v_{cn} - p \left(y_n = c | x_n; \theta_i \right) \right] \tilde{x}_n \ \forall \ i \in \{1 \cdots K\} \qquad (14)$$

where v_{cn} are binary variables indicating membership of nth data to the cth class. The estimation algorithm is summarized as follows:

1. Initialization:
   - set the parameters $\{\mu_i, \alpha_i\}_{i=1}^{K}$ by using the procedures described in [3].
   - choose randomly the parameters $\{\theta_{ic}\}_{ic=11}^{KC}$
2. E-step: Update $\Phi \left(\Omega, \Omega^{(t)} \right)$ via Eqs. (6) and (7)
3. M-step: Maximize $\Phi \left(\Omega, \Omega^{(t)} \right)$ with respect to current value of Ω via:
   - update of $\{\mu_i\}_{i=1}^{K}$ via Eq. (13)
   - update of $\{\theta_{ic}\}_{ic=11}^{KC}$ via Eqs. (9) and (14)
4. Repeat steps 2 and 3 until convergence.

4   Experimental Results

To assess the performance of our model, we have considered both synthetic and real-world datasets. In the case of synthetic datasets, we compare our model to the log-ratio transformation approach by performing ME and LME on the clr-transformation of the experiments datasets. The models resulting from this transformation are called respectively C-ME and C-LME. When in our model each classifier $p(y = c|x; \Theta_i)$ performed a clr-transformation as preprocessed step, we called the model C-DMD. For each classifier $p(y = c|x; \Theta_i)$, the nonlinear optimization method used is the '$minFunc$' implementation [11] of L-BFGS with the convergence tolerance set to 1×10^{-3} over 200 maximum update cycles. We set the EM convergence tolerance to 1×10^{-4} over 50 maximum update cycles. In the case of real-world problem, we compare C-DMD and DMD with two more models named: Random Subspace (RS) and α-Knn [13]. These models are chosen based on the fact that they allow an easy understanding of the classification results. In fact, Random Subspace train a set of Decision trees where each one is trained on a random subspace of dimension d $(d < D)$ and α-Knn is a model proposed in [13] which train a K-nearest neighbor method on 'power-transformation' of a dataset and a specific metric called ES-OV$_\alpha$. We used ten pruned C4.5 trees in RS with $d = D/2$ and for the α-Knn, we set $\alpha = 0.5$ and $K = 2$ as in [13]. All models are evaluated with respect to their accuracy, F-score and the Matthews correlation coefficient (MCC).

4.1 Artificial Datasets

We generate four datasets with binary response ($y \in 0, 1$) where each one is composed of data sampled from a set of three-dimensional Dirichlet distribution. Figure 1 represents these four datasets as well as their parameters. For the validation purpose, we repeat 100 stratified trials of holdout where each one uses a quarter of a dataset for training and the remaining for the test (in this first experiment, $K = 2$). Table 1 reports the classification accuracy, the F-Score and the MCC. The results show that our model achieves at least the same performance when compared to LME and ME using clr-transformation. Moreover, although the clr-transformation allows an interpretation in the transformed space; it does not guarantee a better classification. In fact, ME achieves an average accuracy of 86.84% but this result fells to 83.8%. The opposite fact is made for LME where the accuracy increases from 87.43% to 97.99% after the clr-transformation. In contrast, the results of our model are similar when each classifier either uses a clr-transformation or not. These observations are consistent with those obtained when each one of the four datasets is considered. The good results of C-LME are due to its ability to deal with regions with different covariances but the data need to be transformed first. By contrast, the weight function used by our model deals directly with the compositional nature of the data and allows to interpret the repartition of the tasks among the classifiers without any transformation.

4.2 Water Dataset

This dataset comes from hydrochemical and contains 478 samples of 14 molarities measured along the Llobregat river and its four tributaries in the northeastern Spain during two years [10]. The goal of our study is to classify the

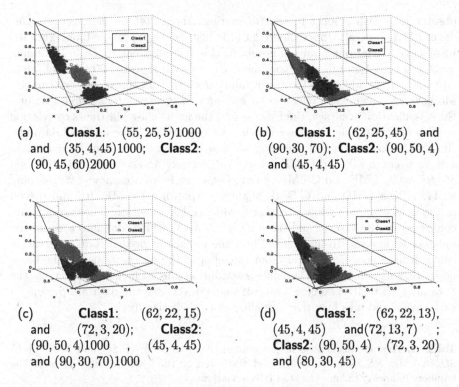

(a)   **Class1**: $(55, 25, 5)1000$
and $(35, 4, 45)1000$;   **Class2**:
$(90, 45, 60)2000$

(b)   **Class1**: $(62, 25, 45)$ and
$(90, 30, 70)$;   **Class2**: $(90, 50, 4)$
and $(45, 4, 45)$

(c)   **Class1**: $(62, 22, 15)$
and $(72, 3, 20)$;   **Class2**:
$(90, 50, 4)1000$, $(45, 4, 45)$
and $(90, 30, 70)1000$

(d)   **Class1**: $(62, 22, 13)$,
$(45, 4, 45)$ and$(72, 13, 7)$;
Class2: $(90, 50, 4)$, $(72, 3, 20)$
and $(80, 30, 45)$

Fig. 1. Artificial datasets used in our experiments: (a) Dataset1; (b) Dataset2; (c) Dataset3; (d) Dataset4. The numbers inside the brackets (.) represent the parameters of the Dirichlet distribution. The number following the brackets (.) denotes the number of instances sampled. If this latter one is not specified, it is equal to 2000.

Table 1. Average accuracies in percentage (Acc), F-Score (F-Sc) and MCC of *C-ME*, *C-LME*, *C-DMD* and *DMD* run on the synthetic datasets. The best scores are set in bold. The numbers following '±' are the standard deviations.

	C-ME			C-LME			C-DMD			DMD		
	Acc	F-Sc	MCC	Acc	F-Sc	MCC	Acc	F-Sc	MCC	Acc	F-Sc	MCC
Dataset1	96.66 ± 0.49	0.96	0.93	99.57 ± 0.11	0.99	0.99	99.7 ± 0.11	0.99	0.99	**99.71** ± 0.1	0.99	0.99
Dataset2	88.25 ± 2.07	0.88	0.76	99.07 ± 0.07	0.99	0.98	99.13 ± 0.07	0.99	0.98	**99.14** ± 0.07	0.99	0.98
Dataset3	87.32 ± 8.25	0.88	0.75	96.2 ± 7.18	0.96	0.92	**99.53** ± 0.15	**0.99**	**0.99**	97.49 ± 4.87	0.97	0.94
Dataset4	74.21 ± 4.64	0.73	0.48	97.94 ± 0.1	0.97	0.95	97.72 ± 0.12	0.97	0.95	**98.08** ± 0.11	**0.98**	**0.96**
Average	83.8	0.84	0.67	97.99	0.98	0.95	**98.77**	0.98	**0.97**	98.40	0.98	0.96

observations with respect to the tributaries: Anoia (145 measurements), Cardener (90 instances), Upper Llobregat (134 instances) and Lower Llobregat (109 instances). Given our above annotation, this is equivalent to $D = 14$ dimensions and $C = 4$ and in this experiment, $K = 6$. We repeated 20 trials of 5-fold stratified cross-validation for the validation of the classification performance. Moreover, we train each classifier by adding a *l2-regularization*. Table 2 reports the classification accuracy, the F-Score and the multi-class Matthews correlation coefficient (MCC) [4]. Contrary to what we observe with the artificial datasets, the results do not vary significantly with or without the clr-transformation. In fact, in average ME and C-ME achieve respectively an accuracy of 85.13% and 85.69% while LME and C-LME achieve respectively an accuracy of 90.19% and 90.44%. Both DMD and C-DMD slightly outperform C-LME, RS and α-Knn in terms of accuracy while ME and C-ME are less accurate. These results are confirmed by the F-score and the multi-class MCC which took into account the unbalanced classes. To understand how the molarities predict if an observation comes from a specific tributary, our model split the observations in 6 homogeneous regions defined by a Dirichlet distribution and each classifier predicts the data affected to it. Therefore, we can understand how the data are present in each region and also how they are affected to each classifier.

Table 2. Average accuracies in percentage (Acc), F-Score (F-Sc) and MCC of *C-ME*, *C-LME*, *RS*, α-Knn, *C-DMD* and *DMD* run on the hydrochemical dataset. The numbers following '$\pm$' are the standard deviations.

Models	Accuracy	F-score	MCC
C-ME	85.69 $\pm$ 3.84	0.84	0.8
C-LME	90.44 $\pm$ 2.75	0.89	0.87
Random Subspace (RS)	90.33 $\pm$ 3.36	0.89	-
α-Knn [13]	90.46 $\pm$ 2.66	0.89	0.87
DMD	91.03 $\pm$ 2.6	0.9	0.88
C-DMD	91.03 $\pm$ 3.23	0.9	0.88

5   Conclusion

In this paper, we proposed a mixture model for discriminative learners (*DMD*) to classify compositional data while allowing to understand how the data are distributed among the classifiers used. The main idea behind our model is to divide a complex task into several sub-tasks through a set of weight functions. These latter are chosen based on a mixture of Dirichlet distribution. When we are in the case where the output of each classifier needs to be explained, one can choose a log-ratio transformation. However, the results conducted on both artificial and real-world data show that the performance is similar either we apply a log-ratio transformation to each classifier or not. The advantage of our method compared to some ensemble methods is the fact that we do not need to transform the data first before applying such methods. This is possible because our weight function deals directly with the compositional nature of the data without any transformation. Future work may include an extension to the hierarchical mixture case and the choice of an appropriate classifier easy to interpret without any transformation.

References

1. Aitchison, J.: The Statistical Analysis of Compositional Data. Chapman and Hall, London, New York (1986)
2. Bouguila, N., Ziou, D., Vaillancourt, J.: Maximum likelihood estimation of the generalized Dirichlet mixture. Dept. Mathematiques et dinformatique, Univ. Sherbrooke, Shrebrooke, QC, Canada (2002)
3. Bouguila, N., Ziou, D., Vaillancourt, J.: Unsupervised learning of a finite mixture model based on the Dirichlet distribution and its application. IEEE Trans. Image Process. **13**(11), 1533–1543 (2004)
4. Gorodkin, J.: Comparing two K-category assignments by a K-category correlation coefficient. Comput. Biol. Chem. **28**(5), 367–374 (2004)
5. Hijazi, R.: Analysis of compositional data using Dirichlet covariate models. The American University, Washington, DC, USA (2003)
6. Jacobs, R., Jordan, M.I., Nowlan, S.J., Hinton, G.: Adaptive mixtures of local experts. Neural Comput. **3**(1), 79–87 (1991)
7. Ksantini, R., Boufama, B.: A novel Bayesian logistic discriminant model with dirichlet distributions: an application to face recognition. In: Kamel, M., Campilho, A. (eds.) ICIAR 2009. LNCS, vol. 5627, pp. 461–470. Springer, Heidelberg (2009). doi:10.1007/978-3-642-02611-9_46
8. Masoudnia, S., Ebrahimpour, R.: Mixture of experts: a literature survey. Artif. Intell. Rev. **42**, 275–293 (2014)
9. Ng, K.W., Tian, G.L., Tang, M.L.: Dirichlet and Related Distributions: Theory, Methods and Applications, vol. 888. Wiley, Hoboken (2011)
10. Otero, N., Tolosana-Delgado, R., Soler, A., Pawlowsky-Glahn, V., Canals, A.: Relative vs. absolute statistical analysis of compositions: 1 comparative study of surface waters of a mediterranean river. Water Res. **39**(7), 1404–1414 (2005)
11. Schmidt, M.: minFunc: unconstrained differentiable multivariate optimization in Matlab (2012). http://www.di.ens.fr/mschmidt/Software/minFunc.html
12. Shi, J.Q., Murray-Smith, R., Titterington, D.: Bayesian regression and classification using mixtures of Gaussian processes. Int. J. Adapt. Control Signal Process. **17**(2), 149–161 (2003)
13. Tsagris, M., Preston, S., Wood, A.T.: Improved classification for compositional data using the α-transformation. J. Classif. **33**(2), 243–261 (2016)
14. Xu, L., Jordan, M.I., Hinton, G.: An Alternative Model for Mixtures of Experts (1995)
15. Yuksel, S., Wilson, J., Gader, P.: Twenty years of mixture of experts. IEEE Trans. Neural Netw. Learn. Syst. **23**(8), 1177–1193 (2012)

Biomedical Image Analysis

Mesh-Based Active Model Initialization for Multiple Organ Segmentation in MR Images

M.R. Mohebpour[✉], F. Guibault, and F. Cheriet

École Polytechnique de Montréal, Montréal, Québec, Canada
majidreza.mohebpour@polymtl.ca

Abstract. Active models are widely used for segmentation of medical images. One of the key issues of active models is the initialization phase which affects significantly the segmentation performance. This paper presents a novel method for an automatic initialization of different types of active models by exploiting an adaptive mesh generation technique which is suitable for automatic detection of multiple organs. This method has been applied on MR images and results show the ability of the proposed method in simultaneously extracting initial approximate boundaries that are close to the exact boundaries of multiple organs. The effect of the proposed initialization algorithm on the segmentation has been tested on a series of arm and thoracic MR images and the results show an improvement in the convergence and speed of active model segmentation of multiple organs with respect to those obtained using manual initialization.

Keywords: Adaptive mesh · Active models initialization · Medical image segmentation · Multiple organ segmentation

1 Introduction and Background

Active models are used in a very large range of applications such as image processing, computer vision, computer animation, etc. Different models can be classified based on their contour representation as continuous or discrete representations. The difference between continuous and discrete representations is that in discrete form, the geometry of contours is only known at finite sets of points. Continuous forms must be discretized for computational needs, but it is possible to compute normal and curvature along the whole curve.

A number of studies have led to the development of methods which applied active models for medical image segmentation [1–6]. There are some advantages in using active models in medical images over other segmentation techniques. They are able to generate closed parametric templates from images in a smooth manner, making them robust to noise and spurious edges, and able to manage complex geometries and topology changes (curve splitting and merging). Moreover, they provide consistent mathematical descriptions, which can be used in subsequent applications.

Active models provide some closed curves or surfaces with the ability to expand and contract to capture desired features by minimizing an energy functional based on

© Springer International Publishing AG 2017
F. Karray et al. (Eds.): ICIAR 2017, LNCS 10317, pp. 429–436, 2017.
DOI: 10.1007/978-3-319-59876-5_47

some constraints. The energy functional contains internal and external energy to control the smoothness and image features respectively. Initialization of active models is a crucial part which affects the ultimate result of the segmentation. A suitable initialization can prevent failure caused by entrapment in local minima and help capture the boundary of objects. It will also reduce the number of iterations that the method needs to converge. Although considerable research has been carried out on active model improvement, only a small portion of research has paid attention to model initialization.

Some of the current methods use manual initialization by selecting some initial points in the image [7, 8]. Although this approach is effective, in the case of volumetric data such as MR images it is very difficult and time consuming to draw a surface manually, and it is also not suitable in an automation context. Some other models employ simple geometric primitives such as circles, which may need many iterations to converge and also may not converge to the desired feature of interest. There have been some investigations on automatic initialization. Ge and Tian [9] proposed a method to start multiple active contours based on some points called Center of Divergence (CoD) computed from a Gradient Vector Flow (GVF) field. Their approach is suitable for simple images, but for complex images, it is difficult to determine the CoD. Tauber et al. [10] generalized the notion of CoD by representing skeleton of divergence, but it needs a starting point to be determined by the user. They also presented a method [11] based on GVF field with a combination of anisotropic diffusion in order to improve the results, but this method is also considered as quasi-automatic and needs user interaction. Bing and Acton [12] proposed a method based on an external energy force field and Dirichlet boundary conditions and automatically initialize parametric active models by solving Poisson's equation. Their method is limited to parametric active models and it may fail in the presence of strong noise. Saha et al. [13] presented a probabilistic quad-tree based method for automatic snake initialization which is computationally inexpensive. It has, however, not been tested on challenging images such as MR images.

In this paper we propose a new automatic initialization method based on adaptive meshes. The suggested method has the following features which make it efficient for automating the active models. It generates an initial boundary close to the exact boundary of the organ in the image; it can capture multiple organs in the image and is able to generate separate active contours for each organ that can evolve simultaneously and iteratively. This model is capable of modeling initial curves for both parametric and geometric active models.

2   Proposed Method

2.1   Metric Tensor Construction and Mesh Adaptation

We use an anisotropic mesh adaptation technique to discretize the image domain in order to satisfy some specifications and characteristics. The result of the discretization process is a mesh with both regular and skinny triangles with adequate size and orientation to fit image features. In the present application, desirable features of an image are considered as all the edges in the image that represent boundaries of different regions (in particular, distinct anatomical tissues) within the image. The edge length of the mesh elements is controlled through the specification of a metric. This metric, which is globally defined as a tensor across space, is constructed directly from image information. In this regard, Hessian matrix composed of second order directional derivatives in the gradient direction is used to construct a Riemannian metric tensor $\mathcal{M}_R$ including the length and stretching factors for mesh adaptation as follows:

$$\mathcal{M}_R = \mathcal{R}\Lambda\mathcal{R}^T = \begin{pmatrix} \vec{e_1} & \vec{e_2} \end{pmatrix} \begin{pmatrix} \lambda_1 & 0 \\ 0 & \lambda_2 \end{pmatrix} \begin{pmatrix} \vec{e_1}^T \\ \vec{e_2}^T \end{pmatrix}$$

where $\vec{e_1}$ and $\vec{e_2}$ are eigenvectors and λ_1 and λ_2 are eigenvalues such that:

$$\lambda_{1,2} = f_{xx} + f_{yy} \pm \sqrt{\left(f_{xx} - f_{yy}\right)^2 + 4f_{xy}^2}$$

where f_{xx}, f_{yy}, and f_{xy} denote second order derivatives from the Hessian matrix.

Figure 1 shows a geometric representation of a metric tensor as an ellipse with its axes displaying the eigenvalues and eigenvectors of the metric tensor.

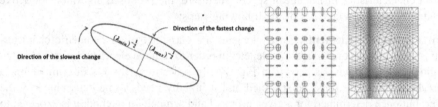

Fig. 1. (left) Geometric representation of a metric tensor; (right) and example of a metric tensor field (center) and a mesh that fits the metric

These eigenvectors and eigenvalues define the stretching and orientation of the elements and specify how the size and distance depend on the orientation. Mesh adaptation algorithms enforce the target size, stretching and orientation prescribed by this control metric as it is shown in Fig. 1.

The edge length between two points (A and B) of the mesh in the Riemannian metric is calculated by the following formula:

$$l_{AB} = \int_0^1 \sqrt{\vec{AB}^T \mathcal{M}_R\left(A + t\vec{AB}\right)\vec{AB}}\, dt$$

with edge AB parametrized by $t \in [0,1]$.

As may be seen in Fig. 2, adaptation based on a thoracic MR image according to the metric tensor yields larger and more regular triangles in regions of fairly uniform intensity of the image, represented as small variations in the metric tensor field. On the contrary, at the boundaries of organs in the image, where the intensity changes rapidly, triangles are refined and stretched as a result of rapid variation in the metric tensor field and the edges of these elements are aligned with the boundaries of the organs.

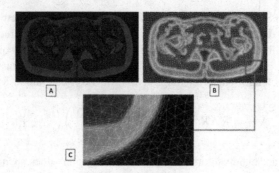

Fig. 2. (a) Thoracic MR image; (b) adaptive mesh of the same image; (c) zoom of the highlighted part in (b)

2.2   Active Models Initialization

The rationale behind the proposed approach is to capture image content using a unit mesh generated in Riemannian space and then to discriminate structures by analyzing mesh characteristics in Euclidean space. Therefore, the proposed algorithm for active model initialization consists of the following steps:

Partitioning the Elements. First, elements are classified using geometric characteristics such as aspect ratio, area, angle, etc. in order to remove elements which belong to regions inside organs, as shown in Fig. 3-b. In order to apply this discrimination, a threshold parameter must be specified and it initially needs to be determined by the user, but once determined for a given protocol and acquisition method, it is expected to be adequate for all similar image sequences. Partitioning the elements produces some holes inside the mesh, each one depicting the presence of an organ in that area. This action also removes the background area of the image from the mesh.

Detecting the Holes. Detecting the holes and identifying the elements around the holes yields a rough representation of the interfaces of the organs, which are very close to their true boundaries. An algorithm has been developed to identify holes by locating boundary edges, linking them together into loops, no matter how complicated the holes might be (Fig. 3-c).

Constructing the Curves. After identifying admissible holes, several active contours are fitted to each organ, based on their interface elements, thereby providing the initialization of active models. An important aspect is that the extracted points for each hole are connected in an order which make it possible to reconstruct curves through interpolation or approximation. Therefore, both discrete and continuous active models are admissible. For continuous models, the edge points can be considered as control points for constructing the curves for initializing the active model as shown in Fig. 3-d. It is possible as well to represent these curves in an explicit parametric form (snakes) or in an implicit form (level sets) in order to support different approaches.

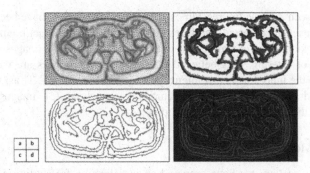

Fig. 3. (a) Adaptive mesh of the image; (b) extracting boundary elements; (c) identifying holes by locating boundary edges; (d) constructing several curves for active model

3 Results and Discussion

In order to validate our method, we applied it to initialize continuous active contours on a series of MR images and those initializations have been utilized in an active contour framework to evaluate the performance of the segmentation results.

3.1 Active Model Initialization

We have applied the proposed method on two series of arm and thoracic MR images, each containing 28 slices. The Object-Oriented Remeshing Toolkit (OORT) adaptive mesh technique [14] has been used to generate adaptive meshes. Then, for each image, after partitioning the elements and detecting the holes, several distinct closed curves were constructed, using a B-Spline parametric representation considering the extracted edge points as control points. Figure 4 illustrates some of the results from our tests on the arm and thoracic MR images.

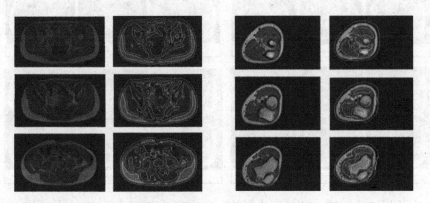

Fig. 4. Original MR image and multiple active model initialization; (left) thoracic section; (right) arm section.

As can be seen in these figures, almost all organs in the image have been segmented and in many cases the detected interfaces are very close to the exact boundaries of the organs. In all our tests, operations were done on the original images without any preprocessing or filtering. It is believed that thoracic MR images are the most challenging due to the inhomogeneous intensity inside the same organ. Thus, the proposed method should be applicable to other modalities such as CT images, where the boundaries of the organs should be well defined.

3.2   Active Model Segmentation

In the previous section, we have prepared an initialization for starting active models which are indispensable starting points in the evolution process. Better starting points help reduce the convergence time and prevent convergence to local minima in order to improve the final segmentation results. In the following, segmentation results using our initial contours as starting points are presented and compared to results obtained using a simpler initialization. In this regard, we have used a snake model based on Gradient Vector Flow (GVF) external force field [15]. The GVF forces are calculated by applying diffusion equations to the gradient of an image edge map. We have applied GVF snake with the same configuration for different initializations; first with our proposed initialization, then with a circle inside and outside the objects. Some of the results are presented in Fig. 5. As may be seen, using a simple initialization, in many cases causes the snake to be trapped in a local minimum and prevents correct attraction to the boundaries. However, using our automatic initialization, we have detected the objects in the region of interest without getting stuck in local minima.

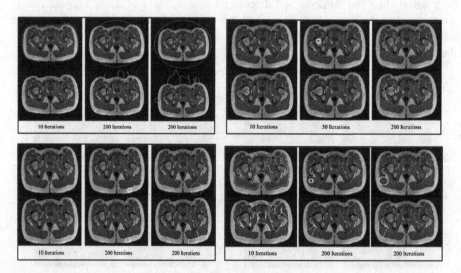

Fig. 5. GVF snake segmentation result with different initialization; (left) proposed initialization; (middle) circle inside organs; (right) circle outside organs

As it is shown in the examples, there are simple and complex structures which have been detected adequately by considering the proposed initialization method. In the sense of convergence speed for all the cases that we have done, in average, our proposed approach needs around 10 iterations to reach the final result, which is considered very rapid compared to other initialization options. The figures also indicate the number of iterations needed for each example. Although the GVF snake has been designed to have an extended capture range in order to find organs that are quite far from the snake's initial position, these examples show that it can easily be distract or get stuck because of careless initialization. These sorts of distractions or entrapments usually happen due to the noise in the image or when the organs are very close to each other, which is so common in medical images. Therefore, we need to initialize more carefully and provide the initial contours as close as possible to the desired organ. Figure 6 shows the final result of extracting multiple regions of the original image.

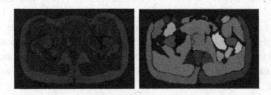

Fig. 6. Original MR image and multiple extracted regions

4 Conclusion

We have proposed a novel automatic initialization method for active contour models by incorporating mesh adaptation technique which offers several advantages. It is applicable for automatically initializing different forms of active models in the category of discrete or continuous approaches with explicit or implicit representation. It is capable of capturing multiple organs in the image simultaneously and constructing several distinct active models without overlapping problems. It provides initial curves very close to the exact boundaries to speed up the evolution process and prevents convergence to local minima. It therefore has a great potential to improve the performance of segmentation results. Initialization results illustrate the performance of the proposed method for multiple organ segmentation. Segmentation results indicate that using the proposed initialization we can automatically segment organs in medical images more efficiently. The only parameter needs to be tuned for current implementation, is a threshold for partitioning the elements which is initially set by the user for each image sequence.

References

1. Kass, M., Witkin, A., Terzopoulos, D.: Snakes: active contour models. Int. J. Comput. Vis. **1** (4), 321–331 (1988)
2. Liu, T., et al.: Improving image segmentation by gradient vector flow and mean shift. Pattern Recogn. Lett. **29**(1), 90–95 (2008)

3. He, L., et al.: A comparative study of deformable contour methods on medical image segmentation. Image Vis. Comput. **26**(2), 141–163 (2008)
4. Li, C., Huang, R., Ding, Z., Gatenby, C., Metaxas, D., Gore, J.: A variational level set approach to segmentation and bias correction of images with intensity inhomogeneity. In: Metaxas, D., Axel, L., Fichtinger, G., Székely, G. (eds.) MICCAI 2008. LNCS, vol. 5242, pp. 1083–1091. Springer, Heidelberg (2008). doi:10.1007/978-3-540-85990-1_130
5. Gao, Y., Tannenbaum, A., Kikinis, R.: Simultaneous multi-object segmentation using local robust statistics and contour interaction. In: Menze, B., Langs, G., Tu, Z., Criminisi, A. (eds.) MCV 2010. LNCS, vol. 6533, pp. 195–203. Springer, Heidelberg (2011). doi:10.1007/978-3-642-18421-5_19
6. Lee, M., et al.: Segmentation of interest region in medical volume images using geometric deformable model. Comput. Biol. Med. **42**(5), 523–537 (2012)
7. Ardon, R., Cohen, L.: Fast constrained surface extraction by minimal paths. Int. J. Comput. Vis. **69**(1), 127–136 (2006)
8. Neuenschwander, W., et al.: Initializing snakes [object delineation]. In: 1994 Proceedings of the IEEE Computer Society Conference on Computer Vision and Pattern Recognition, CVPR 1994 (1994)
9. Ge, X., Tian, J.: An automatic active contour model for multiple objects. In: 2002 Proceedings of the 16th International Conference on Pattern Recognition (2002)
10. Tauber, C., Batatia, H., Ayache, A.: A general quasi-automatic initialization for snakes: application to ultrasound images. In: 2005 IEEE International Conference on Image Processing, ICIP 2005 (2005)
11. Tauber, C., Batatia, H., Ayache, A.: A robust active contour initialization and gradient vector flow for ultrasound image segmentation. In: MVA (2005)
12. Bing, L., Acton, S.T.: Automatic active model initialization via poisson inverse gradient. IEEE Trans. Image Process. **17**(8), 1406–1420 (2008)
13. Saha, B.N., Ray, N., Zhang, H.: Automating snakes for multiple objects detection. In: Kimmel, R., Klette, R., Sugimoto, A. (eds.) ACCV 2010. LNCS, vol. 6494, pp. 39–51. Springer, Heidelberg (2011). doi:10.1007/978-3-642-19318-7_4
14. Courchesne, O., Guibault, F., Dompierre, J., Cheriet, F.: Adaptive mesh generation of MRI images for 3D reconstruction of human trunk. In: Kamel, M., Campilho, A. (eds.) ICIAR 2007. LNCS, vol. 4633, pp. 1040–1051. Springer, Heidelberg (2007). doi:10.1007/978-3-540-74260-9_92
15. Chenyang, X., Prince, J.L.: Snakes, shapes, and gradient vector flow. IEEE Trans. Image Process. **7**(3), 359–369 (1998)

Sperm Flagellum Center-Line Tracing in Fluorescence 3D+t Low SNR Stacks Using an Iterative Minimal Path Method

Paul Hernandez-Herrera[1,2], Fernando Montoya[1], Juan M. Rendón[2],
Alberto Darszon[1], and Gabriel Corkidi[1(✉)]

[1] Instituto de Biotecnología, Universidad Nacional Autónoma de México,
Cuernavaca, Mexico
corkidi@ibt.unam.mx
[2] Centro de Investigación en Ciencias, IICBA, UAEM, Cuernavaca, Mexico

Abstract. Intracellular calcium ($[Ca^{2+}]i$) regulates sperm motility. Visualizing $[Ca^{2+}]i$ in 3D is not a simple matter since it requires complex fluorescence microscopy techniques where the resulting images have very low intensity and consequently low SNR (Signal to Noise Ratio). In 3D+t sequences, this problem is magnified since the flagellum beats (for human sperm) at an average frequency of 15 Hz, making it harder to obtain the three dimensional information. Moreover, 3D holographic techniques do not work for these fluorescence based images. In this paper, an algorithm to extract the flagellum's center-line in 3D+t stacks is presented. For this purpose, an iterative algorithm based on the fast-marching method is proposed to extract the flagellum's center-line. Quantitative and qualitative results are presented in a 3D+t stack to demonstrate the ability of the proposed algorithm to trace the flagellum's center-line. Our method was qualitative and quantitatively compared against state-of-the-art tubular structure center-line extraction algorithms outperforming them and reaching a Precision and Recall of 0.96 as compared with a semi-manual method used as reference. The proposed methodology has proven to solve a major problem related with the analysis of the 3D motility of sperm cells in images with very low intensity.

1 Introduction

Sexual reproduction requires communication between the male and female gametes. The flagellum allows the sperm to swim and fulfil its main objective, to fertilize the egg. Intracellular calcium ($[Ca^{2+}]i$) is a key regulator of sperm's swimming [3]. To understand sperm swimming patterns (typical, helical, hyper-helical, hyper-active or chiral ribbons), it is necessary to study flagellar beating. Hence, measuring how the $[Ca^{2+}]i$ modulates the flagellum movement is a fundamental task. Visualizing $[Ca^{2+}]i$ changes is not a simple matter since it requires complex fluorescence microscopy techniques where the resulting images have very low intensity and low SNR (Signal to Noise Ratio). In 3D+t sequences,

© Springer International Publishing AG 2017
F. Karray et al. (Eds.): ICIAR 2017, LNCS 10317, pp. 437–445, 2017.
DOI: 10.1007/978-3-319-59876-5_48

this problem is magnified since the human sperm flagellum beats at least at an average frequency of 15 Hz.

In order to measure $[Ca^{2+}]i$ from an image, it is necessary to trace the flagellum's center-line which becomes a time consuming task if done manually. Up to date, there are no algorithms developed to automatically trace the flagellum's center-line in this type of 3D+t images. The flagellum resembles a tubular structure, thus, as a first approximation, algorithms developed to trace tubular structures must be suitable to trace the flagellum's center-line. There are many algorithms for the automatic tracing of tubular structures in 3D images such as vessels [5], airways [10], neurons [1], etc. Recently, Hernandez-Herrera *et al.* [4] proposed an algorithm (MESON) to trace the center-line from tubular structures using a one-class classification algorithm to segment tubular structures, and the fast-marching algorithm was used to extract the center-line from the segmented tubular structures. Liu *et al.* [6] proposed an iterative method (RIVULET) based on the fast-marching algorithm to extract the center-line from tubular structures. All of these approaches are designed to extract multi-branch structures, hence spurious branches are usually produced for low contrast images (as the ones used in this work) when a single structure must be detected.

In this paper, a tracing algorithm designed to extract a tubular structure from a 3D+t low contrast and noisy image stack is proposed. First, the MESON [4] algorithm is applied to each 3D stack to enhance tubular structures. Then, the center of the sperm's head is automatically detected and it is used as the starting point of the center-line algorithm. The algorithm searches for the best minimal path over a predefined distance of approximately the maximum flagellum's diameter (using the fast-marching algorithm), just to move a small distance from the starting point and then it stops. The cost to travel across points near the detected minimal path are forced to increase (to avoid detecting the same minimal path), consecutively indexing the terminal point as the starting one; this procedure is iterated to complete the whole flagellum's center-line, avoiding detecting erroneous paths, specifically at the more curved parts. After each iteration, the method measures the quality of the detected center-line, if it has low quality, then the iteration stops and it initializes at the next time point. When compared against to multi-branch methods (MESON and RIVULET) in low light intensity conditions (low SNR), the proposed algorithm exhibits a superior performance. Comparing to others center-line extraction algorithms based on the fast marching method, our approach has important differences. First, some algorithms segment the whole tubular structures and then extract the center-line. The segmentation procedure for this approach may fail for low intensity fluorescence images due to the high level of noise, consequently the center-line extraction fails (fast marching). Second, other methods propagate a fast-marching algorithm from a starting point and stop when the complete image has been processed. Then, the center-lines are extracted from the most distant points and the quality of each center-line is measured. Center-line with high quality are kept while low quality are disregard. These approaches generate erroneous extra branches for noisy images and usually require a fine tuning of the algorithm parameters to produce a good tracing and this would require even more processing time than a semi-manual trace.

2 Methods

2.1 Flagellum Enhancement

Recently, Hernandez-Herrera et al. [4] proposed an algorithm to enhance and segment neurons which could be also suitable to segment other tubular structures such as the sperm flagellum. The enhancement algorithm is based on the one-class classification approach where a training set of points belonging to the background is automatically detected using a Laplacian filter. The training set is used to construct a one-class classifier where points with features similar to the points in the training set are assigned values close to 1 while points with dissimilar features are assigned values close to 0. The dissimilarity value obtained from the one-class classifier is used to enhance the foreground of the 3D image stack. Hence, low values are assigned to the background while high values are assigned to the flagellum. In the algorithm proposed by Hernandez-Herrera et al. [4], the enhanced image R_f is used to segment tubular structures and the skeleton (center-line) is extracted for each segmented structure. In this work, the enhanced image R_f is used to guide a minimal path procedure towards the center of the flagellum. In addition, it is employed to stop an iterative center-line extraction routine.

2.2 Center-Line Extraction

Starting Point Detection. The starting point s_0 used to initialize our algorithm is selected at the center of the sperm's head. The sperm's head is the brightest structure in the 3D image stack, hence it is segmented by applying a binary threshold T. The starting point s_0 is selected as the centroid from the 3D coordinates of the voxels in the largest connected component.

Minimal-Cost Path. The approach for center-line tracing is based on the minimal-cost path problem. The minimum cost path in a 3D image stack to travel from s_0 to t_0 is given by a minimization problem $T(t_0) = \min_C \int_{s_0}^{t_0} F(C(s))ds$ where $C(s)$ is a curve in $\mathbb{R}^3$ (3D image stack), s_0 is the initial point, t_0 the terminal point and $F(x)$ is a function determining the cost of travelling across point x, where x is a point in the 3D image stack. Among all the possible paths to travel from s_0 to t_0 the path with minimum integral is called the minimum cost path. The minimum cost path at each point x can be found using the fast-marching method [8]. The minimum cost path is detected by back-tracing from t_0 along the gradient of the fast-marching solution until s_0 is reached.

In the experiments the cost function F is defined as: $F = \exp^{-40 \cdot G_{\sigma_a} * [I \cdot R_f]}$, where I is the image stack with values in the range $[0, 1]$, R_f is the enhanced flagellum (Sect. 2.1), G_{σ_a} is an anisotropic Gaussian kernel (this depends on the voxel's anisotropy), "·" and "∗" represent the dot product and the convolution, respectively. The value 40 in the cost function was found by parameter optimization. The fast-marching algorithm is initialized at s_0 and it is stopped at the first point t_0 such that $\|s_0 - t_0\| > l_0$. Finally, the minimum cost path from t_0 to s_0 is extracted by back-tracing the fast-marching solution until the point s_0 is reached. Let's call the detected path $c_0(s)$. In our experiments the value l_0 is set to 10 which is a value slightly larger than the maximum diameter of the flagellum.

Increase Cost-Values for Already Visited Point. In order to discourage future iterations to travel across the already detected path $c_0(s)$, the cost function is increased for points close to $c_0(s)$. Recall that the flagellum resembles a tubular structure and the path c_0 should travels across the center of it. Hence, a tube is constructed and the cost value F from any point inside the tube is increased to ∞. The new cost function $F(x)$ discourage the minimal cost path to travel across points in the tube since the new cost value for traveling across those points is infinitely expensive (∞).

2.3   Stopping Criteria

The iterative algorithm has two criteria to stop. The algorithm is stopped if the detected minimum cost path has a low probability of being inside the flagellum. The second criteria is based on the orientation of two connected paths, if there is a sharp change in orientation thus the detected path is disregard and the algorithm is stopped.

Low Possibility to Be Inside the Flagellum. Recall from Sect. 2.1 that the enhanced image gives values close to 1 for points inside the flagellum while low values and close to 0 in the background. Hence, for each $x_i \in c_0(s)$ the value of x_i in the enhanced image can be interpreted as the "probability" of the point x_i to be inside the flagellum. The 50% percentile of the values $\{R_f(x_i)\}$ is used as a measure of the path to be inside the flagellum. If the 50% percentile of the values of the path is higher than 0.5, then the path is detected as belonging to the flagellum. Otherwise, the path $c_0(s)$ is detected to be outside the flagellum.

Orientation. This step is only acceptable starting from the second iteration of the algorithm since it requires at least two detected minimum cost paths. Let's call the consecutive minimum cost paths $c_{j-1}(s)$ and $c_j(s)$. In order to accept the new detected path $c_j(s)$ as a continuation of the previous path, the point at which they connect (terminal point of $c_{j-1}(s)$ and starting point of path $c_j(s)$) must have a small change in orientation since the structure to be detected (flagellum) does not have sharp changes in orientation. The path c_j is detected as a continuation of path c_{j-1} if $orientation < 110°$ (the maximum orientation was set to 110° since changes in orientation larger than this value were not observed among the different stacks analyzed), otherwise the path is detected as an incorrect continuation. If any of the two criteria to stop the algorithm is satisfied, then the algorithm stops. Otherwise, the algorithm returns to step **Minimal-Cost Path** where the new starting point s_0 is set as the terminal point of the previous detected minimal path.

2.4 Post-process

Our algorithm for 3D center-line extraction is applied to each time point (TP) of the 3D+t stack to extract the flagellum's center-line of the 3D stack. In some time points, the algorithm could have stopped prematurely due to large gaps in the image stack. Hence, the algorithm is re-run for stacks with a detected center-line length of less than l_t units. For those stacks, the algorithm is re-run with the parameter l_0 increased to twice its initial value and the sensitivity of the one-class classifier is increased to allow detection of finer structures.

3 Results

In this section, qualitative and quantitative results are presented in a 3D+t stack. The data were acquired using an Olympus IX71 inverted microscope, with an Optronis 5000 camera acquiring 3000 images per second and a resolution of 512×512 pixels. To obtain a 3D image stack, a piezoelectric device (Physic Instruments) attached to a water immersion 60x objective (Olympus) is employed, allowing changes in the focal plane at a frequency of 90 Hz and amplitude of 20 microns. A detailed description of the acquisition protocol is explained in Corkidi et al. [2] and Silva et al. [9]. The dataset is available upon request to the authors.

3.1 Qualitative Results

Our method was qualitatively compared to state-of-the-art tubular structures extraction algorithms (MESON ([4]) and RIVULET ([6])). An exhaustive search for the best parameter configuration for the MESON and RIVULET algorithms was performed. Figure 1(a, f, k) illustrates a maximum intensity projection along the z-axis for three different time points (TP 7, 49 and 103) from a 3D+t stack with high, medium and low signal to noise ratio (SNR), respectively. The head is the brightest structures of the 3D stack and the flagellum is very difficult to visualize because it has very low contrast. Figure 1(b, g, l) depicts the semi-manual trace for each time point as a green line. The semi-manual trace for each time point was extracted using the plugin Simple Neurite Tracer [7] from the FIJI software. Figure 1(c, h, m) depicts the center-line traced with our method as a red line. Note that our method was able to trace most of the flagellum's structure detected by the semi-manual trace in the three time points. Figure 1(d, i, n) depicts the extracted trace by the MESON algorithm. Note that the algorithm produces extra-branches in the three time points because it ignores the a-priori

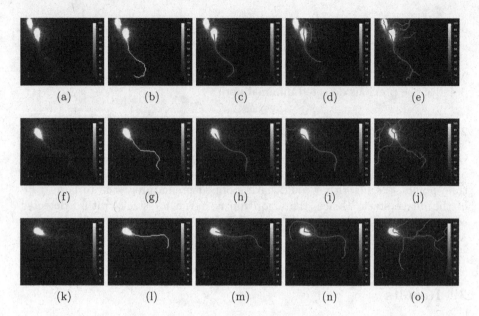

Fig. 1. First column depicts the MIP (Maximum Projection Intensity) from three different time points with high, medium and low SNR. Second, third, fourth and fifth column depicts the traced center-line by the semi-manual trace, our method, MESON and RIVULET. (Color figure online)

knowledge that there is only one structure to trace and the images are very noisy. Furthermore, the algorithm fails to trace the flagellum's tail at TP 7 (Fig. 1(d)). Figure 1(e, j, o) depicts the extracted trace by the RIVULET algorithm. Similar to the MESON algorithm, RIVULET produces extra-segments due to the multi-branch nature of the algorithm. Those multi-branch algorithms can be modified to extract a single structure by selecting only the largest segment. However, this approach may fail since there are time points where the largest segment does not correspond to the flagellum's centerline. Furthermore, selecting a single branch from a given time point by a multi-branch algorithm would require to do an exhaustive search for the best parameter configuration, then manually select the best trace among all the reconstruction and finally select the largest segment. This approach would be required for each time point which is unfeasible since there are usually more than 200 time points for each 3D+t stack. Indeed, the best traces depicted in Fig. 1 for the MESON and RIVULET algorithms have different parameter setting for each time point. Video 1 (http://bit.ly/2js7MFp) depicts the full reconstruction of a 3D+t stack using our method.

3.2   Quantitative Results

In this section, quantitative result are presented for a 3D+t stack. The metrics *Precision* (P), *Recall* (R), and the *Miss-Extra-Score* (MES) [4] are employed

Table 1. Performance evaluation on a 3D+t Stack

Method	Time points	Precision (μ)	Recall (μ)	MES (μ)	ADE (μ)
Our method	108	0.96	0.96	0.92	1.73
	3	0.98	0.97	0.95	1.88
MESON [4]	3	0.51	0.86	0.45	2.64
RIVULET [6]	3	0.34	0.96	0.30	2.54

to measure the quality of the reconstruction, $P = S_C/S_T$, $R = S_C/(S_C + S_{miss})$, $MES = (S_G - S_{miss})/(S_G + S_{extra})$, where S_C is the total length of the automatic trace that was correctly traced, S_T is the total length of the automatic trace, S_G is the total length of the manual trace, S_{miss} and S_{extra} are the total length of the missing and extra segments of the automatic trace, respectively. The values for each metric are in the range $[0, 1]$ with a low value corresponding to a wrong tracing while high values correspond to good tracing. A point p_c in the trace is classified correctly if there is a point p_m in the semi-manual trace such that $\|p_c - p_m\| < C$, where C is a constant defined by the user. In our experiments, C was set to 7.5 since this value is the maximum flagellum's diameter. The last metric is $Average - Displacement - Error$ (ADE) which measures the average distance error between the manual trace and the automatic trace.

Table 1 depicts the average value (μ) for the metrics Precision, Recall, MES and ADE for the evaluated stack. Our method was tested using the 108 time points from the stack (108 flagellum beating cycles). The mean value for the metric Precision is 0.96 which means that the automatic tracing on average detected the 96% of the total path inside the flagellum. Figure 2 depicts the traced center-line (red line) by our algorithm and the semi-manual trace (green line) by different time points from a 3D+t stack. Figure 2(a) illustrates the traced center-line (red line) from the time point 13 which corresponds to the lowest value of the metric Precision (0.78). The extra segment comes from an erroneous detection of the flagellum's initial position (head detection). The algorithm fails to detect the sperm's head because there are two nearby sperm in the 3D stack.

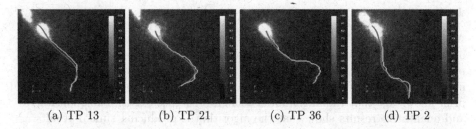

(a) TP 13   (b) TP 21   (c) TP 36   (d) TP 2

Fig. 2. (a-d) Depiction of the traces with the lowest and highest Precision (TP 13 and 21) and Recall (TP 36 and 2) value, respectively. Green line depicts the semi-manual trace while red line depicts the trace by our algorithm. (Color figure online)

Figure 2(b) depicts the traced center-line (red line) by our algorithm from time point 21 which correspond to the highest Precision (1.00) value. A precision value of 1.00 means that for each point p_c from the traced center-line (red line) by our method, exist a point p_m from the semi-manual trace (green line) such that $\|p_c - p_m\| < 7.5$. The metric Recall achieves an average error of 0.96 which means that the proposed algorithm missed only a small part of the flagellum. Figure 2(c) depicts the MIP from time point 36 which correspond to the lowest value for the metric Recall (0.44). The algorithm fails to detect the whole structure because there is a large gap and it stops prematurely. The algorithm was able to detect approximately the initial half part of the flagellum. Figure 2(d) depicts the traced center-line (red line) from time point 2 which correspond to the highest recall (1.00) value. A Recall value of 1.00 means that for each point p_m from the semi-manual trace (green line), exist a point p_c from the traced center-line (red line) by our algorithm such that $\|p_m - p_c\| < 7.5$. The metric MES is a metric penalizing the automatic trace if a segment is missed or has extra segments. Note that the average value 0.92 is near the optimal value, hence on average the algorithm has good tracings. The minimum value for the metric MES was achieved for time point 36 (Fig. 2(c)). Finally, the average ADE was 1.73 which means that the distance error between the manual trace and automatic trace was not even 2 voxels difference which is very close.

Additionally, our algorithm was compared quantitatively against MESON and RIVULET using the three time points (TP 7, 49 and 103) depicted in Fig. 1(a, f, k). Table 1 depicts the average values for the three methods, note that our algorithm has the highest Precision (0.98), Recall (0.97) and MES(0.95) value indicating that it produces better tracings than MESON and RIVULET. In addition the ADE value is the lowest indicating that our tracing is closer to the semi-manual trace than MESON and RIVULET. Note that MESON and RIVULET have very low Precision and MES value, this is because those metrics are penalized if there are extra-segments in the tracing. Note that only three time points were used to compare the algorithms; including more time points would produce similar results (low Precision and MES) due to the multi-branch approach of the algorithms, nevertheless, to evaluate exhaustively our algorithm, we have also used 108 flagellum beatings (see Table 1).

4   Conclusions

An iterative algorithm for the tracing of the sperm's flagellum from a 3D+t stack of very low SNR fluorescence based images has been developed. There are no previous algorithms able to extract the center-line from this type of fluorescent images. Other algorithms usually extract more than one branch even though prior knowledge that the flagellum is a single structure exists. The quantitative and qualitative results show that the algorithm is producing similar results as manual tracing. The proposed methodology has proved to solve a major problem related with the analysis of the 3D motility of sperm cells in fluorescence images with very low intensity and low SNR. Evaluation of the performance of the algorithm using a larger dataset will be investigated in a future work.

Acknowledgments. This work was supported by Consejo Nacional de Ciencia y Tecnología (CONACyT) (grants 253952 to G. Corkidi and Fronteras 71 to 39908-Q to A. Darszon) and Posdoctoral scholarship 291142 (Paul Hernandez Herrera); the Dirección General de Asuntos del Personal Académico by the Universidad Nacional Autónoma de México (DGAPA-UNAM) grants CJIC/CTIC/4898/2016 to F. Montoya and IN205516 to A. Darszon.

References

1. Acciai, L., Soda, P., Iannello, G.: Automated neuron tracing methods: an updated account. Neuroinformatics **14**(4), 353–367 (2016)
2. Corkidi, G., Taboada, B., Wood, C., Guerrero, A., Darszon, A.: Tracking sperm in three-dimensions. Biochem. Biophys. Res. Commun. **373**(1), 125–129 (2008)
3. Darszon, A., Nishigaki, T., Beltran, C., Treviño, C.L.: Calcium channels in the development, maturation, and function of spermatozoa. Physiol. Rev. **91**(4), 1305–1355 (2011)
4. Hernandez-Herrera, P., Papadakis, M., Kakadiaris, I.A.: Multi-scale segmentation of neurons based on one-class classification. J. Neurosci. Methods **266**, 94–106 (2016)
5. Lesage, D., Angelini, E.D., Bloch, I., Funka-Lea, G.: A review of 3D vessel lumen segmentation techniques: models, features and extraction schemes. Med. Image Anal. **13**(6), 819–845 (2009)
6. Liu, S., Zhang, D., Liu, S., Feng, D., Peng, H., Cai, W.: Rivulet: 3D neuron morphology tracing with iterative back-tracking. Neuroinformatics **14**(1), 387–401 (2016)
7. Longair, M.H., Baker, D.A., Armstrong, J.D.: Simple neurite tracer: open source software for reconstruction, visualization and analysis of neuronal processes. Bioinformatics **27**(17), 2453–2454 (2011)
8. Sethian, J.A.: Level Set Methods and Fast Marching Methods: Evolving Interfaces in Computational Geometry, Fluid Mechanics, Computer Vision, and Materials Science, vol. 3. Cambridge University Press, Cambridge (1999)
9. Silva-Villalobos, F., Pimentel, J., Darszon, A., Corkidi, G.: Imaging of the 3D dynamics of flagellar beating in human sperm. In: 36th Annual International Conference of the IEEE Engineering in Medicine and Biology Society, pp. 190–193 (2014)
10. Smistad, E., Falch, T.L., Bozorgi, M., Elster, A.C., Lindseth, F.: Medical image segmentation on GPUs-a comprehensive review. Med. Image Anal. **20**(1), 1–18 (2015)

Curvelet-Based Classification of Brain MRI Images

Rafat Damseh[1(✉)] and M. Omair Ahmad[2]

[1] Institut de génie biomédical, École Polytechnique de Montréal,
Montreal, QC H3T 1J4, Canada
rafat.damseh@polymtl.ca
[2] Department of Electrical and Computer Engineering, Concordia University,
Montreal, QC H3G 1M8, Canada
omair@encs.concordia.ca

Abstract. Classification of brain MRI images is crucial in medical diagnosis. Automatic classification of these images helps in developing effective non-invasive procedures. In this paper, based on curvelet transform, a novel classification scheme of brain MRI images is proposed and a technique for extracting and selecting curvelet features is provided. To study the effectiveness of their use, the proposed features are employed into three different prediction algorithms, namely, K-nearest neighbours, support vector machine and decision tree. The method of K-fold stratified cross validation is used to assess the efficacy of the proposed classification solutions and the results are compared with those of various state-of-the-art classification schemes available in the literature. The experimental results demonstrate the superiority of the proposed decision tree classification scheme in terms of accuracy, generalization capability, and real-time reliability.

Keywords: MRI imaging · Curvelet transform · Feature extraction and classification

1 Introduction

Magnetic resonance imaging (MRI) is an imaging modality that captures high quality anatomical images of the human body, especially the brain. It provides a huge information that facilitate clinical diagnosis and surgical operations. Brain MR images help in the evaluation of neoplasms and plays a critical role in making decisions regarding initial and evolving treatment strategies. Automatic classification of normal and abnormal brain MR images can make one avoid invasive procedures and anticipate the diagnosis without time-consuming histological examinations.

This work was supported in part by the Natural Sciences and Engineering Research Council (NSERC) of Canada and in part by the Regroupement Strategique en Microelectronique du Quebec (ReSMiQ).

© Springer International Publishing AG 2017
F. Karray et al. (Eds.): ICIAR 2017, LNCS 10317, pp. 446–454, 2017.
DOI: 10.1007/978-3-319-59876-5_49

Sparse representation as an approach for feature extraction has been successfully applied to numerous applications, especially in image processing, computer vision and pattern recognition. One well-known simple sparse representation using pre-specified transform functions, namely, discrete wavelet transform (DWT), has been widely used [1–6]. Although the DWTs has an impressive reputation as a good tool for signal processing, it has the drawback of poor directionality, which make its usage limited in many applications. The development of directional wavelets has been investigated in recent years. Ridgelet transform, which is an anisotropic geometric wavelet transform was proposed by Candés and Donoho [7]. To analyze local line or curve singularities, the ridgelet transform is applied to partitions (subsignals) of the original signal. This block ridgelet-based transform is named as curvelet transform (CT) [8]. First-generation curvelet transform is limited because the geometry of ridgelets is unclear. A simpler second-generation curvelet transform based on a frequency partition technique has been proposed later [9]. The second-generation curvelet transform has been shown to be an efficient tool for different applications in signal processing and its discrete version has been proven to be efficient in representing curve-like edges in digital images. Several other developments of directional sparse representations have been proposed with the goal of optimal representation of directional features of signals. However, none of these approaches has reached the publicity of the curvelet transform. In the literature, many classification and detection schemes have exploited the curvelet transform as a tool for extracting meaningful features from signals with applications in medical studies, remote sensing and vehicle verification [10–12].

In this paper, a novel curvelet-based feature extraction scheme for Brain MR images is proposed. Up to the author knowledge, the curvelet coefficients have never been tested as features of the brain MR images. The effectiveness of the proposed feature extraction scheme is examined by combining it with different prediction algorithms and testing on a benchmark dataset consisting of 160 brain MR images to classifiy normal and abnormal brains. The performance of the proposed classification schemes is compared with that of various state-of-the-art classification schemes proposed in the literature. This paper is organized as follows. In Sect. 2, a brief overview of the related work is presented. In Sect. 3, the proposed mechanism of feature extraction is illustrated. In Sect. 4, experiments are performed. Finally, some concluding remarks highlighting the contribution of the paper and scope of further research are provided in Sect. 5.

2 Related Work

Many recent schemes for classification of normal and abnormal brain MR images have been proposed in the literature. These schemes vary in the pre-processing procedures used for constructing the feature descriptors of the data and the learning algorithm applied to build the classifier. Chaplot, et al. [3] used the approximation coefficients obtained by DWT, and employed the self-organizing map (SOM) neural network and support vector machine (SVM). More promising results have been obtained by El-Dahshan, et al. [4]. In [4], the coefficients of 3-level DWT are reduced via principal component analysis (PCA), and a hybrid combination of feed-forward back-propagation artificial neural network (FP-ANN) and K-nearest neighbour (KNN) classifiers has been used. Zhang, et al. [5] proposed another scheme for feature extraction and selection, using DWT and PCA, respectively, and implemented forward neural network (FNN) with scaled chaotic artificial bee colony (SCABC) as a classifier. However, superior results have been obtained in [6] by Zhang and Wu using kernel SVMs; the kernels of homogeneous polynomial, inhomogeneous polynomial, and Gaussian radial basis have been suggested. To reduce the storage memory required to obtain the DWT coefficients, Yang et al. [13] proposed to use the wavelet energy (WE) calculated from the wavelet coefficients as a technique for features extraction. Despite the success in reducing the required memory storage, their solution has limited generalization capability without a big improvement in the classification accuracy. To overcome the limited directionality and non-supportiveness to anisotropy of the DWT, which make it not able to capture the subtle and intrinsic details of the brain MRI images, Das et al. [14] developed a scheme for feature extraction using the Ripplet transform Type-I (RT). On the other hand, to have a translation-invariant feature extraction, Wang et al. [15] employed the stationary wavelet transform (SWT) instead of DWT. However, the two lastly mentioned works have had some improvement on the classification accuracy but at the expense of increasing the prediction computational time due to their complex feature extraction scheme.

From the aforementioned classification schemes recently proposed in the literature, it is noticed that there is still a room for further investigations to develop a more powerful classification scheme that can achieve a better trade-off between obtaining a higher accuracy rate, reducing the prediction time and providing better generalization capability. The aim of this work is to develop a novel classification solution that can meet this concern.

3   Proposed Scheme for Feature Extraction

As known, in the process of developing a solution for a classification problem, one should design a mechanism for feature extraction. In this section, The proposed feature extraction scheme using CT is illustrated. It is to be noted that CT is originally defined in the continuous domain but implemented for the discrete domain. Derived from the spatial mother curvelet function, curvelets spatially differ in scale parameter j, rotation parameter $l \in \mathbb{N}_0$, and translation parameter $k \in \mathbb{Z}^2$. CT coefficients of a function $f \in \mathbb{R}^2$ is given as the inner product of the function with the curvelets $\varphi_{j,l,k}$ as in

$$c_{j,l,k} = \langle f, \varphi_{j,l,k} \rangle = \int_{\mathbb{R}^2} f(x)\overline{\varphi_{j,l,k}(x)}dx \qquad (1)$$

Discrete curvelet tiling in the frequency domain differs from its continuous counterpart as its ring windowing function is defined as square rings and its angular windowing function is defined as shear functions.

In general, the number of CT coefficients of an image is huge, and using all these coefficients as features of the corresponding image can drastically decrease the classification performance. Hence, in the proposed feature extraction scheme, the use of PCA is proposed for reducing the dimension of the feature vector. The full schematic diagram that describes the proposed feature extraction scheme is depicted in Fig. 1. The process of feature extraction is performed by two steps. First, the normalized five-scale curvelet coefficients of an input MRI image is obtained. Second, PCA is applied on these coefficients to produce the final feature vector. In Fig. 1, the input image is in gray-scale format with the size 256×256 and its pixel values are normalized to $[0, 1]$. As suggested in [16], the five-scale CT is selected based on the following formula

$$number\ of\ scales = \log_2(N) - 3 \tag{2}$$

where N defines the size $N \times N$ of an image. It is to be noted that the number of orientations in the curvelet transform depends on the image size. In our case, there are 4×8 orientations at scales 4 and 5, whereas there are 4×16 orientations at scale 3. In the proposed feature extraction scheme, the CT coefficients are partitioned into subsets and then a first-stage PCA is separately applied on the CT coefficients of each subset. Partitioning is done based on the scale and the orientation of the CT coefficients. Specifically, the CT coefficients at scale S and orientation in $O_i, i = 1, 2, 3, 4$, where $O_1 \in [-\pi/2, \pi/2)$, $O_2 \in [\pi/2, 3\pi/2)$, $O_3 \in [3\pi/2, -3\pi/2)$ and, $O_4 \in [-3\pi/2, -\pi/2)$, are clustered as one partition denoted as $P_{(S,i)}$. The approximation coefficients of CT are clustered in a separate partition denoted as P_0.

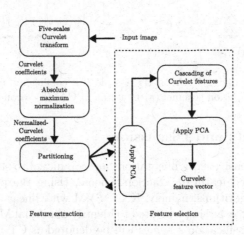

Fig. 1. Proposed curvelet-based feature extraction scheme.

It is to be noted that partitioning of CT coefficients is done in order to reduce the complexity of the PCA computation since it is not feasible to perform PCA on the whole CT coefficients at once. The CT coefficients selected from the various partitions using the first-stage PCA are then aggregated and a second PCA is applied on them. It is to be mentioned that in each stage of PCA, 80% of the data variance is chosen as a threshold to determine the desired number of features. This threshold is chosen based on experimental observations, through which it is demonstrated that a higher threshold, leading into more features, does not improve the classification performance. Based on the dataset on-hand, the result obtained when applying the first-stage PCA on partition $P_{(3,1)}$ (this partition is arbitrary chosen for explanation purpose only) of the CT coefficients is depicted in Fig. 2. It is seen from this figure that 9 principle components (features) are sufficient to represent 80% of the data variance, however, this result could vary when applying PCA on a different partition. Figure 2 shows the result obtained after applying the second-stage PCA. It is seen from this figure that, for an MRI image of size 256×256, the final feature vector obtained using the proposed feature extraction scheme contains only 6 features. Having such a small number of features can substantially reduce the computational burden of the classification solution and make it more feasible. In the next section, the effectiveness of utilizing these features to classify different brain MRI images is studied with implementations of various prediction algorithms.

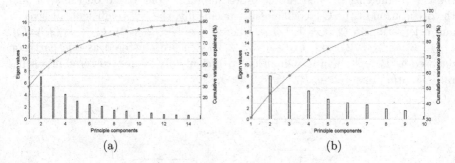

(a)   (b)

Fig. 2. Results of applying PCA on the 5-scale CT coefficients as described in the proposed feature extraction scheme: (a) first-stage PCA; (b) second-stage PCA.

4   Experiments and Discussion

In this section, extensive experimentations are conducted to study the effectiveness of the proposed feature extraction scheme. Using the proposed features, three prediction algorithms, namely, KNN, SVM with linear and gaussian kernels, and decision tree, are implemented to identify abnormal MRI brains images. These proposed classification schemes will be denoted as CT-KNN, CT-LSVM, CT-GSVM and CT-DT, respectively. The performance of the proposed classification schemes is compared with that of other classification schemes available in the literature, namely, the schemes of DWT-GSVM [3], DWT-PCA-KNN [4], DWT-PCA-GSVM [6], Ripplet-PCA-LSSVM [14] and SWT-PCA-FNN [15]. Experiments are carried out on the basis of averaging the results of 10 executions of 2-, 4-, 5-, 10- and 20-fold cross validation.

4.1   Data Collection and Analysis

The dataset contains 160 T2-weighted MR brain images in axial plain with 256 × 256 in-plane resolution, which were down-loaded from the website of Harvard Medical School (URL: http://med.harvard.edu/AANLIB/), OASIS dataset (URL: http://www.oasis-brains.org/), and ADNI dataset (URL: http://adni.loni.uc-la.edu/). The abnormal brain MR images in the dataset are of the following diseases: Alzheimer's disease, meningioma, glioma, Pick's disease, sarcoma, Huntington, and Alzheimer's disease with visual agnosia. Samples of the various disease are shown in Fig. 3. In the dataset, 20 images capture normal brains and the rest capture various brain abnormalities (20 images/each disease). The statistical characteristics and cross validation settings of the dataset are listed in Table 1.

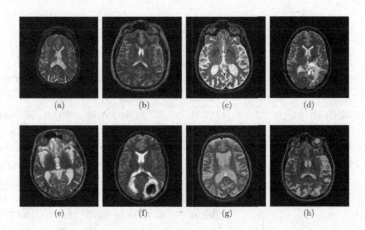

Fig. 3. MRI brain dataset consisting of images representing normal and 7 types of different ubnormal brains: (a) Normal; (b) Alzheimer; (c) Meningioma; (d) Glioma; (e) Pick's disease; (f) Sarcoma; (g) Huntington's disease; (h) Alzheimer with visual agnosia.

Table 1. Cross-validation with various K-fold settings.

K-fold	Total		Training		Validation		Run
	Normal	Abnormal	Normal	Abnormal	Normal	Abnormal	
2	20	140	10	70	10	70	10
4	20	140	15	105	5	35	10
5	20	140	16	112	4	28	10
10	20	140	18	126	2	14	10
20	20	140	19	133	1	7	10

4.2   Classification Performance

Now, after designing the experimental procedure, a comparison between the various classification schemes is carried out in terms of their accuracy rates. The comparison results are shown in Table 2. Some entries of the table are extracted from the corresponding original papers. It is clear from the table that, when the proposed CT-DT scheme is used, the highest average accuracy rate is achieved. The second-best accuracy rate belongs also to the proposed CT-GSVM scheme. Since the experiments have been carried out using different K-fold settings, the average accuracy rates obtained in Table 2 provide a strong indication about the generalization capabilities of the various classification schemes. Thus, the performance of the proposed CT-DT scheme is expected to be superior to those of the other various schemes when predicting new brain MRI datasets.

Table 2. Accuracy rates of the various classification schemes.

Classification schemes	Number of Features	Data partitioning for cross validation					Average accuracy (%)
		2-fold	4-fold	5-fold	10-fold	20-fold	
DWT-GSVM [3]	4761	97.23	97.48	97.33	97.13	96.9	97.214
DWT-PCA-KNN [4]	7	97.77	97.47	97.54	97.73	97.71	97.644
DWT-PCA-GSVM [6]	19	98.52	98.96	99.38	99.07	98.13	98.812
Ripplet-PCA-LSSVM [14]	9	99.08	99.89	100	99.21	99.24	99.484
SWT-PCA-FNN [15]	7	99.46	100	100	99.68	99.18	99.664
CT-KNN	6	98.93	98.42	98.37	98.36	98.13	98.442
CT-LSVM	6	98.35	98.9	99.25	99.1	99.07	98.934
CT-GSVM	6	99.91	99.33	100	99.76	99.77	99.754
CT-DT	6	100	100	100	99.83	100	**99.966**

Since the aim of this paper is to develop a computer-aided diagnosis tool for identifying abnormal brains, the development of such tool implies that it be reliable in real-time environments. Thus, studying the computational-time complexity of the proposed classification scheme is very important. It is known that the classification solution for an input is obtained by firstly training the classifier on a dataset on-hand. Since the training process can be performed off-line, the computational time associated with this process is not a significant factor in assessing the performance of the proposed classification scheme. Hence, only the prediction time-complexity of the proposed classification scheme is studied. By performing the prediction process on all the images in the dataset using MAT-LAB implementation on a 2.9-GHz 8-GB machine, the average time of prediction for one image is recorded as 0.053 s. This time includes the time of feature extraction (0.049 s), feature selection (0.003 s) and classification decision (0.001 s). It is obvious that the proposed classification scheme can successfully meet the time requirement for real-time diagnosis. Despite that the time-complexity for feature extraction in the proposed CT-DT scheme is slightly greater than that of the other various schemes, the use of CT-DT which employs only 6 dimensional feature vector results in lower storage cost. Also, from Table 2, it is proven that the use of this short features vector provides a positive impact on the generalization capability of the classifier.

5 Conclusion

Manual classification of brain MRI images is time consuming and not reliable in some cases. Therefore, the development of automatic diagnostic tools to identify normal and abnormal brains in such images is necessary. Many schemes for automatic classification of normal and abnormal brain MRI images have been proposed in the literature. However, These schemes have limitation in their accuracy or generalization capability. In this paper, a novel scheme for feature extraction using the curvelet transform (CT) has been proposed, and the discrimination ability of the extracted features has been explored by employing them into three prediction algorithms, namely, K-nearest niebours (KNN), support vector machine (SVM) and decision tree. In order to extract the curvelet features from an MRI image, five-scale curvelet transform has been applied and followed by two stages of principle component analysis (PCA). By applying the proposed feature extraction scheme on MRI images of size 256 × 256, the size of the classification problem has been successfully reduced to include only 6 dimensions. The method of K-fold stratified cross validation has been used to assess the performance of the proposed classification schemes. By comparing with other state-of-the-art classification schemes available in the literature, the experimental results have demonstrated the superiority of the proposed DT classification scheme in terms of accuracy and generalization capability, and have proven the reliability of the proposed scheme in real-time environments. One can further extend the research work undertaken in this paper by studying the classification performance of the proposed scheme on other types of datasets that include different MRI modalities, such as T1-weighted, Proton-density weighted, and diffusion weighted images. Also, the behaviour of the proposed scheme could be examined on noisy datasets with different noise characteristics; this examination can provide more insight about the robustness of the proposed classification scheme.

Acknowledgments. The authors would like to thank Yudong Zhang for providing a portion of the MRI dataset.

References

1. Yao, J., Chen, J., Chow, C.: Breast tumor analysis in dynamic contrast enhanced MRI using texture features and wavelet transform. IEEE J. Sel. Top. Sig. Process. **3**(1), 94–100 (2009)
2. Nanthagopal, A.P., Sukanesh, R.: Wavelet statistical texture features-based segmentation and classification of brain computed tomography images. IET Image Process. **7**(1), 25–32 (2013)

3. Chaplot, S., Patnaik, L., Jagannathan, N.: Classification of magnetic resonance brain images using wavelets as input to support vector machine and neural network. Biomed. Sig. Process. Control **1**(1), 86–92 (2006)

4. El-Dahshan, E.S.A., Hosny, T., Salem, A.B.M.: Hybrid intelligent techniques for MRI brain images classification. Digit. Sig. Proc. **20**(2), 433–441 (2010)

5. Zhang, Y., Wang, S., Wu, L.: A novel method for magnetic resonance brain image classification based on adaptive chaotic PSO. Prog. Electromagnet. Res. **109**, 325–343 (2010)

6. Zhang, Y., Wu, L.: An MR brain images classifier via principal component analysis and kernel support vector machine. Prog. Electromagnet. Res. **130**, 369–388 (2012)

7. Candès, E.J., Donoho, D.L.: Ridgelets: a key to higher-dimensional intermittency? Philos. Trans. R. Soc. Lond. A: Math. Phy. Eng. Sci. **357**(1760), 2495–2509 (1999)

8. Candes, E.J., Donoho, D.L.: Curvelets: a surprisingly effective nonadaptive representation for objects with edges. Technical report, DTIC Document (2000)

9. Candes, E.J., Donoho, D.L.: Continuous curvelet transform: II. discretization and frames. Appl. Comput. Harmonic Anal. **19**(2), 198–222 (2005)

10. Bekker, A.J., Shalhon, M., Greenspan, H., Goldberger, J.: Multi-view probabilistic classification of breast microcalcifications. IEEE Trans. Med. Imaging **35**(2), 645–653 (2016)

11. Uslu, E., Albayrak, S.: Curvelet-based synthetic aperture radar image classification. IEEE Geosci. Remote Sens. Lett. **11**(6), 1071–1075 (2014)

12. Guo, J.M., Prasetyo, H., Farfoura, M.E., Lee, H.: Vehicle verification using features from curvelet transform and generalized Gaussian distribution modeling. IEEE Trans. Intell. Transp. Syst. **16**(4), 1989–1998 (2015)

13. Yang, G., Zhang, Y., Yang, J., Ji, G., Dong, Z., Wang, S., Feng, C., Wang, Q.: Automated classification of brain images using wavelet-energy and biogeography-based optimization. Multimedia Tools Appl. **75**(23), 15601–15617 (2015)

14. Das, S., Chowdhury, M., Kundu, M.K.: Brain MR image classification using multiscale geometric analysis of ripplet. Prog. In Electromagn. Res. **137**, 1–17 (2013)

15. Wang, S., Zhang, Y., Dong, Z., Du, S., Ji, G., Yan, J., Yang, J., Wang, Q., Feng, C., Phillips, P.: Feed-forward neural network optimized by hybridization of PSO and ABC for abnormal brain detection. Int. J. Imaging Syst. Technol. **25**(2), 153–164 (2015)

16. Candes, E., Demanet, L., Donoho, D., Ying, L.: Curvelab toolbox, version 2.0. CIT (2005)

A Novel Automatic Method to Evaluate Scoliotic Trunk Shape Changes in Different Postures

Philippe Debanné[1,2(✉)], Ola Ahmad[1,2], Stefan Parent[2], Hubert Labelle[2], and Farida Cheriet[1,2]

[1] Polytechnique Montréal, Montreal, Canada
philippe.debanne@polymtl.ca
[2] CHU Sainte-Justine, Montreal, Canada

Abstract. We present a novel method to evaluate the external trunk shape of Adolescent Idiopathic Scoliosis (AIS) patients. A patient's trunk surface is acquired in different postures (neutral standing, left and right lateral bending) at their preoperative visit and in standing posture at their postoperative visit following spinal deformity corrective surgery with an optical digitizing system. We use spectral shape decomposition to compute the eigenmodes of the trunk surface. This allows us to intrinsically define the principal shape directions robustly with respect to the patient's posture. We then extract a set of contour levels that follow the trunk's deformation from bottom to top, and characterize the trunk shape as a set of multilevel measurements taken at each level. Changes in trunk shape between postures/visits are calculated as differences between the measurement functionals. We performed a study on a small cohort of 14 patients with right thoracic spinal curves to assess the relationship between shape changes induced by the lateral bending positions and those resulting from surgical correction. The proposed method for scoliotic trunk shape evaluation represents a significant improvement over previous ones, as it is completely automatic and it adapts well to the lateral bending posture without the need to manually define control points/curves on the trunk surface.

Keywords: Spectral shape representation · Trunk surface model · Scoliosis

1 Introduction

Adolescent Idiopathic Scoliosis (AIS) is a complex musculoskeletal deformity causing abnormal curvatures of the spine and deformation of the ribcage, inducing a deformation of the whole shape of the human trunk. The deformity can appear in childhood and often worsens during the adolescent growth spurt. In more severe cases, surgery is required to correct the spine shape and by extension the whole trunk shape. Standard surgical planning uses radiographic images of the patient's spine, both in standing and in side bending postures. The radiographic lateral bending test allows the orthopaedist to assess spinal curve

© Springer International Publishing AG 2017
F. Karray et al. (Eds.): ICIAR 2017, LNCS 10317, pp. 455–462, 2017.
DOI: 10.1007/978-3-319-59876-5_50

mobility and predict surgical outcome. However, this test is unable to assess the reducibility of trunk asymmetry, even though external trunk appearance is of primary concern to the patients [2,10]. In light of this, several systems have been developed to scan the surface of the trunk of scoliotic patients, and indices to quantify asymmetry from surface cross-sections or regions have been studied [6,7].

Our research team has for a number of years utilized non-invasive surface topography (ST) to acquire the whole trunk of patients at the Sainte-Justine Hospital scoliosis clinic [9]. Several trunk shape indices have been developed and their reliability evaluated on patients in standing position [8]. More recently, we have demonstrated that the scoliotic trunk surface shape can be reliably documented by a set of multi-level cross-sectional measurements [12].

In a preliminary study on a small group of scoliosis patients, we demonstrated that ST measurements of AIS patients taken in voluntary lateral bending can bring useful information to the clinician in terms of the reducibility of external trunk asymmetry [5]. That work provided evidence that a relationship exists between the changes in trunk shape in side bending (versus neutral standing) and the surgical correction of trunk shape. However, the methodology used had several drawbacks that restricted its reliability. Specifically, the manual selection of a set of landmarks on the digitized trunk surface (in order to establish a patient reference frame and to construct "guiding curves" in order to extract shape cross-sections that followed the trunk's shape in bending) raised issues of repeatability. Another major limitation was that the resulting curves for the different asymmetry measures tended to be unstable. The multilevel shape analysis introduced by [12] addressed the latter shortcoming with its functional data analysis (FDA) modeling of the raw measurement points. Nevertheless, the cross-sectional approach explored thus far does not exploit the *intrinsic* properties of trunk shape in order to identify its deformations, since it relies on manually identified landmarks and must define an axis system prior to asymmetry analysis.

The purpose of this paper is to present a novel method for scoliotic trunk analysis that addresses the limits of previous methods and that exploits the intrinsic properties of shape through a spectral representation of the surface, a framework that we used previously in [1]. We will first describe the different stages of our mesh processing, leading to a multilevel characterization of trunk asymmetry. Then, we will present the results of using this method on a small cohort of AIS patients having undergone corrective surgery, to evaluate the relationship between the changes in trunk shape in side bending versus neutral standing position and those in the post-operative versus the pre-operative visits.

2   Methods

2.1   Trunk Data Acquisition

The scoliotic trunk geometries are acquired with a multi-head optical digitizing system (Capturor II LF, Creaform Inc.) located in the scoliosis clinic of Sainte-Justine Hospital. The system comprises four digitizers that acquire, in a sequence

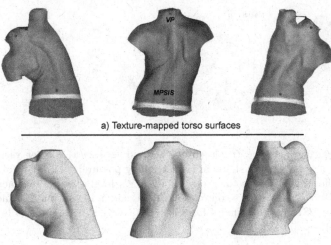

a) Texture-mapped torso surfaces

b) Pre-processed surface meshes

Fig. 1. Trunk mesh data and processed surface meshes in standing, left and right lateral bending postures for a sample patient (preoperative visit). Two landmarks, the vertebral prominens (VP) and the midpoint of the posterior-superior iliac spines (MPSIS), indicate the upper and lower limits of the region of the trunk to be analysed. (Color figure online)

lasting approximately five seconds, different portions of the trunk surface. Using dedicated processing software, the partial trunk surfaces are then registered and merged to produce a complete, textured surface mesh of the trunk containing between 50k and 100k vertices and between 100k and 200k triangles, depending on the patient's size. A previous study of this system demonstrated an accuracy of about 1.4 mm [9]. At the patient's preoperative visit, acquisitions are done in neutral standing posture (with arms in slight abduction by the side) and in maximum voluntary right and left side-bending. At the postoperative visit, the trunk is acquired only in standing posture.

2.2   Mesh Pre-processing

The triangulated meshes obtained by the acquisition system are not homogeneous over the whole torso shape, with triangles of poor quality in certain areas, notably under the arms (due to occlusion) and in surface regions tangential to the digitizers. The mesh data also contain holes at the boundaries of the trunk, i.e., where the head, arms and pelvis are cropped off, and therefore the mesh is not entirely connected. To solve these problems, we pre-process each trunk mesh using the well-known Radial Basis Functions (RBFs) method [3]. Figure 1b (bottom row) shows the processed surface meshes corresponding to the original surfaces in the top row.

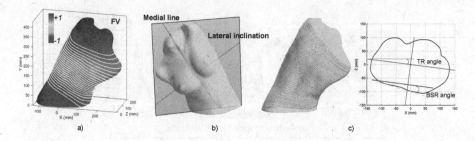

Fig. 2. Computation of the trunk surface contour levels: (a) Fiedler vector and its level sets for a female trunk shape in right bending posture; (b) Best-fit 3D plane (in purple) to the integral curve (in green) of the FV's level sets, and extracted medial curve (in red); (c) *(left)* Surface contour levels (overlaid in red on trunk surface); *(right)* Illustration of BSR and TR angles for a selected contour at the breasts level. (Color figure online)

2.3   Spectral Shape Decomposition

Based on the theory in [4], let $\mathcal{G} = (\mathcal{V}, \mathcal{E})$ be a connected undirected graph where $\mathcal{V}$ represents a set of vertices corresponding to the mesh elements (e.g., triangles), and $\mathcal{E}$ is the set of edges connecting pairs of neighboring vertices. We introduce the graph Laplacian L of the torso mesh points $\mathbf{x}$, together with its $|\mathcal{V}| \times |\mathcal{V}|$ weighted adjacency matrix W, where the edge weights are defined by the Gaussian kernel: $w_{ij} = \exp(-\|\mathbf{x}_i - \mathbf{x}_j\|^2/2\sigma^2)$ if $(i,j) \in \mathcal{E}$ and 0 otherwise. The Laplacian matrix is then defined by $\mathcal{L} = D^{-1}(D - W)$, where D is the diagonal degree matrix with $D_{ii} = \sum_j w_{ij}$. The spectral decomposition of the Laplacian matrix, $\mathcal{L} = \Phi^T \Lambda \Phi$, is defined by the set of eigenvalues $\Lambda = (\lambda_1, \ldots, \lambda_{|\mathcal{V}|})$, in ascending order such that λ_1 is the smallest non-zero eigenvalue, and their associated eigenvectors $\Phi = (\phi_1, \ldots, \phi_{|\mathcal{V}|})$, which capture the coarse to fine geometry of the shape. In this work, we exploit the Fiedler vector (FV), i.e. the 1st eigenvector of the graph Laplacian corresponding to λ_1, which describes the natural vibration of the human body shape along its craniocaudal direction (Fig. 2a).

2.4   Multi-level Characterization of the Trunk Asymmetry

The multi-level characterization of the trunk involves the representation of the 3D shape geometry as a set of contour levels defined between the L5–S1 and C7–T1 vertebral levels and uniformly spaced along the craniocaudal direction of the surface coordinates [5,12]. The vertebral levels are identified prior to acquisition by the VP and MPSIS landmark points (Fig. 1a). To handle the different trunk postures correctly, the surface contour levels must be computed intrinsically. In other words, they should be invariant with respect to the shape's posture. We therefore propose to exploit the spectral shape eigenmodes to efficiently compute the surface contour levels. The lower Laplacian eigenmodes are stable across postural changes [11]; hence they can be used to measure local and integral geometric features that correspond across the shapes. From the FV we compute its integral curve (Fig. 2b), which is constituted of the mean values of the 3D point sets at each level. As can be seen in Fig. 2a, these level sets capture the coarse geometric deformations of the shape resulting mainly from its posture. The integral curve can therefore be used to approximate the global direction of the shape as well as the direction of the medial line of the trunk. To do this, we search for the 3D plane that best fits the integral curve by means of a principal component analysis (PCA) on the set of 3D integral curve points. The PCA gives the position of a central shape point and two orthogonal directions, one corresponding to the medial line and the other to the shape's lateral inclination (Fig. 2b).

The medial line belongs to the left-right plane of symmetry of the trunk, which is orthogonal to the plane shown in Fig. 2b. The intersection of the plane of symmetry with the anterior and posterior halves of the trunk yields two point sets. We then compute a new curve, called the medial curve (Fig. 2b), by taking the midpoints between the corresponding pairs of anterior/posterior intersection points (closest point pairs). In order to have flexibility in determining the final number N of contour levels, we apply a least-square cubic spline approximation to the medial curve. The spline curve is then uniformly sampled at N points to get local positions and tangent vectors; these in turn define 3D cutting planes that intersect the trunk mesh to give the surface contours (Fig. 2c). In the present study, we use 300 contour levels to characterize the trunk asymmetry.

Multi-level Trunk Asymmetry Indices. Using singular value decomposition on the 2D coordinates of a given contour, we obtain its local X and Y axes; the centroid of the 2D coordinates is used as the center point (Fig. 2c). At each contour level, 4 measures are then computed: the back surface rotation (BSR) that measures the angle between the dual-tangent to the back and its local X-axis; the trunk rotation (TR) that measures the angle between the horizontal projection of the local X-axis and the X-axis of the mesh (global) coordinate system; and the coordinates of the center point along the global X and Z axes. These local indices, when computed from all contour levels, define the underlying multi-level asymmetry indices of the trunk. Figure 3 illustrates these characteristic functionals in different postures for the same patient.

3   Results and Discussion

3.1   Patient Cohort and Data Acquisition

Our cohort comprised 14 AIS patients with right thoracic curvatures (Lenke 1A curve type), operated at Sainte-Justine Hospital between 2010 and 2012. The preoperative ST scans occurred between 1 day and 4 months before the surgery date, while the postoperative ST scans occurred between 6 and 20 months after the surgery. These surface meshes (4 per patient) were then processed as described in Sect. 2 to obtain the asymmetry functionals.

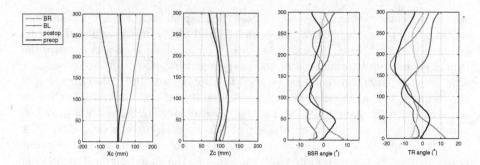

Fig. 3. Example of the torso asymmetry functionals using 300 contour levels. From left to right: Centroid lines (x_c, z_c), BSR and TR for a sample patient in right bending (BR), left bending (BL) and standing (preop, postop) postures. (Color figure online)

3.2 Comparison of Asymmetry Between Postures and Correlations

We want to compare the changes in trunk asymmetry between different postures, these being preop neutral standing (Pre), preoperative left bending (BL) and right bending (BR) and postop neutral standing (Post). We concentrate on the TR and BSR measures described in Sect. 2.4 above. For every surface mesh, we obtain sets of $n = 300$ values for each of these indices. We then compute the differences between postures (taking the indices' absolute values) as follows:

- $\Delta_{BSR}(BL)$: difference between BSR functionals of BL and Pre;
- $\Delta_{BSR}(BR)$: difference between BSR functionals of BR and Pre;
- $\Delta_{BSR}(Post)$: difference between BSR functionals of Post and Pre.

We use a similar notation for the differences in the TR index. To test the *intra-patient* relationships between these differences (Δ curves), we evaluate the correlations between these curves for Post and those for BL and BR, using Pearson's linear correlation coefficient in each case.

Figure 4 shows the results of this test. All the Pearson's r values for both indices are statistically significant (p-value $< 5\%$).

3.3 Discussion

We observe in Fig. 4 that for both asymmetry indices, the correlations are variable from patient to patient. Nonetheless, certain trends are visible. For the BSR, 12 out of 14 patients have at least one of the two r values above 0.5, and 7 of them have an r above 0.7. Of the two bendings, it is $\Delta_{BSR}(BR)$ that most often yields the better correlation with $\Delta_{BSR}(Post)$. Intuitively, this makes sense in that, for patients with right thoracic spinal curves, is it the rightward trunk motion that is more likely to "unfold" the spine curve, in essence artificially correcting the deformity (at least partially) with observable results on the back surface. For the TR, 13 out of 14 patients have at least one of the two r values above 0.5,

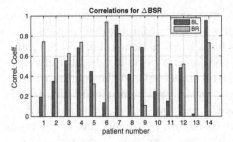
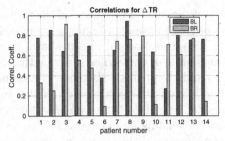

Fig. 4. Correlations obtained for the 14 AIS patients for the TR *(at left)* and BSR *(at right)* measures. For each patient, the Pearson's r values are shown between $\Delta(Post)$ and $\Delta(BL)/\Delta(BR)$. (Color figure online)

and 12 of them have an r above 0.7. Here, however, it is the left bending that more often yields the better correlation with $\Delta_{TR}(Post)$. Further analysis will be necessary to understand why this is the case, but we may hypothesize that it is the leftward trunk motion that is more likely to correct (at least partially) the axial rotation of the trunk, which the TR index evaluates.

Unlike the present study, in the earlier work we examined the correlations between asymmetry changes *across* a small cohort (*inter-patient*), and found moderate to good correlations ($r = 0.52$ for BSR and $r = 0.77$ for TR). Here, we chose instead to look at correlations between Δ curves for each patient. Given the aim of examining how the changes in trunk shape in lateral bending can help predict those resulting from corrective surgery, it seems more useful to analyse how the functional curves evolve *intra-patient*. Here, the best correlations obtained per patient range from moderate ($r = 0.40$ for BSR, $r = 0.38$ for TR) to very strong ($r = 0.95$ for BSR, $r = 0.94$ for TR). The presence of several cases with lower correlation values (e.g. patients 5 and 13 for BSR, patient 6 for TR) suggests that other factors may affect the relationship between the changes in lateral bending versus postop, such as the severity of the spinal curve.

Given the small patient sample size used, further study will be required to understand the complex relationship between the torso shape changes in the different cases. In that context, the proposed method will be a valuable tool as it provides an automated means of obtaining reproducible measures that capture trunk shape changes between different patient visits and postures.

4   Conclusion

In this paper, we have proposed a robust, automatic process to analyse scoliotic trunk shapes in neutral standing and lateral bending postures with the aim of evaluating changes in trunk asymmetry between different acquisitions. This approach presents several significant advantages over previous ones: it automates processing steps that were previously manual or semi-manual, and it adapts well to the lateral bending position without the need for additional control points/curves, unlike in [5]. Future work will focus on a more comprehensive investigation and validation of the medial curve and contour levels computations, in order to improve the method's reliability for more complex deformations.

The experimental results shown here for a small patient cohort corroborate the finding of our earlier study [5] that a relationship exists between the changes in trunk shape in side bending (versus neutral standing) and the surgical correction of trunk shape. However, further investigation will be necessary on a larger patient cohort to better understand how the external deformity is modified in lateral bending and how this relates to its correction by surgery and to the type of deformity of the spine and ribcage.

Acknowledgments. This research was funded by the Canadian Institutes of Health Research (grant number MPO 125875).

References

1. Ahmad, O., Lombaert, H., Parent, S., Labelle, H., Dansereau, J., Cheriet, F.: Longitudinal scoliotic trunk analysis via spectral representation and statistical analysis. In: Reuter, M., Wachinger, C., Lombaert, H. (eds.) SeSAMI 2016. LNCS, vol. 10126, pp. 79–91. Springer, Cham (2016). doi:10.1007/978-3-319-51237-2_7
2. Buchanan, R., Birch, J.G., Morton, A.A., Browne, R.H.: Do you see what I see? looking at scoliosis surgical outcomes through orthopedists' eyes. Spine **28**(24), 2700–2704 (2003)
3. Carr, J.C., Beatson, R.K., Cherrie, J.B., Mitchell, T.J., Fright, W.R., McCallum, B.C., Evans, T.R.: Reconstruction and representation of 3D objects with radial basis functions. In: Proceedings of SIGGRAPH, pp. 67–76. ACM (2001)
4. Chung, F.R.K.: Spectral Graph Theory. American Mathematical Society, Providence (1997)
5. Debanné, P., Pazos, V., Labelle, H., Cheriet, F.: Evaluation of reducibility of trunk asymmetry in lateral bending. Stud. Health Technol. Inf. **158**, 72–77 (2010)
6. Goldberg, C.J., Kaliszer, M., Moore, D.P., Fogarty, E.E., Dowling, F.E.: Surface topography, Cobb angles, and cosmetic change in scoliosis. Spine **26**(4), E55–E63 (2001)
7. Patias, P., Grivas, T.B., Kaspiris, A., Aggouris, C., Drakoutos, E.: A review of the trunk surface metrics used as scoliosis and other deformities evaluation indices. Scoliosis **5**(1), 12 (2010)
8. Pazos, V., Cheriet, F., Dansereau, J., Ronsky, J., Zernicke, R.F., Labelle, H.: Reliability of trunk shape measurements based on 3-D surface reconstructions. Eur. Spine J. **16**(11), 1882–1891 (2007)
9. Pazos, V., Cheriet, F., Song, L., Labelle, H., Dansereau, J.: Accuracy assessment of human trunk surface 3D reconstructions from an optical digitising system. Med. Biol. Eng. Comput. **43**(1), 11–15 (2005)
10. Pratt, R.K., Burwell, R.G., Cole, A.A., Webb, J.K.: Patient and parental perception of adolescent idiopathic scoliosis before and after surgery in comparison with surface and radiographic measurements. Spine **27**(14), 1543–1550 (2002)
11. Reuter, M., Wolter, F.-E., Shenton, M., Niethammer, M.: Laplace-Beltrami eigenvalues and topological features of eigenfunctions for statistical shape analysis. Comput. Aided Des. **41**(10), 739–755 (2009)
12. Seoud, L., Dansereau, J., Labelle, H., Cheriet, F.: Multilevel analysis of trunk surface measurements for noninvasive assessment of scoliosis deformities. Spine **37**(17), E1045–1053 (2012)

Breast Density Classification Using Local Ternary Patterns in Mammograms

Andrik Rampun[1][✉], Philip Morrow[1], Bryan Scotney[1], and John Winder[2]

[1] School of Computing and Information Engineering, Ulster University,
Coleraine BT52 1SA, Northern Ireland, UK
{y.rampun,pj.morrow,bw.scotney}@ulster.ac.uk
[2] School of Health Sciences, Institute of Nursing and Health, Ulster University,
Newtownabbey BT37 0QB, Northern Ireland, UK
rj.winder@ulster.ac.uk

Abstract. This paper presents a method for breast density classification. Local ternary pattern operators are employed to model the appearance of the fibroglandular disk region instead of the whole breast region as the majority of current studies have done. The Support Vector Machine classifier is used to perform the classification and a stratified ten-fold cross-validation scheme is employed to evaluate the performance of the method. The proposed method achieved 82.33% accuracy which is comparable with some of the best methods in the literature based on the same dataset and evaluation scheme.

1 Introduction

In the United States (US), an estimated 246,660 new cases of invasive breast cancer were expected to be diagnosed in the US in 2016 with 40,450 women expected to die [1]. Although dense breasts can be inherited genetically, many studies have indicated that breast density is a strong risk factor for developing breast cancer [2–4,6–12]. Based on the Breast Imaging Reporting and Data System (BI-RADS), there are four major categories used for classifying breast density: (a) predominantly fat, (b) fat with some fibroglandular tissue, (c) heterogeneously dense and (d) extremely dense.

During the last two decades, many breast density classification methods have been proposed in the literature. However, only a few studies have reported accuracy above 80%. The methods of Oliver *et al.* [2] and Parthaláin *et al.* [6] segment the breast region into dense and fatty tissue classes using fuzzy c-means clustering and extract a set of features from each cluster. Feature selection was performed to remove redundant descriptors before feeding them into the classifier. Oliver *et al.* [2] achieved 86% accuracy and Parthaláin *et al.* [6] who used a sophisticated feature selection framework achieved 91.4% accuracy. Bovis and Singh [3] used a combined classifier paradigm and achieved 71.4% correct classification based on a combination of features extracted using the Fourier and Discrete Wavelet transforms in conjunction with first and second-order statistical features. Chen *et al.* [7] compared the performance of local binary patterns (LBP),

© Springer International Publishing AG 2017
F. Karray et al. (Eds.): ICIAR 2017, LNCS 10317, pp. 463–470, 2017.
DOI: 10.1007/978-3-319-59876-5_51

local greylevel appearance (LGA), textons and basic image features (BIF) and achieved accuracies of 59%, 72%, 75% and 70%, respectively. Later, they proposed a method by modelling the distribution of the dense region in topographic representation and reported slightly higher accuracy of 76%. Petroudi *et al.* [12] implemented breast density segmentation using textons based on the Maximum Response 8 (MR8) filter bank and reported 75.5% accuracy. He *et al.* [11] achieved an accuracy of 70% using the relative proportions of the four Tabár's building blocks. Muštra *et al.* [5] captured the characteristics of the breast region using multi-resolution of first and second-order statistical features and reported 79.3% accuracy.

The majority results reported in the literature (achieved below 80%) indicate that breast density classification is a difficult task due to a wide variation of tissue appearance in the mammograms. Most of the proposed methods [2–4,6–12] in the literature extract texture information from the whole breast region whereas most dense tissues are located within the breast fibroglandular disk (FGD_{roi}). In this paper, we propose a method that extracts texture features from the FGD_{roi} only (see Fig. 3 later) to obtain more descriptive information. The reminder of the paper is organised as follows: Sect. 2 presents the technical aspects of our proposed method and; Sect. 3 discuss experimental results and the quantitative evaluation of the proposed method including quantitative comparisons; and finally Sect. 4 presents conclusions of this paper.

2   Methodology

We segment the breast region and extract only the FGD_{roi} followed by a simple median filter for noise reduction and extract features to capture the microstructure information from different orientations. The SVM classifier is employed as a classification approach.

2.1   Pre-Processing

We used our breast segmentation method [17] to disassociate the breast region from the pectoral muscle and the air background, and extract FGD_{roi} from the breast region. The left most image in Fig. 3 shows the estimated fibroglandular disk area. To extract FGD_{roi}, we find B_w which is the longest perpendicular distance between the y-axis and the breast boundary (magenta line). Therefore, the width and the height of the square area of the FGD_{roi} (amber line Fig. 3) can be computed as $B_w \times B_w$ with the center located at the intersection point between B_h and B_w lines. B_h is the height of the breast which is the longest perpendicular distance between the x-axis and the breast boundary. B_h is then relocated in the middle of B_w to get the intersection point. The size of the FGD_{roi} varies depending on the width of the breast. For noise reduction, several techniques have been tested such as mean filter, anisotropic diffusion and Gaussian filter, and we found that the median filter using a 3×3 window size produced better results on average.

2.2 Feature Extraction

The Local Binary Pattern (LBP) operators were first proposed by Ojala *et al.* [13] to encode pixel-wise information based on its different mapping tables namely uniform LBP ('u2'), rotation invariant LBP ('ri') and rotation invariant uniform ('riu2'). The Local Ternary Pattern (LTP) operators are among the variants of the LBP operators which have shown promising results in a study conducted by Nanni *et al.* [14]. Both are similar in terms of architecture as each are defined by using a circle and a number of neighbours (see Fig. 1). The main difference is that the LTP thresholds the neighbouring pixels into three values −1, 0 and 1 (three-value encoding) using a threshold constant set by the user, whereas the LBP thresholds the neighbouring pixels into two values 0 and 1 (two-value encoding). The LTP decimal value of a pixel (i, j) is given by:

$$LTP_{(P,R)}^{pattern}(i,j) = \sum_{p=0}^{(P-1)} s_{pattern}(g_p)2^p \tag{1}$$

where R is the circle radius, P is the number of neighbours, k is the threshold constant, g_c is the grey level value of the center pixel, p is the neighbouring pixel, g_p is the grey level value of the pth neighbour, and $pattern \in \{upper, lower\}$. Once the LTP code is generated, it is split into two binary patterns (upper and lower patterns) by considering its positive, zero and negative components, as illustrated in Fig. 2 using the following conditions

$$s_{upper}(p) = \begin{cases} 1, & \text{if } s(p) > 0 \\ 0, & \text{if } s(p) \leq 0 \end{cases} \tag{2}$$

$$s_{lower}(p) = \begin{cases} 1, & \text{if } s(p) < 0 \\ 0, & \text{if } s(p) \geq 0 \end{cases} \tag{3}$$

The LTP code can be generated using the following conditions

$$s(p) = \begin{cases} -1, & \text{if } p < g_c - k \\ 0, & \text{if } p \geq g_c - k \text{ and } p \leq g_c + k \\ 1, & \text{if } p > g_c + k \end{cases} \tag{4}$$

where $s(p)$ is the pth neighbour containing the LTP code value. In this study we use the uniform LTP ('u2') due to its stability as it is less prone to noise and by considering only uniform patterns making the number of possible LBP labels significantly lower and reliable estimation of their distribution requires fewer samples [16]. To enrich texture information, we extract feature histograms from both upper and lower binary pattern map of the FGD_{roi} at different orientations; $\theta = 0°$, $45°$, $90°$, $135°$, $180°$, $225°$, $270°$, and $315°$ (as shown in Fig. 1) resulting in eight histograms. Subsequently, we concatenate these histograms to be a long histogram (representing the occurrences of edges, corners, spots, lines, etc. within

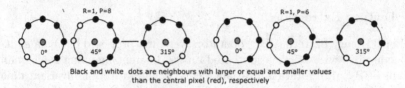

Fig. 1. An illustration of different uniform patterns at different θ and P.

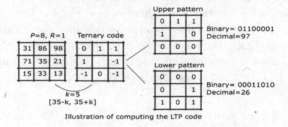

Fig. 2. An illustration of computing the LTP code using $P = 8$ and $R = 1$.

the FGD_{roi}) which will be treated as a feature vector that will be used in the classification phase.

Figure 3 summarises the feature extraction process in this study. Note that, in comparison to the other methods [2–4,6–12] our proposed method applied the LTP operators only within the FGD_{roi} and features were extracted at eight different orientations.

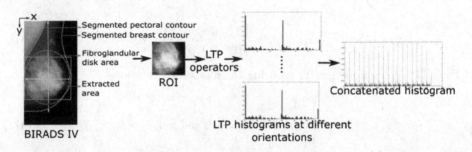

Fig. 3. Summary of feature extraction in our study using the LTP operators.

2.3 Classification

Once the feature extraction was completed, the Support Vector Machine (SVM) was employed as our classification approach. The GridSearch technique was used to explore the best two parameters (complexity (C) and exponent (e)) by testing all possible values of C and e ($C = 1..10$ and $e = 0.1..1.0$) and selecting the best combination based on the highest accuracy. The SVM classifier was trained and in the testing phase, each unseen FGD_{roi} from the testing set is classified as BI-RADS I, II, III or IV.

3 Experimental Results

To test the performance of the proposed method, we used the well known Mammographic Image Analysis Society (MIAS) database [15] which consists of 322 mammograms of 161 women. Each image contains BI-RADS information (e.g. BI-RADS class I, II, III or IV) provided by an expert radiologist. A stratified ten runs 10-fold cross validation scheme was employed, where the patients are randomly split into 90% for training and 10% for testing and repeated 100 times. The metric accuracy (Acc) is used to measure the performance of the method which represents the total number of correctly classified images compared to the total number of images. The following parameter values for the number of neighbours (P), radius (R), threshold (k) and orientations (θ) were tested to evaluate the performance of the proposed method: (a) $P \in \{5, 6, 7, 8\}$, (b) $R \in \{1, 3, 5\}$, (c) $k \in \{3, 4, 5\}$ and (d) $\theta \in \{0°, 45°, 90°, 135°, 180°, 225°, 270°, \text{ and } 315°\}$. In addition, we compare the performance when features are extracted from the whole breast (wb) region versus from the FGD_{roi}.

Table 1 presents the quantitative results when extracting features from the wb region versus from the FGD_{roi}. Note that θ^{ALL} means histograms from the following orientations were concatenated: $0°$, $45°$, $90°$, $135°$, $135°$, $180°$, $225°$, $270°$, and $315°$. Results show that features are more discriminant if extracted only from the FGD_{roi} instead of from the whole breast region. For each case in Table 1, the same parameter settings were used but features were derived from the FGD_{roi} and wb, and at least $3-5\%$ difference in accuracy was observed. Our explanation for these results is that in many cases non-fibroglandular disk areas predominantly contain fatty tissues regardless of its BI-RADS class because most dense tissues start to develop within the fibroglandular disk area. Therefore, capturing micro-structure information outside the fibroglandular disk means extracting similar information resulting in less discriminative features across BI-RADS classes. In cases where the non-fibroglandular disk region is dominated by dense tissue the FGD_{roi} also mostly contains dense tissue. For example, Fig. 4 shows histograms extracted from the whole breast regions (wb) and FGD_{roi} at $\theta = 0°$ for two different breasts (BI-RADS I and IV). To measure the difference quantitatively, we used the χ^2 distance (d) to measure the difference between these histograms and found that $d = 0.0705$ for H_1^{wb} and H_2^{wb} and $d = 0.122$ for $H_1^{FGD_{roi}}$ and $H_2^{FGD_{roi}}$. This means that H_1^{wb} is more similar to H_2^{wb} than $H_1^{FGD_{roi}}$ and $H_2^{FGD_{roi}}$.

Table 2 shows the classification results for the FGD_{roi} when varying the values of k, P, R and θ. The proposed method achieved the best $Acc = 82.33\%$ using $k = 4$, $P = 6$, $R = 5$ and θ^{ALL}. Experimental results suggest that using a single orientation is insufficient to capture the characteristics of dense and fatty tissues due to multidirectional appearance which is complex and uncertain. However,

Table 1. Classification results when extracting features from wb versus FGD_{roi}

k	P	R	θ	Acc(%)
4	6	5	θ^{ALL}	**82.33(FGD_{roi})**, 76.94(wb)
5	6	5	θ^{ALL}	80.07(FGD_{roi}), 76.64(wb)
4	8	5	θ^{ALL}	78.18(FGD_{roi}), 77.43(wb)
4	6	7	θ^{ALL}	77.69(FGD_{roi}), 74.08(wb)
4	6	3	θ^{ALL}	76.85(FGD_{roi}), 71.69(wb)
3	6	3	θ^{ALL}	76.18(FGD_{roi}), 71.67(wb)

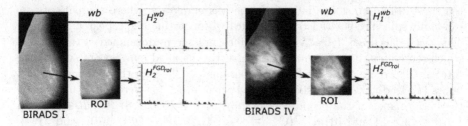

Fig. 4. Histograms extracted from the wb versus FGD_{roi} with BI-RADS class I and IV.

using several orientations enable our method to capture complex appearances of the FGD_{roi}. The number of neighbours affects the performance of the extracted features. Results also suggest that the proposed method is highly dependent on the parameter settings in the LTP operators which is the main drawback of our method. However, it can be seen in Tables 1 and 2 that the LTP operators can extract robust features for breast density classification in mammograms.

Table 2. Classification results when varying the value of k, P, R and θ (features were extracted from FGD_{roi}).

k	P	R	θ	Acc(%)
-	6	5	θ^{ALL}	79.24($k = 3$), **82.33($k = 4$)**, 80.07($k = 5$), 76.59($k = 6$)
4	-	5	θ^{ALL}	76.38($P = 5$), **82.33($P = 6$)**, 77.04($P = 7$), 78.18($P = 8$)
4	6	-	θ^{ALL}	72.75($R = 2$), 78.17($R = 3$), 79.31($R = 4$), **82.33($R = 5$)**
4	6	5	-	**82.33(θ^{ALL})**, 76.67($\theta = 0°$), 74.25($\theta = 45°$), 74.07($\theta = 90°$), 75.44($\theta = 135°$), 76.89($\theta = 180°$), 75.22($\theta = 225°$), 75.17($\theta = 270°$), and 76.61($\theta = 315°$)

In comparison to the other methods in the literature, our method achieved 82.33% accuracy which is better than the methods proposed by Muštra *et al.* [5] (79.3%), Chen *et al.* [7,9] (59%, 70%, 72%, 75% and 76%), Bovis and Singh [3]

(71.4%), and He *et al.* [11] (70%). However, the methods of Parthaláin *et al.* [6] and Oliver *et al.* [2] achieved 91.4% and 86%, respectively. Note that, to minimise bias, these comparisons are based on only those studies that have used the MIAS database [15], four-class classification, and using the same evaluation technique as in this study. Our method outperformed the methods in [3,4,7–12] because (a) robust feature extraction operators are used which are able to capture richer micro-structure information using the three-value encoding technique and are less sensitive to noise, (b) the use of FGD_{roi} minimises the texture similarity representation of the breast region hence resulting in more descriptive features across different BI-RADS classes, and (c) we are able to capture a wider range of texture rotation/variation by extracting features from eight different orientations. In breast imaging, deep learning based approaches are becoming popular due to its capability to learn complex appearances especially in the area of segmentation and classification. Despite deemed as a 'black-box' approach study of Kallenberg *et al.* [18] reported their results have a very strong positive relationship with manual ones. Future work would be to employ more texture futures as investigated by the study of Rampun *et al.* [19].

4 Conclusion

In conclusion, we have presented and developed a breast density classification method using the LTP operators applied only within the fibroglandular disk area which is the most prominent region of the breast instead of the whole breast region as suggested in current studies [2–4,6–12]. By only extracting features from this area, we obtained a set of more discriminative and distinctive texture descriptors across BI-RADS classes. Quantitative comparisons with the existing studies on the same dataset and evaluation technique suggest that the proposed method outperformed most of the current studies in the literature.

Acknowledgments. This research was undertaken as part of the Decision Support and Information Management System for Breast Cancer (DESIREE) project. The project has received funding from the European Union's Horizon 2020 research and innovation programme under grant agreement No. 690238.

References

1. Breast Cancer. 'U.S. Breast Cancer Statistics' (2016). http://www.breastcancer.org/symptoms/understand_bc/statistics. Accessed 6 Jan 2017
2. Oliver, A., Freixenet, J., Martí, R., Pont, J., Perez, E., Denton, E.R.E., Zwiggelaar, R.: A Novel breast tissue density classification methodology. IEEE Trans. Inf Technol. Biomed. **12**(1), 55–65 (2008)
3. Bovis, K., Singh, S.: Classification of mammographic breast density using a combined classifier paradigm. In: 4th International Workshop on Digital Mammography, pp. 177–180 (2002)
4. Oliver, A., Tortajada, M., Lladó, X., Freixenet, J., Ganau, S., Tortajada, L., Vilagran, M., Sentś, M., Martí, R.: Breast density analysis using an automatic density segmentation algorithm. J. Digit. Imaging **28**(5), 604–612 (2015)

5. Muštra, M., Grgić, M., Delać, K.: A novel breast tissue density classification methodology. Breast density classification using multiple feature selection. Automatika **53**(4), 362–372 (2012)
6. Parthaláin, N.M., Jensen, R., Shen, Q., Zwiggelaar, R.: Fuzzy-rough approaches for mammographic risk analysis. Intell. Data Anal. **14**(2), 225–244 (2010)
7. Chen, Z., Denton, E., Zwiggelaar, R.: Local feature based mamographic tissue pattern modelling and breast density classification. In: The 4th International Conference on Biomedical Engineering and Informatics, pp. 351–355 (2011)
8. Bosch, A., Munoz, X., Oliver, A., Martí, J.: Modeling and classifying breast tissue density in mammograms. In: Computer Vision and Pattern Recognition (CVPR 2006), pp. 1552–1558 (2006)
9. Chen, Z., Oliver, A., Denton, E., Zwiggelaar, R.: Automated mammographic risk classification based on breast density estimation. In: Sanches, J.M., Micó, L., Cardoso, J.S. (eds.) IbPRIA 2013. LNCS, vol. 7887, pp. 237–244. Springer, Heidelberg (2013). doi:10.1007/978-3-642-38628-2_28
10. Byng, J.W., Boyd, N.F., Fishell, E., Jong, R.A., Yaffe, M.J.: Automated analysis of mammographic densities. Phys. Med. Biol. **41**(5), 909–923 (1996)
11. He, W., Denton, E., Stafford, K., Zwiggelaar, R.: Mammographic image segmentation and risk classification based on mammographic parenchymal patterns and geometric moments. Biomed. Sig. Process. Control **6**(3), 321–329 (2011)
12. Petroudi, S., Kadir, T., Brady, M.: Automatic classification of mammographic parenchymal patterns: a statistical approach. In: Proceedings of IEEE Conference on Engineering Medicine and Biology Society, vol. 1, pp. 798–801 (2003)
13. Ojala, T., Pietikainen, M., Maenpaa, T.: Multiresolution gray-scale and rotation invariant texture classification with local binary patterns. IEEE Trans. Pattern Anal. Mach. Intell. **24**(7), 971–987 (2002)
14. Nanni, L., Luminia, A., Brahnam, S.: Local binary patterns variants as texture descriptors for medical image analysis. Artif. Intell. Med. **49**(2), 117–125 (2010)
15. Suckling, J., et al.: The mammographic image analysis society digital mammogram database. In: Proceedings of Excerpta Medica Internatinal Congress Series, pp. 375–378 (1994)
16. Hadid, A., Pietikainen, M.K., Zhao, G., Ahonen, T.: Computer Vision Using Local Binary Patterns. Springer, London (2011). pp. 13–47
17. Rampun, A., Winder, R.J., Morrow, P.J., Scotney, B.W.: Fully automated breast boundary and pectoral muscle segmentation in mammograms. Artificial Intelligence in Medicine (2016). (under review)
18. Kallenberg, M., et al.: Unsupervised deep learning applied to breast density segmentation and mammographic risk scoring. IEEE Trans. Med. Imaging **35**(5), 1322–1331 (2016)
19. Rampun, A., et al.: Computer-aided detection of prostate cancer in T2-weighted MRI within the peripheral zone. Phy. Med. Biol. **61**(13), 4796–4825 (2016)

Segmentation of Prostate in Diffusion MR Images via Clustering

Junjie Zhang[1], Sameer Baig[1], Alexander Wong[2], Masoom A. Haider[1], and Farzad Khalvati[1(✉)]

[1] Medical Imaging, Sunnybrook Research Institute, University of Toronto, Toronto, Canada
{junjie.zhang,farzad.khalvati}@sri.utoronto.ca
[2] Systems Design Engineering, University of Waterloo, Waterloo, Canada

Abstract. Automatic segmentation of prostate gland in magnetic resonance (MR) images is a challenging task due to large variations of prostate shapes and indistinct boundaries with adjacent tissues. In this paper, we propose an automatic pipeline to segment prostate gland in diffusion magnetic resonance images (dMRI). The most common approach for segmenting prostate in MR images is based on image registration, which is computationally expensive and solely relies on the pre-segmented images (also known as atlas). In contrast, the proposed method uses a clustering method applied to the dMRI to separate prostate gland from the surrounding tissues followed by a post-processing stage via active contours. The proposed pipeline was validated on prostate MR images of 25 patients and the segmentation results were compared to manually delineated prostate contours. The proposed method achieves an overall accuracy with mean Dice Similarity Coefficient (DSC) of 0.84 ± 0.04, while being the most effective in the middle prostate gland producing a mean DSC of 0.91 ± 0.03. The proposed method has the potential to be integrated into clinical decision support systems that aid radiologists in monitoring prostate cancer.

Keywords: Prostate segmentation · Diffusion magnetic resonance imaging (dMRI) · Gaussian Mixture Model (GMM) · Active contour

1 Introduction

Prostate diseases (e.g., prostate cancer, prostatitis and benign prostate hyperplasia (BPH)) are common afflictions in men. In particular, prostate cancer is the most diagnosed cancer and the third leading cause of cancer death in Canadian men [1]. Accurate estimation of prostate volume and boundary is very useful for treatment planning and other diagnostic and therapeutic procedures for prostate

F. Khalvati—This work was partially supported by Ontario Institute of Cancer Research (OICR) and Cancer Care Ontario (CCO) - Imaging Network of Ontario (CINO).

© Springer International Publishing AG 2017
F. Karray et al. (Eds.): ICIAR 2017, LNCS 10317, pp. 471–478, 2017.
DOI: 10.1007/978-3-319-59876-5_52

cancer as well as other prostate diseases, which requires segmentation of prostate from various imaging modalities.

For instance, prostate volume combined with prostate-specific antigen (PSA) level serves as an important indicator in determining the presence of benign or malignant tumours [2], and robust measurements of the prostate boundary is an important component of radiotherapy planning [3].

Magnetic Resonance (MR) imaging is a good fit for the task of prostate segmentation, since it has been widely used in diagnostic workflow and treatment planning for prostate diseases and provides favourable soft tissue contrast. However, manual segmentation of prostate in 3D MR images is time-consuming and subjective, suffering from inter- and intra-observer variations and personal biases. In this regard, automated segmentation methods are highly demanded in clinical practice.

Automatic segmentation of prostate in MR images, however, is a very challenging task for several reasons [4]. First, different MR images have global inter-scan variability and intra-scan intensity variations due to different MR scan protocols and prostate signal inhomogeneity. Second, prostates exhibit large contrast variation in image response, due to heterogenous anatomic structures of prostate and its surrounding tissue (e.g., blood vessel, bladder, rectum and seminal vessels), which adds up to the difficulty in the determination of prostate boundaries. Third, prostates have a wide range of morphological variation (e.g., shape, size, and volume) among different subjects due to pathology changes or different resolutions of images.

With the growing importance of prostate MRI segmentation, over the past few years, several (semi-)automatic methods have been proposed to meet the challenge, such as atlas-based methods [5], shape-based models [6], and machine learning based methods [7]. A detailed overview of the different segmentation methods and their performance can be found in [8]. Most of these methods rely on training data, a relatively large set of images manually contoured by expert. Nevertheless, creating training data is prohibitively expensive and requires highly trained and paid clinicians to spend a significant amount of time and manually contour the images. Moreover, registration-based techniques are computationally expensive and may not be suitable for a clinical setting with soft or hard real-time performance requirements.

In this paper, we propose a novel hybrid approach to automatically segment prostate gland in diffusion-weighted images (DWIs) that combines an unsupervised clustering method and a deformable method in a pipeline. The performance of our method is validated using MR images of a dataset with 25 cases and the segmentation results are compared to manually delineated prostate contours by an expert. Experimental results show that our proposed method performs reasonably well in segmenting the prostate, especially in the middle prostate gland.

To the best of our knowledge, this is among the first attempts to segment prostate gland solely utilizing DWI modalities, especially high b-value (i.e., $1,000\,s/mm^2$) images, where most methods target segmentation of prostate in T2-weighted images. The proposed method is fully automatic and requires no prior training. This greatly eases the burden of extensive manual contouring and annotation before segmentation, compared to conventional atlas or machine learning based methods. Also, as the proposed method focuses only on the case to be segmented, it is free from inter-subject variation and applicable to other datasets with different scan protocols.

The rest of the paper is organized as follows. The proposed pipeline for automatic prostate segmentation is formulated in Sect. 2 followed by the presentation of quantitative and qualitative evaluation of our method in Sect. 3. We finally draw conclusions in Sect. 4.

2   Methods

The proposed method utilizes different b-value modalities at different steps and incorporates various algorithms in the pipeline to achieve an effective segmentation of the prostate gland. The block diagram of our proposed method is illustrated in Fig. 1.

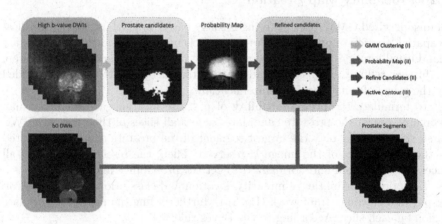

Fig. 1. Schematic representation of the proposed pipeline. Abbreviations used: GMM: Gaussian Mixture Model.

2.1   GMM Clustering

In the first step, a Gaussian Mixture Model (GMM) clustering is used to pre-segment the prostate and background regions from high b-value ($1000\,\mathrm{s/mm^2}$) DWI images to get prostate candidates. We use mixture of Gaussian distributions to model an image, in which the value of each pixel is a linear combination of underlying Gaussian components. The GMM can be solved effectively through expectation-maximization (EM) technique [9]. Given the image as a random variable X, a vector of unknown parameter θ of the Gaussian distributions, and a set of latent unobserved data Y, along with a likelihood function $P(X, Y|\theta)$, the EM algorithm seeks to find the maximum likelihood estimate of the unknown parameters θ by iteratively applying the expectation and maximization steps. In Eq. (1), the expectation step calculates the expected value of the log likelihood with current estimated parameters θ^t; and in Eq. (2), the maximization step finds the parameters that maximizes this quantity.

$$Q(\theta|\theta^t) = E_{Y|X,\theta^t}[logP(X,Y|\theta)] \tag{1}$$

$$\theta^{t+1} = \arg\max_{\theta}[Q(\theta|\theta^t)] \tag{2}$$

The prostate gland appears to be the bright region in the central part in high b-value prostate DWI images. Therefore, in our GMM, the intensity of the MR image is approximated with two classes: the prostate candidates (high-intensity cluster) and the background regions (low intensity cluster). The prostate candidates generated by GMM are not necessarily the exact prostate segment, but give a good idea of the approximate location and shape of prostate on each slice. Figure 2(a) shows an example of GMM segmentation result.

2.2   Probability Map Creation

In unsupervised GMM, the clustering is solely based on pixel intensity regardless of spatial location of prostate in the image. It is natural to find that the prostate gland usually resides in the central part of the prostate MR images. In this step, we further refine the prostate candidates obtained by GMM clustering guided by the prior information of prostate location.

To formulate the prior probability of prostate location, a probability map is created out of the prostate candidates across all slices of the given 3D DWI. First, on each slice, only the largest segment of the prostate candidates located in the central part of the image is reserved. Then, the reserved regions of all slices are aggregated and normalized to get the probability map.

The prior probabilities of intensity, location and class prior probabilities can be put in a Bayesian framework (Eq. 3) to further refine the prostate candidate region on each slice. According to the Bayes rule:

$$\begin{aligned} P(C_i|X) &\propto P(X|C_i)P(C_i) \\ &\propto P(x_{in}|C_i)P(x_{ps}|C_i)P(C_i) \\ &\propto P(C_i|x_{in})P(x_{ps}|C_i) \end{aligned} \tag{3}$$

where $P(C_i|X)$ represents the posterior probability distribution of a class given the prior $P(C_i)$ and the likelihood $P(X|C_i)$. The likelihood $P(X|C_i)$ can be further obtained from the product of the probability of a pixel intensity being the class (e.g. prostate), $P(x_{in}|C_i)$, from the GMM-EM framework and the probability of a pixel location being the class, $P(x_{ps}|C_i)$, as formulated in the probability map. Figure 2 shows an example of refined prostate candidate by the probability map.

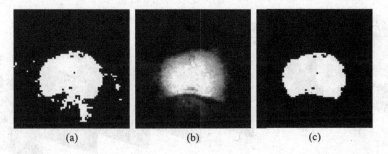

(a)   (b)   (c)

Fig. 2. Probability map for further refinement of prostate candidates. (a) GMM-clustering prostate candidates, (b) Probability map, and (c) Refined prostate candidates.

2.3 Active Contour Segmentation

The refined prostate candidates still suffer from noisy pixels resulting in holes and rough boundaries (see Fig. 2(c)). Thus, a deformable model is required to apply to the refined prostate candidates to get the final prostate segmentation.

The refined prostate candidates are laid on b-0 images in DWI series to carry out the active contour segmentation. B-0 images are used at this stage instead of the high b-value images on which we previously applied the GMM clustering to get initial prostate candidates, as more anatomical structures are shown on low b-value DWI modalities. A geodesic active contour algorithm similar to that developed by Caselles et al. [10] is used in the implementation. The contour evolution of geodesic active contour can be formulated as:

$$C_t(t, \mathbf{u}) = F_{internal} + F_{external} \qquad (4)$$

where $C_t(t, \mathbf{u})$ is the contour at time t parameterized by $\mathbf{u}$ and F are forces acting on the contour in the normal direction. The internal force modulates evolution based on the contour's geometry, which is the mean curvature of the contour, while the external force incorporates information from the image being segmented, computed from the gradient magnitude of the image intensity.

3   Experiments and Results

We have validated the accuracy and robustness of our method on prostate diffusion MR images with a resolution of 144×144 pixels from a dataset with 25 cases. The images were acquired using a Philips Achieva 3.0T machine at Sunnybrook Health Sciences Centre, Toronto, ON, Canada. All data was obtained under the local institutional research ethics board.

The automatic segmentation results were compared with contours manually delineated by an expert on diffusion-weighted MRI data ($b = 0\,\text{s/mm}^2$) using ProCanVAS platform [11]. Figure 3 shows some qualitative results of our method: the automatic segmentation results (Green) and manually delineated contours

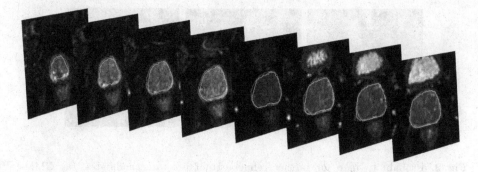

Fig. 3. Automatic segmentation results (Green) and manually delineated contours (Yellow) through several axial slices. (Color figure online)

(Yellow) through several axial slices of a case. It is observed that our method can produce accurate segmentation results.

The quantitative evaluation of our method are shown in Table 1 and Fig. 4. This is done by comparing the automatic segmentation results slice by slice with the manual segmentations via Dice Similarity Coefficient (DSC) [12].

Table 1. Summary of DSC results for the whole prostate gland, the central gland, and apex & base sections of the prostate.

Segmentation	Dice similarity coefficient (DSC)		
	Overall	Mid-section	Apex & Base
GMM	0.81 ± 0.04	0.88 ± 0.04	0.75 ± 0.07
GMM+AC	0.84 ± 0.04	0.91 ± 0.03	0.77 ± 0.07

Table 1 presents the overall average DSCs for the initial GMM generated prostate candidates and the final prostate segmentation produced by active contour (AC). DSCs for middle prostate glands and the apex and base sections are also summarized. The results confirm the effectiveness of our proposed pipeline, that the final prostate segmentations from AC increase DSCs in all cases with respect to the initial prostate candidates by GMM. The unsupervised clustering method applied on high b-value DWI images produces satisfactory overall results with $DSC = 0.81$. AC further refines the segmentations and improves DSC for about 3%. It is also notable that the method achieves good segmentation results especially in middle prostate glands with high DSC value of 0.91, while as expected, the performance is not as good in the apex and base sections of the prostate with $DSC = 0.77$. This is in accordance with the observation that, in the central region, the soft tissue contrast is better, whereas the soft tissue contrast is significantly distorted in the base and the apex section of the prostate, making those sections difficult to segment. Above qualitative analysis of the proposed method is illustrated in Fig. 4.

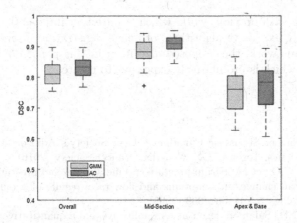

Fig. 4. DSC results for the whole prostate gland (Overall), the central gland (Mid-Section), and two end sections (Apex & Base) of prostate.

In a previous work [13], a semi-automatic registration-based segmentation of prostate was proposed in T2-weighted MRI which required intensive manual segmentation of 3 slices (apex, base, and middle) which yielded a median overall DSC of 0.92 using non-rigid registration algorithm. Another semi-automatic registration-based segmentation algorithm was proposed for prostate in T2-weighted MRI work [14], which again required intensive manual contouring of 3 or more slices with DSC of 0.77 to 0.85 (depending on registration algorithm) when only 5 slices of a given 3D image were left to segment automatically. Recently, an atlas-based segmentation algorithm was proposed [11] for prostate segmentation in DWI images which also required user input as a bounding box around prostate gland in at least 3 (5) slices yielding an average DCS of 0.80 (0.85). Given that in all these cases, heavy user intervention was required and they all relied on time-consuming image registration algorithm, the added value of the proposed algorithm in this paper becomes more clear; in addition to the fact that it does not require training data or atlas, it is fully automatic and it runs in real time.

4 Conclusion

In this work, a fully automatic pipeline was proposed for prostate segmentation in diffusion MR images. This method solely utilizes diffusion weighted images (DWIs) and is comprised of three major steps: (1) Gaussian Mixture Model (GMM) clustering for prostate candidates detection, (2) probability map creation to refine prostate candidate regions, and (3) active contour segmentation to get the final prostate segments. The evaluation of the proposed algorithm performed using MR images of 25 patients showed reasonably accurate results for prostate gland with overall average DSC of 0.84 and average DSCs of 0.91 and 0.77 for middle section, and apex and base sections, respectively. These results are comparable with previous work, which require user intervention and rely on computationally expensive algorithms of image registration. Moreover, in contrast to previous work, the proposed algorithm in this paper does not require training data, which is prohibitively expensive to create.

References

1. Canadian Cancer Statistics: Canadian Cancer Society's Advisory Committee on Cancer Statistics. Toronto, ON: Canadian Cancer Society (2016)
2. Roehrborn, C.G., et al.: Serum prostate-specific antigen and prostate volume predict long-term changes in symptoms and flow rate: results of a four-year, versus placebo. Urology **54**(4), 662–669 (1999)
3. Huyskens, D.P., Salamon, E., et al.: A qualitative and a quantitative analysis of an auto-segmentation module for prostate cancer. Radiother. Oncol. **90**(3), 337–345 (2009)
4. Mahapatra, D., Buhmann, J.: Prostate MRI Segmentation using learned semantic knowledge and graph cuts. IEEE Trans. Biomed. Eng. **61**(3), 756–764 (2014)
5. Klein, S., et al.: Automatic segmentation of the prostate in 3D MR images by atlas matching using localized mutual information. Med. Phys. **35**, 1407–1417 (2008)
6. Toth, R., et al.: Accurate prostate volume estimation using multifeature active shape models on T2-weighted MRI. Acad. Radiol. **18**(6), 745–754 (2011)
7. Moschidis, E., Graham, J.: Automatic differential segmentation of the prostate in 3-D MRI using random forest classification and graph-cuts optimization. In: IEEE ISBI, pp. 1727–1730 (2012)
8. Litjens, G., et al.: Evaluation of prostate segmentation algorithms for MRI: the PROMISE12 challenge. Med. Image Anal. **18**(2), 359–373 (2014)
9. Dempster, A., Larid, N., Rubin, D.: Maximum likelihood from incomplete data via the EM algorithm. J. R. Stat. Soc. **39**(1), 1–38 (1977)
10. Caselles, V., Kimmel, R., Sapiro, G.: Geodesic active contours. Int. J. Comput. Vis. **22**, 61–79 (1997)
11. Zhang, J., Baig, S., Wong, A., Haider, M.A., Khalvati, F.: A local ROI-specific atlas-based segmentation of prostate gland and transitional zone in diffusion MRI. J. Comput. Vis. Imaging Syst. **2**(1) (2016)
12. Bharatha, A., Hirose, M., Hata, N., Warfield, S.K., Ferrant, M., Zou, K.H., Suarez-santana, E., Ruiz-Alzola, J., D'Amico, A., Cormack, R.A., Kikinis, R., Jolesz, F.A., Tempany, C.M.C.: Evaluation of three-dimensional finite element-based deformable registration of pre- and intra-operative prostate imaging. Med. Phys. **28**(12), 2551–2560 (2001)
13. Khalvati, F., Salmanpour, A., Rahnamayan, S., Rodrigues, G., Tizhoosh, H.R.: Inter-slice bidirectional registration-based segmentation of the prostate gland in MR and CT image sequences. Med. Phys. **40**(12), 123503-1-11 (2013)
14. Khalvati, F., Salmanpour, A., Rahnamayan, S., Haider, M.A., Tizhoosh, H.R.: Sequential registration-based segmentation of the prostate gland in MR image volumes. J. Digit. Imaging **29**(2), 254–263 (2016)

Facial Skin Classification Using Convolutional Neural Networks

Jhan S. Alarifi[⊠], Manu Goyal, Adrian K. Davison, Darren Dancey,
Rabia Khan, and Moi Hoon Yap

Manchester Metropolitan University, Manchester M1 5GD, UK
Jhan.s.alarifi@mmu.ac.uk

Abstract. Facial skin assessment is crucial for a number of fields including the make-up industry, dermatology and plastic surgery. This paper addresses skin classification techniques which use conventional machine learning and state-of-the-art Convolutional Neural Networks to classify three types of facial skin patches, namely normal, spots and wrinkles. This study aims to accomplish the pivotal work on the basis of these three classes to provide the collective facial skin quality score. In this work, we collected high quality face images of people from different ethnicities to create a derma dataset. Then, we outlined the skin patches of 100 × 100 resolution in the three pre-decided classes. With extensive parameter tuning, we ran a number of computer vision experiments using both traditional machine learning and deep learning techniques for this 3-class classification. Despite the limited dataset, GoogLeNet outperforms the Support Vector Machine approach with *Accuracy* of 0.899, *F-Measure* of 0.852 and *Matthews Correlation Coefficient* of 0.779. The result shows the potential use of deep learning for non-clinical skin images classification, which will be more promising with a larger dataset.

Keywords: Facial skin · CNNs · Classification · Skin quality assessment

1 Introduction

Data on skin quality properties are often assembled and evaluated by a well-trained expert who allocates noticeable skin samples, both live or from photographs, to a recognised quality grade on a predefined grading scale. However, a machine vision approach to assess skin quality properties is useful in providing an objective analysis [1,2]. This can avoid problems with repeatability and reproducibility, since a professional's experience and knowledge is subjective and can differ amongst graders. This can also potentially result in reduced cost and more effective analysis while providing a consistent assessment of skin quality [3]. There is great importance in providing objectivity to the dermatologist's visual evaluation of skin in order to efficiently develop effective pharmaceutical treatments. Recently, several skin assessment methods have been established; for instance, analysis of the skin appearance around pores on the face [4], evaluation of facial wrinkle improvements over time [5], measuring facial wrinkles using

© Springer International Publishing AG 2017
F. Karray et al. (Eds.): ICIAR 2017, LNCS 10317, pp. 479–485, 2017.
DOI: 10.1007/978-3-319-59876-5_53

quantification methods and automatic detection [6]. Most of these assessments were subjective and revolved around a clinical perspective and a professional's opinion rather than an objective assessment. Further research is required to understand the definition of skin quality based on human perception. In this work, we have experimented with several conventional machine learning (CML) methods and Convolutional Neural Networks (CNNs) with different parameters and settings to classify spots, wrinkles and normal skin patches. This was followed by a comparison of Support Vector Machine (SVM) [7] and GoogLeNet [8] performances.

The rest of the paper is organised as follows: The related work on classification of skin is given in Sect. 2; Sect. 3 elaborates on the dataset and experimental settings; Sect. 4 evaluates the efficiency of the different classifiers that are the best fit for the proposed purpose, for instance, *Sensitivity* and *F-Measure*; in the final section, the conclusion presents the prospects for future work and the limitations of this work.

2   Related Work

Standard machine learning methods have been widely used in several pattern recognition tasks. They have also been used for the detection of skin conditions such as acne [9]. These traditional machine learning methods performed well in many classification tasks. However, they do come with some consequences. For example, ANN (Artificial Neural Network) can be affected by the number of hidden layers, hidden nodes and learning rates. Another disadvantage is that the network has to be extensively trained in order to achieve optimal performance, which is why SVM was chosen for this experiment as a more suitable option. SVM has been used commonly over the last decade [10]. A categorisation of skin texture in early melanoma detection method was implemented using SVM and for skin colour categorisation [11,12]. However, is the Convolutional Neural Networks (CNNs), which is a deep learning framework, has outperformed other method in image classification domain [10].

Recently, with the rapid growth of deep learning algorithms, they become most effective in classification tasks such as in facial recognition and face tracking. The purpose is to understand hierarchical representations of data by using a deep architecture model [13]. Krizhevsky et al. [14] used a deep convolutional neural network to classify high-resolution images in the ImageNet LSVRC-2010. Therefore, including deep learning in the process would provide better performance and more reliable results for the desired output. The network was trained with a total of 1.2 million images and 1000 different classes with error rates of 39.7% for top 1 and 18.9% for top 5. This illustrates the advantage of using this approach. On the other hand, the data used in that approach do not relate to skin attributes. Andre et al. [15] applied successful deep learning approach of skin cancer to dermatologist level by comparing the network performance against 21 dermatologists. Nevertheless, the research focused on clinical use. Therefore, this work will observe the performance of using CNNs in classification of non-clinical skin features such as spots and wrinkles.

3   Methodology

This section will describe in depth the appropriate datasets available, the two sets of experiments and their settings.

3.1   Dataset

Currently, there are limited datasets available for the analysis of facial skin conditions. An available dataset called DermNet consists of a total of 23000 images of various skin diseases. However, this dataset has two limitations. One limitation was that the data collection was not under a controlled environment, which has caused inconsistencies in the images and affected their integrity as well as their accuracy. Another limitation was that the images were not only of facial skin conditions, but also of different diseased body parts, which are unsuitable for this experiment focusing on the classification of common facial skin conditions. To address these limitations, we proposed an ongoing collection of consistent, high-quality images of faces from a wide demographic and from participants who engage in different social habits. These habits can include, but are not limited to, smoking and alcohol consumption. The dataset currently consists of 164 images of participants with a mean age of 48.43 (standard deviation (SD): 21.44, ages between 18 and 92). There are 25 different self-reported ethnicities in the dataset including African, Arabic, Chinese and Malaysian. The ethnic group with most participants is White British with 119 images. The main reported gender was female with a total number of 107 participants; there were also 56 male participants next and 1 transgender participant. To understand how certain habits can affect a person's facial skin properties, participants were asked to complete a questionnaire asking if they consumed alcohol or smoked. Overall, 68 participants never drank alcohol, 88 currently drink and 8 used to drink but had stopped. As for smoking, 85 people never smoked, 21 currently smoke tobacco in some form, 1 smokes electronic cigarettes only, 6 had partaken in smoking a few times in their lives and 51 used to smoke but stopped. The images were taken with a Nikon D5300 at a resolution of 4496 × 3000 to ensure that as much detail as possible on participants' faces was captured.

Firstly, five expressionless images of each participant were captured at different angles to allow for a full view of the face and its profiles. Next, participants were asked to pose with six different facial expressions which were based on Ekmans [16] universal facial expressions: happiness, sadness, surprise, disgust, anger and fear. The replication of these expressions allows for the dataset to include within it some variation in the way each participant's facial skin changes due to natural expressions. Being able to differentiate between actual wrinkles and ridges caused by expression lines would be extremely useful when analysing facial conditions in the future, as would the ability to distinguish between changes caused by natural expressions and deformities caused by other reasons like aging and social habits. The dataset is an on-going project of data collection. Therefore, in the near future, the dataset is likely to increase in size.

Skin patches were also collected. These were of size 100×100 and consisted of three categories: normal skin, skin with spots and skin with wrinkles, as illustrated in (Fig. 1). The spotted skin class has different stages of spots, inflamed and non-inflamed. The wrinkled skin class has two different types of wrinkles: deep and fine wrinkles, which were taken from different parts of the face. The total number of patches is 325.

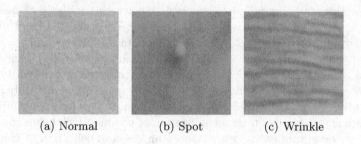

(a) Normal   (b) Spot   (c) Wrinkle

Fig. 1. Sample skin patch from each three classes.

3.2   Traditional Machine Learning

In this section, we used traditional supervised machine learning for the classification of three classes (Normal, Spot, and Wrinkle). Since these three classes of skin have major textural differences amongst them, we investigated popular feature extraction techniques including texture descriptors such as Local Binary Patterns (LBP) [17], and Histogram of Oriented Gradients (HOG) [14]. We did not include mutltifractal as texture descriptor due to it is better in representing face features than skin region [18]. In addition, we also used color descriptors such as Normalized RGB, HSV, and L*u*v features. After the feature extraction from images, we used the machine learning classifier Sequential Minimal Optimization (SMO) to train Support Vector Machine (SVM) for classification task.

3.3   Deep Learning

The Caffe framework [19] was chosen to implement the state-of-the-art CNNs architecture of GoogLeNet. The intention was to provide improvements on the existing model AlexNet when it comes to classifying ImageNet [8]. This model contains 22 layers, compared to AlexNet and CaffeNets [20]. To investigate the best optimisation algorithm for skin patches classification, we tested a number of solvers such as Stochastic Gradient Descent (SGD), Nesterovs Accelerated Gradient (NAG) and Adaptive Gradient (AdaGrad). SGD is one of the most commonly used approaches for large-scale machine learning tasks [21]. AdaGrad presented strong experimental performance on real-world complications, which were tested under different parameters as follows: Each optimizer was tested on the default setting using 30 epochs and 0.01 learning rate. On the second tested set, the number of epochs was increased to 60 and the learning rate was kept the same. For the last tested set, the learning rate was decreased to 0.001 and the number of epochs was kept at 60 [22]. Since the data starts to converge, there is no need to increase the number of epochs.

4 Results and Discussion

In this section, we present the results for various classification experiments on our face dataset of 164 images. These high-resolution images were manually split into the three pre-defined classes of skin patches with 100 × 100 resolution. For this 3-class classification, we divided the dataset of skin patches into 70% for the training set and 30% for the testing set. We adopted the 10-fold cross validation technique to create 10 test cases with a total of 325 images of skin patches. We then divided the number of images equally from the three categories of skin patches. Thus there were 228 skin patches for the training dataset, 97 for testing set. As for evaluation. We chose a number of popular performance measures such as *Sensitivity, False Negative Rate (FNR), F-Measure. Recall, Precision, Matthews Correlation Coefficient (MCC)*, and *Accuracy*. We investigated both traditional machine learning and extensive deep learning techniques to carry out the classification experiments.

Table 1. SVM resuts

Method	Sensitivity	F-Measure	Recall	Precision	MCC	Accuracy
LBP	**0.742**	**0.741**	0.741	**0.740**	**0.597**	**0.815**
LBP and HOG	0.736	0.738	**0.742**	0.742	0.591	0.811
LBP, HOG and Colour Descriptor	0.733	0.735	0.740	0.740	0.586	0.808

Tables 1 and 2 show the classification results achieved with the help of traditional machine learning techniques and deep learning techniques respectively. It is clearly illustrated that with proper parameter tuning, the deep learning techniques outperformed the traditional machine learning ones within the dataset used. Though deep learning techniques usually require a large dataset to train models for classification, this limited dataset was still able to achieve an accuracy rate of 85% and *Sensitivity, Recall, Precision* and *MCC* rates of 0.854, 0.856, 0.856, and 0.779 respectively. This is promising since traditional machine learning techniques are only able to get the best accuracy of approximately 74%. NAG is a first order method and has a distinctive mechanism compared to gradient descent in certain conditions in terms of convergence rate [21]. This predicts the gradient for the next epoch and updates the learning rate for the existing iteration based on the predicted gradient. Therefore, if the gradient is increased for the next set, the learning rate for the present iteration would be higher.

Table 2. GoogLeNet results.

Solver	Epochs	Learning rate	Sensitivity	F-Measure	Recall	Precision	MCC	Accuracy
SGD	30	0.01	0.666	0.661	0.666	0.666	0.472	0.754
	60	0.01	0.833	0.835	0.835	0.835	0.745	0.884
	60	0.001	0.677	0.670	0.671	0.671	0.487	0.761
NAG	30	0.01	0.646	0.639	0.645	0.645	0.439	0.738
	60	**0.01**	**0.854**	**0.852**	**0.856**	**0.856**	**0.779**	**0.899**
	60	0.01	0.729	0.727	0.731	0.732	0.579	0.856
AdaGrad	30	0.01	0.521	0.425	0.375	0.375	0.192	0.624
	60	0.01	0.646	0.650	0.667	0.667	0.449	0.739
	60	0.001	0.708	0.703	0.707	0.707	0.545	0.790

Conversely, if the gradient is low, it would slow down the learning rate. In this experiment, the solver received the highest accuracy with 60 epochs and default learning rate.

5   Conclusion

We presented a dataset that is suitable for facial skin analysis. Our experiments showed the potential for using CNNs in classifying skin attributes. Thus far, GoogLeNet using NAG outperforms the other optimisers used in the experiments. Although the data collection was under a controlled environment and had high-resolution images, it is limited to three categories. Therefore, an expansion of the data is needed.

To improve the classification accuracy for non-clinical skin images, future research involves conducting experiment to understand human perception in classifying skin types and collect more data. We are also interested in comparing the performance of the experts and non-experts [23], in this case, the differences between dermatologists to non-dermatologists.

References

1. Ng, C.-C., Yap, M.H., Costen, N., Li, B.: Automatic wrinkle detection using hybrid Hessian filter. In: Cremers, D., Reid, I., Saito, H., Yang, M.-H. (eds.) ACCV 2014. LNCS, vol. 9005, pp. 609–622. Springer, Cham (2015). doi:10.1007/978-3-319-16811-1_40
2. Ng, C.-C., Yap, M.H., Costen, N., Li, B.: Wrinkle detection using Hessian line tracking. IEEE Access **3**, 1079–1088 (2015)
3. Prats-Montalbán, J.M., Ferrer, A., Bro, R., Hancewicz, T.: Prediction of skin quality properties by different multivariate image analysis methodologies. Chemometr. Intell. Lab. Syst. **96**(1), 6–13 (2009)
4. Mizukoshi, K., Takahashi, K.: Analysis of the skin surface and inner structure around pores on the face. Skin Res. Technol. **20**(1), 23–29 (2014)

5. Luebberding, S., Krueger, N., Kerscher, M.: Comparison of validated assessment scales and 3D digital fringe projection method to assess lifetime development of wrinkles in men. Skin Res. Technol. **20**(1), 30–36 (2014)

6. Cula, G.O., Bargo, P.R., Nkengne, A., Kollias, N.: Assessing facial wrinkles: automatic detection and quantification. Skin Res. Technol. **19**(1), e243–e251 (2013)

7. Wang, L.: Support Vector Machines: Theory and Applications, vol. 177. Springer Science & Business Media, Heidelberg (2005)

8. Szegedy, C., Liu, W., Jia, Y., Sermanet, P., Reed, S., Anguelov, D., Erhan, D., Vanhoucke, V., Rabinovich, A.: Going deeper with convolutions. In: Proceedings of the IEEE Conference on Computer Vision and Pattern Recognition, pp. 1–9 (2015)

9. Liao, H.: A deep learning approach to universal skin disease classification

10. Schmidhuber, J.: Deep learning in neural networks: an overview. Neural Netw. **61**, 85–117 (2015)

11. Yuan, X., Yang, Z., Zouridakis, G., Mullani, N.: SVM-based texture classification and application to early melanoma detection. In: 28th Annual International Conference of the IEEE, Engineering in Medicine and Biology Society, EMBS 2006, pp. 4775–4778. IEEE (2006)

12. Khan, R., Hanbury, A., Stöttinger, J., Bais, A.: Color based skin classification. Pattern Recogn. Lett. **33**(2), 157–163 (2012)

13. Wang, L., Sng, D.: Deep learning algorithms with applications to video analytics for a smart city: a survey. arxiv preprint (2015). arXiv:1512.03131

14. Krizhevsky, A., Sutskever, I., Hinton, G.E.: Imagenet classification with deep convolutional neural networks. In: Advances in neural information processing systems, pp. 1097–1105 (2012)

15. Esteva, A., Kuprel, B., Novoa, R.A., Ko, J., Swetter, S.M., Blau, H.M., Thrun, S.: Dermatologist-level classification of skin cancer with deep neural networks. Nature **542**, 115–118 (2017)

16. Ekman, P.: Facial expressions. Handb. Cogn. Emot. **16**, 301–320 (1999)

17. Guo, Z., Zhang, L., Zhang, D.: A completed modeling of local binary pattern operator for texture classification. IEEE Trans. Image Process. **19**(6), 1657–1663 (2010)

18. Yap, M.H., Ugail, H., Zwiggelaar, R., Rajoub, B., Doherty, V., Appleyard, S., Hurdy, G.: A short review of methods for face detection and multifractal analysis. In: International Conference on CyberWorlds, CW 2009, pp. 231–236. IEEE (2009)

19. Jia, Y., Shelhamer, E., Donahue, J., Karayev, S., Long, J., Girshick, R., Guadarrama, S., Darrell, T.: Caffe: convolutional architecture for fast feature embedding. In: Proceedings of the 22nd ACM International Conference on Multimedia, pp. 675–678. ACM (2014)

20. Liu, D., Wang, Y.: Monza: image classification of vehicle make and model using convolutional neural networks and transfer learning

21. Singh, B., De, S., Zhang, Y., Goldstein, T., Taylor, G.: Layer-specific adaptive learning rates for deep networks. In: 2015 IEEE 14th International Conference on Machine Learning and Applications (ICMLA), pp. 364–368. IEEE (2015)

22. Duchi, J., Hazan, E., Singer, Y.: Adaptive subgradient methods for online learning and stochastic optimization. J. Mach. Learn. Res. **12**, 2121–2159 (2011)

23. Yap, M.H., Edirisinghe, E., Bez, H.: Processed images in human perception: a case study in ultrasound breast imaging. Eur. J. Radiol. **73**(3), 682–687 (2010)

Automatic Detection of Globules, Streaks and Pigment Network Based on Texture and Color Analysis in Dermoscopic Images

Amaya Jiménez[1], Carmen Serrano[2(✉)], and Begoña Acha[2]

[1] Dpto. Teoría de la Señal y Comunicaciones,
Universidad Carlos III Madrid, Getafe, Spain
ajimenez@tsc.uc3m.es
[2] Dpto. Teoría de la Señal y Comunicaciones, Universidad de Sevilla,
Seville, Spain
{cserrano,bacha}@us.es

Abstract. Melanoma diagnosis in early stages is a difficult task, which requires highly qualified and trained staff. Therefore, a computer aided diagnosis tool to assist non-specialized physicians in the assessment of pigmented lesions would be desirable. In this paper a method to detect streaks, globules and pigment network, which are very important features to evaluate the malignancy of a lesion, is presented. The algorithm calculates the texton histograms of color and texture features extracted from a filter bank, that feed a Support Vector Machine. The method has been tested with 176 images attaining an accuracy of 80%, outperfoming the benchmark techniques used as comparison.

Keywords: Pigmented lesion · Streak · Globules · Pigment network · Texture · Color

1 Introduction

Malignant melanoma is a form of skin cancer with an increasing incidence and a potential for metastases, making it one of the most serious skin cancers [18]. The early diagnosis of melanoma is a key tool that improves the prognosis and the survival rate [13]. For these reasons, the noninvasive techniques, as dermoscopy, are very popular tools to assist physicians in the diagnosis.

Dermoscopy is an epiluminescence microscopy light that allows a better visualization of the skin structures [3]. Trained dermatologists can use this imaging tool to examine the pigmented skin lesions down to the dermoepidermal junction. Numerous studies have proven that the accuracy of a trained dermatologist in melanome diagnosis may reach 69% in early stages whereas this accuracy is decreased to 12% for non-specialist physicians. Thus, the main objective of this

This work have been funded by Junta de Andalucía, Spain, Project no. P11-TIC-7727.

© Springer International Publishing AG 2017
F. Karray et al. (Eds.): ICIAR 2017, LNCS 10317, pp. 486–493, 2017.
DOI: 10.1007/978-3-319-59876-5_54

paper is to assist dermatologists and, mainly, physicians at primary health care centres in the task of analysing the malignancy in pigmented lesions.

Dermatologists employ different methods to diagnose pigmented lesions from dermoscopic images: pattern analysis [16], ABCD rule [23], 7-point checklist [4], or Menzies method [14]. The first one is the preferred method by the dermatologists [4]; however, the primary care practitioners usually apply one of the three simplified methods (ABCD rule, 7-point checklist, or Menzies method) to detect suspicious lesions. Both pattern analysis and the simplified methods require the detection of globules, streaks and pigment network to score a pigmented lesion in order to evaluate its suspicion of being melanoma.

Thus, the aim of this work is to distinguish among streaks, globules, pigment network and homogeneous pattern. This method could help the physicians to identify some very important features which can evaluate the malignancy of the lesion.

The automatic detection of irregularly distributed streaks is a difficult problem and there are only few works trying to solve this problem (e.g., [8,12,15,19]). Sadegui et al. [20] present a method for recognizing radial streaming patterns based on a clinically inspired feature set. Mirzaalian et al. [15] propose a machine learning approach for identifying streaks and distinguish among absent, regular, and irregular streaks. Delibasis et al. [8] present an algorithm that detects streaks based on local image curvature information obtained by the Hessian matrix. Kropidlowski et al. [12] detect not only streaks but also pigment network. They use histogram-based features and an artificial neural network.

The automatic detection of irregularly distributed globules has been more widely faced (e.g., [7,11,17,26]), but only few methods have addressed the detection of both features in the same work, just those related with the implementation of the complete 7-point checklist algorithm (e.g., [9]).

Different authors have devoted their research to the detection of pigment network. Recently, García-Arroyo and García-Zapirain [10] applied machine learning to the detection of pigment network. Anantha et al. [1] applied Laws mask and neighborhood gray-level dependence matrix to detect this feature. Shrestha et al. [22] extracted ten texture features to detect atypical pigment network in dermoscopic images. Sadeghi et al. [21] detect and analyse cyclic graphs in the pigmented lesion in order to find possible pigment networks. Barata et al. [5] employed directional filters for the detection of pigment network.

The main novelty of the paper is that the problem of detecting streaks, globules and pigment networks in pigmented lesions is solved in a unified manner. To this aim a new methodology based on color and texture analysis is proposed. A dictionary of textons with the output of a filter bank and color features is constructed to characterize the above mentioned patterns.

2   Methodology

Following the approach in [25], textures are modeled with the responses obtained with a filter bank. However, in this paper color information is added and a

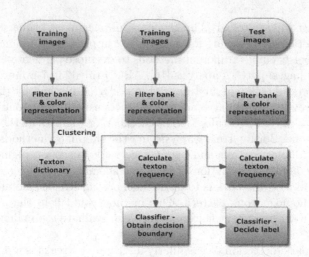

Fig. 1. Flowchart of the proposed method based on filter bank.

different filter bank than in [25] is implemented. In such a way, the feature vector for each pixel consists in the filter responses and the color representation.

Next, the most representative vectors of each pattern (streaks, globules, pigment network and other (considered as homogeneous)) are calculated by a clustering algorithm, obtaining the so-called textons. All these textons, obtained from images with different patterns, are grouped in a texton dictionary that represents both texture and color information. Finally, to characterize the texture and color of a new image, the distribution of the textons is calculated as their normalized histogram.

In a training stage, the filter responses and color representations are obtained for every training image, conforming a set of vectors that are labeled with the textons (from the dictionary) more similar to them. Thus, the frequency of occurrence of every texton can be obtained from the normalized histogram. In such a way, the texton frequencies determine a texture and color model associated to each image. All the models calculated from the training images are used in a SVM classifier to obtain the decision boundary.

In the test stage, the image to be classified goes through the filter bank; also, the color representation is calculated. As in the previous phase, the model, with the frequency of occurrence of every texton, is derived. Finally, the obtained model is compared to the decision boundary and labeled accordingly.

Figure 1 shows a block diagram with the main steps of this algorithm. In the next subsections, every stage of this approach will be explained in detail.

2.1   Filter Bank

In first place, the image to be analyzed is converted to $CIEL^*a^*b$ color space. The L component, representing the lightness of the pixels, is the input to the filter bank.

The filter bank used in our proposal is motivated by the maximum response (MR) set presented in [25], but with some slight differences. In our method, we use 48 anisotropic filters (first and second derivative Gaussian filters at 3 scales and 8 orientations, which are an edge and bar detector, respectively), along with 5 Laplacian of Gaussian (LoG) isotropic filters with $\sigma = 1, \ldots, 5$ pixels, making a total of 53 filters, all of them normalized to get unit L_1 norm. We reduce the obtained filter outputs by maximizing the responses along the 8 orientations. In this way, we get a total of 11 filter responses with rotation invariance.

The reasons to use this set of filters are: first, the anisotropic filters are oriented to identify streaks, as they detect sharp changes with linear shape in the brightness of the image. As the detection of streaks is a difficult problem, we want to allow enough scales and orientations to properly locate streaks in any position of the pigmented lesion. Second, the globules are defined as darker spots than the surrounding lesion, so the LoG filters are the most adequate to locate them. We allow a set of σ values selected as the most suitable for the size of the considered block images. Moreover, these values are able to capture the variability in the sizes of the globules. In addition, as holes of the pigment network show lighter intensities than pigmented lines, with the LoG filters we avoid to highlight them and wrongly label them as globules.

To include color information, hue and saturation of the *HSV* color space are used. To avoid the hue discontinuity for the red color the following simple transformation is applied: $H' = cos(H)$.

Therefore, every pixel in the image is represented by a vector containing the 11 filter responses, representing the texture, and the 2 color values, H' and S. Each component of the vector is normalized to get zero mean and unit variance. In the following subsection we explain how to obtain the texton dictionary from these 13-dimensional vectors that characterize the image.

2.2 Building the Texton Dictionary

For each considered pattern, a set of randomly selected images is analyzed as it was previously explained in the previous subsection. All the obtained 13-dimensional vectors, with texture and color information, are collected from this set of images. Then, we apply a clustering algorithm to obtain the most representative vectors, referred as textons. This process is repeated with the four patterns, obtaining a set of textons for each one. All of them form the texton dictionary.

We calculate 10 textons for each pattern, obtaining a dictionary with a total of 40 textons, i.e., it contains the 10 most representative 13-dimensional vectors of the streaks, globules, homogeneous, and pigment network patterns. The texton dictionary is the basis to model statistically the textures and colors in the images.

Every image (both in training and classification phase) is represented as the texton frequency. To this end, the filter responses and color representations for each image are calculated, and the obtained 13-dimensional vectors are labeled with the most similar textons from the dictionary (as similarity measure we use the Euclidean distance). From these labeled vectors, the texton histogram is

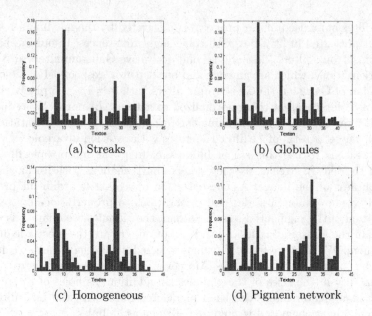

(a) Streaks

(b) Globules

(c) Homogeneous

(d) Pigment network

Fig. 2. Texton frequencies obtained with an image of each pattern. In the x-axis, the 40 textons of the dictionary; in the y-axis, the frequency.

calculated and normalized, obtaining the texton frequencies. In this way, each image is represented by a 40-dimensional vector, where each bin represents the frequency of a texton from the dictionary. Figure 2 shows an example of the texton frequencies obtained for an image of each pattern; as it can be seen, these frequencies are different for each one, helping to correctly label the image.

To carry out the classification, we use the SVM classifier. In the training phase, the obtained models with the texton frequencies are used to build the decision boundaries, which are applied in the test phase to classify each new image. The classification method will be explained in detail in Sect. 3.2.

3   Experimental Results

3.1   Image Database

The image database used in the experimental evaluation to test our proposal, consists of 44 images of each considered class (i.e., streaks, globules, homogeneous, and pigment network), making a total of 176 images, with size of 81×81 pixels. All images were extracted from the Interactive Atlas of Dermoscopy, published by Edra Medical Publishing New Media [2].

The ground truth has been taken according to the diagnosis described in the Interactive Atlas of Dermoscopy and the recommendation of an expert physician.

3.2 Classification Method Based on SVM

SVMs [6] is used as classifier whose input are the frequency of the 40 texton in the dictionary. To evaluate the proposed methods a *k-fold* cross validation technique is used. Specifically, we consider a *4-fold* strategy; in such a way, a quarter of the data (44 images, 11 per pattern) are used in the test phase and the remaining data (132 images, 33 per pattern) are for the training.

3.3 Evaluation of the Proposed Method

To evaluate the performance of our proposal, the accuracy in the classification is measured. We decide to compare our method with another approximation that has proven to be very useful in the texture classification problem, the proposal in [24]. To make a fair comparison, we also include the color information in the method presented in [24], following the same reasoning described in Sect. 2.1. Moreover, we also compare the performance of different classifiers. Our method is developed with a SVM classifier but, as the k-nearest neighbors (KNN) is usually applied when texton histograms characterize the images, we include it in our evaluation.

In Table 1 the classification accuracy for all the evaluated methods is shown. **[24]+Col+KNN** and **[24]+Col+SMV** refer to the method presented in [24] along with the color information and the two considered classifiers. **FB+KNN** and **FB+SVM** refer to the filter bank-based method (Sect. 2) with the two classifiers. As it can be seen, the proposed method clearly outperform the widely used approach [24], improving the accuracy in the classification more than a 20%. Also, to apply a SVM classifier implies an improvement of nearly 8%. In addition, Table 1 shows the classification accuracy specified by each pattern and it shows that this accuracy improvement is attained when detecting each of the four different patterns.

Table 1. Classification results, specified by each considered pattern, for the proposed methods compared with [24] (%).

Method	Streaks	Globules	Homogeneous	Pigmented network	Average
[24]+Col+KNN	66.0	47.7	52.3	52.3	54.4
[24]+Col+SVM	75.0	45.5	68.2	63.6	63.0
FB+KNN	72.7	63.6	77.3	77.3	72.6
FB+SVM	**79.5**	**70.5**	**90.9**	**79.5**	**80.2**

4 Discussion and Conclusions

In this paper a new method to detect streaks, globules and pigment network in dermoscopic images is presented. The detection of these three features is required to score 4 out of 9 points in Menzies scoring method and to score 3 out of 7 points in the 7-point checklist method.

A new methodology based on color and texture analysis is proposed. Similarly to [25], a dictionary of textons with the output of a filter bank is constructed to characterize texture. However, color information is included and a different filter bank, that takes into account the different scales that textures in pigmented lesion can present, is proposed. In addition, the texton histograms are classified with a SVM, instead of with the k-nearest neighbor method.

The proposed method have been compared to [24], also based on textons, when the final classification is performed with k-nearest neighbor, as proposed originally in [24], and with an SVM, as proposed in this paper. As shown in Sect. 3.3, the methodology based on the proposed filter bank outperforms [24] and the classification based on SVM outperforms the k-nearest neighbor proposed in [24].

References

1. Anantha, M., Moss, R.H., Stoecker, W.V.: Detection of pigment network in dermoscopy images using texture analysis. Comput. Med. Imaging Graph. **28**, 225–234 (2004)
2. Argenziano, G., Soyer, H., et al.: Interactive atlas of dermoscopy. In: EDRA-Medical Publishing New Media (2000)
3. Argenziano, G., Soyer, H.P.: Dermoscopy of pigmented skin lesions: a valuable tool or early diagnosis of melanoma. Lancet Oncol. **2**(7), 443–449 (2016)
4. Argenziano, G., Soyer, H.P., Chimenti, S., et al.: Dermoscopy of pigmented skin lesions: Results of a consensus meeting via the internet. J. Am. Acad. Dermatol. **48**(5), 679–693 (2003)
5. Barata, C., Marques, J.S., Rozeira, J.: A system for the detection of pigment network in dermoscopy images using directional filters. IEEE Trans. Biomed. Eng. **59**(10), 2744–2754 (2012)
6. Chang, C.C., Lin, C.J.: LIBSVM: a library for support vector machines. ACM Trans. Intell. Syst. Technol. **2**, 1–27 (2011)
7. De Vita, V., Di Leo, G., Fabbrocini, G., Liguori, C., Paolillo, A., Sommella, P.: Statistical techniques applied to the automatic diagnosis of dermoscopic images. Acta Imeko **1**(1), 7–18 (2012)
8. Delibasis, K., Kottari, K., Maglogiannis, I.: Automated detection of streaks in dermoscopy images. In: Chbeir, R., Manolopoulos, Y., Maglogiannis, I., Alhajj, R. (eds.) AIAI 2015. IAICT, vol. 458, pp. 45–60. Springer, Cham (2015). doi:10.1007/978-3-319-23868-5_4
9. Di Leo, G., Paolillo, A., Sommella, P., Fabbrocini, G., Rescigno, O.: A software tool for the diagnosis of melanomas. In: 2010 IEEE Instrumentation and Measurement Technology Conference (I2MTC), pp. 886–891, May 2010
10. Garcia-Arroyo, J.L., Garcia Zapirain, B.: Detection of pigment network in dermoscopy images using supervised machine learning and structural analysis. Comput. Biol. Med. **44**, 144–157 (2014)
11. Jaworek-Korjakowska, J., Tadeusiewicz, R.: Assessment of dots and globules in dermoscopic color images as one of the 7-point check list criteria. In: 2013 20th IEEE International Conference on Image Processing (ICIP), pp. 1456–1460, September 2013

12. Kropidlowski, K., Kociolek, M., Strzelecki, M., Czubinski, D.: Nevus atypical pigment network distinction and irregular streaks detection in skin lesions images. In: 2015 Signal Processing: Algorithms, Architectures, Arrangements, and Applications (SPA), pp. 66–70, September 2015

13. Leiter, U., Buettner, P.G., Eigentler, T.K., Garbe, C.: Prognostic factors of thin cutaneous melanoma: an analysis of the central malignant melanoma registry of the German Dermatological Society. J. Clin. Oncol. **22**(18), 3660–3667 (2004)

14. Menzies, S.W., Ingvar, C., Crotty, K.A., McCarthy, W.H.: Frequency and morphologic characteristics of invasive melanomas lacking specific surface microscopic features. Arch. Dermatol. **132**(10), 1178–1182 (1996)

15. Mirzaalian, H., Lee, T., Hamarneh, G.: Streak-detection in dermoscopic color images using localized radial flux of principal intensity curvature (Chap. 7). In: Dermoscopy Image Analysis, pp. 211–229 (2015)

16. Pehamberger, H., Steiner, A., Wolff, K.: In vivo epiluminescence microscopy of pigmented skin lesions. i. pattern analysis of pigmented skin lesions. J. Am. Acad. Dermatol. **17**(4), 571–583 (1987)

17. Pellacani, G., Grana, C., Cucchiara, R., Seidenari, S.: Automated extraction and description of dark areas in surface microscopy melanocytic lesion images. Dermatology **208**, 21–26 (2004). http://www.karger.com/DOI/10.1159/000075041

18. Rigel, D.S., Carucci, J.A.: Malignant melanoma: prevention, early detection, and treatment in the 21st century. CA Cancer J. Clin. **50**(4), 215–236 (2000)

19. Sadeghi, M., Lee, T.K., McLean, D., Lui, H., Atkins, M.S.: Detection and analysis of irregular streaks in dermoscopic images of skin lesions. IEEE Trans. Med. Imaging **32**(5), 849–861 (2013)

20. Sadeghi, M., Lee, T., Mclean, D., Lui, H., Atkins, M.: Detection and analysis of irregular streaks in dermoscopic images of skin lesions. IEEE Trans. Med. Imaging **32**(5), 849–861 (2013)

21. Sadeghi, M., Razmara, M., Lee, T.K., Atkins, M.: A novel method for detection of pigment network in dermoscopic images using graphs. Comput. Med. Imaging Graph. **35**(2), 137–143 (2011). Advances in Skin Cancer Image Analysis

22. Shrestha, B., Bishop, J., Kam, K., Chen, X., Moss, R.H., Stoecker, W.V., Umbaugh, S., Stanley, R.J., Celebi, M.E., Marghoob, A.A., Argenziano, G., Soyer, H.P.: Detection of atypical texture features in early malignant melanoma. Skin Res. Technol. **16**(1), 60–65 (2010)

23. Stolz, W., Riemann, A., Cognetta, A.B., et al.: ABCD rule of dermatoscopy: a new practical method for early recognition of malignant melanoma. Eur. J. Dermatol. **4**(7), 521–527 (1994)

24. Varma, M., Zisserman, A.: A statistical approach to material classification using image patch exemplars. IEEE Trans. Pattern Anal. Mach. Intell. **31**(11), 2032–2047 (2009)

25. Varma, M., Zisserman, A.: A statistical approach to texture classification from single images. Int. J. Comput. Vis. **62**(1–2), 61–81 (2005)

26. Yoshino, S., Tanaka, T., Tanaka, M., Oka, H.: Application of morphology for detection of dots in tumor. In: SICE 2004 Annual Conference, vol. 1, pp. 591–594, August 2004

Image Analysis in Ophthalmology

Learning to Deblur Adaptive Optics Retinal Images

Anfisa Lazareva[1][(⊠)], Muhammad Asad[2], and Greg Slabaugh[2]

[1] Department of Electrical and Electronic Engineering,
City, University of London, London, UK
anfisa.lazareva.1@city.ac.uk
[2] Department of Computer Science, City, University of London, London, UK

Abstract. In this paper we propose a blind deconvolution approach for reconstruction of Adaptive Optics (AO) high-resolution retinal images. The framework employs Random Forest to learn the mapping of retinal images onto the space of blur kernels expressed in terms of Zernike coefficients. A specially designed feature extraction technique allows inference of blur kernels for retinal images of various quality, taken at different locations of the retina. This model is validated on synthetically generated images as well as real AO high-resolution retinal images. The obtained results on the synthetic data showed an average root-mean-square error of 0.0051 for the predicted blur kernels and 0.0464 for the reconstructed images, compared to the ground truth (GT). The assessment of the reconstructed AO retinal images demonstrated that the contrast, sharpness and visual quality of the images have been significantly improved.

Keywords: Adaptive optics imaging · Deconvolution · Image restoration · Regression · Random forest

1 Introduction

Direct observation of the retina suffers from various optical aberrations of the eye. A wavefront sensor in an AO instrument performs calculation and compensation of high-order ocular aberrations thus providing a high level of resolution when imaging the retina. Despite this, due to hardware limitations of the wavefront corrector, this correction is not perfect. Therefore, the acquired retinal images are still corrupted by residual aberrations resulting in blur. Enhancement of the retinal images facilitates better distinction of photoreceptor cells and thereby assists clinicians in the examination of living retina, allowing for more accurate quantitative analysis of photoreceptor cell packing density. Additional improvements in the contrast and resolution of retinal images can be obtained *a posteriori*, by using an image restoration technique such as image deconvolution. In AO, the degradation function of the system can be estimated partially from wavefront sensing (WFS), assuming that the measurements of the deformable mirror are accurate [1]. Based on the information about the residual errors after AO correction, the system's point spread function (PSF) can be reconstructed and employed in the deconvolution process [2]. However, the WFS is not a reliable source of data due to multiple types of noise in the AO imaging system as well as the unsynchronized process

© Springer International Publishing AG 2017
F. Karray et al. (Eds.): ICIAR 2017, LNCS 10317, pp. 497–506, 2017.
DOI: 10.1007/978-3-319-59876-5_55

of image capturing and wavefront calculation. As a result, the obtained WFS data does not always correspond to the acquired set of frames. Therefore, ordinary deconvolution from WFS is not a suitable method for post-processing of AO images [1].

When the PSF is not available, 'blind' deconvolution, a more generalized technique, can be applied to the images. This type of image deconvolution allows for recovery of the object and the PSF distributions simultaneously from a series of measurements. This is made by the use of physical constraints about the target and knowledge of the imaging system [3]. A few blind deconvolution methods have been reported in the literature for restoring AO high-resolution retinal images [4–6]. However, conventional blind deconvolution has a drawback of getting trapped in local minima that makes it hard to find a unique solution, especially when there is only a single blurred image to be restored [1].

In this paper, we propose an image deconvolution method based on a multi-variate Random Forest regressor. Random Forests are fast and efficient learners, particularly suitable for high dimensional problems. Randomness induced into the trees during the training makes them highly resistant to overfitting. Although a number of learning-based techniques have been proposed in the literature for the purposes of image deconvolution [7, 8], these methods rely on generalized models and therefore their accuracy is limited to specific types of blur. In addition, in most of the reported methods the achieved resolution of the recovered blur kernel is often found to be restricted by size. In our work, the proposed framework is specifically designed for deconvolution of retinal images acquired with a commercially available flood-illuminated AO instrument (rtx1, Imagine Eyes, Orsay, France). The blur kernel is modeled by the physics/optics of the AO system and thus constrained as a member in a class of parametric functions. This allows to significantly reduce the space of valid PSFs. A convolution kernel is estimated through non-linear regression of retinal images onto the space of PSFs expressed in terms of Zernike coefficients. By performing regression on a compact representation of the PSF, we are able to infer a convolution blur kernel for AO retinal images without compromising the resolution of the PSF. The feature extraction technique is specially developed to allow for better generalization on a large set of retinal images. To our knowledge, learning-based methods have not been previously used for AO retinal images.

2   Method

2.1   Image Model

Many tasks in image processing can be formulated as a regression problem where we learn a mapping function $f_w : X \rightarrow Y$ from the input space X to the output space Y, which is parametrized by a learned parameter w [7]. The task of image deconvolution requires estimation of original image X from its degraded observations Y obtained as a result of convolution with the system's PSF w and an additive noise n. In that sense, the imaging process can be expressed as follows:

$$Y = X * \mathrm{w} + n. \tag{1}$$

Given that, the GT is provided in the form of training data T composed of input and output image pairs (X_i, Y_i^{GT}), learning the optimal convolution kernel w^* can be generally formulated using the principle of empirical risk minimization:

$$w^* = \underset{w}{\operatorname{argmin}} \sum_{i=1}^{N} \left\| Y_i^{GT} - X_i * w \right\|^2, \qquad (2)$$

where N is a total number of samples.

We propose solving this problem with Random Forest, where optimal convolution kernel w^* is found through non-linear regression of blurred images $\{Y_i^{GT}\}$ onto the space of system's PSFs $\{w_i\}$. In this work, the PSFs of AO system $\{w_i\}$ are defined by the vectors of Zernike coefficients $\{a_i\}$ and the blurred images are represented by Histograms of oriented Gradients (HoG) $\{H_i\}$ [9]. Then, with the use of the Random Forest we learn a mapping function $f_a : \{H_i\} \rightarrow \{a_i\}$.

The proposed deconvolution approach was developed as one of the stages in the image processing framework for enhancement of high-resolution retinal images [10]. In this framework, the system's noise is filtered prior to the image deconvolution stage. Therefore, here, we neglect the noise term n and assume that images are corrupted only by convolution blur kernel w.

2.2   Multi-variate Random Forest

The forest is a collection of T decision trees which are trained independently using a training dataset $U = \{H_i, a_i\}$. Each tree consists of non-terminal split nodes and terminal leaf nodes. The split nodes are responsible for performing a binary split on the input dataset, whereas the leaf nodes store the probability distribution of data arriving at their terminal position. At the jth split node, splitting function $f(U_j, \Theta)$ learns the optimal parameter $\Theta = (k, \tau)$, where k is the index of the test image feature and τ is its corresponding learned threshold defining the split. The optimal parameter Θ, that maximizes the information gain $Q(U_j, \Theta)$, is selected from a pool of randomly generated parameters. Training continues until a maximum depth D is reached or the data arriving at the jth node contains a minimum number of samples required for creating a leaf node.

In this work, Random Forest consisted of $T = 80$ trees with a maximum depth $D = 15$. These parameters were established by performing greedy optimization, where the root-mean-square error (RMSE) between the training $\{a_i\}$ and predicted PSFs a^* was evaluated.

In order to find an optimal split parameter Θ at j^{th} split node, an objective function is defined as the information gain $Q(U_j, \Theta)$:

$$Q(U_j, \Theta) = E(U_j) - \sum_{q \in \{left, right\}} \frac{|U_j^q|}{|U_j|} E\left(U_j^q\right), \qquad (3)$$

where $E(U_j) = \log\left(\frac{1}{|U_j|}\sum_{i=1}^{N}\left(a_i^j - \overline{a^j}\right)^2\right)$ is the multi-variate differential entropy for the target vector of Zernike coefficients a_i^j with mean $\overline{a^j}$ of data U_j and q defines the data for child nodes.

2.3   Generation of Training Data

Since no GT data is available for AO retinal images, the Random Forest was trained on synthetically generated retinal images and blur kernels replicating the PSFs of the flood-illuminated AO system. For a training dataset U, a set of convolutional blur kernels $\{w_i\}$ was generated so as to simulate optical aberrations of the eye. Low-order aberrations such as astigmatism, defocus and prism are usually well compensated with the Badal system embedded in the AO instrument [11]. These aberrations are represented by the first six Zernike polynomials (as defined by Noll [12]). Thus, the pupil phase of the PSF was expanded on Zernike polynomials of higher order aberrations, retaining the first 15 Zernike terms. An additional term corresponding to defocus was added so as to account for residual blur coming from different layers of the retina [6]. Mathematically this model of PSF is presented as:

$$w(y, \varphi) = \left|\Im\left\{P(u)\exp\left(-j\left(\frac{2\pi}{\lambda l}\right)y\,u\right)\exp(j(\varphi(u) + \varphi_d(u)))\right\}\right|^2, \qquad (4)$$

where P is the pupil function, λ is the central wavelength of imaging beam; l is the focal length of the optical system; y defines the coordinates of two-dimensional focal plane, $\varphi(x, y) = \sum_{m=7}^{15} a_m Z_m(x, y)$ is the wavefront phase error, $\varphi_d(x, y) = a_4 Z_4(x, y)$ is the defocus phase, a_m and Z_m are Zernike coefficients and Zernike polynomial.

The values of the Zernike coefficients were sampled from a statistical model of the wavefront aberrations in healthy eyes reported in [13]. The range of Zernike coefficients was quantized between $[mean - \frac{std}{2}, mean + \frac{std}{2}]$ with the step size of 0.01 μm. In order to account for partial compensation of aberrations with the AO system, these values were scaled by 0.42 [14]. Imaging wavelength and focal plane sampling were set according to the specifications of AO instrument, rtx1 (750 nm and 1.6 μm). A pupil diameter was assigned to 6 mm and axial length to 24 mm. Since all parameters defining the PSF were fixed to constant values, the PSF of the AO system can be represented by a vector of Zernike coefficients only, i.e. $\{a_i\}$.

In order to generate a set of synthetic blurred retinal images $\{Y_i\}$, we firstly created an ideal retinal image X, using the algorithm described in [15]. The obtained synthetic image was convolved with each PSF from the set $\{w_i\}$. HoG feature vectors were extracted around the strongest corners in small regions of size 10×10 pixels centered at photoreceptor cell locations. As the cone coordinates are known through the process of synthetic image generation, locating windows with photoreceptor cells is straightforward. In addition to distinct variations caused by different types of optical blur, the retinal mosaic has a unique pattern, varying across the retina as well as human eyes. By extracting features from small windows containing a single cone, we limit the

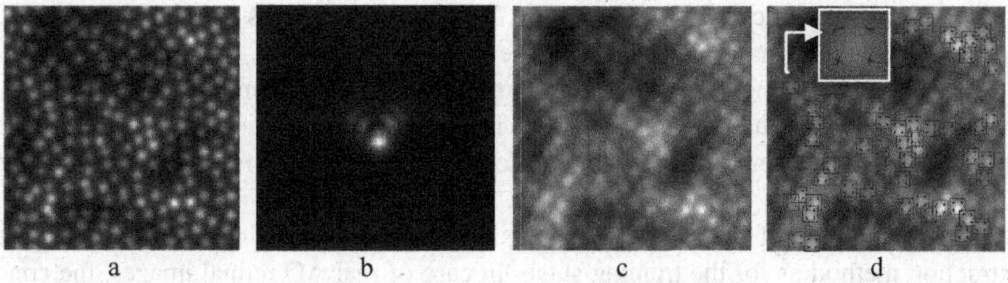

Fig. 1. Results of synthetic data generation and feature extraction, showing (a) ideal retinal image, (b) generated PSF, (c) synthetic blurred retinal image, (d) HoG features extracted from cone windows

nature of variations down to the corruption of cone shape due to blur, thus assuring the inference of PSFs for any retinal image, taken at different locations of the retina. Moreover, due light scattering and the angle of incident light, photoreceptor cells appear with different intensity levels in the acquired image. To eliminate these variations, for each blurred image HoG features were extracted from 50 windows containing the brightest cones only. The resulting vector $\{\boldsymbol{H}_i\}$ of size 36×1 was obtained by averaging HoG features across 50 windows. Figure 1 shows the example of synthetic data generation and image feature extraction.

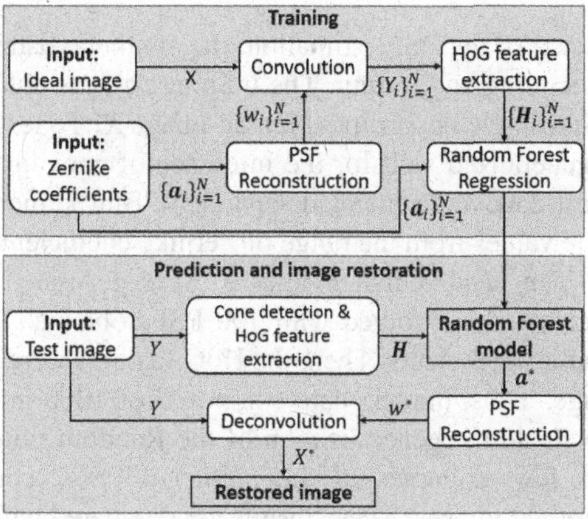

Fig. 2. Flowchart showing the process of training, prediction and image restoration

2.4 Training, Prediction and Image Restoration

Training is done offline using the training dataset: HoG feature vectors $\{\boldsymbol{H}_i\}$, extracted from a set of blurred images $\{Y_i\}$, and Zernike coefficients $\{\boldsymbol{a}_i\}$, where each vector represents a combination of different types of optical aberrations in the eye. Based on the range of values for each of the Zernike coefficients, we obtained 134,400 combinations of optical PSFs. Thus, the Random Forest was trained on 134,400 synthetic blurred retinal images and blur kernels.

During the training stage, Random Forest learns the optimal split function by maximizing the information gain as defined in Eq. 3. The branches in the tree terminate with leaf nodes that contain the vectors of Zernike coefficients arriving as a result of the splitting process.

The prediction of Zernike coefficients is done online using the same feature extraction method as for the training stage. In case of real AO retinal images, the cone coordinates are found automatically, using the algorithm described in [16]. During the prediction, a given image feature vector propagates down the branches of each tree where a leaf node gives a posterior probability and the corresponding data. Kernel density estimation was used on the data aggregated from all the leaf nodes to find an optimal vector of Zernike coefficients a^* with the highest posterior probability.

After prediction, the estimated optimal vector of Zernike coefficients a^* was used to reconstruct the corresponding PSF w^*, using Eq. 4. Then, the obtained PSF was employed in the restoration of the blurred retinal image with Lucy-Richardson deconvolution algorithm [17, 18]. Figure 2 illustrates the process of training, prediction and image restoration.

3   Results and Evaluation

3.1   Experimental Validation Using Synthetic Data

To evaluate the accuracy of the PSF estimation, the synthetic dataset was divided into two subsets used for training and testing. The training set was obtained by convolving blur kernels $\{Y_i\}$ with a single uncorrupted retinal image X. To test whether the trained Random Forest has generalized well for the inference of convolution blur kernels for any retinal images, test data was generated separately. Blur kernels were produced by taking the intermediate values from the range of Zernike coefficients and reconstructing corresponding PSFs. Ten ideal retinal images X were generated so as to reproduce different retinal mosaics and convolved with 100 PSFs obtained from randomly generated vectors of Zernike coefficients. Then, the HoG features were extracted and stored for the prediction stage. Thus, the test data composed of 1000 image pairs represents the unseen data to evaluate the generalization of the Random Forest.

Figure 3 presents few examples of the predicted PSFs, corresponding blurred images and restored retinal images. These results are compared with GT data: the PSFs obtained from training vectors of Zernike coefficients, corresponding blurred images and uncorrupted synthetic retinal images X.

To evaluate the performance of the proposed method, quantitative assessment was performed on 1000 synthetically blurred retinal images. All images were normalized from the original range of intensities to [0, 1]. The mean RMSE between the predicted convolutional blur kernels and the GT PSFs was found to be 0.0051 across 100 samples of each synthetic image X. The mean RMSE between the restored retinal images and original synthetic images across 100 samples of each test data was 0.0464. This represents 0.5% and 4.6% of generalization error correspondingly.

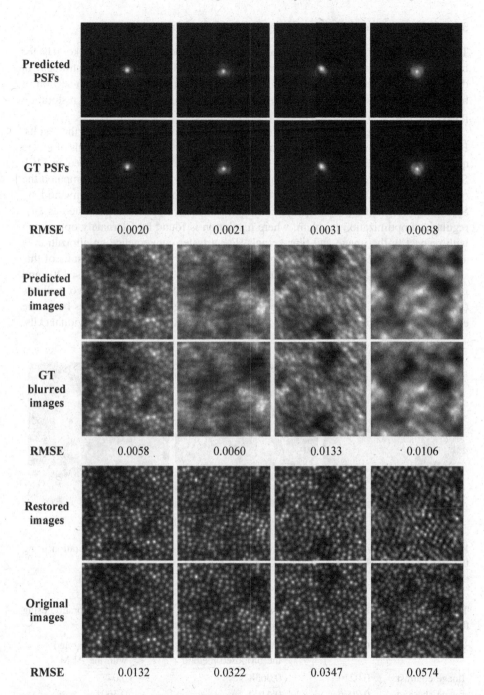

Fig. 3. Deconvolution of four representative synthetic images imitating retina at different eccentricities, showing the predicted PSFs, corresponding blurred images, images restored with the estimated PSFs and the GT data

3.2   Experimental Validation Using Real AO Retinal Images

The trained Random Forest was used for predicting convolutional blur kernels for high-resolution retinal images, acquired with the flood-illuminated AO instrument in different subjects and at various eccentricities. Figure 4-(a) and (b) shows a section of high-resolution retinal image before and after applying the proposed restoration process.

In the case of real retinal images, there is no GT data for evaluation of the results. For this reason, we assessed the performance of the proposed method in terms of image quality metrics in 25 high-resolution retinal images processed as described in [10]. Since the WFS data is not available in a commercial AO instrument, we compared the obtained results with the images restored using a blind deconvolution method by Sroubek and Milanfar [19]. In this method, blind deconvolution is represented as a l_1-regularized optimization problem, where a solution is found by alternately optimizing with respect to the image and blur kernels. For a faster convergence, minimization is addressed with an augmented Lagrangian method (ALM). Based on the results of this study, the proposed method outperforms the ALM in terms of the contrast [20] and image sharpness [21] (Table 1). From retinal images processed with the two methods (Fig. 4-(b) and (c)), it becomes apparent that the proposed method preserves better the edges of photoreceptor cells as well as achieves better differentiation of individual cells.

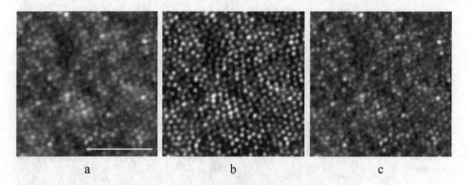

a   b   c

Fig. 4. Representative AO high-resolution retinal image before (a) and after restoration using the proposed approach (b) and ALM (c). The scale bar is 81 μm

Table 1. Quality assessment of the original AO retinal images before and after restoration using the proposed approach and ALM averaged for 25 images from different subjects.

	Original image	Images restored with the proposed method	Images restored with the ALM
Image contrast	0.0216	0.0638	0.0479
Image sharpness	0.2076	0.4100	0.3601

4 Discussion and Conclusion

In this paper, we demonstrated that Random Forest regression can be used for estimating PSF of an AO imaging system. The Random Forest was trained on a large dataset by learning the mapping of HoG features, extracted from synthetically blurred retinal images onto the space of corresponding PSFs represented by Zernike coefficients. A mathematical model for the PSF was parameterized through the pupil phase, thereby significantly reducing the number of unknowns in the regression target. By extracting the information about the object shape from a single cone with the use of HoG features averaged across the image, we limited the nature of variations present in retinal images. This reduced the generalization error and allowed for the inference of PSFs from unseen images of various quality, acquired at different locations.

The proposed method has been tested on 1000 synthetic retinal images and 25 AO high-resolution retinal images. The validation study on synthetic data showed an average error of 0.0051 for the predicted blur kernels and 0.0464 for the reconstructed images, compared to the GT. Qualitative analysis of the results indicated that most of the errors come from images that were significantly distorted (Fig. 3, last column). In clinical practice, images with such poor quality would be rarely used for quantitative assessment of cone photoreceptor distribution. Retinal images where photoreceptor cells cannot be resolved are usually discarded from analysis or attributed to eye pathologies. While in this study the Random Forest was trained on the model of healthy eyes, in case of pathological retinas a different model might be required for setting the range of values of Zernike coefficients.

The obtained results proved that the proposed approach is applicable for the enhancement of real retinal images. For validation purposes, a comparison analysis was performed using 25 retinal images restored with the proposed method and ALM. The results demonstrated that the method based on Random Forest regressor provides higher image contrast and sharpness than the ALM (Table 1) as well as achieves a better differentiation of photoreceptor cells. However, the proposed approach could not always restore the regularity of photoreceptor cell shape. This can be attributed to the limitations of the proposed method, such as fixed axial length and pupil diameter as well as compensation of aberrations up to the 4th radial order of the Zernike polynomial expansion. Despite these limitations, the Random Forest is still able to generalize to most cases. This shows the promise of the proposed method, where more complex blur kernels can be modeled. We aim to address this in our future work.

References

1. Rao, C., Yu, T., Hua, B.: Topics in adaptive optics. AO-based high resolution image post-processing. In: Tyson, R.K. (eds.) Topics in Adaptive Optics, pp. 69–94. InTech (2012)
2. Arines, J.: Partially compensated deconvolution from wavefront sensing images of the eye fundus. Opt. Commun. **284**(6), 1548–1552 (2011)
3. Christou, J.C., Roorda, A., Williams, D.R.: Deconvolution of adaptive optics retinal images. J. Opt. Soc. Am. A. Opt. Image Sci. Vis. **21**(8), 1393–1401 (2004)

4. Blanco, L., Mugnier, L.M.: Marginal blind deconvolution of adaptive optics retinal images. Opt. Express **19**(23), 23227 (2011)
5. Li, H., Lu, J., Shi, G., Zhang, Y.: Real-time blind deconvolution of retinal images in adaptive optics scanning laser ophthalmoscopy. Opt. Commun. **284**(13), 3258–3263 (2011)
6. Chenegros, G., Mugnier, L.M., Lacombe, F., Glanc, M.: 3D phase diversity: a myopic deconvolution method for short-exposure images: application to retinal imaging. J. Opt. Soc. Am. A **24**(5), 1349 (2007)
7. Fanello, S.R., Keskin, C., Kohli, P., Izadi, S., Shotton, J., Criminisi, A., Pattacini, U., Paek T.: Filter forests for learning data-dependent convolutional kernels. In: IEEE CVPR, pp. 1709–1716 (2014)
8. Schuler, C.J., Burger, H.C., Harmeling, S., Scholkopf, B.: A machine learning approach for non-blind image deconvolution. In: IEEE CVPR, pp. 1067–1074 (2013)
9. Dalal, N., Triggs, B.: Histograms of oriented gradients for human detection. CVPR **1**, 886–893 (2005)
10. Lazareva, A., Liatsis, P., Rauscher, F.G.: An automated image processing system for the detection of photoreceptor cells in adaptive optics retinal images. In: IWSSIP, pp. 196–199 (2015)
11. Atchison, D.A., Bradley, A., Thibos, L.N., Smith, G.: Useful variations of the Badal Optometer. Optom. Vis. Sci. **72**(4), 279–284 (1995)
12. Noll, R.J.: Zernike polynomials and atmospheric turbulence. J. Opt. Soc. Am. **66**(3), 207 (1976)
13. Thibos, L.N., Bradley, A., Hong, X.: A statistical model of the aberration structure of normal, well-corrected eyes. Ophthalmic Physiol. Opt. **22**(5), 427–433 (2002)
14. Valeshabad, A.K., Wanek, J., Grant, P., Lim, J.I., Chau, F.Y., Zelkha, R., Camardo, N., Shahidi, M.: Wavefront error correction with adaptive optics in diabetic retinopathy. Optom. Vis. Sci. **91**(10), 1238–1243 (2014)
15. Mariotti, L., Devaney, N.: Performance analysis of cone detection algorithms. J. Opt. Soc. Am. A **32**(4), 497 (2015)
16. Lazareva, A., Liatsis, P., Rauscher, F.G.: Hessian-LoG filtering for enhancement and detection of photoreceptor cells in adaptive optics retinal images. J. Opt. Soc. Am. A **33**(1), 84 (2015)
17. Lucy, L.B.: An iterative technique for the rectification of observed distributions. Astron. J. **79**, 745 (1974)
18. Richardson, W.H.: Bayesian-based iterative method of image restoration. J. Opt. Soc. Am. **62**(1), 55 (1972)
19. Sroubek, F., Milanfar, P.: Robust multichannel blind deconvolution via fast alternating minimization. IEEE Trans. Image Process. **21**(4), 1687–1700 (2012)
20. Peli, E.: Contrast in complex images. J. Opt. Soc. Am. A **7**(10), 2032 (1990)
21. Kanjar, D., Masilamani, V.: A new no-reference image quality measure for blurred images in spatial domain. J. Image Graph. **1**(1), 39–42 (2013)

A Deep Neural Network for Vessel Segmentation of Scanning Laser Ophthalmoscopy Images

Maria Ines Meyer[1]([✉]), Pedro Costa[1], Adrian Galdran[1],
Ana Maria Mendonça[1,2], and Aurélio Campilho[1,2]

[1] INESC-TEC - Institute for Systems and Computer Engineering,
Technology and Science, Porto, Portugal
{maria.i.meyer,pedro.costa,adrian.galdran}@inesctec.pt
[2] Faculdade de Engenharia da Universidade do Porto, Porto, Portugal
{amendon,campilho}@fe.up.pt

Abstract. Retinal vessel segmentation is a fundamental and well-studied problem in the retinal image analysis field. The standard images in this context are color photographs acquired with standard fundus cameras. Several vessel segmentation techniques have been proposed in the literature that perform successfully on this class of images. However, for other retinal imaging modalities, blood vessel extraction has not been thoroughly explored. In this paper, we propose a vessel segmentation technique for Scanning Laser Opthalmoscopy (SLO) retinal images. Our method adapts a Deep Neural Network (DNN) architecture initially devised for segmentation of biological images (U-Net), to perform the task of vessel segmentation. The model was trained on a recent public dataset of SLO images. Results show that our approach efficiently segments the vessel network, achieving a performance that outperforms the current state-of-the-art on this particular class of images.

Keywords: Scanning Laser Ophthalmoscopy · Retinal vessel segmentation

1 Introduction

The retina offers a unique and noninvasive insight into the vascular system. A number of diseases cause visible alterations to the retina. Cardiovascular disease, for example, causes changes in the ratio between the diameter of arteries and veins, and can be related to anatomical changes, such as increased vessel tortuosity and alterations in the branching angles at bifurcation points [10]. Complications deriving from diabetes, namely Diabetic Retinopathy (DR), can also be detected through retinal alterations. DR is the main cause of blindness in most developed countries [9]. When detected at an early stage it can be controlled and blindness prevented. DR manifests itself in a number of different ways, depending on the progression stage of the disease. Some of the changes

© Springer International Publishing AG 2017
F. Karray et al. (Eds.): ICIAR 2017, LNCS 10317, pp. 507–515, 2017.
DOI: 10.1007/978-3-319-59876-5_56

that can be associated with DR are retinal hemorrhages and the formation of new vessels, among others.

As a consequence, the retina is routinely examined in medical settings as a means of diagnosis, through the acquisition of retinal images. The standard image modality in this context is color fundus photography. Other types of techniques applied to retinal imaging include Optical Coherence Tomography (OCT), which produces high-resolution cross sectional images of tissues, and Scanning Laser Ophthalmoscopy (SLO), which also produces high-quality images of the eye fundus by using laser beams to acquire the image [15]. In particular, retinal SLO images present some important advantages over standard fundus photography, namely a stronger contrast and definition of the structures. Nevertheless, this class of images has been less studied in the past, due to the lack of publicly available databases.

Among the existing computational retinal image analysis tasks, the segmentation of the retinal vasculature is of particular relevance. Manual retinal vessel segmentation is a costly process that can greatly benefit from an automated approach. For this reason, several methods have been proposed to approach this problem [4,5,8,11]. However, most of the available techniques are designed for standard color retinal images, and may fail to generalize correctly to SLO images.

In this work, we build on an existing Deep Neural Network (DNN) architecture designed for segmentation of cells and neuronal structures in microscopy images [13], and suitably modify the architecture to adapt it to the segmentation of retinal vessel trees. The method is tested on the newly available IOSTAR dataset [16], which contains SLO images and manually segmented vessel networks. We also validate the approach on the RC-SLO dataset, which comprises a selection of region samples displaying hard to segment cases, such as bifurcations or crossings. In both cases, our approach achieves state-of-the-art results, reaching high values for accuracy and for AUC (area under the ROC curve).

In the following sections, we present a brief overview on the particularities of SLO retinal imaging. Next, we give a description of the employed DNN architecture and the training procedure. Finally, we provide an analysis of the obtained results, and we conclude with some remarks on future research.

2   Scanning Laser Ophthalmoscopy

Scanning Laser Ophthalmoscopy (SLO) was first proposed in [15]. In short, the eye fundus is scanned by a laser beam and light is collected only from one retinal point at a time. Different wavelengths enable the imaging of deeper structures in the retina. As a result, this type of imaging has a number of advantages over color fundus photography, namely an improved contrast, lower light exposure, finer detail, and direct digital acquisition. Nevertheless, there are also associated problems, e.g. a large sensitivity to movement artifacts, or an expensive acquisition that cannot be performed in true color. A first approach to obtain color SLO images was proposed in [6], by imaging the eye fundus with different laser wavelengths and combining them to form a single image.

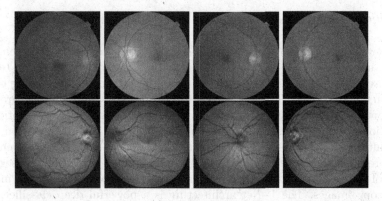

Fig. 1. Comparison between fundus photographs (top) and SLO images (bottom). (Color figure online)

The aforementioned drawbacks contributed to a more widespread use of color fundus photography, which remains the standard in retinal imaging, both in clinical settings and in research. For comparison, some examples of color fundus photographs and SLO images are shown in Fig. 1. An immediate observation is that in SLO images the anatomical structures are represented with higher contrast and detail, and the outer rim of the optic disk has a strong dark green coloring, which corresponds to the neuroretinal rim. These and other differences between both image types may lead to an incorrect generalization of current vessel segmentation methods, that were developed for color fundus photographs.

Few works propose retinal vessel segmentation techniques specifically designed for SLO images [1,12,16]. In [12] a supervised method is presented where the SLO image is pre-processed to enhance the contrast and illumination of the vascular network. The maximum intensity value at each pixel location is extracted at different scales to form an intensity map. Several other parametric maps are extracted, reflecting the width range, standard deviation and local orientation of the vessels, from the combination of the different scales in the previous step. These alternative representations are used as input to a neural network classifier. Similarly to [12], the technique presented in [1] consists of a combination of retinal image enhancement and feature extraction procedures. In the first stage, the SLO image is denoised, and its blood vessels are enhanced by lifting the image to an associated space of positions and orientations. Following, several visual features are computed, including intensities, wavelet responses and an extension of Gaussian derivatives. These features are then supplied to a neural network to produce vessel likelihoods for each pixel. The approach in [16] is related to [1], since the SLO images undergo a similar transformation to a joint positions and orientations space. However, after enhancing the vessel tree, the image is backprojected to the 2D space, and the vasculature is segmented by means of a global thresholding, turning the method into an unsupervised technique.

3   Deep Neural Network for Vessel Segmentation

To achieve vessel segmentation, we propose to consider the problem as a semantic segmentation task. This approach differs from a standard segmentation task in the sense that the aim is to classify each pixel in an image as belonging to one class. In our particular case, the classes to consider are vessel and background.

Deep learning architectures have been widely applied to the semantic segmentation problem. A relevant technique was proposed in [14], based on Fully Convolutional Networks (FCNs). Building on the ideas of FCNs, Ronneberger *et al.* proposed the U-Net architecture [13], which was designed for the segmentation of neuronal structures in electron microscopic recordings and cells in light microscopy images. The U-Net architecture is a powerful deep classifier, that can be trained with a low quantity of images to produce precise segmentations.

The architecture of the U-Net is represented in Fig. 2. The network has two main sections, or paths - a first contracting path, followed by a dimensionally symmetric expanding path. The contracting path consists of consecutive convolutional layers, with stride two and followed by Rectified Linear Unit (ReLU) activations. Assuming that the dimension of the input is a power of two, the stride is selected so that the dimension of the output feature map of the contracting path layers decreases until a dimension of $1 \times 1 \times nf$ (nf: number of filters) is reached. This point in the network marks the beginning of the expanding path. The output of the layer is upsampled to have the same dimension as the previous layer in the contracting path. To compensate for the loss in spacial resolution that results from the multiple downsampling operations, the upsampled feature map is concatenated with the feature map of the corresponding layer in the contracting path. The new feature map serves as input for a new convolutional layer, followed by ReLU activation. This procedure is repeated for the output of each convolutional layer until the output of the expanding path layers reaches the same dimension as the first layer of the network. Finally, the feature vectors are mapped into vessel/background classes by a sigmoid activation function.

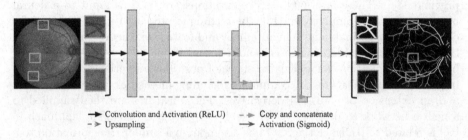

→ Convolution and Activation (ReLU)   ⇢ Copy and concatenate
⇢ Upsampling   → Activation (Sigmoid)

Fig. 2. Overview of the U-net model. Each box represents a multi-channel feature map.

4   Datasets and Implementation

The method was trained using the IOSTAR dataset [16], which includes 30 SLO images taken with an EasyScan camera (provided by i-Optics B.V., the Netherlands). The images have a resolution of 1024 × 1024 pixels with a 45° field of view (FOV), and were annotated and corrected by two different experts. Annotations in the optic disk region are not available, for which reason an optic disk mask is provided, along with the FOV mask. Further information about the dataset is available in [1] and [16].

From the initial dataset, the first 20 images were employed for training, and the remaining 10 were used to build an independent test set. Images in the training set were divided into overlapping patches of 128 × 128 pixels. The resulting set of patches was randomized and 10% of this set was used as a validation set.

The method trained on IOSTAR was tested with no further modifications on the RC-SLO vessel patch dataset [16], which is composed of 40 images with dimension 360 × 320 pixels. This dataset provides a good test because it comprises cases that are difficult to segment, such as central vessel reflex, crossings, bifurcations, background artifacts and high curvature changes. These images were also divided into smaller overlapping patches of 128 × 128 pixels.

In order to build a vessel tree segmentation of the entire retinal image, we first need to combine the predictions generated locally for each image patch. To achieve this, at the time each patch is produced, the coordinates translating it to its corresponding location in the input image are stored. This allows for a simple reconstruction step to be performed once the predictions for every patch are available. When there is overlap between patches, the resulting local probability maps are simply averaged pixel-wise.

The described DNN was optimized by minimizing the pixel mis-classification error, according to the cross-entropy loss function. The minimization followed a stochastic mini-batch gradient descent, with the gradients computed by standard backpropagation. The applied optimization technique was the Adam optimizer [3] with a learning rate of 1×10^{-4}. The model was trained on the training set, while the loss was monitored on the validation set. Training took approximately 12 h on a NVIDIA GeForce GTX 1080 GPU. Once trained, the model can generate a vessel segmentation for one 128 × 128 patch in approximately 0.05 s. The implementation was done in Python using Keras [2].

5   Experimental Results and Evaluation

The model was evaluated on the IOSTAR test set. The test images were normalized following the same process as at training time, i.e., they were linearly transformed so that they had zero mean and unit standard deviation, and divided into overlapping patches of 128 × 128 pixels. The vessel segmentation was predicted for each patch, and the image was then reconstructed. The pixel probabilities for overlapping patches were averaged, in order to get the final probability map of the retinal segmentation, as described in Sect. 4.

The method was then also evaluated on the RC-SLO dataset. Each 360×320 pixel patch was normalized to have unit standard deviation and mean zero, and divided into smaller patches of 128×128 pixel, to be fed into our pre-trained network. The same reconstruction protocol as described above was used.

The resulting soft segmentations were subjected to ROC analysis. The Mathews Correlation Coefficient (MCC) [7] was computed at different thresholds. MCC is a commonly used score to assess the performance of binary classifiers in the presence of skewed classes, as is our case since there is a more predominant presence of background pixels than vessel pixels. The threshold maximizing the MCC was employed to generate binary segmentations, from which Accuracy (Acc), Sensitivity (Se) and Specificity (Sp) were calculated.

For both datasets the values achieved outperformed the previously existing approaches, with AUC scores of 0.9771 on IOSTAR and 0.9807 on RC-SLO. The derived metrics also reflect a state-of-the-art performance. Table 1 summarizes the results, and compares our method with previous approaches that reported results on the same datasets. Figures 3 and 4 present example results of applying the proposed technique to IOSTAR and RC-SLO images, respectively.

Table 1. Summary of our results and comparison with other previously proposed techniques that also reported performance on the IOSTAR and RC-SLO datasets. (Best values are displayed in bold).

Methods	Approach	Dataset	Se	Sp	Acc	AUC	MCC
BIMSO [1]	Supervised	IOSTAR	0.7863	0.9747	0.9501	0.9615	0.7752
LAD-OS [16]	Unsupervised	IOSTAR	0.7545	0.9740	0.9514	0.9626	0.7318
LAD-OS [16]	Unsupervised	RC-SLO	0.7787	0.9710	0.9512	0.9626	0.7327
OURS	Supervised	IOSTAR	**0.8038**	**0.9801**	**0.9695**	**0.9771**	**0.7920**
OURS	Supervised	RC-SLO	**0.8090**	**0.9794**	**0.9623**	**0.9807**	**0.7905**

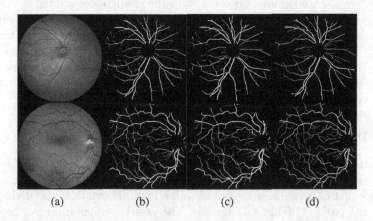

(a)   (b)   (c)   (d)

Fig. 3. Segmentation results for IOSTAR dataset. (a) Real image. (b) Output probability map. (c) Segmentation of the probability map at threshold that optimizes MCC. (d) Ground truth annotation. (Color figure online)

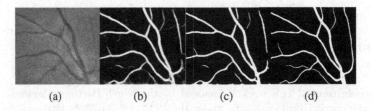

(a)   (b)   (c)   (d)

Fig. 4. Example of segmentation result for RC-SLO dataset. (a) Real image; (b) Output probability map. Note the artifacts at the bottom of the image; (c) Segmentation of the probability map at the threshold that optimizes MCC; (d) Ground truth annotation.

Visual inspection of the results in Fig. 3 shows that our technique sometimes under-segments difficult cases, such as thin vessels. We also observe some vessel discontinuities, indicating that the method can still be improved. For the RC-SLO dataset some results present artifacts at the bottom part, as exemplified in Fig. 4(b) and (c). This could be due to the model being trained on patches extracted from entire images, which contained the round FOV borders.

6   Conclusions and Future Work

In this work we proposed a supervised method to segment the retinal vessel tree of Scanning Laser Opthalmoscopy images. Our approach is based on the U-Net DNN architecture, designed for image segmentation. The method was trained on the IOSTAR dataset and achieved above state-of-the-art results on both IOSTAR and RC-SLO datasets.

An important limitation of the present work is its inability to reliably segment vessels within the optic disk, which is a direct result of missing ground truth segmentations in this region, where the background is very different from the remaining image. We are optimistic that if the annotated ground truth contained information regarding these vessels our method could have resolved well. An issue to address is the appearance of artifacts at the bottom part of some of the RC-SLO segmentations. It is noteworthy that even with the presence of such artifacts the performance of the method was very good.

Another limitation of the proposed technique is that it was trained for SLO images, and we can expect it to generalize badly to standard retinal images. A future line of research will be to perform transfer learning from one problem to the other. Visually, vessel segmentation from SLO and from standard retinal images are similar problems. For this reason, after training the method on SLO images, we can take advantage of the representations learned by the neural network, fine-tuning its weights so that it performs well also on standard retinal images. This would avoid the necessity of carrying out a complete re-training. Training the same model simultaneously on both image types is also appealing, since this could confer the neural network more generalization capability.

Finally, the proposed technique could be improved to segment patches of any size and resolution. For this, the method should be modified to account for multiple scales. This is a well known problem in semantic segmentation, and could be overcome by altering the architecture so that it could support images of different resolutions and patch sizes. This could be relevant to avoid re-training the network each time it is presented with images of different resolutions.

Acknowledgments. This work is financed by the North Portugal Regional Operational Programme (NORTE 2020), under the PORTUGAL 2020 Partnership Agreement, and the European Regional Development Fund (ERDF), within the project "NanoSTIMA: Macro-to-Nano Human Sensing: Towards Integrated Multimodal Health Monitoring and Analytics/NORTE-01-0145-FEDER-000016".

References

1. Abbasi-Sureshjani, S., Smit-Ockeloen, I., Zhang, J., Ter Haar Romeny, B.: Biologically-Inspired supervised vasculature segmentation in SLO retinal fundus images. In: Kamel, M., Campilho, A. (eds.) ICIAR 2015. LNCS, vol. 9164, pp. 325–334. Springer, Cham (2015). doi:10.1007/978-3-319-20801-5_35
2. Chollet, F.: Keras (2015). https://github.com/fchollet/keras
3. Kingma, D., Ba, J.: Adam: a method for stochastic optimization. In: International Conference on Learning Representations, pp. 1–13 (2014)
4. Liskowski, P., Krawiec, K.: Segmenting retinal blood vessels with deep neural networks. IEEE Trans. Med. Imaging **35**(11), 2369–2380 (2016)
5. Maninis, K.-K., Pont-Tuset, J., Arbeláez, P., Gool, L.: Deep retinal image understanding. In: Ourselin, S., Joskowicz, L., Sabuncu, M.R., Unal, G., Wells, W. (eds.) MICCAI 2016. LNCS, vol. 9901, pp. 140–148. Springer, Cham (2016). doi:10.1007/978-3-319-46723-8_17
6. Manivannan, A., Kirkpatrick, J.N., Sharp, P.F., Forrester, J.V.: Novel approach towards colour imaging using a scanning laser ophthalmoscope. Br. J. Ophthalmol. **82**(4), 342–345 (1998)
7. Matthews, B.: Comparison of the predicted and observed secondary structure of T4 phage lysozyme. Biochimica et Biophysica Acta (BBA) - Protein Structure **405**(2), 442–451 (1975)
8. Mendonça, A.M., Campilho, A.: Segmentation of retinal blood vessels by combining the detection of centerlines and morphological reconstruction. IEEE Trans. Med. Imaging **25**(9), 1200–1213 (2006)
9. Mohamed, Q., Gillies, M., Wong, T.: Management of diabetic retinopathy: a systematic review. JAMA **298**(8), 902–916 (2007)
10. Nguyen, T.T., Wong, T.Y.: Retinal vascular changes and diabetic retinopathy. Curr. Diab. Rep. **9**(4), 277–283 (2009)
11. Orlando, J.I., Blaschko, M.: Learning fully-connected CRFs for blood vessel segmentation in retinal images. In: Golland, P., Hata, N., Barillot, C., Hornegger, J., Howe, R. (eds.) MICCAI 2014. LNCS, vol. 8673, pp. 634–641. Springer, Cham (2014). doi:10.1007/978-3-319-10404-1_79
12. Pellegrini, E., Robertson, G., Trucco, E., MacGillivray, T.J., Lupascu, C., van Hemert, J., Williams, M.C., Newby, D.E., van Beek, E., Houston, G.: Blood vessel segmentation and width estimation in ultra-wide field scanning laser ophthalmoscopy. Biomed. Opt. Express **5**(12), 4329–4337 (2014)

13. Ronneberger, O., Fischer, P., Brox, T.: U-Net: convolutional networks for biomedical image segmentation. In: Navab, N., Hornegger, J., Wells, W.M., Frangi, A.F. (eds.) MICCAI 2015. LNCS, vol. 9351, pp. 234–241. Springer, Cham (2015). doi:10.1007/978-3-319-24574-4_28

14. Shelhamer, E., Long, J., Darrell, T.: Fully convolutional networks for semantic segmentation. In: Proceedings of the IEEE Conference on Computer Vision and Pattern Recognition (2015)

15. Webb, R.H., Hughes, G.W., Pomerantzeff, O.: Flying spot TV ophthalmoscope. Appl. Opt. **19**(17), 2991–2997 (1980)

16. Zhang, J., Dashtbozorg, B., Bekkers, E., Pluim, J.P.W., Duits, R., ter Haar Romeny, B.M.: Robust retinal vessel segmentation via locally adaptive derivative frames in orientation scores. IEEE Trans. Med. Imaging **35**(12), 2631–2644 (2016)

Adversarial Synthesis of Retinal Images from Vessel Trees

Pedro Costa[1(✉)], Adrian Galdran[1], Maria Ines Meyer[1],
Ana Maria Mendonça[1,2], and Aurélio Campilho[1,2]

[1] INESC TEC - Institute for Systems and Computer Engineering,
Technology and Science, Porto, Portugal
{pvcosta,adrian.galdran,maria.i.meyer}@inesctec.pt
[2] Faculdade de Engenharia da Universidade do Porto, Porto, Portugal
{amendon,campilho}@fe.up.pt

Abstract. Synthesizing images of the eye fundus is a challenging task that has been previously approached by formulating complex models of the anatomy of the eye. New images can then be generated by sampling a suitable parameter space. Here we propose a method that learns to synthesize eye fundus images directly from data. For that, we pair true eye fundus images with their respective vessel trees, by means of a vessel segmentation technique. These pairs are then used to learn a mapping from a binary vessel tree to a new retinal image. For this purpose, we use a recent image-to-image translation technique, based on the idea of adversarial learning. Experimental results show that the original and the generated images are visually different in terms of their global appearance, in spite of sharing the same vessel tree. Additionally, a quantitative quality analysis of the synthetic retinal images confirms that the produced images retain a high proportion of the true image set quality.

Keywords: Retinal image synthesis · Generative adversarial learning

1 Introduction

Modern machine learning methods require large amounts of training data. This data is rarely available in the field of medical image analysis, since obtaining clinical annotations is often a costly process. Therefore, the possibility of synthetically generating medical visual data is greatly appealing, and has been explored for years. However, the realistic generation of high-quality medical imagery still remains a complex unsolved challenge for current computer vision methods.

Early methods for medical image generation consisted of digital phantoms, following simplified mathematical models of human anatomy [2]. These models slowly evolved to more complex techniques, able to reliably model relevant aspects of the different acquisition devices. When combined with anatomical and physiological information arising from expert medical knowledge, realistic images can be produced [4]. These are useful to validate image analysis techniques, for medical training, therapy planning, and a wide range of applications [6,11].

© Springer International Publishing AG 2017
F. Karray et al. (Eds.): ICIAR 2017, LNCS 10317, pp. 516–523, 2017.
DOI: 10.1007/978-3-319-59876-5_57

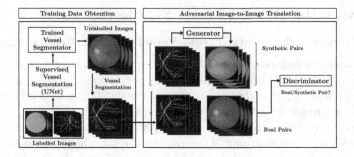

Fig. 1. Overview of the proposed retinal image generation method.

However, the traditional top-down approach of observing the available data and formulating mathematical models that explain it (*image simulation*) implies modeling complex natural laws by unavoidably simplifying assumptions. More recently, a new paradigm has arisen in the field of medical image generation, exploiting the bottom-up approach of directly learning from the data the relevant information. This is achieved with machine learning systems able to automatically learn the inner variability on a large training dataset [18]. Once trained, the same system can be sampled to output a new but plausible image (*image synthesis*).

In the general computer vision field, the synthesis of natural images has recently experimented a dramatic progress, based on the general idea of adversarial learning [5]. In this context, a generator component synthesizes images from random noise, and an auxiliary discriminator system trained on real data is assigned the task of discerning whether the generated data is real or not. In the training process, the generator is expected to learn to produce images that pose an increasingly more difficult classification problem for the discriminator.

Although adversarial techniques have achieved a great success in natural image generation, medical imaging applications are still incipient. This is partially due to the lack of large amounts of training data, and partially to the difficulty of finely controlling the output of the adversarial generator. In this work, we propose to apply the adversarial learning framework to retinal images. Notably, instead of generating images from scratch, we propose to generate new plausible images from binary retinal vessel trees. Therefore, the task of the generator remains achievable, as it only needs to learn how to generate part of the retinal content, such as the optical disk, or the background's texture (Fig. 1).

The remaining of this work is organized as follows: we first describe a recent generative adversarial framework [7] that can be employed on pairs of vessel trees and retinal images to learn how to map the former to the latter. Then, we briefly review U-Net, a Deep Convolutional Neural Network designed for image segmentation, which allows us to generate pairs of retinal images and corresponding binary vessel trees. This model provides us with a dataset of vessel trees and corresponding retinal images that we then use to train an adversarial model, producing new good-quality retinal images out of a new vessel tree. Finally, the quality of the generated images is evaluated qualitatively and quantitatively, and a description of potential future research directions is presented.

2   Adversarial Retinal Image Synthesis

2.1   Adversarial Translation from Vessel Trees to Retinal Images

Image-to-image translation is a relatively recent computer vision task in which the goal is to learn a mapping G, called *Generator*, from an image x into another representation y [7]. Once the model has been trained, it is able to predict the most likely representation $G(x_{new})$ for a previously unseen image x_{new}.

However, for many problems a single input image can correspond to many different correct representations. If we consider the mapping G between a retinal vessel tree v and a corresponding retinal fundus image r, variations in color or illumination may produce many acceptable retinal images that correspond to the same vessel tree, i.e. $G(v) = \{r_1, r_2, \ldots, r_n\}$. Directly related to this is the choice of the objective function to be minimized while learning G, which turns out to be critical. Training a model to naively minimize the $L2$ distance between $G(v_i)$ and r_i for a collection of training pairs given by $\{(r_1, v_1), \ldots, (r_n, v_n)\}$ is known to produce low-quality results with lack of detail [12], due to the model selecting an average of many equally valid representations.

Instead of explicitly defining a particular loss function for each task, it is possible to employ Generative Adversarial Networks to implicitly build a more appropriate loss [7]. In this case, the learning process attempts to maximize the misclassification error of a neural network (called *Discriminator*, D) that is trained jointly with G, but with the goal of discriminating between real and generated images. This way, not only G but also the loss are progressively learned from examples, and adapt to each other: while G tries to generate increasingly more plausible representations $G(v_i)$ that can deceive D, D becomes better at its task, thereby improving the ability of G to generate high-quality samples. Specifically, the adversarial loss is defined by:

$$\mathcal{L}_{adv}(G, D) = \mathbb{E}_{v, r \sim p_{data}(v, r)}[log D(v, r)] + \mathbb{E}_{v \sim p_{data}(v)}[log(1 - D(v, G(v)))], \quad (1)$$

where $\mathbb{E}_{v, r \sim p_{data}}$ represents the expectation of the log-likelihood of the pair (v, r) being sampled from the underlying probability distribution of real pairs $p_{data}(v, r)$, while $p_{data}(v)$ corresponds to the distribution of real vessel trees. An overview of this process is shown in Fig. 2.

To generate realistic retinal images from binary vessel trees, we follow recent ideas from [7,15], which propose to combine the adversarial loss with a global $L1$ loss to produce sharper results. Thus, the loss function to minimize becomes:

$$\mathcal{L}(G, D) = \mathcal{L}_{adv}(G, D) + \lambda \mathbb{E}_{v, r \sim p_{data}(v, r)} (\|r - G(v)\|_1), \quad (2)$$

where λ balances the contribution of the two losses. The goal of the learning process is thus to find an equilibrium of this expression. The discriminator D attempts to maximize Eq. (2) by classifying each $N \times N$ patch of a retinal image, deciding if it comes from a real or synthetic image, while the generator aims at minimizing it. The $L1$ loss controls low-frequency information in images generated by G in order to produce globally consistent results, while the adversarial loss promotes sharp results. Once G is trained, it is able to produce a realistic retinal image r from a new binary vessel tree v.

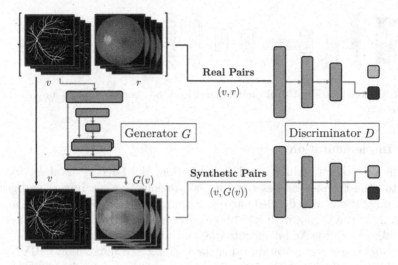

Fig. 2. Overview of the generative model mapping vessel trees to retinal images.

2.2 Obtaining Training Data

The model described above requires training data in the form of pairs of binary retinal vessel trees and corresponding retinal images. Since such a large scale manually annotated database is not available, we apply a state-of-the-art retinal vessel segmentation algorithm to obtain enough data for the model to learn the mapping from vessel trees to retinal images. There exist a large number of methods capable of providing reliable retinal vessel segmentations. Here we employ a supervised method based on Convolutional Neural Networks (CNNs), namely the U-Net architecture, first proposed in [13] for the segmentation of biomedical images. This technique is an extension of the idea of Fully-Convolutional Network (FCNs), introduced in [14], adapted to be trained with a low number of images and produce more precise segmentations.

The architecture of the U-Net consists of a contracting and an expanding part. The first half of the network follows a typical CNN architecture, with stacked convolutional layers of stride two and Rectified Linear Unit (ReLU) activations. The second part of the architecture is an expanding path, symmetric to the contracting path. The output feature map of the last layer of the contracting path is upsampled so that it has the same dimension of the second last layer. The result is concatenated with the feature map of the corresponding layer in the contracting path, and this new feature map undergoes convolution and activation. This is repeated until the expanding path layers reach the same dimensions as the first layer of the network.

The final layer is a convolution followed by a sigmoid activation in order to map each feature vector into vessel/non-vessel classes. The concatenation operation allows for very precise spatial localization, while preserving the coarse-level features learned during the contracting path. A representation of this architecture as used in the present work is represented in Fig. 3.

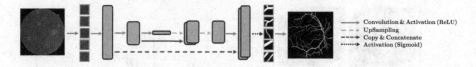

Fig. 3. Overview of the U-Net architecture. Each box corresponds to a multi-channel feature map.

2.3   Implementation

For the purpose of retinal vessel segmentation, the DRIVE database [16] was used to train the method described in the previous Section. Images and ground truth annotations were divided into overlapping patches of 64 × 64 pixels and fed randomly to the U-Net, with 10% of the patches used for validation. The network was trained with the Adam optimizer [8] and a binary crossentropy loss function.

Retinal vessel segmentation using the U-Net was evaluated on DRIVE's test set, achieving a 0.9755 AUC, aligned with state-of-the-art results [10]. The optimal binarization threshold maximizing the Youden index [19] was selected. Messidor [3] images were cropped, in order to only display the field of view, and downscaled to 512 × 512. Then, the segmentation method was applied to these images. Messidor contains 1200 images annotated with the corresponding diabetic retinopathy grade, and displays more color and texture variability than DRIVE's 20 training images. Due to the U-Net being trained and tested in different databases, some of the produced segmentations were not entirely correct. This may be related to DRIVE only containing 7 examples of images with signs of mild diabetic retinopathy (grade 1). For this reason, we retained only pairs of images and vessel trees in which the corresponding image had grade 0, 1, and 2.

The final dataset collected for training our adversarial model consisted of 946 Messidor image pairs. This dataset was further randomly divided into training (614 pairs), validation (155 pairs) and test (177 pairs) sets. Regarding image resolution, the original model in [7] used pairs of 256 × 256 images, with a U-Net-like generator G. We modified the architecture to handle 512 × 512 pairs, which is closer to the resolution of DRIVE images. For that, we added one layer to the contracting part and another to the expanding part of G. The discriminator D classifies 16 × 16 overlapping patches of size 63 × 63. The implementation was developed in Python using Keras[1] [1]. The learning process starts by training D with real (v, r) and generated pairs $(v, G(v))$. Then, G is trained with real (v, r) pairs. This process was repeated iteratively until the losses of D and G stabilized.

3   Experimental Evaluation

For subjective evaluation of the images generated by our model, we show in Fig. 4 some visual results. The first row depicts a random sample of vessel trees extracted from the held-out test set, which was not used during training. The

---

[1] Code to reproduce our results is available at https://github.com/costapt/vess2ret.

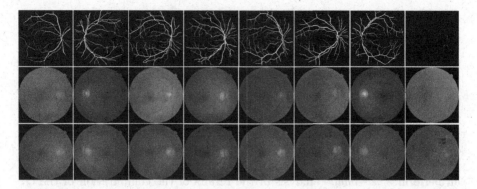

Fig. 4. Results of our model. First row: Vessel trees not used during training. Second row: True retinal images corresponding to the above vessel trees. Third row: Corresponding retinal images generated by our model. All images have 512 × 512 resolution.

second row shows the real images from which those vessel trees were segmented with the method outlined in Sect. 2.2, and the bottom row shows the synthetic retinal images produced by the proposed technique. We see that the original and the generated images share some global geometric characteristics. This is natural, since they approximately share the same vascular structure. However, the synthetic images have markedly different high-level visual features, such as the color and tone of the image, or the illumination. This information was extracted by our model from the training set, and effectively applied to the input vessel trees in order to produce realistic retinal images.

The last column in Fig. 4 shows a failure case of the proposed technique. Therein, the segmentation technique described in Sect. 2.2 failed to produce a meaningful vessel network out of the original image. This is probably due to the high degree of defocus that the input image had. In this situation, the binary vessel tree supplied to the generator contained too few information, and it reacted by creating spurious artifacts and chromatic noise in the synthetic image. Fortunately, the amount of cases in which this happened was relatively low: from our test set of 177 images, 7 were found to suffer from artifacts.

Regarding objective image quality verification, this is a hard challenge when no reference is available. In addition, for generative models it has been recently observed that specialized evaluation should be performed for each problem [17]. In our case, to achieve a meaningful objective quantitative evaluation of the quality of the generated images, we apply the no-reference retinal image quality assessment technique proposed in [9]. This score, denoted Q_v, is derived by calculating a local degree of vesselness around each pixel, computing a local estimate of anisotropy on regions that are good candidates for containing vessels, and averaging the results, see [9] for the technical details. The results of computing the Q_v metric on both sets of real and synthetic images are shown in Table 1.

The first two columns on Table 1 show the mean and standard deviation of the Q_v scores computed from the original and synthetic images. We can see that

Table 1. Result of computing the Q_v quality measure on real/synthetic images.

	Mean Q_v score	Std. dev	Avg. *per-image* variation
Real images	0.1234	0.0207	100%
Synthetic images	0.1040	0.0131	87.55%

the mean Q_v score obtained for the synthetic images was relatively close to the score computed from the dataset of true images. Furthermore, since from each vessel tree we have the corresponding true and synthetic images available, we can perform a *per-image* analysis of the results of the computation of the Q_v measure. For that, we considered the quality of the true retinal fundus images to be 100%, and for each synthetic image we computed the percentage of quality variation observed. Results of this analysis are shown in the third column, where we see that, on average, 87.55% of the true images quality was preserved. A more detailed analysis revealed that, from the 177 test binary vessel trees, the corresponding synthetically generated images achieved a better Q_v scores than the true images in 30 cases.

4   Conclusions and Future Work

The above results demonstrate the feasibility of learning to synthesize new retinal images from a dataset of pairs of retinal vessel trees and corresponding retinal images, applying current generative adversarial models. In addition, the dimension of the produced images was 512×512, which is greater than commonly generated images on general computer vision problems. We believe that achieving this resolution was only possible due to the constrained class of images in which the method was applied: contrarily to generic natural images, retinal images show a repetitive geometry, where high-level structures such as the field of view, the optical disc, or the macula, are usually present in the image, and act as a guide for the model to learn how to produce new texture and background intensities.

The main limitation of the presented method is its dependence on a pre-existing vessel tree in order to generate a new image. Furthermore, if the vessel tree comes from the application of a segmentation technique to the original image, the potential weaknesses of the segmentation algorithm will be inherited by the synthesized image. We are currently working on overcoming these challenges.

Acknowledgments. This work is funded by the ERDF - European Regional Development Fund through the Operational Programme for Competitiveness and Internationalisation - COMPETE 2020 Programme, by the FCT - Fundação para a Ciência e a Tecnologia within project CMUP-ERI/TIC/0028/2014 and by the North Portugal Regional Operational Programme (NORTE 2020), under the PORTUGAL 2020 Partnership Agreement within the project "NanoSTIMA: Macro-to-Nano Human Sensing: Towards Integrated Multimodal Health Monitoring and Analytics/NORTE-01-0145-FEDER-000016".

References

1. Chollet, F.: Keras (2015). https://github.com/fchollet/keras
2. Collins, D.L., Zijdenbos, A.P., Kollokian, V., Sled, J.G., Kabani, N.J., Holmes, C.J., Evans, A.C.: Design and construction of a realistic digital brain phantom. IEEE Trans. Med. Imaging **17**(3), 463–468 (1998)
3. Decencière, E., Zhang, X., Cazuguel, G., Lay, B., Cochener, B., Trone, C., Gain, P., Ordonez, R., Massin, P., Erginay, A., Charton, B., Klein, J.C.: Feedback on a publicly distributed database: the Messidor database. Image Anal. Stereol. **33**(3), 231–234 (2014)
4. Fiorini, S., Ballerini, L., Trucco, E., Ruggeri, A.: Automatic generation of synthetic retinal fundus images. In: Reyes-Aldasoro, C.C., Slabaugh, G. (eds.) Medical Image Understanding and Analysis 2014, pp. 7–12. BMVA Press (2014)
5. Goodfellow, I., Pouget-Abadie, J., Mirza, M., Xu, B., Warde-Farley, D., Ozair, S., Courville, A., Bengio, Y.: Generative adversarial nets. In: Advances in Neural Information Processing Systems, pp. 2672–2680 (2014)
6. Hodneland, E., Hanson, E., Munthe-Kaas, A.Z., Lundervold, A., Nordbotten, J.M.: Physical models for simulation and reconstruction of human tissue deformation fields in dynamic MRI. IEEE Trans. Bio-Med. Eng. **63**(10), 2200–2210 (2016)
7. Isola, P., Zhu, J.Y., Zhou, T., Efros, A.A.: Image-to-image translation with conditional adversarial networks, November 2016. arXiv.org, arXiv: 1611.07004
8. Kingma, D., Ba, J.: Adam: A method for stochastic optimization. In: International Conference on Learning Representations, pp. 1–13 (2014)
9. Köhler, T., Budai, A., Kraus, M.F., Odstrcilik, J., Michelson, G., Hornegger, J.: Automatic no-reference quality assessment for retinal fundus images using vessel segmentation. In: Proceedings of the 26th IEEE CMBS, pp. 95–100, June 2013
10. Liskowski, P., Krawiec, K.: Segmenting retinal blood vessels with deep neural networks. IEEE Trans. Med. Imaging **35**(11), 2369–2380 (2016)
11. Liu, X., Liu, H., Hao, A., Zhao, Q.: Simulation of blood vessels for surgery simulators. In: 2010 International Conference on Machine Vision and Human-Machine Interface, pp. 377–380, April 2010
12. Lotter, W., Kreiman, G., Cox, D.: Unsupervised learning of visual structure using predictive generative networks. arxiv preprint (2015). arXiv:1511.06380
13. Ronneberger, O., Fischer, P., Brox, T.: U-Net: convolutional networks for biomedical image segmentation. In: Navab, N., Hornegger, J., Wells, W.M., Frangi, A.F. (eds.) MICCAI 2015. LNCS, vol. 9351, pp. 234–241. Springer, Cham (2015). doi:10.1007/978-3-319-24574-4_28
14. Shelhamer, E., Long, J., Darrell, T.: Fully convolutional networks for semantic segmentation. In: Proceedings of the IEEE CVPR, pp. 3431–3440 (2015)
15. Shrivastava, A., Pfister, T., Tuzel, O., Susskind, J., Wang, W., Webb, R.: Learning from simulated and unsupervised images through adversarial training. arxiv preprint (2016). arXiv:1612.07828
16. Staal, J.J., Abramoff, M.D., Niemeijer, M., Viergever, M.A., van Ginneken, B.: Ridge based vessel segmentation in color images of the retina. IEEE Trans. Med. Imaging **23**(4), 501–509 (2004)
17. Theis, L., Oord, A.v.d., Bethge, M.: A note on the evaluation of generative models. In: International Conference on Learning Representations (2016)
18. Tulder, G., Bruijne, M.: Why does synthesized data improve multi-sequence classification? In: Navab, N., Hornegger, J., Wells, W.M., Frangi, A.F. (eds.) MICCAI 2015. LNCS, vol. 9349, pp. 531–538. Springer, Cham (2015). doi:10.1007/978-3-319-24553-9_65
19. Youden, W.J.: Index for rating diagnostic tests. Cancer **3**(1), 32–35 (1950)

Automated Analysis of Directional Optical Coherence Tomography Images

Florence Rossant[1(✉)], Kate Grieve[2], and Michel Paques[2]

[1] Institut Supérieur d'Electronique de Paris (ISEP), Paris, France
Florence.rossant@isep.fr
[2] Clinical Investigation Center 1423, Quinze-Vingts Hospital, Paris, France
kategrieve@gmail.com, michel.paques@gmail.com

Abstract. Directional optical coherence tomography (D-OCT) reveals reflectance properties of retinal structures by changing the incidence angle of the light beam. As no commercially available OCT device has been designed for such use, image processing is required to homogenize the grey levels between off-axis images before differential analysis. We describe here a method for automated analysis of D-OCT images and propose a color representation to highlight angle-dependent structures. Clinical results show that the proposed approach is robust and helpful for clinical interpretation.

Keywords: Optical coherence tomography · Directional OCT (D-OCT) · Retina · Ophthalmology · Differential analysis · Markov random fields

1 Introduction

Photoreceptors have unique optical properties, one of which is the angular dependence of their absorbance and reflectance, known as the Stiles-Crawford effect (SCE) [1, 2]. As a result, the distribution of backscattered light through the pupil is modulated by its incidence. The angle-dependent absorbance and reflectance of individual cone photoreceptors follows a Gaussian curve, with a peak whose orientation defines the photoreceptor pointing and whose acceptance angle correlates with the span of photon capture. Cones account for most of the SCE.

Gao et al. first used a custom OCT setup to image off-axis macular photoreceptors and measure the contributions of macular photoreceptor substructures to the optical SCE (oSCE) [3]. They suggested that directional reflectance of the cone outer segment tip (COST) line and, to a lesser extent, of the inner/outer segment junction (IS/OS), accounts for most of the oSCE of macular photoreceptors. This was supported by the findings of Miloudi et al. who documented the directional reflectance of photoreceptors using a combined approach by en face adaptive optics imaging and D-OCT [4]. The potential medical interest of D-OCT was suggested by Lujan et al. who used D-OCT to delineate the Henle fiber layer, that is, the photoreceptor axons, and subsequently extract the thickness of the outer nuclear layer [5]. In addition, our group and others have recently demonstrated differences due to pathology in the photoreceptor layers using D-OCT in patients (Pedinielli et al. IOVS 2016: ARVO E-Abstract 4248, Lujan et al. IOVS 2016: ARVO E-Abstract 4250). Hence, D-OCT appears to be an interesting

© Springer International Publishing AG 2017
F. Karray et al. (Eds.): ICIAR 2017, LNCS 10317, pp. 524–532, 2017.
DOI: 10.1007/978-3-319-59876-5_58

approach to document subtle retinal changes. We describe here a method for automated analysis of D-OCT images acquired with commercial OCT systems. We propose an efficient visualization of D-OCT data that helps develop the clinical interpretation by highlighting angle-dependent structures.

2   Methods

D-OCT scans were acquired using a commercially available OCT apparatus (Spectralis® SD-OCT). A reference scan was acquired in the standard mode, with a light beam parallel to the visual axis. Two off-axis images were also acquired, based on a manual procedure, at approximately equal $2°$ to $3°$ angles α to the right and left of the visual axis. All images were aligned by the acquisition system (Fig. 1).

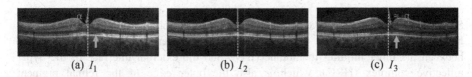

(a) I_1 (b) I_2 (c) I_3

Fig. 1. A set of two off-axis images (a) (c) along with a standard on-axis acquisition (b), all aligned. Red and green arrows show evidence of anisotropy. (Color figure online)

Anisotropy properties of the reflectance can be brought out by computing the difference between the two off-axis images. Retinal structures that have similar reflectance responses whatever the light incidence angle are considered isotropic and lead to values near zero. On the contrary, anisotropic areas lead to positive or negative values, depending on the local orientation of the retinal structures with respect to the incidence angle. However the simple difference of the two angled images is not sufficient here. Indeed, the global brightness of the image is no longer homogeneous over the entire image when the operator turns the light beam away from the visual axis (Fig. 1(a) and (c)). Note that the left side is darker than the right side for the angle α and conversely for the opposite angle $-\alpha$. So the basic difference of the two off-axis images cannot provide suitable enhancement of anisotropic features (Fig. 2).

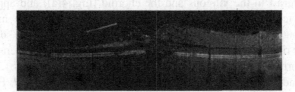

Fig. 2. The absolute difference of the two input off-axis images.

Consequently, we have developed dedicated image processing methods to compensate for illumination inhomogeneity, before performing differential analysis. The proposed method relies on the segmentation of the retinal area (Sect. 3), followed by an illumination correction applied on this area (Sect. 4). We do not have prior knowledge of the physical model underlying this non-uniform illumination of the off-axis images, as we use the OCT acquisition system in an unconventional manner. However, we observe that the overall illumination basically varies along the horizontal axis. Therefore, our correction models rely on functions of the horizontal x-coordinate. We also propose post-processing algorithms to achieve a colored representation of aniso-tropic areas (Sect. 5). To the best of our knowledge only one article [6] presents image processing methods for D-OCT analysis. However, a global correction of the overall brightness is applied on each image separately, which does not allow one to deal with cases like the one presented in Fig. 2.

Let us denote by $I_1(x,y)$ and $I_3(x,y)$ the two off-axis images and by $I_2(x,y)$ the standard image. The grayscale values are coded by floating point numbers in [0,1].

3   Retinal Segmentation and Estimation of the Incidence Angle

Segmentation is necessary to determine the area over which the illumination correction has to be applied. We apply the method of [7, 8] to determine the region comprised between the inner limiting membrane (ILM) and the external interface of the hyper reflective complex (HRC), which is also the outer edge of the retinal pigment epithelium (RPE). Horizontally, we restrict the region of interest (ROI) to the interval $[x_L,x_R]$, 6 mm wide, centered on the foveola x_F (Fig. 3).

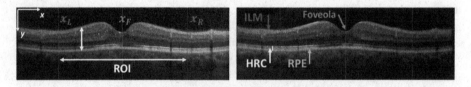

Fig. 3. Segmentation of the retinal data and definition of the ROI.

The beam incidence is manually and approximately set by the operator, as no commercially available OCT apparatus has been designed for such use. Hence, it is interesting to calculate the beam angle α from the image itself. We dramatically increase the contrast in the vitreous and the choroid (Fig. 4(a)) and apply a morpho-logical closing to simplify the image. These operations reveal the main edges, in particular the contours of the rectangular image that is tilted off by the sought for angle α (see the upper right black triangle). We extract the edges (Canny operator) and apply a Hough transform so as to obtain the main linear segments (Fig. 4(b)). We compute the angle $\theta(j)$ and the length $l(j)$ of each extracted segment $j = 1,...J$. For each angle $\theta(j)$, we calculate the sum of the lengths of all segments having the same orientation up to $2°$ (1). The angle with the highest cumulative length L is the incident angle α.

$$L(j) = \sum_{i \in E(j)} l(i), \; E(j) = \{i \in \{1,...,J\}/|\theta(i) - \theta(j)| < 2°\} \tag{1}$$

(a)   (b)

Fig. 4. Estimation of the incident light beam angle from the saturated image (a); linear segments (in green) extracted from the contour image through the Hough transform (b). (Color figure online)

4 Illumination Correction

4.1 Individual Illumination Correction

A first the illumination correction is independently applied to both off-axis images to better balance their overall intensity. We divide the input image $I(x, y)$ (actually $I_1(x, y)$ or $I_3(x, y)$, Fig. 5(a)) into two parts, by slicing it vertically at the foveola x_F. We classify the pixels of the retinal area into 3 classes with a k-means algorithm ($k = 3$ classes), applied to both sub-images independently (Fig. 5(b)). Let us denote by $c_k^{(L)}$, $c_k^{(R)}$, $k = 1,2,3$, the centers of the clusters for the left (L) and right (R) sub-images. These values represent the mean intensity of each cluster. We use them to define straight lines that represent a linear estimation of the intensity variation of the actual value $I_k^{(0)}$ of cluster k with respect to the x coordinate (Fig. 5(b)).

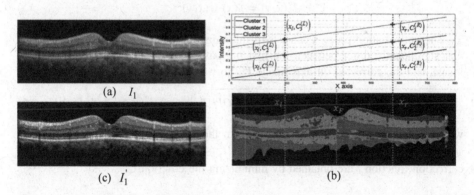

(a) I_1

(c) I_1' (b)

Fig. 5. Individual correction of the intensity of off-axis images; (b) K-means classification applied to input image (a), and linear estimation of the intensity variation; (c) Corrected image.

The two points defining each of the three lines are set at the middle x coordinate of every half image, and vertically take the intensity value of the cluster. We propose modeling the variation of the intensity $I_k^{(0)}$ along the horizontal axis as follows:

$$I_k(x,y) = I_k^{(0)} + a \frac{x - x_F}{x_r - x_l} I_k^{(0)} + b \frac{x - x_F}{x_r - x_l} \tag{2}$$

The parameters a and b are estimated from points $\left(x_l, c_k^{(L)}\right)$ and $\left(x_r, c_k^{(R)}\right)$, $k = 1, 2, 3$, by minimizing the mean quadratic error at these six points. We deduce from (2) the transform to apply to each pixel (x, y) of intensity $I(x, y)$:

$$I'(x,y) = \left[I(x,y) - b \frac{x - x_F}{x_r - x_l}\right] \Big/ \left[1 + a \frac{x - x_F}{x_r - x_l}\right] \tag{3}$$

So, the overall intensity is better balanced over the corrected off-axis images, I_1' and I_3', which are then normalized in $[0,1]$ by linear contrast stretching (Fig. 5(c)).

4.2 Joint Illumination Correction

The corrected images are then jointly processed in the ROI. We propose modeling the illumination correction as 3rd order polynomial function of the x-coordinate:

$$f(x) = ax^3 + bx^2 + cx + d \tag{4}$$

The purpose of the optimization is to define the parameters a, b, c, d of this function. The corrected image is defined by

$$I_3''(x,y) = f(x)\, I_3'(x,y) \tag{5}$$

The optimization process consists of minimizing the number of pixels of the ROI that significantly differ between the first off-axis image I_1', viewed as the reference image, and the corrected off-axis image I_3''. So we define the following criterion:

$$C(a,b,c,d) = card\left\{(x,y) \in ROI / \left|I_1'(x,y) - I_3''(x,y)\right| \geq T\right\} \tag{6}$$

where T is a threshold dynamically set from the source images, equal to twice the standard deviation of the difference image $\left|I_1' - I_3'\right|$. The optimal parameters of the correction function f are obtained by minimizing the criterion C:

$$\left(a_{opt}, b_{opt}, c_{opt}, d_{opt}\right) = Arg \min_{a,b,c,d}\{C(a,b,c,d)\} \tag{7}$$

The image I_3' is corrected by Eq. (5) with the optimal parameters of the polynomial function obtained through (7) (Fig. 6). It is worth noticing that the RPE layer gives a zero response in the difference image, which proves retrospectively that our correction model is valid, as this retinal layer is known to be isotropic.

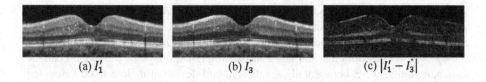

(a) I_1' (b) I_3'' (c) $|I_1' - I_3''|$

Fig. 6. The absolute difference (c) of the two pre-processed off-axis images (a), (b).

5 Differential Analysis

5.1 Color Representation

A color representation of the anisotropic retinal structures is proposed to indicate the angle of maximal response. We calculate the difference image I_D on the ROI:

$$I_D(x,y) = \frac{I_{D0}(x,y)}{\max_{(x,y)}\{|I_{D0}(x,y)|\}}, \text{ where } I_{D0}(x,y) = \begin{cases} I_1'(x,y) - I_3''(x,y) & \text{if } (x,y) \in ROI \\ 0 & \text{otherwise} \end{cases} \tag{8}$$

The operator sets a threshold $S \in [0,1]$ that classifies the pixels (x,y) in two sets: isotropic if $|I_D(x,y)| < S$ and anisotropic otherwise. Two colored images, coded in the RGB color space, are finally derived from this classification. Let us denote by $c \in \{1,2,3\}$ the third coordinate coding the color channel (1 for red, 2 for green and 3 for blue). The first output image I_{RG} shows the anisotropic regions in red and green, with the color coding the angle of maximal response (9). The second output image I_{RGB} provides the same information superimposed on a fusion image, which is defined as the minimum of I_1' and I_3''. Figure 7 shows the output color images obtained with our example, for $S = 0.1$, which is generally a good choice.

$$I_{RG}(x,y,1) = I_D(x,y) \text{ if } I_D(x,y) > S, \quad 0 \text{ otherwise}$$
$$I_{RG}(x,y,2) = -I_D(x,y) \text{ if } I_D(x,y) < -S, \ 0 \text{ otherwise} \tag{9}$$
$$I_{RG}(x,y,3) = 0$$

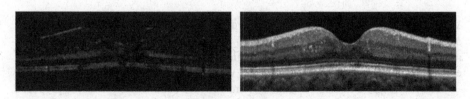

Fig. 7. Color representation of anisotropy: I_{RG} (left), I_{RGB} (right). (Color figure online)

5.2 Regularization

The results presented in Fig. 7 can be regularized based on Markov random field (MRF) techniques [9], since the Markovian hypothesis allows us to take into account spatial interactions between connected pixels. We define $K = 3$ classes corresponding to the regions in red, black and green respectively in I_{RG} (9). The center μ_k of every cluster $k \in \{1, 2, 3\}$ is initialized to the mean intensity in I_D of the pixels belonging to class k. We also assume that the noise in the difference image I_D follows a Gaussian distribution, and we experimentally set its standard deviation σ_k to a fixed value, 0.1. So we have defined the probability distributions $P(I_{D_s}|\omega_s = k)$ of the pixel intensities (denoted by I_{D_s} where s is referred to as a site, i.e. a pixel), conditionally to every class k. We refine the classification according to the Bayesian maximum a posteriori (MAP) criterion, i.e. by looking for the label configuration that maximizes the probability of the class field (the labels) conditionally to the observation field (the intensity image I_D). This optimal configuration corresponds to a minimum state of an energy function U, defined as follows.

$$U(\omega|I_D) = \sum_s \left(\frac{(I_{Ds} - \mu_{\omega_s})^2}{2\sigma_{\omega_s}^2} + \ln\left(\sqrt{2\pi}\sigma_{\omega_s}\right) \right) + \beta \sum_{(s,t)} \varphi(\omega_s, \omega_t) \qquad (10)$$

In this equation, $\varphi(\omega_s, \omega_t)$ refers to the Potts model, expressing interactions between the 8-connected sites s and t. The first term of (10) is related to the image data, while the second term is a regularization term. The parameter β, empirically set ($\beta = 0.9$), weights the relative influence of each. The energy function is minimized by running the simulated annealing (SA) algorithm. Figure 8 shows the resulting images.

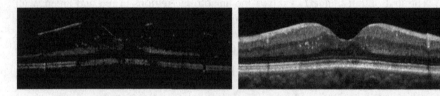

Fig. 8. Final images (I_{RG} and I_{RGB}) obtained after RMF based regularization, less noisy than in Fig. 7, showing abnormal distribution of anisotropy of outer retinal structures.

6   Experimental Results and Conclusion

We have applied the proposed method on a database of 60 sets of D-OCT images, from 33 patients, including normal cases (31%) and pathological cases: 27% of resolved macular edema, 42% with macular telangiectasia. The study followed international ethical requirements and informed consent was obtained from all patients.

The results presented below (Fig. 9) were obtained automatically with the proposed method, with $S = 0.1$. Overall, we observed a good robustness of our algorithm, with about 15% failure, revealed by noisy results or an anisotropy detected in the RPE. The probable causes of failure are saturation of an off-axis image to black or white, off-axis images that are too noisy, or a significant shift in the position of the vertical or horizontal slice, due to eye movement during acquisition, and causing artefacts.

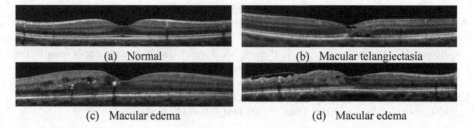

(a) Normal   (b) Macular telangiectasia

(c) Macular edema   (d) Macular edema

Fig. 9. Some clinical results. (a) Control patient, with uniform color in each hemiretinas; (b), (c), (d) Pathological cases: disruption of the coloration suggests abnormal distribution of anisotropy. (Color figure online)

Highlighting the anisotropy of outer retinal layers on OCT scans may help to better interpret some abnormalities seen in the photoreceptor layers. A common finding in retinal diseases is indeed the observation of discontinuities of outer retinal layers, which is usually attributed to loss of photoreceptors. Identifying directional variability of the photoreceptor reflectance helps to identify "hidden" photoreceptor outer structures. We have observed areas of outer retinal layers showing a profile of directional reflectance which is clearly different from normal eyes. In particular, color coding enabled detection of adjacent areas with different directions of peak reflectivity (Figs. 8 and 9). This suggests the presence of photoreceptor disarray, that is, changes in the pointing of photoreceptors. The fact that such disarray was detected in very different clinical situations suggests that it is a common process in retinal diseases. D-OCT may help therefore to disambiguate missing from misaligned cones.

We plan now to refine our acquisition protocol and to increase the number of incidence angles in order to study more accurately reflectance properties.

References

1. Stiles, W.S., Crawford, B.S.: The luminous efficiency of rays entering the eye pupil at different points. Proc. R. Soc. Lond. **112**, 428–450 (1933)
2. Westheimer, G.: Directional sensitivity of the retina: 75 years of Stiles-Crawford effect. Proc. Biol. Sci. **275**, 2777–2786 (2008)
3. Gao, W., Cense, B., Zhang, Y., et al.: Measuring retinal contributions to the optical Stiles-Crawford effect with optical coherence tomography. Opt. Express **16**, 6486–6501 (2008)
4. Miloudi, C., Rossant, F., Bloch, I., et al.: The negative cone mosaic: a new manifestation of the optical stiles-crawford effect in normal eyes. Invest. Ophthalmol. Vis. Sci. **56**, 7043–7050 (2015)
5. Lujan, B.J., Roorda, A., Croskrey, J.A., et al.: Directional optical coherence tomography provides accurate outer nuclear layer and Henle fiber layer measurements. Retina **35**, 1511–1520 (2015)
6. Makhijani, V.S., Roorda, A., Bayabo, J.-K., et al.: Chromatic visualization of reflectivity variance within hybridized directional OCT images. In: Proceedings of SPIE 8571, Optical Coherence Tomography and Coherence Domain Optical Methods in Biomedicine XVII, 857105, 20 March, 2013. doi:10.1117/12.2007141

7. Ghorbel, I., Rossant, F., Bloch, I., Tick, S., Pâques, M.: Automated segmentation of macular layers in OCT images and quantitative evaluation of performances. Pattern Recogn. **44**(8), 1590–1603 (2011)
8. Rossant, F., Bloch, I., Ghorbel, I., Pâques, M.: Parallel double snakes. Application to the segmentation of retinal layers in 2D-OCT for pathological subjects. Pattern Recogn. **48**(12), 3857–3870 (2015)
9. Geman, S., Geman, D.: Stochastic relaxation, Gibbs distribution, and the Bayesian restauration of images. IEEE Trans. PAMI **6**(6), 721–741 (1984)

Contrast Enhancement by Top-Hat and Bottom-Hat Transform with Optimal Structuring Element: Application to Retinal Vessel Segmentation

Rafsanjany Kushol[(✉)], Md. Hasanul Kabir, Md Sirajus Salekin,
and A.B.M. Ashikur Rahman

Department of Computer Science and Engineering,
Islamic University of Technology, Dhaka, Bangladesh
{kushol,hasanul,salekin,ashikiut}@iut-dhaka.edu

Abstract. Automatic detection of the retinal blood vessel can be used in biometric identification, computer assisted laser surgery, and diagnosis of many eye related diseases. Early detection of retinal blood vessel helps people to take proper treatment against diseases such as diabetic retinopathy, hypertension which can significantly reduce possible vision loss. This paper presents an efficient and simple contrast enhancement technique where morphological operations like top-hat and bottom-hat are applied to enhance the image. Edge Content-based contrast matrix is measured for selecting the optimal structuring element size and simple straightforward steps are applied for completely extracting the vessels from the enhanced retinal image. The proposed method acquires an average accuracy rate of 0.9379 and 0.9504 on two publicly available DRIVE and STARE benchmark dataset respectively.

Keywords: Retina · Top-hat · Bottom-hat · Vessel segmentation

1 Introduction

Medical image processing has become one of the most significant and essential parts among the modern medical health care community. Automatic detection and analysis of the retinal vascular network is helpful for computer assisted laser surgery as well as speed up the diagnosis of comprehensive retinal vessel related diseases [1]. The evaluation and diagnosis of various ophthalmologic diseases like diabetes, arteriosclerosis, and choroidal neovascularization can be performed by measuring morphological characteristics of retinal vessel such as width, length, branching pattern, and angles [2]. Doctors workload can be reduced by accurate and automatic vessel segmentation system since it provides several useful features for the treatment of various eye related diseases. However, retinal blood vessel structure plays a vital role in the detection of diabetic retinopathy which gradually leads to retinal revascularization and vision loss. Furthermore, there

F. Karray et al. (Eds.): ICIAR 2017, LNCS 10317, pp. 533–540, 2017.
DOI: 10.1007/978-3-319-59876-5_59

are some other diseases like hypertensive retinopathy which can be identified based on the changes in reflectivity of the arterioles and tortuosity of retinal blood vessels [1]. Isolation of the blood vessel from the background is not an easy task. Different challenges such as variation in the vessel width, branching pattern, image resolution, and shape need to be overcome for a proper solution approach. There can be some similar types of objects in the retina like exudates, lesions, hemorrhage, optic disk, and cotton wool spots which usually produce false positive results in the output. Moreover, vessel crossing, central light reflex, bifurcation, and less variation in the contrast between vessel tree and surrounding tissues make it difficult to properly identify the thin vessels.

This paper presents a new blood vessel enhancement technique using top-hat and bottom-hat transformation with optimal structuring element which can efficiently enhance the retinal blood vessels. The enhanced image helps to separate the blood vessels from the background image more accurately.

2   Literature Review

A substantial amount of methods has been developed to accurately separate the blood vessels from retinal images. The first significant concept for detecting blood vessels was the matched filter. Chaudhuri et al. [3] used a two-dimensional Gaussian linear kernel for separating retinal blood vessels as the intensity profile along the cross section of a vessel has the Gaussian-shaped curve. To improve the efficiency of this concept some variants [4–6] have been proposed. Hoover et al. [4] introduced a threshold probing technique on a matched filter response image. The methodology described by Zhang et al. [5] combines a pair of filters to detect the vessels, the zero-mean Gaussian filter (MF) and the first-order derivative of the Gaussian (FDOG). A self-adaptive matched filter is proposed by Chakraborti et al. [6] where the matched filter with high specificity and the vesselness filter with high sensitivity is obtained using orientation histogram.

Multiscale approaches are introduced based on the idea of varying width of the vessels. Vlachos and Dermatas [7] proposed multiscale line tracking method for retinal blood vessel segmentation which starts its procedure from a small group of pixels, derived from a brightness selection rule and terminates when a cross-sectional profile condition becomes invalid.

Morphological processing is a collection of non-linear operations to represent shape or morphological attributes of an image. Mendonca and Campilho [8] presented an algorithm which combines directional differential filters for centerline extraction and iterative region growing method for filling vessel segments. Miri and Mahloojifar [9] introduced curvelet transform coefficients to enhance the retinal image edges with morphological operators by reconstruction.

Pattern classification and machine learning based methods try to extract features from the images which are trained with either supervised or unsupervised classifier. The methods produce segmentation by classifying each image pixel as vessel or non-vessel, based on the pixels feature vector. Soares et al. [10] composed feature vectors from the pixels intensity and 2-D Morlet wavelet transform responses taken at multiple scales with a Bayesian classifier. A 9-D feature vector based on the orientation analysis of gradient vector field, morphological transformation, line strength measures, and Gabor filter responses is designed by Fraz et al. [11]. Marin et al. [12] presented a supervised method for blood vessel detection by using a neural network (NN) classifier where a 7-D feature vector is composed of gray-level and moment invariants-based features.

Some works [13,14] focus on image contrast enhancement. Lidong et al. [13] combined Contrast limited adaptive histogram equalization and discrete wavelet transform for vessel enhancement. By using morphological filters Hassanpour et al. [14] introduced another enhancement technique for medical images.

3 Proposed Method

The overall idea of our proposed method for retinal blood vessel segmentation is depicted in Fig. 1. It uses the following steps: (1) Pre-processing, (2) Retinal Vessel Enhancement, (3) Background Exclusion and Classification, (4) Post Processing.

3.1 Pre-processing

Among the three channels of the retinal RGB image, the green channel appears the best contrast for vessel background. So the RGB color image is converted to the Gray-scale image by extracting the green channel only. Some false gap inside the vessel can be generated by central light reflex. Morphological opening is applied to take aside this problem with a structuring element of three-pixel radius disk. Finally, the border of the inverted green plane image is extended by the pre-processing step used in [10] to dispel undesired border effects.

3.2 Retinal Vessel Enhancement

In image processing, morphological operators are mainly used for characterization of region shapes, such as boundaries and skeletons. The most commonly used operators are dilation and erosion which employ a structuring element (SE).

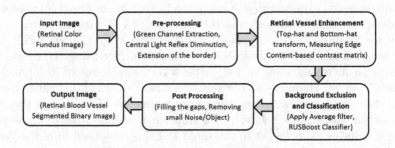

Fig. 1. Flow chart of the proposed method

Dilation is used for filling gaps and connecting disjoint regions whereas erosion is performed for eliminating irrelevant detail. The other two main operations are opening and closing. Opening is performed by an erosion followed by a dilation whereas closing is executed by a dilation followed by an erosion. Following equations represent opening and closing respectively where A is an image matrix and B is the SE.

$$A \circ B = (A \ominus B) \oplus B \tag{1}$$

$$A \bullet B = (A \oplus B) \ominus B \tag{2}$$

For image enhancement, two morphological transformation Top-hat and Bottom-hat are used frequently. They are very effective tool in medical imaging for enhancing detail in presence of shading or dark areas. The top-hat is defined as the difference between the input image and its opening by a SE whereas the bottom-hat is the difference between the closing and the input image. By applying top-hat we can retrieve those objects or elements which are smaller than the SE as well as brighter than their surroundings. On the other hand, bottom-hat produces those objects which are smaller than the SE and darker than their surroundings. So we can take the advantage of these two operators by adding the top-hat result and subtracting the bottom-hat result. Therefore, by adding the bright regions and by removing the dark areas we can obtain an enhanced image. Top-hat, bottom-hat, and enhanced image can be expressed as follows.

$$A_{top} = A - (A \circ B) \tag{3}$$

$$A_{bot} = (A \bullet B) - A \tag{4}$$

$$A_{enhance} = A + A_{top} - A_{bot} \tag{5}$$

Among four types of shape, we have chosen disk type of SE as they offer the rotation invariance property. However, depending on the SE size, the contrast of the image may vary a lot. For this reason, to get the most contrast enhanced image we need a dynamic procedure for selecting the SE size. Gupta et al. [15] discussed 15 contrast enhancement measures for brain and breast cancer images. Among those methods, they found Edge Content-based contrast matrix [16] provides the best response. In our study, some conventional contrast measurement techniques are also experimented. Finally, the Edge Content-based contrast value is applied for dynamically selecting the disk size. An exhaustive search is performed for finding the best response by calculating Edge Content (EC) value where the radius size starts from 5 and ends at 60. Increasing the radius also results in an increased value in EC. After a certain amount of iterations, the value of EC is not changing with a significant amount which indicates increasing the disk size is not enhancing the contrast anymore. Therefore, a threshold T is used for finding this behavior. We stop our searching if three consecutive increased value of EC is less than threshold T. EC is computed from the magnitude of the gradient vector. The changes in contrast of different shapes inside an area can be assessed by EC and it has the ability to capture the changes even if the pixels are closely connected. The gradient vector of an image $A(x, y)$ at any pixel position (x, y) and for an image block of size $m \times n$ where $1 \leq x \leq m$ and $1 \leq y \leq n$ the EC is computed as follows.

$$\triangledown A(x, y) = \begin{bmatrix} G_x \\ G_y \end{bmatrix} = \begin{bmatrix} \frac{\partial}{\partial x} A(x, y) \\ \frac{\partial}{\partial y} A(x, y) \end{bmatrix} \tag{6}$$

$$EC = \frac{1}{(m \times n)} \sum_x \sum_y |\triangledown A(x, y)| \tag{7}$$

Table 1 provides the result of different SE and corresponding EC value for two images. The optimal SE size found for these two are 31 and 27 (1st image of STARE and DRIVE). Figure 2 shows the comparison among traditional methods with our proposed idea where we can visualize the improvement of our approach.

Table 1. Statistics of different SE size and corresponding EC value for two images of STARE and DRIVE dataset

1st Image of STARE		1st Image of DRIVE	
SE size	EC value	SE size	EC value
5	0.0859	5	0.0876
10	0.1016	10	0.1007
15	0.1073	15	0.1052
20	0.1107	20	0.1086
25	0.1126	25	0.1100
30	0.1141	30	0.1110
35	0.1152	35	0.1116
40	0.1162	40	0.1119
45	0.1172	45	0.1123
50	0.1179	50	0.1124
55	0.1185	55	0.1126
60	0.1191	60	0.1128

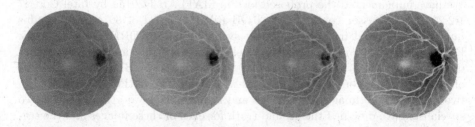

Fig. 2. Output of Gray-scale image, Green channel extraction, Contrast Limited Adaptive Histogram Equalization, and Proposed method respectively (from left to right)

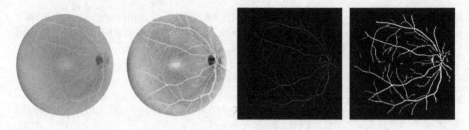

Fig. 3. Output of Original image, Enhanced image, Background subtracted image, and RUSBoost classification performed image respectively (from left to right)

3.3 Background Exclusion and Classification

The background of the retinal image is removed by applying an averaging filter of size 12×12 over the enhanced image. Then the enhanced image is subtracted from the average filtered image. The output of background subtracted image is shown in Fig. 3. Finally, RUSBoost [17] classifier is applied on train images of DRIVE dataset. By combining data sampling and boosting, RUSBoost provides an efficient method for improving classification performance. The binary segmented output image is generated from the classification result.

3.4 Post Processing

Two things are considered for post-processing step. First one is filling the gaps of inside vessel tree and the second one is removing small noise or object. The concept of Marin et al. [12] is applied for filling the gaps where an iterative filling operation is performed. Some non-vessel pixels are classified as vessel points if at least six neighbors are also vessel pixels. Furthermore, the area of each connected region is computed and small regions are removed by morphological area open operation. All connected components with a region below 40 pixels are eliminated in this final step. The outputs of the above-mentioned steps are shown in Fig. 3.

4   Experimental Results and Performance Evaluation

We have implemented the proposed idea in MATLAB R2015a by Intel Core i7 @2.20 GHz processor containing 8.00 GB RAM. A total of 60 colorful fundus images is evaluated from two publicly available dataset DRIVE and STARE. DRIVE (Digital Retinal Images for Vessel Extraction) [18] consists of 40 eye fundus images with 45° field of view (FOV) whereas STARE (STructured Analysis of the Retina) [4] consists of 20 images with 35° FOV. DRIVE is organized in two parts, first 20 images are test set and last 20 images are training set. Two specialists have designed the ground truth for each of these images and the segmentation of the first specialist is considered as ground truth in our experimental outcomes. In the case of STARE, each image has a resolution of $700 * 605$ pixels with 8 bits per color plane in PPM format. The database contains two sets of

Table 2. Performance comparison of vessel segmentation methods on DRIVE dataset

No	Techniques	Year	Average Sn	Average Sp	Average Acc
1	Chaudhuri et al. [3]	1989	0.2663	0.9901	0.8773
2	Mendonca and Campilho [8]	2006	0.7344	0.9764	0.9452
3	Soares et al. [10]	2006	0.7332	0.9782	0.9461
4	Zhang et al. [5]	2010	0.7120	0.9724	0.9382
5	Vlachos and Dermatas [7]	2010	0.747	0.955	0.929
6	Miri and Mahloojifar [9]	2011	0.7352	0.9795	0.9458
7	Marin et al. [12]	2011	0.7067	0.9801	0.9452
8	Fraz et al. [11]	2012	0.7406	0.9807	0.9480
9	Chakraborti et al. [6]	2015	0.7205	0.9579	0.9370
10	**Proposed Method**	**2017**	**0.6296**	**0.9830**	**0.9379**

manual segmentations made by two different experts from which segmentation of the first expert is considered as ground truth in our experimental analysis.

Common statistical measures Sensitivity (Sn), Specificity (Sp), and Accuracy (Acc) are calculated for the performance evaluation. The performance comparison of our approach with some existing works is shown in Table 2. Our method also achieves an average Sn, Sp, and Acc rate of 0.6495, 0.9847, and 0.9504 respectively on STARE dataset. As our focus is not segmentation part, that's why the Acc especially the Sn is little bit lower than most of the methods. The Acc can be improved by applying the prominent segmentation approaches over our enhanced images. Most of the methods fail to detect the thin vessels. But our method can increase the contrast throughout the whole vessel tree. Thus it provides better information for detecting thin vessels and achieves better Acc.

5   Discussion and Conclusion

An efficient blood vessel enhancement technique can accelerate the further processing more accurately. Our proposed image enhancement technique can be applied with most of the current methods as a pre-processing step to improve the accuracy. It can also be utilized for visualizing and detecting different important elements of medical images more clearly. Our future work will be introducing a robust blood vessel segmentation technique from our proposed enhanced image where the segmentation and classification can retrieve the blood vessels more accurately. Wavelet-based approaches already proved their efficiency for separating blood vessels from the background. By applying those techniques over our enhanced image will help to detect thin vessels more accurately.

References

1. Joussen, A.M., Gardner, T.W., Kirchhof, B., Ryan, S.J.: Retinal Vascular Disease. Springer, Heidelberg (2007)

2. Fraz, M.M., Remagnino, P., Hoppe, A., Uyyanonvara, B., Rudnicka, A.R., Owen, C.G., Barman, S.A.: Blood vessel segmentation methodologies in retinal images a survey. Comput. Methods Prog. Biomed. **108**(1), 407–433 (2012)
3. Chaudhuri, S., Chatterjee, S., Katz, N., Nelson, M., Goldbaum, M.: Detection of blood vessels in retinal images using two-dimensional matched filters. IEEE Trans. Med. Imaging **8**(3), 263–269 (1989)
4. Hoover, A., Kouznetsova, V., Goldbaum, M.: Locating blood vessels in retinal images by piecewise threshold probing of a matched filter response. IEEE Trans. Med. Imaging **19**(3), 203–210 (2000)
5. Zhang, B., Zhang, L., Zhang, L., Karray, F.: Retinal vessel extraction by matched filter with first-order derivative of Gaussian. Comput. Biol. Med. **40**(4), 438–445 (2010)
6. Chakraborti, T., Jha, D.K., Chowdhury, A.S., Jiang, X.: A selfadaptive matched filter for retinal blood vessel detection. Mach. Vis. Appl. **26**(1), 55–68 (2015)
7. Vlachos, M., Dermatas, E.: Multi-scale retinal vessel segmentation using line tracking. Comput. Med. Imaging Graph. **34**(3), 213–227 (2010)
8. Mendonca, A.M., Campilho, A.: Segmentation of retinal blood vessels by combining the detection of centerlines and morphological reconstruction. IEEE Trans. Med. Imaging **25**(9), 1200–1213 (2006)
9. Miri, M.S., Mahloojifar, A.: Retinal image analysis using curvelet transform and multistructure elements morphology by reconstruction. IEEE Trans. Biomed. Eng. **58**(5), 1183–1192 (2011)
10. Soares, J.V., Leandro, J.J., Cesar, R.M., Jelinek, H.F., Cree, M.J.: Retinal vessel segmentation using the 2-D Gabor wavelet and supervised classification. IEEE Trans. Med. Imaging **25**(9), 1214–1222 (2006)
11. Fraz, M.M., Remagnino, P., Hoppe, A., Uyyanonvara, B., Rudnicka, A.R., Owen, C.G., Barman, S.A.: An ensemble classification-based approach applied to retinal blood vessel segmentation. IEEE Trans. Biomed. Eng. **59**(9), 2538–2548 (2012)
12. Marin, D., Aquino, A., Gegundez-Arias, M.E., Bravo, J.M.: A new supervised method for blood vessel segmentation in retinal images by using gray-level and moment invariants-based features. IEEE Trans. Med. Imaging **30**(1), 146–158 (2011)
13. Lidong, H., Wei, Z., Jun, W., Zebin, S.: Combination of contrast limited adaptive histogram equalisation and discrete wavelet transform for image enhancement. IET Image Process. **9**(10), 908–915 (2015)
14. Hassanpour, H., Samadiani, N., Salehi, S.M.: Using morphological transforms to enhance the contrast of medical images. Egypt. J. Radiol. Nucl. Med. **46**(2), 481–489 (2015)
15. Gupta, S., Porwal, R.: Appropriate contrast enhancement measures for brain and breast cancer images. Int. J. Biomed. Imaging **2016**, 8 p. (2016). Article ID 4710842. doi:10.1155/2016/4710842
16. Saleem, A., Beghdadi, A., Boashash, B.: Image fusion-based contrast enhancement. EURASIP J. Image Video Process. **2012**(1), 1 (2012)
17. Seiffert, C., Khoshgoftaar, T.M., Van Hulse, J., Napolitano, A.: Rusboost: a hybrid approach to alleviating class imbalance. IEEE Trans. Syst. Man Cybern.-Part: A Syst. Hum. **40**(1), 185–197 (2010)
18. Niemeijer, M., Staal, J., Ginneken, B.V., Loog, M., Abramoff, M.D.: Comparative study of retinal vessel segmentation methods on a new publicly available database. In: International Society for Optics and Photonics, Medical Imaging 2004, pp. 648–656 (2004)

Retinal Biomarkers of Alzheimer's Disease: Insights from Transgenic Mouse Models

Rui Bernardes[1,2]([⊠]), Gilberto Silva[1], Samuel Chiquita[1], Pedro Serranho[1,2,3], and António Francisco Ambrósio[1]

[1] Institute for Biomedical Imaging and Life Sciences (IBILI), Faculty of Medicine, University of Coimbra, Coimbra, Portugal
rmbernardes@fmed.uc.pt
[2] Coimbra Institute for Biomedical Imaging and Translational Research (CIBIT), ICNAS, University of Coimbra, Coimbra, Portugal
[3] Department of Sciences and Technology, Universidade Aberta, Lisboa, Portugal

Abstract. In this paper, we use the retina as a window into the central nervous system and in particular to assess changes in the retinal tissue associated with the Alzheimer's disease. We imaged the retina of wild-type (WT) and transgenic mouse model (TMM) of Alzheimer's disease with optical coherence tomography and classify retinas into the WT and TMM groups using support vector machines with the radial basis function kernel. Predictions reached an accuracy over 80% at the age of 4 months and over 90% at the age of 8 months. Texture analysis of computed fundus reference images suggests a more heterogeneous organization of the retina in transgenic mice at the age of 8 months in comparison to controls.

Keywords: Alzheimer's disease · 3xTg mouse model · Optical coherence tomography · Retina · Classification

1 Introduction

In this work, we explore the use of statistical descriptors of Optical Coherence Tomography (OCT) data of wild-type (WT) and transgenic mouse model (TMM) of Alzheimer's Disease (AD). Our objective is two-fold. First, we aim to test the classification approach previously applied to OCT data of human retinas to classify those into the healthy control and disease groups [1]. Second, we aim to clarify if changes in the retina are present and can be detected in the early stages of the disease.

According to the WHO (World Health Organization) report [2], projections from 2013 estimate a total of over 47 million people living with dementia (including AD) in 2015. Projections for 2030 are of 66 million, resulting in the indisputable burden for patients and caregivers. Although these figures account for all types of dementia, AD is the most common type and estimated to represent 60% to 80% of the cases [3].

© Springer International Publishing AG 2017
F. Karray et al. (Eds.): ICIAR 2017, LNCS 10317, pp. 541–550, 2017.
DOI: 10.1007/978-3-319-59876-5_60

One of the fundamental problems tackling AD is the lack of early diagnostic biomarkers, the reason why AD may develop undiagnosed for over a decade [4]. While the definite diagnosis of AD can only be "confirmed after a postmortem examination of the brain" [4], up to now the "probable" diagnosis is possible only when "significant neurological damage has already occurred" [4].

A particular effective diagnostic tool requires positron emission tomography (PET) imaging facilities and the use of the Pittsburgh compound-B (PiB) agent, therefore severely limiting the access to this biomarker due to the (high) associated costs and due to the limited number of PET facilities [4].

Ideally, a biomarker should be easy to collect and measure in addition to the objective and reproducible mandatory characteristics [5]. The definition of biomarker clearly points to the need for a different approach when it comes to AD diagnosis. The use of the retina as a window into the brain is being followed by an increasing number of research groups and appears as the natural choice. The retina is actually the visible part of the central nervous system (CNS). There is also cumulating evidence indicating that the retina can also be affected in neurodegenerative diseases [6].

OCT is a widespread imaging technique in the field of ophthalmology and is an important tool to diagnose a variety of diseases. OCT provides the absorption and scattering properties of the tissues and thus inherently conveys pertinent information on the structure of the retina. In addition to the basic information on the thickness of the retina, previous studies conducted by our research group demonstrate that information gathered by the OCT allows to classify eyes, automatically, into the Parkinson, Diabetic Retinopathy and Multiple Sclerosis groups [7,8]. These works paved the way for the work herewith reported.

2   Methodology

The use of transgenic mouse models of AD allows overcoming a fundamental problem when trying to collect data in the early stages of the disease in humans, that the diagnosis of AD is possible only after a significant neurological damage occurred [4]. Furthermore, it allows to follow-up the natural development and aging of the retina for WT controls and TMM and to compare the two groups at the same age at regular intervals during their lifespan.

The TMM used in this study is the 3xTg-AD (triple-transgenic mouse model). This model presents characteristics of the human form of the disease, namely the aggregation of amyloid-β (Aβ) protein accumulating into plaques and the hyperphosphorylation of tau protein [9].

In this work, we report data gathered at the ages of 4 and 8 months.

2.1   OCT Imaging

OCT is a noninvasive diagnostic imaging technology able to provide cross-sectional images (B-scans) of the retinal tissue *in vivo* and *in situ* (Fig. 1). Its working principle is analogous to that of ultrasonography with light taking the role of sound [10]. Furthermore, OCT terms such as A-scan and B-scan match the definition of those in ultrasound. The principle, based on the backscattering of low-coherence light has been extensively described in the literature [11,12]. OCT readings convey information on refractive index changes along the light path. Consequently, any change in the content or structural organization of the retina, on the healthy condition, is captured in the statistics of the signal.

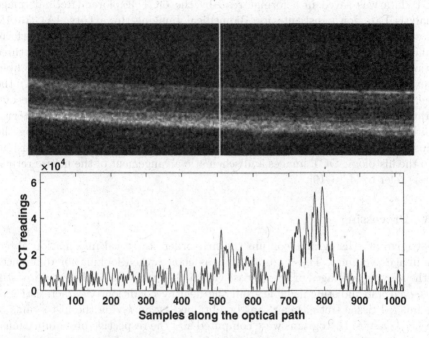

Fig. 1. Optical coherence tomography. Top: B-scan of the right eye of a wild-type (control) mice at the age of 8 months. Bottom: Plot of the A-scan earmarked in yellow (B-scan above). A-scan values up to sample ~500 correspond to OCT readings within the vitreous. A-scan values from sample ~500 to ~800 correspond to OCT readings within the retina and A-scan values from sample ~800 to 1024 correspond to OCT readings within the choroid. The light travels from top to bottom (B-scan) and left to right (A-scan). (Color figure online)

The system used in this study is the Micron IV OCT (Phoenix Research Labs, Pleasanton, CA, USA). It is tailored to image mouse and rat retinas and allows to scan the posterior pole of the eye gathering 512 by 512 by 1024 voxels over 50° field of view.

The animals were anesthetized and pupils dilated. Animals were then placed in front of the OCT for data acquisition with all scans being performed at the same retinal location, horizontally centered on the optic disc and vertically above the optic disc border.

A total of 512 B-scans, of 512 A-scans each, were produced per eye scan. Each B-scan is saved in the computer disc as an uncompressed TIFF file of 512 by 1024 pixels. The 512 images were then read using Matlab 9.1.0 (R2016b) (The MathWorks Inc., Natick, MA, USA) for processing and analysis.

2.2   Pre-processing

OCT data was saved in a format read by the OCT Explorer (Retinal Image Analysis Lab, Iowa Institute for Biomedical Imaging, Iowa City, IA) [13–15] software for segmentation. In particular, we take advantage of the ease to perform manual segmentation/correction using this software interface to segment three major layers. The first major layer (Layer 1) comprises the retinal nerve fiber layer (RNFL), the ganglion cell layer (GCL), the inner plexiform layer (IPL), the inner nuclear layer (INL) and the outer plexiform layer (OPL) layers. The second major layer (Layer 2) comprises the outer nuclear layer (ONL) and the external limiting membrane (ELM), and; the third major layer (Layer 3) comprises the ellipsoid zone and retinal pigment epithelium (RPE) layers. For a detailed insight into the histology, OCT images and schematic arrangement of the mouse retina, please refer to [4, 16–19].

2.3   Processing

Histograms. Histograms contain the first-order statistical information about the images/volumes. These are an obvious choice when looking for differences in the very early stages of the disease, when the structure of the retina is still preserved and indistinguishable from the healthy condition. Characteristics of the imaged tissue (retina) can thus be obtained by analyzing the histograms of the OCT values. Histograms were computed and the respective probability mass function (pmf) determined separately for Layer 1 ($j = 1$), Layer 2 ($j = 2$), Layer 3 ($j = 3$) and the whole retina ($j = 4$).

Parameters defining the shape of the pmf are then computed by fitting the function defined by (1)

$$f(x) = \sum_{i=0}^{2} A_i e^{-(x-x_i)^2/(\sigma_i^2)} \tag{1}$$

to the pmf, from where a vector of parameters ($\mathbf{v}_j = [A_0, x_0, \sigma_0, A_1, x_1, \sigma_1, A_2, x_2, \sigma_2]$) ($j = \{1, 2, 3, 4\}$) is computed conveying information on the characteristics of the tissue.

Each of the four individual vectors ($\mathbf{v}_j$) is transformed into the vector $\mathbf{w}_j = [A_0, \sigma_0, A_1, \sigma_1, A_2, \sigma_2, x_0-x_1, x_0-x_2, A_0/A_1, A_0/A_2, \sigma_0/\sigma_1, \sigma_0/\sigma_2, \text{RMSE}]$, where RMSE stands for the root mean squared error of the fit.

Mean Value Fundus Images. Different approaches can be used to compute a fundus reference image from an OCT volume data. These approaches share the idea of using depth-wise averaging (total or partial) of each A-scan [20]. In this work, for each of the three layers (Layer 1, Layer 2 and Layer 3) a mean value fundus (MVF) image [21] is computed by (2) as the average of the A-scan values within the respective layer.

$$MVF_i(x,y) = \frac{1}{Z_i^2(x,y) - Z_i^1(x,y) + 1} \sum_{z=Z_i^1(x,y)}^{Z_i^2(x,y)} V(x,y,z), \qquad (2)$$

where V is the OCT volume (of size $512 \times 512 \times 1024$ voxels), $Z_i^1(x,y)$ and $Z_i^2(x,y)$ are the limits of Layer i ($i = \{1,2,3\}$) at coordinates (x,y) and the coordinate system for the OCT data is defined as: x is the nasal-temporal direction, y is the superior-inferior direction, and z is the anterior-posterior (depth) direction [21].

Texture Analysis. Texture analysis is an image analysis technique used in a wide range of applications with special emphasis on pattern recognition.

In this work, we resort to common texture analysis techniques [22] to find whether the structural arrangement of the retina may be different in mice model of AD in comparison to those of controls. In addition, it may provide further insight into the natural aging process through the analysis of data from WT animals imaged over time.

In this exploratory approach, we make use of the Energy (3) and Contrast (4) statistical properties of the image derived from the gray-level co-occurrence matrix (GLCM). In particular, the number of gray levels of computed fundus reference images was reduced to 16 to limit the size of the GLCM. Furthermore, we split each image into 7 by 7 blocks (of equal size) with an overlap of 10% and analyze each independently of each other, in different directions and at different scales. Besides, with respect to the RNFL, different orientations are expected because fibers converge to the optic disc. Moreover, when considering the human retina, where the OCT scans are performed centered on the fovea, fibers are radially oriented away from it. The maximum energy and maximum contrast across the combination of directions (0°, 45°, 90° and 135°) and scales (1, 2, 3 and 4) were chosen as the value for the block, respectively, for energy and contrast.

$$E_\theta^\alpha = \sum_{i,j} p(i,j)^2, \qquad (3)$$

$$C_\theta^\alpha = \sum_{i,j} |i-j|^2 p(i,j), \qquad (4)$$

where $p(i,j)$ is the joint probability occurrence of pixel pairs having values i and j in the image and α and θ are, respectively, the scale and direction.

2.4 Classification

Each eye scan is characterized by a vector of 52 parameters and is labeled as "C", if belonging to the group of controls (WT), and as "D" otherwise (TMM).

Classification is performed using support vector machines (SVM) with radial basis function (RBF) kernel [23]. The SVM is a supervised learning algorithm able to infer a function from labeled training data, a set of training examples for which the classification is known beforehand. A model (discriminating function) is thus created and later applied to classify new cases.

As not all the features may be required, the backward elimination algorithm was used to increase the performance (i.e., accuracy in classifying unknown cases) of the system. From the initial set of parameters, a subset (of dimension 6) was determined to allow for an increased accuracy. The system accuracy was determined using the k-fold cross-validation procedure, where the cases were split into k sets. $k - 1$ sets were used for training and the set left out was then classified using the established model. The accuracy was determined by comparing the obtained classifications into the "C" and "D" classes with the actual ones. The process repeats k times.

The working hypothesis, that OCT embeds information on the status of the CNS, allowing to discriminate between controls and AD cases, will prove to be correct should the system be able to classify eyes, in the WT and TMM groups, with an accuracy over the simple randomization one (equivalent to tossing a coin).

3   Results

Two eyes per animal were imaged by the OCT system, at the ages of 4 and 8 months, and the respective OCT volume data exported and processed. The number of eyes and scans performed can be found in Table 1. These do not account for the 13 out of 90 (14.4%) scans dropped due to acquisition and segmentation errors.

Table 1. Number of eyes and valid scans performed at the ages of 4 and 8 months. WT – wild-type mice (controls), and; TMM – Transgenic mouse model of Alzheimer's disease.

Age	WT		TMM	
(months)	Eyes	Scans	Eyes	Scans
4	20	20	12	12
8	26	20	32	25

The number of cases in each class ("C" and "D") for every classification performed was matched every time, not to bias results. Cases were randomly

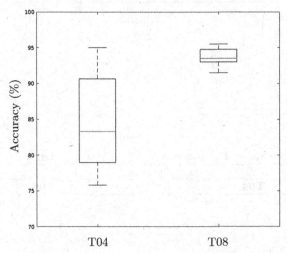

Fig. 2. Classification performance in discriminating wild-type mice (WT) from triple-transgenic mouse model (3xTg-AD) (TMM) of Alzheimer's Disease at time points T04 (4 months) and T08 (8 months), respectively at the ages of 4 and 8 months. Minimum/Median/Maximum accuracy values (%) are of 76/83/95 and 92/94/96, respectively for time points T04 and T08.

selected, and the process repeated multiple times. In consequence, accuracy results are shown as a distribution as opposed to a single value (Fig. 2).

Achieved results strongly suggest that retinas of transgenic mice are actually different from that of healthy controls at the same age. Additionally, results demonstrate that differences do exist at the age of 4 months and that the differences are even clearer at the age of 8 months, as shown by the increase from the minimum 76% to a minimum of 92% of classification accuracy (Fig. 2).

Results from the texture analysis can be found in Fig. 3. At the age of 4 months, WT and TMM mice do show similar values for both the energy and contrast at all layers. At the age of 8 months, the contrast is reduced for the TMM group in comparison to that of the WT group. For the energy, differences become clearer. In particular, for Layer 1 the energy distribution is similar but with an increase in the mean value for the TMM group. For Layers 2 and 3, the distributions show that energy values of WT mice do have a much more homogeneous distribution than that of the TMM group. These findings suggest a difference in tissue properties between controls and mouse model of disease which is in agreement with results from the classification using the SVM algorithm.

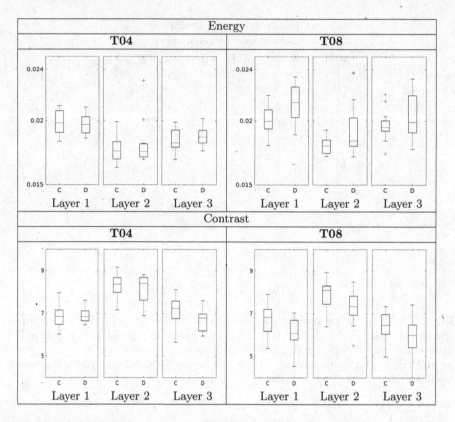

Fig. 3. Texture analysis: energy and contrast from the gray-level co-occurrence matrix (GLCM) of mean value fundus (MVF) images for Layers 1 to 3 and time points T04 and T08, respectively mice at the ages of 4 and 8 months. C – wild-type (WT) mice group; D – transgenic mouse model (TMM) group.

4   Conclusions

In this paper, we have tested the approach followed previously for the classification of human retinas to find it possible to apply it to OCT scans of mice retinas with success.

Our data show that not only it is possible to use the OCT to get an insight on changes occurring in the retina of mouse model of Alzheimer's disease, but that differences are sufficient to allow to discriminate eyes of controls from that of the transgenic group, with an accuracy in the prediction of over 80% at the age of 4 months and over 90% at the age of 8 months (median values).

Finally, found differences in texture of fundus images computed for the layers considered in this work suggest a more heterogeneous organization of the retina in transgenic mice at the age of 8 months, while at the age of 4 months results are similar.

Acknowledgements. This study was supported by the Neuroscience Mantero Belard Prize 2015 (Santa Casa da Misericórdia)(MB-1049-2015), by The Portuguese Foundation for Science and Technology (PEst-UID/NEU/04539/2013), by FEDER-COMPETE (POCI-01-0145-FEDER-007440) and Centro 2020 Regional Operational Programme (CENTRO-01-0145-FEDER-000008: BrainHealth 2020).

References

1. Bernardes, R., Serranho, P., Santos, T., Gonçalves, V., Cunha-Vaz, J.: Optical coherence tomography: automatic retina classification through support vector machines. Eur. Ophthalmic Rev. **6**(4), 200–203 (2012)
2. Prince, M., Guerchet, M., Prina, M.: The Epidemiology and Impact of Dementia: Current State and Future Trends. World Health Organization, Geneva (2015)
3. Alzheimer's Association. 2010 Alzheimer's disease facts and figures. Alzheimer's & Dementia **6**(2), 158–194 (2010). doi:10.1016/j.jalz.2010.01.009
4. Krantic, S., Torriglia, A.: Retina: source of the earliest biomarkers for Alzheimer's disease? J. Alzheimers Dis. **40**(2), 237–243 (2014). doi:10.3233/JAD-132105
5. Strimbu, K., Tavel, J.A.: What are biomarkers? Curr. Opin. HIV and AIDS **5**(6), 463 (2010). doi:10.1097/COH.0b013e32833ed177
6. Cordeiro, M.F.: Eyeing the brain. Acta Neuropathol. **132**(6), 765–766 (2016). doi:10.1007/s00401-016-1628-z
7. Duque, C., Januário, C., Lemos, J., Fonseca, P., Correia, A., Ribeiro, L., Bernardes, R., Freire, A: Optical coherence tomography in LRRK2-associated Parkinson Disease. Neurology, **84**(14), Supplement P2.147 (2015)
8. Bernardes, R., Correia, A., D'Almeida, O.C., Batista, S., Sousa, L., Castelo-Branco, M.: Optical properties of the human retina as a window into systemic and brain diseases. Invest. Ophthalmol. Vis. Sci. **55**, 3367 (2014)
9. Sterniczuk, R., Antle, M.C., LaFerla, F.M., Dyck, R.H.: Characterization of the 3xTg-AD mouse model of Alzheimer's disease: part 2. Behavioral and cognitive changes. Brain Res. **1348**, 149–155 (2010). doi:10.1016/j.brainres.2010.06.011
10. Bouma, B., Tearney, G.: Handbook of Optical Coherence Tomography. Marcel Dekker Inc, New York (2002)
11. Huang, D., Swanson, E.A., Lin, C.P., Schuman, J.S., Stinson, W.G., Chang, W., Hee, M.R., Flotte, T., Gregory, K., Puliafito, C.A., Fujimoto, J.G.: Optical coherence tomography. Science **254**(5035), 1178–1181 (1991)
12. Fujimoto, J.G., Drexler, W.: Introduction to optical coherence tomography. In: Drexler, W., Fujimoto, J.G. (eds.) Optical Coherence Tomography: Technology and Applications. Springer, Berlin, Heidelberg (2008)
13. Abramoff, M.D., Garvin, M., Sonka, M.: Retinal imaging and image analysis. IEEE Rev. Biomed. Eng. **3**, 169–208 (2010). doi:10.1109/RBME.2010.2084567
14. Kang, K., Wu, X., Chen, D.Z., Sonka, M.: Optimal surface segmentation in volumetric images - a graph-theoretic approach. IEEE Trans. Pattern Anal. Mach. Intell. **28**, 119–134 (2006). doi:10.1109/TPAMI.2006.19
15. Garvin, M.K., Abramoff, M.D., Wu, X., Burns, T.K., Russell, S.R., Sonka, M.: Automated 3-D intraretinal layer segmentation of macular spectral-domain optical coherence tomography images. IEEE Trans. Med. Imaging **28**(9), 1436–1447 (2009). doi:10.1109/TMI.2009.2016958

16. Hasegawa, T., Ikeda, H.O., Nakano, N., Muraoka, Y., Tsuruyama, T., Okamoto-Furuta, K., Kohda, H., Yoshimura, N.: Changes in morphology and visual function over time in mouse models of retinal degeneration: an SD-OCT, histology, and electroretinography study. Jpn. J. Ophthalmol. **60**, 111 (2016). doi:10.1007/s10384-015-0422-0

17. Knott, E.J., Sheets, K.G., Zhou, Y., Gordon, W.C., Bazan, N.G.: Spatial correlation of mouse photoreceptor-RPE thickness between SD-OCT and histology. Exp. Eye Res. **92**(2), 155–160 (2011). doi:10.1016/j.exer.2010.10.009

18. Berger, A., Cavallero, S., Dominguez, E., Barbe, P., Simonutti, M., Sahel, J.A., Sennlaub, F., Raoul, W., Paques, M., Bemelmans, A.P.: Spectral-domain optical coherence tomography of the rodent eye: highlighting layers of the outer retina using signal averaging and comparison with histology. PLoS ONE **9**(5), e96494 (2014). doi:10.1371/journal.pone.0096494

19. Dutescu, R.M., Li, Q.X., Crowston, J., Masters, C.L., Baird, P.N., Culvenor, J.G.: Amyloid precursor protein processing and retinal pathology in mouse models of Alzheimer's disease. Graefes Arch. Clin. Exp. Ophthalmol. **247**, 1213–1221 (2009). doi:10.1007/s00417-009-1060-3

20. Guimarães, P., Rodrigues, P., Lobo, C., Leal, S., Figueira, J., Serranho, P., Bernardes, R.: Ocular fundus reference images from optical coherence tomography. Comput. Med. Imaging Graph. **38**, 381–389 (2014)

21. Rodrigues, P., Guimarães, P., Santos, T., Simão, S., Miranda, T., Serranho, P., Bernardes, R.: Two-dimensional segmentation of the retinal vascular network from optical coherence tomography. J. Biomed. Opt. **18**(12), 126011 (2013)

22. Gonzalez, R.C., Woods, R.E.: Digital Image Processing, 2nd edn. Prentice Hall, New Jersey (2002)

23. Duda, R.O., Hart, P.E., Stork, D.G.: Pattern Classification. Wiley-Interscience, Chichester, UK (2000)

Particle Swarm Optimization Approach for the Segmentation of Retinal Vessels from Fundus Images

Bilal Khomri[1,2(✉)], Argyrios Christodoulidis[1], Leila Djerou[2], Mohamed Chaouki Babahenini[2], and Farida Cheriet[1]

[1] LIV4D Laboratory, Polytechnique of Montréal, Montréal, QC, Canada
bilal.khomri@polymtl.ca
[2] LESIA Laboratory, University of Biskra, Biskra, Algeria

Abstract. In this paper, we propose to use the Particle Swarm Optimization (PSO) algorithm to improve the Multi-Scale Line Detection (MSLD) method for the retinal blood vessel segmentation problem. The PSO algorithm is applied to find the best arrangement of scales in the basic line detector method. The segmentation performance was validated using a public high-resolution fundus images database containing healthy subjects. The optimized MSLD method demonstrates fast convergence to the optimal solution reducing the execution time by approximately 35%. For the same level of specificity, the proposed approach improves the sensitivity rate by 3.1% compared to the original MSLD method. The proposed method will allow to reduce the amount of missing vessels segments that might lead to false positives of red lesions detection in CAD systems used for diabetic retinopathy diagnosis.

Keywords: Retinal blood vessel segmentation · Optimization · Particle Swarm Optimization Algorithm · Multi-scale Line Detection

1 Introduction

Retinal vessels constitute an important anatomical structures for the diagnosis and for the severity evaluation of ocular pathologies. Regular follow-up of the signs of these pathologies by the physician on fundus images is required, as these diseases are sight-threatening, and can potentially lead to blindness if left unmonitored or untreated. The detection of the various pathologies is based on the observation by the expert of different lesions on the retina. This task is tedious and time consuming and with the increasing prevalence of diabetes and an aging population, it is expected that by 2025, 333 million diabetic patients worldwide will require retinal examination each year, greatly increasing the burden on ophthalmologists. Thus, the development of automated methods for the diagnosis and grading of ocular diseases will allow the management of patients to be better distributed between different eye care specialists depending on the type of treatment required and the patient/ophthalmologist ratio to be reduced. The segmentation of the retinal vessel network is a prerequisite for the

© Springer International Publishing AG 2017
F. Karray et al. (Eds.): ICIAR 2017, LNCS 10317, pp. 551–558, 2017.
DOI: 10.1007/978-3-319-59876-5_61

automatic detection of red lesions. Indeed, after removing the retinal vessel network, the rest of red structures identified on fundus images are considered as red lesions which may be associated with a diabetic retinopathy [1]. Therefore, the development of automatic methods for vessel segmentation could be incorporated into CAD systems that would promote the screening programs to larger populations for the early diagnosis of sight-threatening pathologies.

There has been a substantial effort to propose automatic vessel segmentation methods in the literature [2]. The existing methods can be classified into two broad categories, supervised [3–5] or unsupervised [6–11] approaches. Supervised methods rely on an extracted set of descriptors that can characterize the pixels to belong to vessels or background. Different kinds of classifiers have been proposed in the literature for the classification of the different regions of interest from the set of extracted features. Among these, the artificial neural network (ANN) [3], the support vector machine (SVM) [4], or the Gaussian mixture model (GMM) [5] were proposed. Unsupervised approaches are further categorized into clustering and intensity-based techniques. In clustering approaches, the pixels are assigned to either vessel or background based on their relative distance to the two clusters based on a specially designed feature space. The features are extracted from intensity-based characteristics such as the grayscale level, the derivatives, or the local texture. Intensity-based methods include the matched filtering [6], the morphology operators [7], the multi-scale line detector (MSLD) [8], the vessel tracking [9], the deformable models [10], and the bio-inspired algorithms/heuristics-metaheuristics [11].

Line detectors is a family of methods devised for identifying linear structures [12] in biomedical images. Recently, Nguyen et al. [8] adapted the basic line detector to accommodate for the various vessel sizes that appear in the retina by considering a range of scales in MSLD. The method is effective in segmenting large to medium-sized vessels compared to existing methods in the literature. However, the method is implemented by an exhaustive use of all the available scales affecting the execution time. Additionally, the method does not take into account the fact that vessels generally yield the optimal line response in a particular scale, but may be not optimal for higher or lower scales. The response also depends on the size of the averaging window. In the recombination phase, small vessels' line response decreases when higher scales are used because a large averaging area is taken into account. Equally, large vessels' central line response decreases when lower scales are used because of the small averaging area.

To overcome the drawbacks of the MSLD algorithm, we optimize the arrangement of the available scales by exploiting the flexibility of the PSO algorithm. Our optimization scheme uses the sensitivity and the specificity measures as objectives functions to evaluate the result of each iteration of the PSO algorithm. To the best of our knowledge, this is the first time where metaheuristics are combined with the line detector method.

2 Methods

The proposed methodology section consists of three parts. First, we review the multi-scale line detection method. Then, we present the basic concepts of the particle swarm optimization algorithm. Finally, we explain and describe the proposed method used for finding the best scales using the PSO algorithm.

2.1 Multi-scale Line Detection Method

The MSLD method as was proposed by Nguyen et al. [8], uses a straight sampling segment of variable length that is rotated around a central pixel to maximize the mean intensity along its length. The winning line's response I_{max}^L is then compared to the average gray level I_{avg}^w of a neighborhood with size equal to the highest chosen scale W. The following equation gives the line response for a single scale:

$$R_{max}^L = I_{max}^L - I_{avg}^w,$$ (1)

According to the authors, the highest chosen scale W, or length of the line, is set as twice the diameter of a typical vessel in the images. Images coming from high resolution databases require a value close to $W = 40$ pixels, while for lower resolution images half the previous value suffices. The multi-scale analysis is achieved by computing Eq. 1 for L at $1 \leq L \leq W$. The final MSLD response is the linear combination of the line responses of different scales, defined as:

$$R_{Combined} = \frac{1}{n_L + 1}\left(\sum_L R_W^L + I_{igc}\right)$$ (2)

where n_L is the number of scales, R_W^L is the response of the line detector at scale L, and I_{igc} is the value of the inverted green channel at the corresponding pixel.

2.2 Particle Swarm Optimization Algorithm

The original PSO algorithm was developed by Eberhart and Kennedy [13]. The PSO takes advantage of the swarm intelligence idea, in which the algorithm mimics the social behavior of a flock of birds known as swarm [14]. In the PSO algorithm, the particle candidates are called solutions. To find an optimal solution, these particles travel through the search space by interacting and sharing information with other particles, namely their individual best solution (local best). In each step of the PSO procedure, the global best solution $\overline{gbest}_i$ is obtained when the entire swarm is updated. For each iteration of the algorithm, the new position of a particle is computed according to their previous position $\overline{X}_i$, velocity $\overline{V}_i$, and $\overline{gbest}_i$ values. These computations are repeated until the convergence criterion is met.

2.3 Process of Finding the Best Scales by PSO Algorithm

In this work, we optimize the arrangement of the available scales in the MSLD method by exploiting the flexibility of the PSO algorithm. In order to reduce the number of scales to compensate for the problems of the MSLD due to the combination of all the available scales, only those contributing to the optimization of the objective function will be retained. The proposed algorithm is applied on the green channel of the retinal images, because it provides the maximum contrast between the background and vessels. The subsequent stages of the proposed vessel extraction method are given by Algorithm 1.

According to the algorithm and at the initialization step, we assign each particle ($\overline{X}_i$) one scale among three because in the retinal image three calibers of vessels can be considered: (1) the small, (2) the large, and (3) the medium range vessels (see Fig. 1). The first scale represents the smallest possible vessel, while the last represents the largest possible vessel. In the initialization step, we fixed the small scale at 3 pixels and the large scale at 41 pixels, respectively (the value of the large scale was chosen as twice the diameter of a typical vessel). At the next steps of the algorithm, we search only the best scales between the smallest and the largest scales in order to detect the medium vessels.

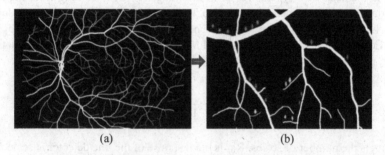

(a)   (b)

Fig. 1. Example of the three different calibers of vessels in the retinal vessels. (a) High-resolution manual reference retinal image, (b) Zoomed region corresponding to the red square inset in (a); red arrows indicate the large vessel, yellow arrows show the small vessels, while blue arrows indicate the medium caliber vessels. (Color figure online)

At the second step and for each iteration of the proposed optimisation algorithm, we add a new scale for each particle in order to search and detect new vessels, then we compare the new one with the previous particle according to their position $\overline{X}_i$ and velocity $\overline{V}_i$ by using the objective functions, where, the used objective functions (Eqs. 3 and 4) represent the sensitivity and the specificity of the segmented image. To sort the particules and keep only the best for each iteration, we use multi-objective optimization strategies, as fast non-dominated sorting method and the population selection strategy [15]. At the end of the process of the proposed algorithm, we keep just the best particle that gave the best scales.

Algorithm 1. PSO algorithm for improving the MSLD algorithm

```
MSLD-PSO (Param1, Param2,…, ParamN)
Begin: MSLD-PSO
For Each particle i do
        Initialize X̄ᵢ and V̄ᵢ randomly.
        Calculate Fitness f̄ᵢ (Eq.3 and Eq.4).
        p̄best̄ᵢ←X̄ᵢ.
End
        ḡbest̄ᵢ←Max (p̄best̄ᵢ); i: 1…s
While  The stopping criterion is not reached do
        For Each particle i do
                Move the particle based on X̄ᵢ and V̄ᵢ.
                Calculate f̄ᵢof the new position(Eq.3 and Eq.4).
                If f̄ᵢ(X̄ᵢ) > f̄ᵢ(p̄best̄ᵢ) then
                        p̄best̄ᵢ←X̄ᵢ
                End
                Add a new scale
                If one scale from ḡbest̄ᵢ ≠all scale in p̄best̄ᵢ then
                        Add a new scale from ḡbest̄ᵢ to X̄ᵢ.
                Else
                        Add randomly new scale to X̄ᵢ.
                End
                If the new f̄ᵢ(X̄ᵢ) > f̄ᵢ(p̄best̄ᵢ) then
                        p̄best̄ᵢ←X̄ᵢ
                End
        End
End
```

---

3 Experimental Results and Discussion

3.1 Datasets and Performance Measures

To assess the effectiveness of the proposed approach, healthy group of a public High-Resolution Fundus database [16] is used. This database contains 15 retinal images, which includes also the manual segmentations of an expert considered as the gold standard. Based on this gold standard we calculated the sensitivity, specificity, and accuracy of the proposed method. The Sensitivity reflects the ability of the algorithm to detect the vessel pixels, and the Specificity is the ability of the algorithm to identify non-vessel pixels. Accuracy is the ratio of the total number of correctly classified pixels (sum of true positives and true negatives) to the number of pixels in the image FOV.

$$\text{Sensitivity} = \frac{TP}{TP + FN} \qquad (3)$$

$$\text{Specificity} = \frac{TN}{TN + FP} \qquad (4)$$

$$Accuracy = \frac{TP + TN}{TP + FN + FP + TN} \qquad (5)$$

where, TP, FP, TN and FN represent the total number of pixels that are true positives, false positives, true negatives and false negatives, respectively.

3.2   Results Analysis

The optimized MSLD method works well on the considered database. We evaluate how well the compared methods segment the vessels given the same background noise, or specificity rate ($T \geq 70$). We show that the performance on the different considered measures in Table 1 is higher for the optimized version compared to standard MSLD method ($p < 0.05$, two sample t-test). The improvement in the sensitivity rate approaches 3%. The higher sensitivity rate indicates that the optimized method segment more retinal vessels. Figure 2 shows the segmentation of the retinal vessel image using the MSLD and the optimized MSLD methods. Figure 3 shows an example of an isolated region, where the two methods are compared. Qualitatively, we can notice that the optimized algorithm detects more vessels than the original method. These vessels include small vessels that are located in the periphery of the image.

Table 1. Performance evaluation on the healthy subjects of HRF database

Method	Sensitivity ($\mu \pm \sigma$)	Specificity ($\mu \pm \sigma$)	Accuracy ($\mu \pm \sigma$)	Time (min)
Original MSLD	84.5% ± 3.78%	97.0% ± 0.44%	95.6% ± 0.50%	3.15
Optimized MSLD	87.3% ± 2.92%	97.7% ± 0.49%	96.2% ± 0.53	2.01

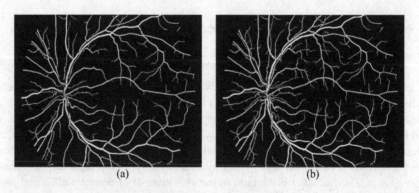

(a)   (b)

Fig. 2. Example of the application of the two compared methods in a retinal image in the (a) original MSLD, and in the (b) proposed optimized MSLD.

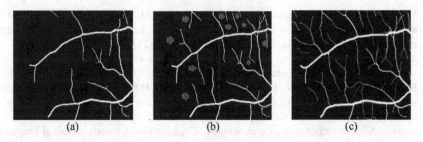

      (a)   (b)   (c)

Fig. 3. Example of an isolated segmented region from a healthy image under the (a) original MSLD, (b) optimized MSLD method. The red arrows indicate small vessels missed by the original MSLD method, while (c) the corresponding region in the reference image. (Color figure online)

Furthermore, the optimized MSLD uses less scales to segment the vessels than the standard MSLD, reducing overall the computational resources (3.01 min vs 2.15 min, system 2.6 GHz, 4-GB RAM). The number of utilized scales is reduced from 21 for the standard MSLD to 15 for the proposed scheme. The optimization chooses the following scales $L_{optimized}$ = {3, 5, 7, 11, 13, 15, 17, 19, 21, 23, 25, 29, 33, 37, and 41}. Scales L = {1, 9, 27, 31, 35, and 39} affect negatively the segmentation performance thus the optimization process excludes them from the set.

4 Conclusion

In this paper, an optimized MSLD method was presented. The particle swarm optimisation (PSO) algorithm is used as a solution to find the optimal number as well as the arrangement of scales in the line detector. For the first time, MSLD is optimized by means of metaheuristics. Evaluated on a healthy images database, our proposed method achieves better vessels segmentation using less scales, ultimately reducing the computation time. The increased sensitivity and speed make the system a suitable substitution to the original MSLD. In the future work, we will further investigate the problems of scales response recombination, weighting, as well as the search space of the PSO algorithm. We also plan to compare the proposed scheme against other already existing methods in retinal databases spanning all the available image resolutions from low to high.

References

1. Ciulla, T.A., Amador, A.G., Zinman, B.: Diabetic retinopathy and diabetic macular edema pathophysiology, screening, and novel therapies. Diab. Care **26**(9), 2653–2664 (2003)
2. Fraz, M.M., Remagnino, P., Hoppe, A., Uyyanonvara, B., Rudnicka, A.R., Owen, C.G., Barman, S.A.: Blood vessel segmentation methodologies in retinal images - a survey. Comput. Methods Programs Biomed. **108**(1), 407–433 (2012)

3. Marin, D., Aquino, A., Gegundez-Arias, M.E., Bravo, J.M.: A new supervised method for blood vessel segmentation in retinal images by using gray-level and moment invariants-based features. IEEE Trans. Med. Imaging 30(1), 146–158 (2011)
4. Ricci, E., Perfetti, R.: Retinal blood vessel segmentation using line operators and support vector classication. IEEE Trans. Med. Imaging 26(10), 1357–1365 (2007)
5. Soares, J.V., Leandro, J.J., Cesar Jr., R.M., Jelinek, H.F., Cree, M.J.: Retinal vessel segmentation using the 2-D gabor wavelet and supervised classification. IEEE Trans. Med. Imaging 25(9), 1214–1222 (2006)
6. Hoover, A., Kouznetsova, V., Goldbaum, M.: Locating blood vessels in retinal images by piecewise thresh-old probing of a matched fillter response. IEEE Trans. Med. Imaging 19(3), 203–210 (2000)
7. Mendonca, A.M., Campilho, A.: Segmentation of retinal blood vessels by combining the detection of centerlines and morphological reconstruction. IEEE Trans. Med. Imaging 25(9), 1200–1213 (2006)
8. Nguyen, U.T., Bhuiyan, A., Park, L.A., Ramamohanarao, K.: An effective retinal blood vessel segmentation method using multi-scale line detection. Pattern Recogn. 46(3), 703–715 (2013)
9. Vlachos, M., Dermatas, E.: Multi-scale retinal vessel segmentation using line tracking. Comput. Med. Imaging Graph. 34, 213–227 (2010)
10. Zhao, Y., Rada, L., Chen, K., Harding, S.P., Zheng, Y.: Automated vessel segmentation using infinite perimeter active contour model with hybrid region information with application to retinal images. IEEE Trans. Med. Imaging 34(9), 1797–1807 (2015)
11. Emary, E., Zawbaa, H.M., Hassanien, A.E., Schaefer, G., Azar, A.T.: Retinal vessel segmentation based on possibilistic fuzzy c-means clustering optimised with cuckoo search. In: International Joint Conference on Neural Networks (IJCNN), pp. 1792–1796. IEEE (2014)
12. Zwiggelaar, R., Astley, S.M., Boggis, C.R., Taylor, C.J.: Linear structures in mammographic images: detection and classification. IEEE Trans. Med. Imaging 23(9), 1077–1086 (2004)
13. Kennedy, J., Eberhart, R.: A new optimizer using particle swarm theory. In: Proceedings of the IEEE Sixth International Symposium on Micro Machine and Human Science, pp. 39–43 (1995)
14. Del Valle, Y., Venayagamoorthy, G.K., Mohagheghi, S., Hernandez, J.C., Harley, R.G.: Particle swarm optimization: basic concepts, variants and applications in power systems. IEEE Trans. Evol. Comput. 12(2), 171–195 (2008)
15. Huo, Y., Zhuang, Y., Gu, J., Ni, S.: Elite-guided multi-objective artificial bee colony algorithm. Appl. Soft Comput. 32, 199–210 (2015)
16. Odstrcilik, J., Kolar, R., Budai, A., Hornegger, J., Jan, J., Gazarek, J., Kubena, T., Cernosek, P., Svoboda, O., Angelopoulou, E.: Retinal vessel segmentationby improved matched filtering: evaluation on a new high-resolution fundusimage database. IET Image Process. 7 (4), 373–383 (2013)

Retinal Vessel Segmentation
from a Hyperspectral Camera Images

Rana Farah[1(✉)], Samuel Belanger[2], Reza Jafari[3], Claudia Chevrefils[3],
Jean-Philippe Sylvestre[3], Frédéric Lesage[2], and Farida Cheriet[1]

[1] Département de génie informatique et génie logiciel,
École Polytechnique de Montréal, 2900, boul. Édouard-Montpetit,
Montréal, QC H3T 1J4, Canada
rana.farah@polymtl.ca
[2] Département de Génie électrique, École Polytechnique de Montréal,
2900, boul. Édouard-Montpetit, Montréal, QC H3T 1J4, Canada
[3] Optina Diagnostics, 7405 Rte Transcanadienne,
Suite #330, St-Laurent, QC H4T 1Z2, Canada

Abstract. In this paper, a vessel segmentation method from hyperspectral retinal images based on the Multi-Scale Line Detection algorithm is proposed. The method consists in combining segmentation information from several consecutive images obtained at specific wavelengths around the green channel to produce an accurate segmentation of the retinal vessel network. Images obtained from six subjects were used to evaluate the performance of the proposed method. Preliminary results suggest a potential advantage of combining multispectral information instead of using only the green channel in segmenting retinal blood vessels.

Keywords: Hyperspectral images · Vessel segmentation · Retinal vessels

1 Introduction

The eye allows the observation of human blood circulation in-vivo. Studies [1, 2] have shown that the arterio-venule diameter ratio (AVR) is associated with different risk factors such as hypertension, cardiovascular diseases, and diabetes. For that reason, assessing the AVR is of paramount importance in ocular examination.

Automatic assessment of AVR, or any other biomarker that manifests in the morphology, the tortuosity or the spatial configuration of the retinal vessels, require accurate segmentation of the whole retinal vascular network including veins and arteries.

Standard segmentation methods use the green channel in fundus images to extract the retinal vessel topology. The reason the green channel is usually used is that it provides the best contrast between the vessel structures and the background of the retina. However, most of these methods lack the capacity to segment the small vessels given that the contrast in the green channel is not sufficient for such segmentation. The width of these vessels and possible discontinuities in their curvature are also a major hindrance to their segmentation.

© Springer International Publishing AG 2017
F. Karray et al. (Eds.): ICIAR 2017, LNCS 10317, pp. 559–566, 2017.
DOI: 10.1007/978-3-319-59876-5_62

Hyperspectral imaging can potentially improve the segmentation of retinal blood vessels using images obtained at different wavelengths around the specific wavelength corresponding to the green channel. The hyperspectral imaging not only captures images of the retina at different wavelengths but also takes advantage of oximetry information which allows the varying oxyhemoglobin concentration or lack of it in a region of interest to be captured differently at each wavelength. For instance 569 nm captures a better contrast between the vessels and the retinal background while 600 nm is more sensitive to the higher level of HbO2 concentration level in the arteries.

Several methods have been proposed to automatically segment the retinal vessels from fundus images. These methods can be classified in two main categories. The first category of retinal vessel segmentation methods uses supervised algorithms [3–6]. The general idea is to train a classifier on local or global extracted features. The classifiers used in the literature include, among others, Bayesian classifiers [3], Neural Networks [4, 5] and Support Vector Machine [6]. These methods gained popularity due to their better performance when compared to their unsupervised counterparts. Their main drawback is the requirement of large datasets for the training phase. In this article, we investigate the potential advantage of using multispectral images for vessel segmentation and only a small dataset is currently available as the hyperspectral imaging is a new modality, which makes the use of supervised methods unpractical.

The second category consists of unsupervised methods [7–14] spanning a wide range of basic schemes. Jiang and Mojon [7] used multi-thresholding at different scales and combined the results of the segmentation of each level. Zana and Klein [8] used mathematical morphological operators to enhance and segment retinal vessels. Chaudhuri et al. [9] proposed a retinal vessel segmentation method based on matched filters. The authors assumed that the intensity distribution along the cross section of the vessels is Gaussian. For this purpose they applied a set of rotated Gaussian filters to detect the vessels. Several extensions were proposed to the initial matched filter segmentation method [10, 11]. However, all these methods assume that the cross-section intensity distribution follows a Gaussian profile, which is not always the case. Other works [12, 13] used tracking to trace vessels from a seed point to an extremity using local information to trace the vessels. The seed points can be either manually or automatically identified. The drawback of these approaches is that they tend to terminate at bifurcations and return an incomplete segmentation. In [14] the authors extended the vessel tracking method by using Tensor Voting to connect the fragmented vessel segments and reach the smaller sections of the vessels. Line detection was also proposed by Ricci and Perfetti [6]. Measuring the average intensity along a rotating line in a window centered at a given pixel, one can associate window size to the assumed range of vessel widths. A pixel is classified as a vessel pixel based on the line detector response. Nguyen et al. [15] expanded on this method and proposed a more generalized version. This version is dubbed the Multi-Scale Line Detector (MSLD). The MSLD algorithm proved to be a very robust algorithm when tested on the STARE [10] and DRIVE [4] datasets. Our work below is based on an extension of this approach. We chose this method for its simplicity and shorter execution time when compared to other highly achieving methods as in [14]. However, we should keep in mind that setting the parameters of the MSLD algorithm is not a trivial task that mostly affects the segmentation of the thinner blood vessels.

To the best of our knowledge, only one work in the literature [16] proposed a dual-wavelength retinal vessel segmentation. The main purpose of the article is to classify arteries and veins. For this purpose, the vessel structure was segmented using the algorithm described in [13]. However, no performance evaluation was provided for the segmentation algorithm.

In this study, we propose a method that extends the Multi-Scale Line Detector (MSLD) method [15] to segment retinal vessels in multispectral retinal images. Our contribution consists in demonstrating that the limitations of the MSLD segmentation algorithm in detecting small vessels could be compensated when multispectral images are available by combining the information from each of the segmentation results into a single vessel map.

2 Methodology Description

Retinal vessels are usually segmented from the green channel (one wavelength) in fundus image. Our method assumes that better segmentation can be achieved when more than one image obtained at different wavelengths are combined. For this reason, we use a sub-band of a hyperspectral retinal image. A hyperspectral retinal image is a 3D image that consists of several 2D images taken at different wavelengths of the same retina in the span of a second.

Each of the wavelength images was preprocessed to enhance its contrast. The vessel structure was then segmented in each of these images using the MSLD algorithm. Finally, a global segmentation map is reconstructed by adding the individual segmented vessel maps from each individual wavelength. The value of a pixel in the reconstructed image is the summation of the pixel intensities from each of the individual wavelength segmentation maps that are at the same position of that pixel. The reconstructed image is then thresholded in such a way that any value that is greater than zero is set to one. The regions wrongly detected as retinal vessels in that reconstructed image are cleared by deleting any connected region with a number of pixels smaller than a given threshold α.

3 Experimental Settings and Results

Our dataset was captured using a Metabolic Hyperspectral Retinal Camera (MHRC) developed by Optina Diagnostic (Montreal, Canada). An early prototype is described in details in [17]. Figure 1(a) shows the camera setup and Fig. 1(b) gives a rough description on the functionality of the apparatus. The camera sends a beam of monochromatic light through a tunable filter. The reflected light is collected on a 2D sCMOS sensor. The 2D images from different wavelength beams are stacked into one 3D cube to form the hyperspectral image. Compared to the previously described MHRC in [17], this version of the system included a sCMOS sensor which permitted acquisitions at a rate up to 100 frames per second.

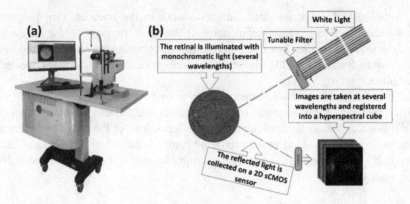

Fig. 1. (a) Shows the MHRC set up (b) Shows the functionality of the system. This picture was adapted from [17]

The processed images are a sub-band of the original multispectral images that span the interval of 450 nm to 900 nm with a step of 5 nm. For our purposes, we considered the interval between 495 and 570 nm that spans the green region of the spectrum and presents the best contrast between the vessel structures and their environment. Thus, the dimensions of the processed images are $1536 \times 1536 \times 16$ pixels.

The raw reflectance images were preprocessed using an in-house Matlab (The Mathworks, Natick, MA) tool. Preprocessing consists in normalization and registration of the acquired images. Normalization was necessary to account for spatial and spectral variations in the light source intensity and the system optics [17]. Registration was used to account for slight eye motion.

We set the threshold α on the small connected regions size to 800. This value was determined empirically and kept constant for all the subjects.

We processed multispectral images from six subjects including healthy controls and subjects followed at the ophthalmology clinic. We qualitatively compared the MSLD results for each of the individual wavelength images and the corresponding MSLD reconstructed image. Figure 2 shows a detailed case for subject 1. Only one case was detailed in this paper due to space constraints. In Fig. 2, the regions shown in the right columns are the enlarged detailed version of the contoured regions in one of the images that belongs to the same row. Each two corresponding regions are contoured with the same color.

To quantitatively evaluate the results of the proposed segmentation we built a ground truth by manually segmenting the vessels. Given that the higher wavelength images showed better contrast, the image at 570 nm was used to build the representative ground truth map for a given multispectral image. It is true that the degree of contrast in that image was not enough to permit an automatic segmentation of the thinner blood vessels. However, it was enough to allow a more precise manual segmentation. For instance, manual segmentation was able to account for discontinuities in the vessel structure in that image.

For each of the individual wavelength images and the reconstructed image, a score was calculated. The score is the normalized sum of the intersection foreground pixels in both the ground truth map and the automatically segmented map as described in Eq. (1).

$$score = \frac{sum\_pixels(GT_{foreground} \cap AS_{foreground})}{sum\_pixels(GT_{foreground})} \tag{1}$$

Where GT refers to the ground truth map and AS refers to an automatically segmented map. The calculated scores for all 6 subjects are shown in Table 1. In this case the higher the score the more successful is the segmentation. Also the optical disk region is excluded from the score calculation.

Table 1. The calculated score for the reconstructed image (Recon.) and each of the wavelength images for the six subjects. The captured wavelength in subject 3 did not extend to 510 nm or lower.

	Subject 1	Subject 2	Subject 3	Subject 4	Subject 5	Subject 6
Recon.	0,90	0,98	0,92	0,89	0,92	0,80
570 nm	0,83	0,95	0,80	0,72	0,87	0,87
565 nm	0,76	0,93	0,73	0,76	0,85	0,84
560 nm	0,73	0,93	0,71	0,75	0,81	0,82
555 nm	0,70	0,92	0,68	0,75	0,80	0,81
550 nm	0,71	0,91	0,72	0,74	0,82	0,83
545 nm	0,74	0,91	0,75	0,74	0,83	0,82
540 nm	0,71	0,87	0,72	0,73	0,81	0,81
535 nm	0,66	0,80	0,73	0,67	0,77	0,76
530 nm	0,57	0,71	0,64	0,61	0,74	0,73
525 nm	0,46	0,64	0,57	0,58	0,67	0,64
520 nm	0,37	0,53	0,51	0,55	0,59	0,55
515 nm	0,31	0,45	0,46	0,51	0,53	0,47
510 nm	0,33	0,41		0,50	0,54	0,44
505 nm	0,30	0,37		0,51	0,52	0,43
500 nm	0,30	0,34		0,50	0,51	0,43
495 nm	0,31	0,31		0,50	0,53	0,42

When examined, the high wavelength individual images appear to have most of the information. The vessels in these images are less fragmented than in the lower wavelength images. However, even the higher wavelength individual images, when taken separately, lack the complete information gathered in the reconstructed image. The fact that the vessels are less fragmented in the reconstructed image makes it easier to clean the final results of noisy patches. In fact, the noise was more likely to be distinguished from any fragmented vessel part. Furthermore, in the MSLD method the choice of the threshold value that provide an optimal vessel segmentation is not obvious and may depend on the intensity distribution in each image. The impact of this effect was minimized using several wavelength images. Even though the threshold value may not be optimal for some of the images, the combination of multiple images resulted in a better reconstructed segmentation map.

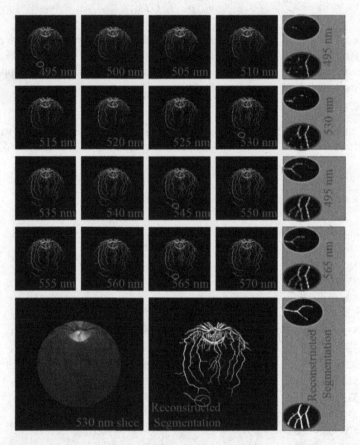

Fig. 2. The upper 4 rows and left side 4 columns show the MSLD segmentation result for the individual images captured at the different identified wavelengths. The image in the lower left column is the preprocessed raw image at wavelength 530 nm. The middle image at the lowest row is the reconstructed segmented image. The right most column shows the enlarged contoured details at the image in the corresponding row. The yellow contour marks the same region in all involved images and the same is true for the green one. (Color figure online)

Figure 3 shows the results for subject 6 where the segmentation was not as successful as for the other subjects. The algorithm wrongly identified vessels in regions where anatomical features related to age-related macular degeneration (ARMD) such as the large hypopigmented region (clear zone) and the drusen (smaller white patches) were present.

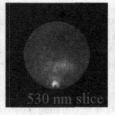

Fig. 3. These are representative images from subject 6. The first and second images show the MSLD segmentation results at wavelength 525 nm and 565 nm respectively. The third image shows the preprocessed images captured at wavelength 530 nm where a multiple drusen and a large hypopigmented region are visible. The last image shows the reconstructed segmentation map.

4 Conclusion

We proposed a multispectral retinal vessel segmentation technique that is based on the MSLD method. The proposed method combined information from the individual segmentation of each spectral image into a global reconstructed segmentation image. We tested our method on six different subjects. The preliminary results show the advantage of combining multispectral images to exploit more information than when using only one-wavelength image.

The current approach used to combine the information is not sufficient when signs of ARMD are present in the images. Thus, a combination method that is less sensitive to these confounding features is needed. Otherwise, a mid-processing step can be applied to the results of the individual MSLD images to exclude the wrongly-identified vessels before combination.

It would also be interesting to investigate the contribution of each wavelength image to the final reconstructed image. This can be used to determine if all the images are needed for the reconstruction.

An extensive validation is needed to understand better why the higher wavelength images gave more accurate segmentation results. We suspect the better contrast obtained for the hemoglobin and the choice of the MSLD threshold. Thus, it could be useful to adapt the threshold value for each wavelength. Also we will enhance the robustness of the method used to set the parameters of the algorithm once applied on a larger database.

References

1. Wong, T.Y., Klein, R., Klein, B.E., Tielsch, J.M., Hubbard, L., Nieto, F.J.: Retinal microvascular abnormalities and their relationship with hypertension, cardiovascular disease, and mortality. Surv. Ophthalmol. **46**(1), 59–80 (2001)
2. Klein, R., Myers, C.E., Lee, K.E., Gangnon, R., Klein, B.E.: Changes in retinal vessel diameter and incidence and progression of diabeticretinopathy. Arch. Ophthalmol. **130**, 749–755 (2012)

3. Soares, J.V.B., Leandro, J.J.G., Cesar, R.M., Jelinek, H.F., Cree, M.J.: Retinal vessel segmentation using the 2-D Gabor wavelet and supervised classification. IEEE Trans. Med. Imag. 25(9), 1214–1222 (2006)
4. Staal, J., Abramoff, M.D., Niemeijer, M., Viergever, M.A., van Ginneken, B.: Ridge-based vessel segmentation in color images of the retina. IEEE Trans. Med. Imag. 23(4), 501–509 (2004)
5. Annunziata, R., Trucco, E.: Accelerating convolutional sparse coding for curvilinear structures segmentation by refining SCIRD-TS filter banks. IEEE Trans. Med. Imaging 35(11), 2381–2392 (2016)
6. Ricci, E., Perfetti, R.: Retinal blood vessel segmentation using line operators and support vector classification. IEEE Trans. Med. Imaging 26(10), 1357–1365 (2007)
7. Jiang, X., Mojon, D.: Adaptive local thresholding by verification based multithreshold probing with application to vessel detection in retinal images. IEEE Trans. Pattern Anal. Mach. Intell. 25(1), 131–137 (2003)
8. Zana, F., Klein, J.C.: Segmentation of vessel-like patterns using mathematical morphology and curvature evaluation. IEEE Trans. Image Process. 10(7), 1010–1019 (2001)
9. Chaudhuri, S., Chatterjee, S., Katz, N., Nelson, M., Goldbaum, M.: Detection of blood vessels in retinal images using two-dimensional matched filters. IEEE Trans. Med. Imaging 8(3), 263–269 (1989)
10. Hoover, A., Kouznetsova, V., Goldbaum, M.: Locating blood vessels in retinal images by piecewise threshold probing of a matched filter response. IEEE Trans. Med. Imaging 19(3), 203–210 (2000)
11. Zhang, B., Zhang, L., Karray, F.: Retinal vessel extraction by matched filter with first-order derivative of Gaussian. Comput. Biol. Med. 40(4), 438–445 (2010)
12. Liu, I., Sun, Y.: Recursive tracking of vascular networks in angiograms based on the detection-deletion scheme. IEEE Trans. Med. Imaging 12(2), 334–341 (1993)
13. Can, A., Shen, H., Turner, J.N., Tanenbaum, H.L., Roysam, B.: Rapid automated tracing and feature extraction from retinal fundus images using direct exploratory algorithms. IEEE Trans. Inf. Tech. Biomed. 3(2), 125–138 (1999)
14. Christodoulidis, A., Hurtut, T., Ben Tahar, H., Cheriet, F.: A multi-scale tensor voting approach for small retinal vessel segmentation in high resolution fundus images. Comput. Med. Imaging Graph. 52, 28–43 (2016)
15. Nguyena, U.T.V., Bhuiyana, A., Park, L.A.F., Ramamohanaraoa, K.: An effective retinal blood vessel segmentation method using multi-scale line detection. Pattern Recogn. 46(3), 703–715 (2013)
16. Narasimha-Iyer, H., Beach, J.M., Khoobehi, B., Roysam, B.: Automatic identification of retinal arteries and veins from dual-wavelength images using structural and functional features. IEEE Trans. Biomed. Eng. 54(8), 1427–1435 (2007)
17. Desjardins, M., Sylvestre, J.P., Jafari, R., Kulasekara, S., Rose, K., Trussart, R., Arbour, J.D., Hudson, C., Lesage, F.: Preliminary investigation of multispectral retinal tissue oximetry mapping using a hyperspectral retinal camera. Exp. Eye Res. 146, 330–340 (2016)

Remote Sensing

The Potential of Deep Features for Small Object Class Identification in Very High Resolution Remote Sensing Imagery

M. Dahmane[1]([✉]), S. Foucher[1], M. Beaulieu[1], Y. Bouroubi[2], and M. Benoit[2]

[1] Computer Research Institute of Montreal, 405, avenue Ogilvy, bureau 101, Montreal, QC H3N 1M3, Canada
{dahmanmo,samuel.foucher}@crim.ca
[2] Effigis Geo-Solutions inc., Montreal, Canada
info@effigis.com

Abstract. Various generative and discriminative methods have been transferred from the computer vision field to remote sensing applications using different low and high semantic level descriptors. However, as classical approaches have shown their limits in representation learning and are not intended to deal with the great variability of the data. With the emergence of large-scale annotated datasets in vision, the convolutional deep approaches represent the most winning solutions by supporting this variability with spatial context integration through different semantic abstraction levels. In the lack of annotated remote sensing data, in this paper, we are comparing the performances of deep features produced by six different CNNs that have been trained on well established computer vision datasets with respect to the detection of small objects (cars) in very high resolution Pleiades imagery.

Our findings show good generalization performance and are very encouraging for future applications.

Keywords: Object detection · Very high resolution · Deep learning

1 Introduction

Traffic research and intelligent transportation systems acquire data from cameras as well as different conductive and wireless sensors that are not covering properly large road network. The increase availability of remote sensing data combined with lower acquisition costs can provide massive road data that can be used in complement to traditional in-situ sensors. Despite the great amount of data, object detection in Very High Resolution (VHR) imagery is not yet mature especially for small objects mainly due to the lack of annotated data.

In computer vision, object detection has been studied for more than four decades making available to the research community large corpus of annotated data that can be used to build more reliable detection algorithms that must be able to decide whether or not the image contains the object of interest in a

© Springer International Publishing AG 2017
F. Karray et al. (Eds.): ICIAR 2017, LNCS 10317, pp. 569–577, 2017.
DOI: 10.1007/978-3-319-59876-5_63

given position and at a particular scale. A detection system has to determine whether the image belongs to the object class or non-object class. The difficulty is to assess at different scales its presence in every position of the image. The problem is formulated as a binary classification with a higher degree of difficulty as the class 'non-object' defining negative examples is almost infinite hence the problem of the imbalanced classes. Besides the problem of the object scale and its real size, we can also identify other properties related to the object or image itself [13] including the number of instances of the object in the image, the degree of occlusions, the deformability of the object to detect, the amount of contrasting texture and color variability in the image.

In remote sensing imagery, small objects such as vehicles present less differences in appearance, size and external boundaries but the challenge is more acute since space imagery data are more complex. Saikat *et al.* [1] used the concept of intrinsic dimension to compare the dimensionality of remote sensing data to established datasets in vision. The authors found that the remote sensing images lie on a much higher dimensional manifold (7 times greater) than the images from vision datasets.

In this article which presents an extension of our previous work [5], we developed a detector of small objects from high resolution Pleiades satellite imagery. We conducted an extensive comparison of the most popular CNNs and end-to-end CNN architecture to determine the scalability limits and the learning transfer performance from computer vision to remote sensing.

2   Related Work

With meaningful and fair evaluations on large datasets, the wining methods of the PASCAL VOC[1] 2008–2012 and ILSVRC[2] 2010–2015 computer vision competitions for object detection and recognition were mainly based on convolution neural networks.

In remote sensing applications, multilayer neural networks have been widely applied [14]. The appearance of deep learning approaches is now opening up a new deep vision in remote sensing. Firat *et al.* [7] used a CNN to detect objects in high resolution satellite images. A deep learning method merging spectral and spatial data has been proposed in [4]. Chen *et al.* [3] studied a set of popular descriptors in comparison to a CNN classifier trained directly on grayscale images from Google Earth. The availability of pre-trained models in computer vision (Caffe model zoo [9]) allows the remote sensing community to apprehend the application of CNNs irrespective of the availability of large annotated space imagery databases. Razavian *et al.* [12] demonstrated the effective capacity of generalization of attributes produced by a CNN which even if it was originally trained for 1000 classes it was able to produce more generic attributes. The work of Penatti *et al.* [11] was conducted in this perspective to evaluate two pre-trained networks (OverFeat [15] and Caffe [9]) on a classification task from

---

[1] http://host.robots.ox.ac.uk/pascal/VOC/.

[2] http://www.image-net.org/challenges/LSVRC/.

remote sensing data. In contrast to Penatti *et al.* [11] which attempted to classify meta-class images such as 'parking lots', the contribution of the present work is to evaluate the performance of learned deep features at different levels of abstraction on a detection task of small objects in 0.5 m resolution Pleiades imagery. The main challenge is that cars are very small objects with only a few pixels across at this scale.

The paper is organized as follow, first, we briefly give some background on deep learning then we describe some winning methods in computer vision, after that we describe and evaluate our approach. Finally, some conclusions are drawn with some intent for future works.

3   Convolutional Neural Net Architecture

The convolutional neural net architectures use three-dimensional volumes of neurons and stack convolution and ReLu blocks with pooling layers followed by fully connected layers. The ever increasing performance of modern CNN in recognition and detection tasks is principally due to the non-saturating neurons and the dropout regularization that enhance the non-linearity and prevent overfitting.

In the literature, to detect objects CNNs can implement end-to-end architecture [3] or serve as a generic deep feature extractor from a high-level layer [11]. In this work, we will focus on the following well known architectures: CIFARNet, AlexNet, GoogleNet and VGG.

CIFAR10Net and CIFAR100Net: The CIFAR10Net is a small model of convolution, pooling, rectified linear unit, non-linearity, and local contrast normalization layers with a linear classifier on top. It was trained on the CIFAR-10[3] dataset which consists of $60K$ 32×32 color images in 10 classes (airplane, automobile, bird, cat, deer, dog, frog, horse, ship, truck).

CIFAR100 is similar to CIFAR10 except it has 100 classes containing 600 images, each one comes with two labels: a "fine" label (the class to which it belongs) and a "coarse" label (its superclass). The 100 classes are grouped into 20 superclasses. The CIFAR100Net architecture is more sophisticated than CIFAR10Net (e.g. using Xavier initialisation vs. Gaussian).

AlexNet: AlexNet is the baseline of the new generation of layered models [10] that expanded the learning capacity of CNNs with efficient GPU implementation of convolution operations to tune up to $60M$ parameters. Two implementations of this network can be used for features extraction CaffeNet and Places205Net.

CaffeNet: The Caffe [9] reference convolutional network (CaffeNet) is a replication of AlexNet architecture with two main differences (i) training without data-augmentation (ii) the order of pooling and normalization layers is reversed. The network parameters were trained on a subset of ImageNet[4] with 1000 categories and $1.2M$ images. The penultimate fully connected layer "FC7" can be used as a 4096-dimension feature set extractor.

---

[3] https://www.cs.toronto.edu/~kriz/cifar.html.
[4] http://www.image-net.org/.

Places205Net: Places205Net [18] is an AlexNet CNN trained on 205 scene categories of Places Database[5] with $2.5M$ images. Feature extractors can be associated with the "FC7" layer.

GoogLeNet: The runner-up in the classification task of the ILSVRC-2014 was GoogLeNet [17] with a deeper network. The use of the average pooling layers instead of fully connected layers at the top of the network, and the introduction of an inception module reduced significantly the number of parameters of the network (GoogLeNet 5M vs. AlexNet 60M). The 1024 outputs of the last pooling layer "$pool5/7 \times 7\_s1$" can be used as a deep generic feature set. The downside of GoogLeNet architecture is the depth-level of abstraction using inception layers which yield more memory usage.

VGGNet: The object localization task winner of the ILSVRC-2014 contest was VGGNet [16]. The very deep network (up to 19 layers with $140M$ parameters) with more small (3×3) convolutional layers obtained the second place in the classification task. The outputs of the fully connected layer "FC7" can be used as a 4096-dimensional descriptor.

4   Approaches and Evaluation

4.1   Generic Deep Features

To test the ability of transferring CNN features to remote sensing, Penatti *et al.* [11] assessed two CNNs, OverFeat [15] and CaffeNet for a classification task of 21 classes delineated as large patches (256×256). Similarly, Castelluccio *et al.* [2] evaluated deep features from GoogLeNet on the same dataset. In the same way as in [2,11], Hu *et al.* [8] extracted deep features of some pre-trained CNNs to perform a classification task on the same dataset. These models have been trained on the ImageNet dataset [6] with $1.2M$ images distributed in 1000 classes.

Differently, in this paper we evaluate the learning transfer from CNNs to remote sensing on different levels of abstraction to detect small objects instead of classifying large images. Our goal is to be able to decide whether the image contains objects as small as cars and at which positions. Furthermore, an end-to-end architecture was evaluated for this task. In addition, we make use of both visible and near-infrared bands. Figure 1 gives some training examples from the Pleiades red band images at a resolution of 0.5 m.

Pleiades images have 4 multispectral bands (blue, green, red and NIR) and a panchromatic band. The multispectral bands were pan-sharpened using the panchromatic band. Two scenes are used in this study: (a) A scene from *Vancouver-East area* that will be divided in two sub-regions one for training and the other for testing; and (b) A completely different scene from *Quebec area* that will be used only for testing.

Fig. 1. Subsets of the 32 × 32 patches used for training (only the red band is shown here, the positive examples are delimited in green). (Color figure online)

Our target objects was the "car" class including individual cars and cars in parking lots. A relatively small patch size (32 × 32) is chosen to cover such small objects (Fig. 1). The training database is formed of 6,307 patches chosen in the *Vancouver scene*. The test set is composed of 4,300 patches sampled from a *Quebec scene* of (9735 × 19947) pixels. The size of the training set is relatively modest and is based on sub-region from a single (15655 × 12273) image (*Vancouver image*). To inject variability in the training data and increase the number of images for each patch, we generated seven other patches corresponding to different orientations. The patches were upsampled using a bilinear kernel for the models that have a data layer with different input size (Table 1).

Table 1. Car/non-car database characteristics.

Data	Car	Non-car
Train-Vancouver	6,944	33,344
Test-Vancouver	872	4,176
Test-Quebec	2,565	5,670

Results were improved when using the band combination R+G+NIR (i.e. the blue band is more noisy and less informative). A SVM 2-class classifier (linear kernel) is used for the final classification (cars against all). The ROC curves on the test set for car objects are shown on Fig. 2 with an auc of 99.98% for VGGNet and 99.96% for Places205Net.

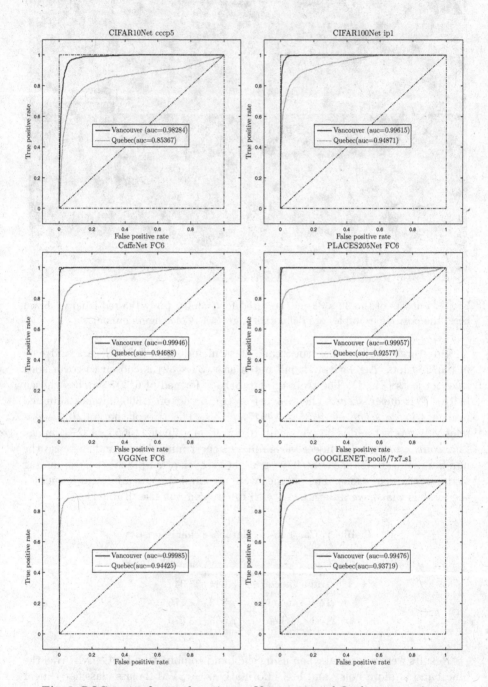

Fig. 2. ROC curves for car detection on Vancouver- and Quebec-test regions.

CIFAR10 led to poor performance (auc = 98.28%) despite the fact that the size of the input layer is the same as our "car" image patches (32×32). One reason could be that the CIFAR10 model has been trained on a restricted set of classes (10), thus the produced features are too specific. CIFAR100Net that consider up to 100 fine classes improved the ratio to (99.61%) on the Vancouver-test area. A noticeable improvement in performance was observed on the Quebec-test area (from 85.38% to 94.87%).

The results in Fig. 2 show fairly a good generalization performance on Vancouver-test with an average of (99.54%) with regard to the mean performance obtained from Quebec-test (92.61%). Except CIFAR10Net, all architectures tend to have aucs > 99% on the Vancouver-test. On the second dataset, CIFAR100Net, CaffeNet and VGGNet gave more than 94%. This small drop in accuracy for Quebec-test region is almost justifiable since we have observed that cars in the Vancouver-area are likely to be more bigger and oversized than those in the Quebec-test region.

4.2   End-to-End Learning

An end-to-end CNN architecture can achieve the optimal balance between robustness and discriminative aspects. From this perspective, we evaluated a convolutional neural network model that we have directly trained from scratch using the same Vancouver-train subset.

Unfortunately, as we can see from Fig. 3, training the network from scratch led to a decline in auc values with only 90% for Vancouver-test images, and a ratio as low as 67% on the Quebec-test subset compared to deep features which seem to be more generic (Table 2). These findings were totally unexpected since, theoretically, it is known that a complete training of a CNN should significantly increase the performance.

A specific training of a CNN network on the Pleiades data could have many advantages. The input layer can be easily modified to take into account all

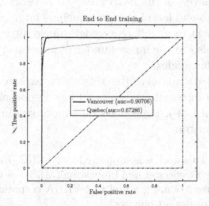

Fig. 3. ROC curves of the end-to-end learning architecture.

Table 2. Models are ranked from left to right by decreasing average AUC performance (%).

Data	CaffeNet	CIFAR100Net	VGGNet	GoogLeNet	Places205Net	CIFAR10Net	End to End
Test-Vancouver	99.95	99.62	**99.99**	99.48	99.96	98.28	90.71
Test-Quebec	94.69	**94.87**	94.43	93.72	92.58	85.34	67.29
Avg	**97.32**	97.24	97.21	96.60	96.27	91.83	79.0

spectral bands (R, G, B, and NIR) with a context-specific architecture which best suits to the type of data; the size of the input images could be adjusted. However, a successful training requires a large number (millions) of images with high variability.

5 Conclusions

In this work we extended the evaluation to different CNNs including GoogLeNet, 205Places, CIFAR10 and VGGNet. Deep features generated by CaffeNet obtained the best global performance for car detection in Pleiade VHR imagery. This model was trained on very large computer vision database where images have a priori little in common with VHR images. CIFAR10 led to poor performance despite the fact that the size of the input layer is the same as the "car" image patches (32×32). Notice that, relatively, CaffeNet has a very large input layer (256×256). Compared to an end-to-end learning, we found that deep features are powerful in representation learning and seem to capture intuitively the crude shapes of vehicles.

As a future work, we plan to investigate the construction of a larger training and test set with greater variability between scenes in order to add more diversity. We intend also to investigate other options as using 4 instead of only 3 channels and training with initialized weights from CaffeNet while the weights for the additional channel could be initialized randomly, an option could be to use late fusion -at the feature level- of the RGB and NIR networks.

In the case of end-to-end learning we can include large inter-season variations (e.g. winter vs. summer) from different geographical sites (e.g. North America vs Europe). We plan also to use images under different acquisition conditions to ensure variation in solar angles.

A large annotated database will enable us to explore in depth the CNNs whose implementation in remote sensing still has a good margin of progression. However, in the absence of a large mass of remote sensing data deep features remain the robust descriptors.

Acknowledgements. This work was supported by NSERC (Engage Grant), the Ministère de l'Économie, des Sciences et de l'Innovation (MESI) of the province of Québec and Effigis Géo Solutions. We are grateful to NVIDIA corporation for the Tesla K40 GPU Hardware Grant to support our work.

References

1. Basu, S., Ganguly, S., Mukhopadhyay, S., DiBiano, R., Karki, M., Nemani, R.R.: DeepSat: a learning framework for satellite imagery. In: Proceedings of the 23rd SIGSPATIAL International Conference on Advances in Geographic Information Systems, Bellevue, WA, USA, 3–6 November 2015, p. 37 (2015)
2. Castelluccio, M., et al.: Land Use Classification in Remote Sensing Images by Convolutional Neural Networks. CoRR, 1508.00092 (2015)
3. Chen, X., Xiang, S., Liu, C.-L., Pan, C.-H.: Vehicle detection in satellite images by hybrid deep convolutional neural networks. Geosci. Remote Sens. Lett. **11**(10), 1797–1801 (2014)
4. Chen, Y., et al.: Deep learning-based classification of hyperspectral data. IEEE Sel. Topics Appl. Earth Observations Remote Sens. **7**(6), 2094–2108 (2014)
5. Dahmane, M., Foucher, S., Beaulieu, M., Riendeau, F., Bouroubi, Y., Benoit, M.: Object detection in pleiades images using deep features. In: 2016 IEEE International Geoscience and Remote Sensing Symposium, IGARSS 2016, Beijing, China, 10–15 July 2016, pp. 1552–1555 (2016)
6. Russakovsky, O., et al.: ImageNet large scale visual recognition challenge. IJCV **115**(3), 211–252 (2015)
7. Firat, O., Can, G., Vural, F.Y.: Representation learning for contextual object and region detection in remote sensing. In: ICPR, pp. 3708–3713 (2014)
8. Hu, F., Xia, G.-S., Hu, J., Zhang, L.: Transferring deep convolutional neural networks for the scene classification of high-resolution remote sensing imagery. Remote Sens. **7**(11), 14680 (2015)
9. Jia, Y., Shelhamer, E., Donahue, J., Karayev, S., Long, J., Girshick, R., Guadarrama, S., Darrell, T.: Caffe: convolutional architecture for fast feature embedding. arXiv:1408.5093 (2014)
10. Krizhevsky, A., Sutskever, I., Hinton, G.E.: Imagenet classification with deep convolutional neural networks. In: Advances in Neural Information Processing Systems, vol. 25, pp. 1097–1105 (2012)
11. Penatti, O., Nogueira, K., Dos Santos, J.: Do deep features generalize from everyday objects to remote sensing and aerial scenes domains? In: CVPRW, pp. 44–51, June 2015
12. Razavian, A.-S., Azizpour, H., Sullivan, J., Carlsson, S.: CNN features off-the-shelf: an astounding baseline for recognition. CoRR, 1403.6382 (2014)
13. Russakovsky, O., Deng, J., Huang, Z., Berg, A.C., Fei-Fei, L.: Detecting avocados to zucchinis: what have we done, and where are we going? In: International Conference on Computer Vision (ICCV) (2013)
14. Sandham, W., Leggett, M., Aminzadeh, F.: Geophysical Applications of Artificial Neural Networks and Fuzzy Logic. Modern Approaches in Geophysics, vol. 21. Kluwer Academic Publishers, Berlin (2004)
15. Sermanet, P., et al.: Overfeat: integrated recognition, localization and detection using convolutional networks. arXiv preprint: 1312.6229v4 (2014)
16. Simonyan, K., Zisserman, A.: Very deep convolutional networks for large-scale image recognition. CoRR, abs/1409.1556 (2014)
17. Szegedy, C., et al.: Going deeper with convolutions. In: CVPR 2015 (2015)
18. Zhou, B., Lapedriza, A., Xiao, J., Torralba, A., Oliva, A.: Learning deep features for scene recognition using places database. In: NIPS (2014)

Segmentation of LiDAR Intensity Using CNN Feature Based on Weighted Voting

Masaki Umemura[1(✉)], Kazuhiro Hotta[1], Hideki Nonaka[2], and Kazuo Oda[2]

[1] Meijo University, 1-501 Shiogamaguchi, Tempaku-ku, Nagoya, Japan
163433007@ccalumini.meijo-u.ac.jp,
kazuhotta@meijo-u.ac.jp
[2] Asia Air Survey Co., Ltd., 1-2-2 Manpukuji, Asao-ku, Kawasaki,
Kanagawa, Japan
{hdk.nonaka,kzo.oda}@ajiko.co.jp

Abstract. We propose an image labeling method for LiDAR intensity image obtained by Mobile Mapping System (MMS). Conventional segmentation method using CNN and KNN could give high accuracy but the accuracies of objects with small area are much lower than other classes with large area. We solve this issue by using voting cost. The first cost is determined from a local region. Another cost is determined from surrounding regions of the local region. Those costs become large when labeling result corresponds to class label of the region. In experiments, we use 36 LIDAR intensity images with ground truth labels. We divide 36 images into training (28 images) and test sets (8 images). We use class average accuracy as evaluation measures. Our proposed method gain 84.75% on class average accuracy, and it is 9.22% higher than our conventional method. We demonstrated that the proposed costs are effective to improve the accuracy.

1 Introduction

Fundamental Geospatial Data of road (FGD) [1] is road map which used for road construction, detailed road mapping and so on. FGD is made by human now. However, manual process has some problems. Human cannot treat a large amount of data, and there is the possibility of human error. In addition, since many people are required to make the FGD, a lot of costs are required. Thus, automatic creation of the FGD is required to reduce human burden and cost.

An automatic classification method from LiDAR intensity images has been proposed [2]. In the method, features are extracted by the CNN from local regions cropped from LiDAR intensity images [3]. Since the similarity between features reflects the similarity of contents of local regions, we search top K similar regions cropped from training samples with a test region. Since regions in training images have manually-annotated ground truth labels, we vote the labels attached to top K similar regions are voted to the test region. The class label with the maximum vote is assigned to each pixel in the test image. The method was able to assign the same label as human beings

F. Karray et al. (Eds.): ICIAR 2017, LNCS 10317, pp. 578–585, 2017.
DOI: 10.1007/978-3-319-59876-5_64

in 97.8% of all pixels in test LiDAR intensity images. However, this approach has a serious issue about low accuracy of objects with small area. For example, the accuracy of road class is 98.56% but the accuracy of gutter is only 19.46%. The goal of our research is to create road maps automatically. Therefore, objects with small area are more important than objects with large area. We want to improve the class average accuracy.

In order to solve this issue, we propose an improved method based on weighted voting. We define unary cost that obtained from the information of a target region and pair-wise cost that obtained from the information of peripheral regions of the target region. By using two costs, we confirmed the great improvement for objects with small area such as catchment basins and gutter.

In experiments, we use 36 LiDAR intensity images obtained by the MMS with ground truth labels. Those images include 9 categories. We use both class average accuracy and pixel-wise accuracy as evaluation measures. Our proposed method achieves 96.77% in pixel-wise accuracy and 84.75% in class average accuracy. Surprisingly, the improvement is 9.22% on class average accuracy.

This paper is organized as follows. We explain the conventional method using KNN of feature obtained by CNN in Sect. 2. Proposed method using two voting costs explained are in Sect. 3. Evaluation and comparison results of our method are shown in Sect. 4. Section 5 is for conclusions and future works.

2   Related Works

We used a segmentation method to classify objects in LiDAR intensity images for automatic road map creation because it can visualize object information (e.g. the number of object, object position and object shape).

Recently, segmentation method can be roughly divided into two approaches; Convolutional neural network (CNN) based or not. The approaches without CNN [4, 5] are faster and use smaller amount of dataset than CNN based method. But, their accuracies are not so high. Since we want to create road map automatically, we must get high accuracy as much as possible. Almost of these methods use color information to improve accuracy but we use LiDAR intensity images which are gray scale images.

On the other hand, CNN based approaches [6–9] are successful on segmentation problem. Those methods gave high accuracy using large scale training dataset. But, we have only 36 LiDAR intensity images and the size of small object is 10×10 pixels in 1500×2000 pixels. Those small objects may be lost by encoder and decoder or convolution and deconvolution processing.

To overcome the problem, we proposed a segmentation method using local regions cropped from LiDAR intensity images. By using local features obtained from the Caffenet [11], segmentation is carried out from small number of LiDAR Intensity images. However, this approach gave low accuracy of objects with small area than other objects. Therefore, we want to improve accuracy of those classes.

3 Proposed Method

We automatically recognize objects in LIDAR intensity images. Figure 1 shows the overview of our method. We use the features extracted by CNN from local regions (e.g. 64 × 64 pixels) cropped from LIDAR intensity images. Since similarity between features represents similarity between local regions, we can search top K local regions which is similar to test region from training samples. Since regions in training images have manually-annotated ground truth labels, we vote the labels which are attached to K similar region to output image.

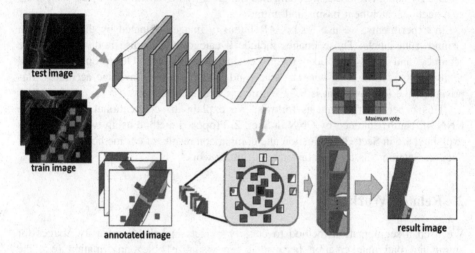

Fig. 1. Overview of our proposed method

Furthermore, we apply new voting approach to conventional method in order to improve class average accuracy. In this approach, we add use costs when we vote labels attached to similar training regions to a test region. We denote a local region as x_i and the set of x_i is denoted as X. Each local region has ground truth label y_i in terms of all pixels in x_i. y_i is N × N dimensional vector if we crop a local region with N × N pixels from a ground truth image. We define ground truth labels at the k-th pixel as $y_{i,k}$ which takes one label in C classes. Before using voting cost, we assign class label l_i to a local region x_i. The class label of surrounding region x_j of x_i is denoted as l_j.

The first cost which called unary cost is determined from the information of local region of x_i. If ground truth label of the k-th pixel $y_{i,k}$ is the same as the class label l_i, we enlarge voting weight for l_i. The second cost which called surrounding cost is determined from the information is determined from the information of surrounding local region x_j. If the number of label l_j corresponds to ground truth label $y_{i,k}$, we enlarge the voting weight for l_j.

We explain the class label which is essential for our costs in Sect. 3.1. The first cost determined from the information of local region explained are in Sect. 3.2. The second cost determined from the information of surrounding local region are shown in Sect. 3.3.

3.1 Setting Class Label of Each Local Region

In order to use two voting costs, we assign the class label to each local region. The size of the objects in LiDAR intensity images is clearly defined. Thus, we can assign appropriate class label to each local region x_i automatically by the area of an object. We define threshold according to the area and the frequency an object. The class which has the largest area in classes exceed threshold is defined as the most appropriate label l_i.

3.2 Unary Cost

The unary cost γ_i is determined from a local region x_i. If ground truth label $y_{i,k}$ corresponds to the class label l_i attached in previous section, we enlarge the voting weight of $y_{i,k}$. The unary cost γ_i is defined as

$$\gamma_i(y_{i,k}) = \begin{cases} \frac{d_k - d_m}{d_k - d_1} & \text{if } y_{i,k} = l_i \\ (1 - \delta) * \frac{d_k - d_m}{d_k - d_1} & \text{if } y_{i,k} \neq l_i \text{ and } y_{i,k} \in c_i \\ 0 & \text{otherwise} \end{cases} \tag{1}$$

where d_m is the distance of the m-th nearest neighbour, d_1 is the distance of the most similar (smallest distance in K regions) region and d_k is the distance of the K-th nearest neighbor. Namely, the weight of the most similar region is 1 and the most unsimilar region is 0. $0 \le \delta \le 1$ is the parameter to define how much we focus on the label l_i. Lager δ, more focus on it.

In the first condition, if ground truth label $y_{i,k}$ attached to a local region x_i corresponds to class label l_i, then weight according to similarity is voted. In the second condition, c_i is the set of class labels which appear in y_i. If ground truth label $y_{i,k}$ attached to a local region x_i is not the same class label l_i, then $(1 - \delta) * d_k - d_m / d_k - d_1$ is voted. For the other cases, voting is not performed. Thus, we can vote larger value to appropriate class of the local region x_i than other labels.

Figure 2 shows the unary cost. The local region x_i is shown as a red square. In the region, road, catchment biases, roadside tree and road shoulder are included, and these classes are c_i. The class l_i of the region is catchment biases. The voting weights are determined automatically.

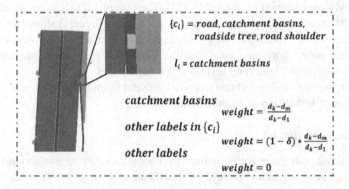

Fig. 2. Example of unary cost (Color figure online)

3.3   Surrounding Cost

The surrounding cost $\gamma_{i,j}$ is determined from surrounding region x_j of x_i. Since we cropped local regions with overlapped manner, the class label l_i of surrounding region x_j should be the same to the class label l_i of the local region x_i. The pairwise cost $\gamma_{i,j}$ is defined as

$$\gamma_{i,j}(y_{i,k}) = \frac{\sum_j \left[y_{i,k} = l_j \right]}{n} * \frac{d_k - d_m}{d_k - d_1} \tag{2}$$

where $[\cdot]$ is 1 if the argument is true and 0 for otherwise and n is the number of surrounding regions x_j. The weight becomes high if the number of label corresponds to l_j is large. We improve the accuracy of objects with small area by using the costs.

Figure 3 show the surrounding cost. Red square shows the local region x_i, 4 nearest surrounding regions of x_i are denoted as $x_1 - x_4$ and n = 4. The class labels $l_1 - l_4$. Since 3 surrounding regions are catchment basins, the weight for the class becomes large.

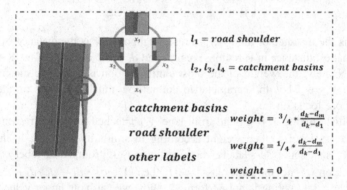

Fig. 3. Example of surrounding cost (Color figure online)

4   Experiments

In experiments, we use 36 LiDAR intensity images with manually annotated ground truth labels. They were obtained by the MMS [12]. The size of LiDAR intensity images are 2000 × 1500 pixels, and an image represents 80 × 60 m. Figure 3 shows the examples of our dataset. Those images include 9 categories; pedestrian crossing, catchment basins, roadside tree, gutter, gore area, road, median, pedestrian path and road shoulder. We divide 36 images into 28 training images and 8 test images.

By using our method, we obtain over 40 thousand training samples and 1 thousand local regions of test image from only 36 images.

4.1   Evaluation Method

We use both class average accuracy and pixel-wise accuracy as evaluation measures. Class average accuracy is the average of classification accuracy of each class, and it is influenced by the classes with small area such as catchment basins. Pixel-wise accuracy is the percent of correctly labeled pixels in all pixels of test images, and it is influenced by the classes with large area such as road.

We consider that class average accuracy is more important than pixel-wise accuracy because the purpose of this study is for making the Fundamental Geospatial Data of road automatically. Thus fore, it is necessary to improve the accuracy of classes with small areas such as catchment basins.

4.2 Evaluation Results

In Table 1, we compare our proposed method with conventional method. The second and sixth rows represent the results which combined features extracted different region sizes; 64 × 64 pixels and 48 × 48 pixels. This approach improve power pf expression, because represent the same object with different sizes. We find top K similar regions separately for each region size, and voting is performed independently.

The pixel-wise accuracy of our proposed method with both unary and surrounding costs is 96.77%. The class average accuracy is 84.74%, and it is 9.21% higher than conventional method. We confirmed a large improvement without a major change in conventional method. Conventional method with only unary cost and surrounding cost have the accuracy of 81.16% and 80.36% in class average accuracy. They are 5.63% and 4.77% better than conventional method. The results demonstrate that each cost is effective to improve performance. Also, class average accuracy changes by the value δ. Higher δ gives better class average accuracy. Thus, the parameter δ in Eq. (1) is set to 0.9. $\delta = 1$ means that we vote only one label in ground truth label y_i. No voting area is generated by this method. Since we want to create a road map automatically, we do not allow no voting area and we do not use $\delta = 1$.

Table 1. Accuracy comparison of segmentation approach

	Class average	Pixel-wise
Conventional method	74.96	97.63
Conventional method + different region sizes	75.53	97.78
Proposed method	82.19	96.89
Proposed method with only unary cost ($\delta = 0.6$)	79.50	97.51
Proposed method with only unary cost ($\delta = 0.9$)	81.16	96.77
Proposed method with only surrounding cost (4 neighborhood)	80.30	96.88
Surrounding cost (8 neighborhood)	79.44	97.10
Proposed method + different region sizes	**84.74**	96.77

Table 2 shows the accuracy of each class by conventional method and the proposed method. Our method much improved the accuracy of classes with small area such as catchment basins and gutter.

Figure 4 shows the results by our proposed method. Since pixel-wise accuracy is not so changed, similar segmentation results are obtained. But we confirmed that voting values for the correct class increase. Figure 5 shows results by the proposed method and conventional method for objects with small area. Our proposed method are obviously better than conventional method.

Table 2. Accuracy of each class

	Pedestrian	Catchment basins	Garden plant	Gutter	Gore area	Road	Median	Pedestrian path	Road shoulder
Conventional	76.43%	29.54%	93.62%	19.46%	76.90%	**98.56%**	**97.14%**	98.65%	**89.53%**
Proposed	**80.01%**	**59.50%**	**98.70%**	**54.26%**	**91.59%**	96.83%	95.11%	**99.17%**	87.63%

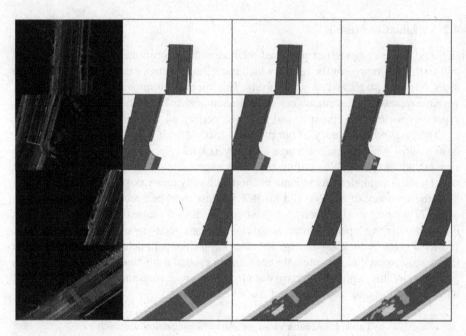

Fig. 4. Results by our method. The first column is input LiDAR intensity image, the second column is ground truth, the third column is results by conventional approach and the forth column is results by proposed method.

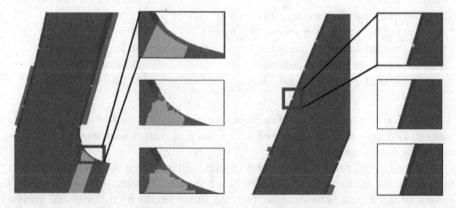

Fig. 5. Results of objects with small area. The first row is region of ground truth, the second row is result by conventional approach and the third row is result by the proposed method.

5   Conclusion

In this paper, we proposed the improved approach for labeling road map using voting costs. Unary cost is the weight for a local region. Surrounding cost is the weight for using surrounding information of the target region effectively. Our method gave 9.22% higher accuracy in comparison with conventional method. We also confirm that both costs are effective to improve the accuracy.

However, the accuracy of object class with small area is improved significantly but the accuracy is not sufficient yet. Thus, we consider to add the weight to the appropriate region size of each object class when we combine the results using different region sizes. We would like to use relationship between classes like CRF.

Acknowledgements. This work is partially supported by MEXT KAKENHI 15K00252.

References

1. Hasegawa, H., Ishiyama, N.: Publication of The Digital Maps (Basic Geospatial Information). Geospatial Inf. Authority Jpn **60**, 19–24 (2013)
2. Umemura, M., Hotta, K., Nonaka, H., Oda, K.: Image labeling for lidar intensity image using K-NN of feature obtained by convolutional neural network, Int. Arch. Photogramm. Remote Sens. Spatial Inf. Sci. XLI-B3, 931–935 (2016)
3. Yan, W.Y., Shaker, A., Habib, A., Kersting, A.P.: Improving classification accuracy of airborne LiDAR intensity data by geometric calibration and radiometric correction. ISPRS J. Photogram. Remote Sens. **67**, 35–44 (2012)
4. Tighe, J., Lazebnik, S.: SuperParsing: scalable nonparametric image parsing with superpixels. In: Daniilidis, K., Maragos, P., Paragios, N. (eds.) ECCV 2010. LNCS, vol. 6315, pp. 352–365. Springer, Heidelberg (2010). doi:10.1007/978-3-642-15555-0_26
5. Kohli, P., Osokin, A., Jegelka, S.: A principled deep random field model for image segmentation. In: Proceedings of the IEEE Conference on Computer Vision and Pattern Recognition, pp. 1971–1978 (2013)
6. Hariharan, B., Arbeláez, P., Girshick, R., Malik, J.: Simultaneous detection and segmentation. In: Fleet, D., Pajdla, T., Schiele, B., Tuytelaars, T. (eds.) ECCV 2014. LNCS, vol. 8695, pp. 297–312. Springer, Cham (2014). doi:10.1007/978-3-319-10584-0_20
7. Girshick, R., Donahue, J., Darrell, T., Malik, J.: Region-based convolutional networks for accurate object detection and segmentation. IEEE Trans. Pattern Anal. Mach. Intell. **38**(1), 142–158 (2016)
8. Jonathan, L., Shelhamer, E., Darrell, T.: Fully convolutional networks for semantic segmentation. In: Proceedings of the IEEE Conference on Computer Vision and Pattern Recognition, Boston, USA, pp. 3431–3440 (2015)
9. Badrinarayanan, V., Kendall, A., Cipolla, R.: Segnet: a deep convolutional encoder-decoder architecture for image segmentation. arXiv preprint arXiv:1511.00561 (2015)
10. Jia, Y., Shelhamer, E., Donahue, J., Karayev, S., Long, J., Girshick, R., Guadarrama, S., Darrell, T.: Caffe: convolutional architecture for fast feature embedding. In: Proceedings of the ACM International Conference on Multimedia, Florida, USA, pp. 675–678 (2014)
11. Novak, K.: Mobile Mapping System: new tools for the fast collection of GIS information. In: Proceedings of the SPIE, vol. 1943 (1993)

A Lattice-Theoretic Approach
for Segmentation of Truss-Like Porous
Objects in Outdoor Aerial Scenes

Hrishikesh Sharma[✉], Tom Sebastian, and Balamuralidhar Purushothaman

Embedded Systems and Robotics Lab, TCS Research and Innovation,
Bangalore, India
{hrishikesh.sharma,tom.sebastian,balamurali.p}@tcs.com

Abstract. Remote video surveillance of vast outdoor systems for structural health monitoring using e.g. drones is gaining rapid popularity. Many such systems are designed as truss structures, due to well-known mechanical reasons. A truss structure has interstices inherently porous, and hence no closed region or contour really represents useful properties or features of just foreground or just background. In this paper, we present a novel approach to segment and detect porous objects of truss-like structures in videos. Our approach is primarily based on modeling of such objects as composite shapes, organized in a structure called geometric lattices. We define a novel feature called shape density to classify and segment the truss region. The segmented region is then analyzed for various surveillance goals such as bending. The algorithm was tested against video data captured for many transmission towers along two different power grid corridors. We believe that our algorithm will be very useful for analysis of truss-like structures in many a outdoor vision applications.

1 Introduction

In most nations, many vast outdoor man-made systems are installed, signifying national utilities, for example dams, bridges, oil and gas supply pipelines, power grid etc. All of these systems have a robust *mechanical design aspect*, done primarily to ensure that they withstand extreme weather condition and have long life. Maintenance of such systems is an important and costly legal responsibility, mainly for public safety, of the organization that owns them. Periodic inspections and subsequent maintenance acts help to minimize the number and duration of their outages in daily usage. The combined use of Unmanned Aerial Vehicles (UAVs) and computer vision techniques for **structural health monitoring** of such vast systems, especially long linear infrastructures such as power grid, is rapidly emerging as a popular option [7].

A specific category of vast outdoor systems are systems having truss as their mechanical model. Such structure is used for systems having requirement for very long spans, such as in bridges, airport terminals, power transmission towers, stadium roof etc. For automation of monitoring of such systems, they need

© Springer International Publishing AG 2017
F. Karray et al. (Eds.): ICIAR 2017, LNCS 10317, pp. 586–595, 2017.
DOI: 10.1007/978-3-319-59876-5_65

to be first segmented in the corresponding aerial imagery collected by a UAV. Only a handful of research works exist in literature, for truss identification from aerial images using machine vision [5]. While analysis of in-situ sensor data using e.g. vibration sensor is promoted [6,8], it has not been used in practice due to well-known reason of problem of collecting sensor data over vast sensor networks. In [12], electricity towers are detected using straight lines in each video frame 2-D IIR filter, and Hough transformation. In [10], a supervised learning app-roach is used for classification of towers. However, our classification performance surpasses the performance of their *shallow* classifier, as well our own prototype deep classifier, as is shown in Sect. 5. One **challenge** for segmentation of truss-like structures in images is that such structure is inherently porous, and hence no closed region or contour within the frame really represents useful properties or features of just foreground or just background. Another challenge is that in aerial images, the background is *varying and complex*. Further, most of these background artefacts are texture-rich, while the foreground objects of interest (the set of GS beams that form the pylon structure) are almost texture-less (other than being fairly thin). This makes the segmentation problem challenging.

In this paper, for identification of *keyframes* and subsequent segmentation within them for locating truss regions, we propose a four-stage **unsupervised** classification algorithm, which has following properties.

- As one novelty, we represent a truss by a region marked by high presence of overlapping planar polygons.
- As another novelty, we organize these polygons in a **geometric lattice** struc-ture, and then use reachability analysis over the lattice to derive the set of atoms, or the *canonical* polygons,
- As another novelty, using canonical polygons, we define a novel feature called **shape density** to distinctly classify broad class of truss-like objects.

The rest of the paper is organized as follows. We first describe our detection algorithm in Sect. 2. It is followed by analysis of certain faults over the segmented region, in Sect. 3. We then describe the nature of our experiments in Sect. 4, followed by the Sect. 5 on results and analysis, before we conclude.

2   Proposed Detection Algorithm

The algorithm consists of four stages, as shown in Fig. 1. The overall intuition behind classification is that the edges of beams or girders of trusses are lines that intersect or join each other in both 2D projection and 3D. Such criss-crossing lines, at multiple angles not just 0° or 90°, participate in many **abutting** polygon

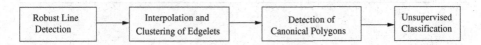

Fig. 1. Block diagram of truss detection

formations. Further, in a typical aerial *sideways* inspection, almost all such joints will be visible and not occluded by projection of other beams. As can be seen from a wide-angle aerial frame in Fig. 2a, a region (2D) or a volume(3D) of any truss structure will have very high number of such polygons *per unit* area or volume. Hence we can estimate 2D or 3D density of such shapes to distinctly classify such regions. Since we work with heavily porous objects, having extensive shape concavities, such measure gives a very useful classification.

In the first two stages of our algorithm, the focus is on detecting beams maximally. This ensures that not only that we detect maximal number of beams (true positive in terms of number of beams), but also maximal length of each beam perimeter. Else, we miss out on certain intersections of lines and hence some count of polygons. The next two stages focus on detection of canonical polygons (those that do not subsume any other polygon), as well as their spatial clustering. The spatially clustered region/s corresponding to abnormally high density of polygon are marked as truss regions. Some analysis is carried out on such regions, to showcase utility of such segmentation in real deployments.

2.1 Robust Line Detection in Outdoor Scenes

There are plenty of edge and line detection algorithms known in machine vision literature. However, when it comes to outdoor images, there are multiple challenges including illumination variance given time of day, presence of diffuse edges and contour leaks etc. It is obvious, and also cross-verified through our experiments, that no single thresholding will be able to detect maximal number of truss beam edges. Hence we employed pyramid method to robustly detect edges by reducing the *variance* of edge gradients using smoothening to multiple levels. Towards that, we used the recent and popular line detection scheme, LSD [4]. The **typical** output of detection for Fig. 2a is shown in Fig. 2b.

Algorithm 1. Detection of truss structures in aerial images

Fast line segmentation based on pyramid method to detect maximal edges of maximal number of beams

Representation of beams by medial lines

   Interpolation of gaps between detected line segments

   Clustering of line segment of each beam face around medial line

Detection of Polygons

   Enumeration of all polygons based on line intersection

   Shortlisting of Canonical Polygons

Shape Density based Classification

   Estimation of Shape Density using Canonical Polygons

   Density-based Removal of False Positives

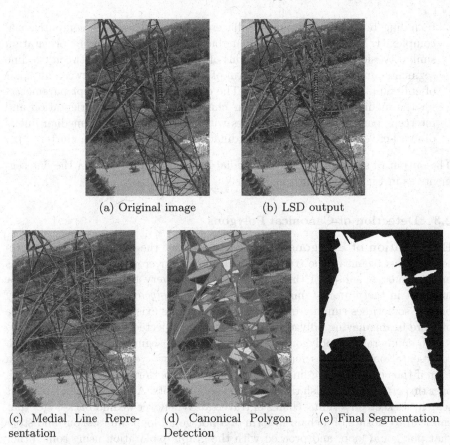

(a) Original image   (b) LSD output

(c) Medial Line Repre-   (d) Canonical Polygon   (e) Final Segmentation
sentation   Detection

Fig. 2. Example stage-wise segmentation for first truss. Zoom for better view.

2.2   Interpolation and Clustering of Line Segments

It can be seen, as illustrated in Fig. 2b, that using LSD, we were able to maximize
the number of beams for which lines at their edges are detected. However, for
each such beam, the full length of edges is not detected in general. Such breakage
of edges occurs very often in localities of extreme diffusion [9,11]. To be able to
maximize length of detected beam edges, we perform two operations.

1. We **interpolate** and cover short gaps between two line segments that (ideally) have same orientation, overlap the hypothetical line created by infinitely
   extending either of the line segment, and have small gap over the hypothetical
   line, once overlaid so.
2. Many of the stronger beams are actually two flat beams welded and abutted
   along the major length (L-angle beam). Hence ideally, we detect upto 3 edges
   in their projection. Practically, for each edge, we get different edgelets at
   different locations on the **same** beam. We group these edgelets and represent them using a single, longer, *medial line*, simply *interpolated* within the

bounding box of such group of adjacent edgelets, to reduce computational complexity. We perform the grouping based on three parameters: orientation similarity, sideways intercept in terms of minimum distance between two line segments, and front intercept in form of minimum distance between any pair of endpoints of two line segments. The orientation and intercept parameters for the median line for a particular cluster are the average orientation and intercept parameters of all the lines within that cluster. The median line is drawn between two extreme x coordinates of the lines within a cluster.

The output of such clustering, and medial line representation, over the line segments as in Fig. 2b, is shown in Fig. 2c.

2.3   Detection of Canonical Polygons

Enumeration of Polygons. It is easy to see that the medial lines that represent various beams of the truss object either join or cross each other at various orientations. A subset of three or more beams, every two of which concur or intersect in the *projected* image, form a *irregular polygon*. Since there are many beams, sometimes running into few hundreds, there exist many subsets of three of more beams having polygonal closure in the projected image.

To enumerate all polygons, we take all the line segments and make a graph representation. The lines represent vertices, and two vertices are joined by an edge if the corresponding lines intersect. We then perform a **depth-first** search over this graph to establish the connected components. As we perform depth first search, we keep on forming chains of traversal. Whenever we find back-edges that can lead to closure of a chain in form of a polygonal loop, we take out and store that polygonal loop, and proceed with the graph exploration using some other edge of the current vertex. This procedure results in detection of **forest** of all connected/abutting polygons, which in term represent the *true positive* as well as *false positive* truss regions.

Geometric Lattice of Polygons. We state here, without proof for lack of space, that these polygonal shapes can be organized into a mathematical structure called **geometric lattice** [1]. To recall that we define **canonical polygons** as those polygons within each connected component of lines, which do not subsume any other polygon inside.

Theorem 1. *The set of all polygons within each connected component of polygons above is homomorphic to the sublattice of a complete lattice formed having all the lines as the elements of the partially ordered set.*

Any geometric lattice can be organized into levels/flats of its elements. Hence to represent a truss region, we detect the (canonical) polygons at level 1, to be used in next section for estimation of shape density of truss-like regions (Fig. 3).

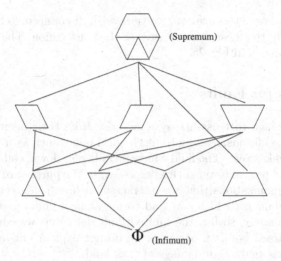

Fig. 3. Geometric lattice of polygon subsumptions

Listing the Canonical Polygons. To establish subsumption, we note that the proof of above theorem is a constructive proof. Hence we take the set of lines and assign them labels from a set. We then **sort** the polygons by the number of lines it is made up of. Put this way, the smallest sized polygon is a triangle. For a n-gon within truss, polygons ranging from triangle upto n-gons are expected to be subsumed within, if true. Starting from smallest polygons i.e. label subsets of size 3, 4, $\ldots$, we **greedily** look for representing the n-gon (for a fixed n) as a union of known label subsets of smaller or same size (i.e. from 3 to n), that abut i.e. share some labels. If we fail to represent a n-gon as a union of smaller or same-sized polygons, then we deem that specific n-gon as a canonical polygon and make it as part of the level-1 list of polygons. We repeat this exercise for each of the connected component, and make many such lists.

The output of this iterative process is a "forest" of a few merged segments as shown in Fig. 2d.

2.4   Unsupervised Classification Using Shape Density Estimation

As has already been explained, the region corresponding to a truss is distinctively represented by a high presence of polygonal shapes that abut each other. Hence if we estimate region density of polygons in a truss region, then even if we use number of canonical polygons **only** to estimate the density, the estimate will be very high when compared to any other, even similar if not random, *false positive* object/region in the background. The value of shape density is defined as the number of abutting *canonical* polygons in a particular region, divided by the number of points that are in the interior of the region. Usage of such measure simultaneously gives controlled **precision** measure, and very less **recall** measure i.e. degree of missed detections during classification.

To perform above classification, we work with disconnected clusters of polygons, and pick up the most dense cluster as the truss region. The output of this (final) stage is shown in Fig. 2e.

3 Analysis for Faults

The biggest damage to truss-like systems arise from their ageing under open environmental conditions, leading to mechanical issues such as fracture of panel joints, brace breakage etc. The final effects include pull-down, collapse and bending/breakage of a part of tower such as cross-arms. The purpose of aerial surveys, as desired by organizations that we interacted with, is to detect and estimate degree of such damage, which can lead to mechanical faults and disruption in service. Unfortunately, so far, the limited fault data that we could get relates to damaged trusses. For lack of space, we do not explain but only sketch our approach towards analysis of damage of such kind.

As shown in Fig. 2e, we are able to segment the exact silhouette of the truss. We take past data as shape template of such kind, taken from fixed camera point of view. Such fixing is possible, and in fact required, if the same planned path is taken for multiple surveys. We then do a template matching of past and current shapes, and calculate the dissimilarity measure using polyline similarity matching described in [2]. For a more precise 3D estimation, we are currently working on registration of line models using the estimated depth map.

4 Experiments and Data Collection

So far, we were allowed to perform aerial survey to gather data related to transmission towers only. It was captured using a 11 MP f/2.8 wide lens RGB camera, GoPro Hero3, which was mounted on a mini-UAV. The image size captured was 3000×2250, which was resized to 1280×960 for testing purposes. Two test sites provided by *Hot Line Training Center*, outside our city of residence, were used for imaging two power grids. A quadcopter provided by our collaborators was flown so as to have a sideways view of the power grids. For such view, the pitch of the camera mount was fixed to around 60°, while yaw was azimuth-facing and roll angle towards horizon. Enough length of power grids was imaged, giving us around **115** different keyframes having towers within the video. The heterogeneous background through the truss was varying all through, typically consisting of vegetation, sky, unpaved roads and houses.

5 Results and Performance Analysis

While the testing and analysis has been carried out on all **115** tower frames, we have only illustrated results for two sample images for sake of brevity. The two images have been chosen with different angle of view, for different sized pylons. One of the aerial image has a different, urban background. The other image has only *partial* true positive: only a part of the truss is visible. Further, it has a false positive as well in the background: a grid in the power substation at the back. Also, the background itself is totally different, a rural terrain.

Before arriving at an unsupervised approach, we first did quick experiments with supervised deep learning approach via TensorFlow library. Since there are too many beams in a truss, labeling the foreground by labeling only the beams was not timely possible for us. Hence we labeled the bounding box of entire truss and partitioned it into 32 × 32 blocks for training. Problem with such trained classifier was many false negatives (missed detection of certain ground truth blocks during testing) due to high amount of noise in each bounding box (only <20% pixels in each 32 × 32 box belong to foreground, on an average). Hence we switched to unclassified and actually got better results.

It can be seen that our algorithm works quite well for such **extreme** segmentation cases, with varying outdoor background conditions. For all the frames having varying degree of presence of towers (i.e. varying region size), we detect hundreds of **connected** polygons. Hence we get shape density figures which range from 50 to 80 for our dataset. In other regions that are present as false positive till the penultimate stage, the worst-case shape density was found to be upto 20. Thus, even when a equivalent-sized false positive is present in the background, and the foreground itself is partial, we are able to classify robustly and detect the right true positive. Importantly, we did not require any specific tuning of thresholds for any specific frame. Rather, same set of parameters worked well to discriminate the truss region in entire tower dataset. We benchmarked our algorithm's performance against a very recent work in [3]. We used the same dataset and the corresponding annotated ground truth. We used a pixel-level annotation tool called LabelMe for ground truth generation. We use F1 score as metric rather, which is more meaningful for our work (minimization of both false positives and false negatives). We investigated three datasets pertaining to transmission tower (collected as per Sect. 4), bridges over rivers and scaffolding, the last two of which were sourced from internet. Our algorithm performed better by giving a F1 measure of **86%**, **88%** and **79%** for dataset 1, 2 and 3 respectively. Corresponding F1 measure for first type of data in [3] is **80%**, which highlights significant improvement when we use our algorithm. Reason for relatively less recall value over scaffolding dataset is that its beams were rounded, and do not show sharp bends like an L-joint, like in all other datasets. In a general scenario, this also alludes to the failure mode of our algorithm, where the bends in beams are not sharp. This leads to weak edges, picked only partially or not picked at all by unsupervised, threshold-driven methods, and corresponding loss in formation and detection of canonical polygons. Shadows, bad lighting conditions, occlusion from the leaves and rusting are the other reasons which will reduce the quality of beam detection stage of the algorithm and eventually will reduce the overall efficiency significantly. We have used the LSD line segment detector for finding the line segments, which works well in normal conditions but it is not fully robust against all the above factors. For improving the accuracy of the algorithm, better line detection approaches need to be investigated.

In few cases, false positives get detected for the reason that the set of transmission lines meet/hang by the tower. Hence their projection in the image, along with certain beams, forms a false cluster of polygons. This can be improved by considering another feature that represents the *heterogeneity* of the set of detected polygons, in terms of the variance of set of their areas. In limited cases, we found that some beams went undetected in initial cases, due to which certain part (overlap) of the ground truth did not get detected in the true positive. For lack of space, we are not able to depict the ground truth in this paper. Such loss of overlap, happening in <**3%** frames, is <**10%** of the ground truth region, and can be further limited by improving the degree of line segment detection in the first stage. Also, for the target of semi-automated processing, minor partial false negative or partial false positive is not a serious problem, though.

5.1   Computational Complexity

For the first stage, it has been shown in [4] that the computational complexity of LSD is $O(n^2)$. The interpolation and clustering of line segments is proportional to number of line segments detected in previous stage, which at maximum are bounded from above by $O(n^2)$ again. Similarly, formation of polygons in third stage is again bounded from above by $O(n^2)$. The final stage of calculation of shape densities for different clusters depends on the number of polygons present, to be accumulated in the numerator. Hence it is bounded from above by $O(n^2)$ as well.

In summary, by taking the maximum of above complexity orders, it can be clearly seen that the overall complexity of our truss segmentation algorithm is $O(n^2)$. This estimation is actually an upper bound, since we work with only edges and not linear shapes in all stages, not regions. Hence in practice, as we have seen during experiments, our algorithm is much faster.

6   Conclusion

Automatic segmentation and analysis of truss-like objects in aerial images is a practically useful research problem. In this paper, we have proposed an algorithm for their segmentation in complex and heterogeneous outdoor surroundings. To do so, we first robustly detect line segments along the edges of beams and girders of the truss. The ideal edges of beams are riveted and joined to each other, and hence in the image, they appear to be criss-crossing, forming a region of dense irregular polygonal shapes that abut each other. Since edges of most beams break while detection due to various reasons, we interpolate their gaps and cluster them along medial line for representation. Then we detect all the polygonal shapes and also count the number of *canonical* polygons that are not subsumed in any other polygon. The region density of such shapes gives a useful classifier that is then used to select the truss region and weed out any false positive regions. The entire algorithm was tested on 100+ truss images collected using a mini-UAV, over two power grid corridors. It was found to exhibit minimum presence of both false positives and false negatives. The work has the potential to be cross-applied to other applications, e.g. identification of a freckled or scaled skin surface of humans/animals, identification and counting of stacked or stuffed inventory in a warehouse etc. We believe that our algorithm for truss segmentation and analysis is robust enough with good performance, and hence can be applicable for detecting various types of trusses in various outdoor surroundings.

References

1. Bárány, I.: Discrete and convex geometry. In: Horváth, J. (ed.) A Panorama of Hungarian Mathematics in the Twentieth Century I, pp. 427–454. Springer, Heidelberg (2006)
2. Cohen, S.D., Guibas, L.J.: Partial matching of planar polylines under similarity transformations. In: ACM-SIAM Symposium on Discrete Algorithms, pp. 777–786 (1997)
3. Dutta, T., Sharma, H., Vellaiappan, A., Balamuralidhar, P.: Image analysis-based automatic detection of transmission towers using aerial imagery. In: Paredes, R., Cardoso, J.S., Pardo, X.M. (eds.) IbPRIA 2015. LNCS, vol. 9117, pp. 641–651. Springer, Cham (2015). doi:10.1007/978-3-319-19390-8_72
4. von Grompone, G.R., Jakubowicz, J., Morel, J.M., Randall, G.: LSD: a fast line segment detector with a false detection control. IEEE Trans. Pattern Anal. Mach. Intell. **32**(4), 722–732 (2010)
5. Jahanshahi, M.R., Kelly, J.S., Masri, S.F., Sukhatme, G.S.: A survey and evaluation of promising approaches for automatic image-based defect detection of bridge structures. Struct. Infrastruct. Eng. **5**(6), 455–486 (2009)
6. Kim, H.M., Bartkowicz, T.J.: An experimental study for damage detection using a hexagonal truss. Springer J. Comput. Struct. **79**(2), 173–182 (2001)
7. Li, Z., Liu, Y., Hayward, R., Zhang, J., Cai, J.: Knowledge-based power line detection for UAV surveillance and inspection systems. In: IEEE International Conference on Image and Vision Computing, pp. 1–6 (2008)
8. Mehrjoo, M., Khaji, N., Moharrami, H., et al.: Damage detection of Truss bridge joints using Artificial Neural Networks. Springer J. Expert Syst. Appl. **35**(3), 1122–1131 (2008)
9. Ravanbakhsh, M., Heipke, C., Pakzad, K.: Road junction extraction from high-resolution aerial imagery. Photogram. Rec. **23**(124), 405–423 (2008)
10. Sampedro, C., Martinez, C., Chauhan, A., Campoy, P.: A supervised approach to electric tower detection and classification for power line inspection. In: International Joint Conference on Neural Networks, pp. 1970–1977 (2014)
11. Sirmacek, B., Unsalan, C.: Building detection from aerial images using invariant color features and shadow information. In: IEEE International Symposium on Computer and Information Sciences, pp. 1–5. IEEE (2008)
12. Tilawat, J., Theera, U.N., Auephanwiriyakul, S.: Automatic detection of electricity pylons in aerial video sequences. In: International Conference on Electronics and Information Engineering, pp. 342–346 (2010)

Non-dictionary Aided Sparse Unmixing of Hyperspectral Images via Weighted Nonnegative Matrix Factorization

Yaser Esmaeili Salehani[(✉)] and Mohamed Cheriet

Synchromedia Laboratory,
University of Quebec's École de technologie supérieure (ÉTS),
Montreal, Canada
yaser.esmaeili@gmail.com

Abstract. In this paper, we propose a method of blind (non-dictionary aided) sparse hyperspectral unmixing for the linear mixing model (LMM). In this method, both the spectral signatures of materials (endmembers) (SSoM) and their fractional abundances (FAs) are supposed to be unknown and the goal is to find the matrices represent SSoM and FAs. The proposed method employs a weighted version of the *non-negative matrix factorization* (WNMF) in order to mitigate the impact of pixels that suffer from a certain level of noise (i.e., low signal-to-noise-ratio (SNR) values). We formulate the WNMF problem thorough the regularized sparsity terms of FAs and use the multiplicative updating rules to solve the acquired optimization problem. The effectiveness of proposed method is shown through the simulations over real hyperspectral data set and compared with several competitive unmixing methods.

Keywords: Hyperspectral images · Unmixing · Weighted nonnegative matrix factorization (WNMF) · Sparse recovery · Non-dictionary aided

1 Introduction

Hyperspectral sensors are able to measure the electromagnetic spectrum of various materials and collect such information as a datacube that contains two spatial dimension and one spectral dimension. In a typical data cube, the mixed pixels often occur due to the insufficient spatial resolution of hyperspectral camera sensors and mixing of the materials in a microscopic scale. Thus, the hyperspectral unmixing (HU) methods are needed to analyse these mixed pixels. In recent years, different HU methods have been proposed to characterize the mixed pixels into a set of spectral signatures called *endmembers* as well as a set of fractional abundances (FAs) of these endmembers [2,3,20]. In the linear mixing model (LMM), which is common in practice, the measured (observed) spectral signatures of materials (SSoMs) are a linear combination of the existing endmembers in the scene of hyperspectral data cube [3]. We can consider the HU methods of LMM either dictionary aided (semi-blind) or non-dictionary aided

© Springer International Publishing AG 2017
F. Karray et al. (Eds.): ICIAR 2017, LNCS 10317, pp. 596–604, 2017.
DOI: 10.1007/978-3-319-59876-5_66

(blind) methods depending on the availability of the library of SSoMs. The main goal of the dictionary aided HU methods is to find the FAs of the materials for a given dictionary of SSoMs. However, the spectral library is not often known in practice. Hence, the non-dictionary aided HU methods are needed and one-step methods such as nonnegative matrix factorization (NMF) [18,22] based app-roach is highly interested due to its noticeable supports. These may contain the nonnegativity constraints of both SSoMs and their corresponding FAs which are automatically included in the NMF-based methods [6,23]. Furthermore, it can make decomposition matrices to be more intractable because of a part-based representation of the data, e.g. [4,6]. Recently, various HU methods through the classic NMF problem [18] have been proposed such as the $\ell_{1/2}$-norm sparsity constrained [23], the manifold regularization into the sparsity constraint [19], the substance dependence constrained [26], the collaborative properties of FAs [7] and the multilayer NMF through the ℓ_p-norm [8,24].

In this paper, the main objective is to alleviate the role of hyperspectral pixels that corrupted due to a certain level of noise (i.e., low signal-to-noise-ratios (SNRs)) for the purpose of HU in order to enhance the results of estimated SSoMs and their corresponding FAs. Therefore, we assign the zero weights to the set of this kind of observed pixels such that the normalized ℓ_2-norms of their corresponding spectral bands are smaller than a constant threshold. We propose to use a weighted NMF (WNMF) approach [12,16] to achieve this goal. Then, we introduce a new minimization problem that considers the WNMF to measure the ℓ_2-norm of modeling error and the regularized ℓ_p-norm term to use the sparse property of FAs. We apply the multiplicative updating rules to solve the acquired minimization problem and show that the proposed updating rules will be guarantee to reach a local minimum. Finally, we evaluate the proposed method over the real hyperspectral data set and show the effectiveness of the proposed method through simulations.

The rest of the paper is organized as follows. The linear mixing model (LMM) and the classic NMF problem are reviewed in Sect. 2. We introduce our method of sparse WNMF in Sect. 3 and report the simulation results in Sect. 4. Section 5 concludes the paper and discuss the future works.

2   System Model and Problem Formulation

2.1   LMM: Linear Mixing Model

The measured spectra of hyperspectral pixels in the LMM is a linear combination of the SSoMs and their corresponding FAs that can be considered as $\mathbf{y} = \mathbf{A}\mathbf{x} + \mathbf{e}$, where $\mathbf{y} \in \mathbb{R}^{L \times 1}$ is the measured spectral signatures of a pixel with L spectral band, $\mathbf{A} \in \mathbb{R}_+^{L \times N}$ is the library of endmemebrs contains N pure SSoMs, $\mathbf{x} \in \mathbb{R}_+^{N \times 1}$ is the corresponding FAs of the endmembers and $\mathbf{e} \in \mathbb{R}^{L \times 1}$ is an $L \times 1$ additive noise vector. There are two physically meaningful constraints for the FAs called the *abundance nonnegativity constraint* (ANC), $\mathbf{x} \geq 0$, and the *abundance sum-to-one constraint* (ASC), $\mathbf{1}^T \mathbf{x} = 1$. Also, we can present a matrix compact of P hyperspectral pixels as follows

$$\mathbf{Y} = \mathbf{AX} + \mathbf{E}, \tag{1}$$

where, $\mathbf{Y}$ and $\mathbf{X}$ are the $L \times P$ and $N \times P$ matrices, respectively. Since the number of endmembers present at each mixed pixel of the hyperspectral scene is much smaller than the dimension of available spectral libraries in practice, the HU problem is considered as the sparse regression problem, e.g., [6,15,23] where $L \gg N$.

2.2 NMF-Based Methods: An Overview

Given the data matrix $\mathbf{Y}$, the NMF method seeks two nonnegative matrices $\mathbf{A}$ and $\mathbf{X}$ to approximate as $\mathbf{Y} \approx \mathbf{AX}$, where $N < \min(L, P)$. The cost function of the classic NMF problem is $\frac{1}{2}\|\mathbf{Y} - \mathbf{AX}\|_2^2$ where finding the global optimal solution for the corresponding minimization problem is NP-hard problem due to the nonconvexity of the problem with respect to both $\mathbf{A}$ and $\mathbf{X}$ [25]. Indeed, one can easily show that for any nonnegative invertible square matrix $\mathbf{B} \in \mathbb{R}_+^{N \times N}$, there are different values for $\mathbf{AX} = (\mathbf{AB})(\mathbf{B}^{-1}\mathbf{X})$ [23]. The regularization method is used to overcome this problem and to add the desired constraints (e.g., sparsity and smoothness) into the cost function. To consider the sparse property of FAs, adding the ℓ_0-norm term is of interest. However, the ℓ_0-norm problem which is defined by the number of nonzero components of a vector is itself NP-hard problem due to the combinatorial exhaustive search of solution. Therefore, using approximation terms such as ℓ_p-norm $(0 < p < 1)$ term is the alternative approach to tackle this problem. Several sparse HU method through the NMF approach when $p = \frac{1}{2}$ have been proposed in [7,19,23,26]. The corresponding minimization problem is as follows:

$$\min_{\mathbf{A} \geq 0, \mathbf{X} \geq 0} f(\mathbf{A}, \mathbf{X}), \tag{2a}$$

$$f(\mathbf{A}, \mathbf{X}) = \frac{1}{2}\|\mathbf{Y} - \mathbf{AX}\|_F^2 + \lambda\|\mathbf{X}\|_{\frac{1}{2}}, \tag{2b}$$

where $\|.\|_F$ denotes the Frobenius norm and $\|\mathbf{X}\|_{\frac{1}{2}} = \sum_{i=1}^{N} \sum_{j=1}^{P} x_{ij}^{\frac{1}{2}}$ and x_{ij} is the corresponding abundance for the i-th endmember at the j-th pixel and $\lambda > 0$ is the Lagrangian parameter. The multiplicative updating rules [18] can be used to solve the problem (2a) which are [23]:

$$\mathbf{A} \leftarrow \mathbf{A}. * \mathbf{YX}^T. / \mathbf{AXX}^T \tag{3}$$

$$\mathbf{X} \leftarrow \mathbf{X}. * \mathbf{A}^T\mathbf{Y}. / (\mathbf{A}^T\mathbf{AX} + \frac{\lambda}{2}\mathbf{X}^{-\frac{1}{2}}) \tag{4}$$

where $.*$ and $./$ are the element-wise multiplication and division respectively, $(.)^T$ denotes the transpose of a matrix, and the power in $\mathbf{X}^{-\frac{1}{2}}$ operates element-wisely as well.

3   The Proposed WNMF Method

The weighted NMF method have been used recently in some applications such as network communication where some entries of the data matrix are missing [16], data mining that used collaborative filtering through WNMG [10] and face feature extraction [12]. To best our knowledge, we are using the idea of WNMF for the first time for the application of HU. Moreover, we consider the feature of sparsity of FAs jointly with the WNMF.

Now, we define the following minimization problem

$$\min_{\mathbf{A} \geq 0, \mathbf{X} \geq 0} c(\mathbf{A}, \mathbf{X}), \tag{5a}$$

$$c(\mathbf{A}, \mathbf{X}) = \frac{1}{2} \|\mathbf{W}^{\frac{1}{2}} . * (\mathbf{Y} - \mathbf{AX})\|_F^2 + \lambda g(\mathbf{X}), \tag{5b}$$

where $g(\mathbf{X}) = \|\mathbf{X}\|_1$ or $\|\mathbf{X}\|_{\frac{1}{2}}$, and $\mathbf{W} = (\mathbf{W}_1, \mathbf{W}_2, \dots, \mathbf{W}_P)$ are the weights defined as follows:

$$\mathbf{W}_i = \begin{cases} \mathbf{1}_L & \text{if } \frac{\|\mathbf{Y}_i\|_2}{E[\|\mathbf{Y}_i\|_2]} \leq \tau \\ \mathbf{0}_L & \text{otherwise,} \end{cases} \tag{6}$$

where $\mathbf{1}_L$ and $\mathbf{0}_L$ are the column vectors of 1's and 0's, respectively, with the size of L, $E[.]$ denotes the mean and τ is the constant threshold.

We use the multiplicative updating rules [18] to solve the problem (5). Thus, we have the following updating rules by applying the *Karush-Kuhn-Tucker* (KKT) conditions, the transposition and division:

$$\mathbf{A} \leftarrow \mathbf{A} . * ((\mathbf{W} . * \mathbf{Y})\mathbf{X}^T) . / ((\mathbf{W} . * (\mathbf{AX}))\mathbf{X}^T) \tag{7}$$

$$\mathbf{X} \leftarrow \mathbf{X} . * (\mathbf{A}^T (\mathbf{W} . * \mathbf{Y})) . / (\mathbf{A}^T (\mathbf{W} . * (\mathbf{AX})) + \frac{\lambda}{2} \frac{\partial g(\mathbf{X})}{\partial \mathbf{X}}) \tag{8}$$

where $.*$ and $./$ are the element-wise multiplication and division respectively and the power in $\mathbf{X}^{-\frac{1}{2}}$ operates element-wisely as well. Also, $\frac{\partial g(\mathbf{X})}{\partial \mathbf{X}} = \mathbf{1}_{N \times P}$ or $\mathbf{X}^{-\frac{1}{2}}$.

3.1 Parameter Setting and Implementation Issues

The regularization parameter λ balances the degree of sparsness of FAs and the squared error of the estimated SSoMs and FAs. The larger λ gives the sparser solution as well as the larger squared error and vice versa. Hence, the parameter λ must be chosen to trade-off between the sparsity and the error. Some methods have been studied in [9]. In our simulations, we used the following estimation method based on the observed data that used in different literature, e.g. [7,9,14]:

$$\lambda = \frac{1}{\sqrt{L}} \sum_{k=1}^{L} \frac{\sqrt{P} - \frac{\|\mathbf{Y}^{(k)}\|_1}{\|\mathbf{Y}^{(k)}\|_2}}{\sqrt{P} - 1}, \tag{9}$$

where $\mathbf{Y}^{(K)}$ denotes the k-th band of the observed hyperspectral data.

We impose the ASC over the FAs through the updating rules by appending $\delta \mathbf{1}^T$ to $\mathbf{Y}$ and $\mathbf{A}$ as follows:

$$\bar{\mathbf{Y}} = \begin{bmatrix} \mathbf{Y} \\ \delta \mathbf{1}_P^T \end{bmatrix}, \quad \bar{\mathbf{A}} = \begin{bmatrix} \mathbf{A} \\ \delta \mathbf{1}_N^T \end{bmatrix} \tag{10}$$

where δ is a constant value and $\mathbf{1}_l$ denotes the $l \times 1$ column vector with all elements equal to one.

3.2   The Convergence and Steps of the Proposed Method

The following proposition shows that our proposed method is converging.

Proposition 1. *The cost function in* (5b) *is nonincreasing under* (7) *and* (8).

Proof. The convergence of the proposed method under the updating rule for (7) is similar to the Theorem 4 in [12] (by setting $\lambda = 0$) since the last term of (5b) is independent of $\mathbf{A}$. For the updating rule of $\mathbf{X}$ in (5b), we can follow up the same procedure used in [23] with the additional weights and shows the cost function in (5b) is nonincreasing under (8).

Algorithm 1 summarizes our proposed Sparse-WNMF method as follows:

Algorithm 1. Pseudocode of the proposed method: weighted NMF (WNMF) unmixing method with the sparsity constraint (Sparse-WNMF)

INPUT   : The observed matrix data $\mathbf{Y}$
- Set parameters λ, δ, ϵ, I_{max} and claculate $\mathbf{W}$ using (6).
OUTPUT: Matrices of SSoMs ($\mathbf{A}$) and FAs ($\mathbf{X}$).
:
- Initialize $\mathbf{A}$ using the VCA method [21] or randomly selected from the interval $[0,1]$.
- Initialize $\mathbf{X}$ using $(\mathbf{A}^T\mathbf{A})^{-1}\mathbf{A}^T\mathbf{Y}$.
REPEAT :
- Replace $\mathbf{A}$ and $\mathbf{Y}$ by (10).
- Compute $c_{old} = c(\bar{\mathbf{A}}, \mathbf{X})$ using (5b).
- Update $\mathbf{A}$ using (7).
- Update $\mathbf{X}$ using (8).
- Replace $\mathbf{A}$ and $\mathbf{Y}$ by (10).
- Compute $c_{new} = f(\bar{\mathbf{A}}, \mathbf{X})$ using (5b).
- Continue if the iteration number is less than I_{max} or $|c_{new} - c_{old}| > \epsilon$.

---

4   Experimental Results and Discussion

We evaluate our proposed Sparse-WNMF method over the Samson hyperspectral data set [1]. This data set contains 952×952 pixels over 156 spectral bands for

Table 1. Comparison of the SAD results and their standard deviation of various methods over the Samson hyperspectral data set.

Method	Soil	Tree	Water	Mean
Sparse-WNMF ($\ell_{\frac{1}{2}}$-norm)	0.1028 ± 0.1505	0.0660 ± 0.0176	0.1387 ± 0.0613	**0.1025 ± 0.0764**
$\ell_{1/2}$-NMF, ($I_{max} = 3000$) [23]	0.0768 ± 0.1489	0.0622 ± 0.0261	0.1761 ± 0.1150	0.1050 ± 0.0967
Sparse-WNMF (ℓ_1-norm)	0.0934 ± 0.1358	0.0683 ± 0.0192	0.1471 ± 0.0725	*0.1029 ± 0.0758*
ℓ_1-NMF, ($I_{max} = 3000$) [13]	0.0840 ± 0.1515	0.0645 ± 0.0245	0.1770 ± 0.1165	0.1085 ± 0.0975
Classic-WNMF ($\lambda = 0$)	0.0937 ± 0.1509	0.0923 ± 0.0301	0.1835 ± 0.0478	0.1232 ± 0.0763
Classic-NMF (CNMF), ($I_{max} = 3000$) [17]	0.0960 ± 0.1879	0.0716 ± 0.0273	0.2442 ± 0.0532	0.1373 ± 0.0895
VCA-Fully Constrained Least Square (FCLS) [11,21]	0.1372 ± 0.2137	0.0491 ± 0.0026	0.1287 ± 0.0049	0.1050 ± 0.0737

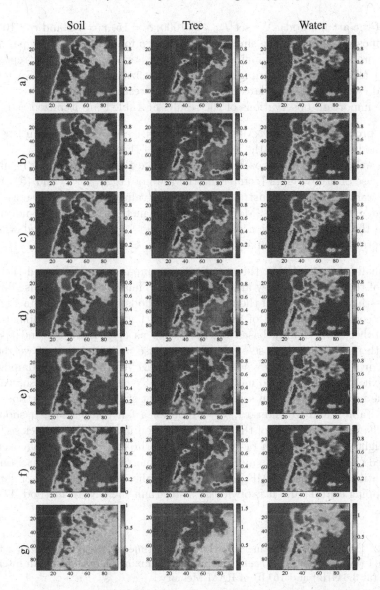

Fig. 1. A visualization of FA maps of endmembers estimated by different unmixing methods over Samson hyperspectral image: (a) Sparse-WNMF ($\ell_{\frac{1}{2}}$) (b) $\ell_{1/2}$-NMF [23] (c) Sparse-WNMF (ℓ_1) (d) ℓ_1-NMF [13] (e) Classic-WNMF ($\lambda = 0$) (f) CNMF [17] (g) VCA-FCLS

the range of 0.4–0.89 μm. For the simulation, we use the similar subset of the original data set with 95×95 pixels similar to one studied in [8, 27]. Moreover, we consider the same ground truth values of SSoMs provided by Zhu *et al.* in [27]. Table 1 compares the SAD results of the Sparse-WNMF method with several state-of-the-art methods. We set $I_{max} = 1000$, $\delta = 15$, $\tau = 0.2$ and $\epsilon = 10^{-3}$ for our proposed method in the simulations and keep following the same parameter setting proposed in the original mentioned competitive methods. We repeat 200 runs for all the methods through the same initialization with the VCA method [21] and calculate the standard deviation of the results as well.

The important observations of the results of Table 1 are itemized as follows:

- The proposed Sparse-WNMF methods (rows 1 and 3) outperform the other methods in the sense of the average results of the SAD.
- The weights in the proposed WNMF enhance the corresponding results for the case of non-weights (comparing the rows of (1, 2), (3, 4) and (5, 6)).
- Generally, the sparse constraints for smaller values of p in the ℓ_p-norm term improve the SAD results. More precisely, the methods through $\ell_{\frac{1}{2}}$-norm term have the better estimation of the endmembers' spectra compared with the ℓ_1-norm term or non sparsity constraint.

It shall be pointed that the quantitative comparison of FAs need the truth values of them in which they are not available. Thus, we provide a visualization of the estimated FAs for all the mentioned methods of the martials in Fig. 1.

The computational cost of the proposed method is basically similar to that of the other NMF family based methods such as $\ell_{1/2}$-NMF [23] and ℓ_1-NMF [13] with the additional cost of the element-wise multiplication of weights **W** present in (7) and (8). They add the order of LP to the original computation complexity of the mentioned methods, i.e. $O(LPN + (NP)^2)$ [23]. Also, we compare the processing times in seconds for different state-of-the-art methods with ours in Table 2 and we choose $I_{max} = 1000$ for a fair comparison. It should be noted the processing time of the weighted version is approximately close to the non-weighted ones for both $\ell_{\frac{1}{2}}$-norm and ℓ_1-norm sparse terms. Moreover, the weighted classic NMF needs more processing time, 2.7161 s, in comparison with the mentioned weighted NMF methods due to the convergence issue related to the stopping criteria and it is obviously faster than the CNMF method [17] with 7.6220 s.

Table 2. Comparison of the measured processing time for the Samson data set used in Table 1 (in Seconds) per 10 runs for different unmixing algorithms on an Intel Core i7-4790 (at 3.6 GHz) and 16 GB of RAM Memory.

Method	Time (seconds)
Sparse-WNMF ($\ell_{\frac{1}{2}}$-norm)	0.2211
Sparse-WNMF (ℓ_1-norm)	0.1837
Classic-WNMF ($\lambda = 0$)	2.7161
$\ell_{1/2}$-NMF, ($I_{max} = 3000$) [23]	0.1673
ℓ_1-NMF, ($I_{max} = 3000$) [13]	0.1640
Classic-WNMF (CNMF) [17]	7.6220

5   Conclusion and Future Works

We proposed a method of non-dictionary aided HU through the weighted NMF under sparsity constraints of FAs. We formulated the sparse unmixing approach via a new minimization problem and utilized the multiplicative updating rules to derive the required steps of our proposed method. We showed that our proposed sparse WNMF method converges. Then, we evaluated it over the real hyperspectral data set and compared the simulation results with several competitive methods. We observed that our proposed method outperformed the other methods in terms of overall SAD result. Although the current experimental results are promising over one type of hyperspectral data set, we need more experiments over the synthetic data as well as the other types of real hyperspectral data set. Furthermore, estimating the threshold value in Eq. (6) to assign 0 and 1 for the weights is the other future plan. Finally, generalizing the idea of weighted NMF with the other kinds of pixel's feature such as collaborative property of FAs or smooth property of the SSoM [5] are of interest for the future work.

References

1. http://opticks.org/confluence/display/opticks/sample+data
2. Bioucas-Dias, J., Plaza, A., Camps-Valls, G., Scheunders, P., Nasrabadi, N., Chanussot, J.: Hyperspectral remote sensing data analysis and future challenges. IEEE Geosci. Remote Sens. Mag. 1(2), 6–36 (2013)
3. Bioucas-Dias, J., Plaza, A., Dobigeon, N., Parente, M., Du, Q., Gader, P., Chanussot, J.: Hyperspectral unmixing overview: geometrical, statistical, and sparse regression-based approaches. IEEE J. Sel. Top. Appl. Earth Obs. Remote Sens. 5(2), 354–379 (2012)
4. Cichocki, A., Zdunek, R., Phan, A.H., Amari, S.: Nonnegative Matrix and Tensor Factorizations Applications to Exploratory Multiway Data Analysis and Blind Source Separation. The Atrium, Southern Gate, Chichester (2009)
5. Esmaeili Salehani, Y., Gazor, S.: Smooth and sparse regularization for NMF hyperspectral unmixing. IEEE J. Sel. Top. Appl. Earth Obs. Remote Sen. (2017, accepted for the publication)
6. Esmaeili Salehani, Y.: Linear hyperspectral unmixing using l0-norm approximations and nonnegative matrix factorization, Ph.D. thesis, October 2016
7. Esmaeili Salehani, Y., Gazor, S.: Collaborative unmixing hyperspectral imagery via nonnegative matrix factorization. In: Mansouri, A., Nouboud, F., Chalifour, A., Mammass, D., Meunier, J., ElMoataz, A. (eds.) ICISP 2016. LNCS, vol. 9680, pp. 118–126. Springer, Cham (2016). doi:10.1007/978-3-319-33618-3_13
8. Esmaeili Salehani, Y., Gazor, S.: Sparse data reconstruction via adaptive ℓ_p-norm and multilayer NMF. In: 2016 IEEE 7th Annual Information Technology, Electronics and Mobile Communication Conference (IEMCON), pp. 1–6. IEEE, October 2016
9. Esmaeili Salehani, Y., Gazor, S., Kim, I.M., Yousefi, S.: ℓ_0-norm sparse hyperspectral unmixing using arctan smoothing. Remote Sens. 8(3), 187 (2016)
10. Gu, Q., Zhou, J., Ding, C.: Collaborative filtering: weighted nonnegative matrix factorization incorporating user and item graphs. In: Proceedings of the 2010 SIAM International Conference on Data Mining, pp. 199–210. SIAM (2010)

11. Heinz, D.C., Chang, C.I.: Fully constrained least squares linear mixture analysis for material quantification in hyperspectral imagery. IEEE Trans. Geosci. Remote Sens. **39**(3), 529–545 (2001)
12. Ho, N., Van Dooren, P., Blondel, V.: Weighted nonnegative matrix factorization and face feature extraction. Submitted to Image and Vision Computing (2007)
13. Hoyer, P.O.: Non-negative sparse coding. In: Proceedings of IEEE Workshop Neural Network Signal Process. XII, Martigny, Switzerland, pp. 557–565 (2002)
14. Hoyer, P.O.: Non-negative matrix factorization with sparseness constraints. J. Mach. Learn. Res. **5**, 1457–1469 (2004)
15. Iordache, M.D., Bioucas-Dias, J., Plaza, A.: Sparse unmixing of hyperspectral data. IEEE Trans. Geosci. Remote Sens. **49**(6), 2014–2039 (2011)
16. Kim, Y.D., Choi, S.: Weighted nonnegative matrix factorization. In: IEEE International Conference on Acoustics, Speech and Signal Processing, ICASSP 2009, pp. 1541–1544. IEEE (2009)
17. Lee, D.D., Seung, H.S.: Algorithms for non-negative matrix factorization. In: Advances in Neural Information Processing Systems, pp. 556–562 (2001)
18. Lee, D.D., Seung, H.: Learning the parts of objects with nonnegative matrix factorization. Nature **401**(6755), 788–791 (1999)
19. Lu, X., Wu, H., Yuan, Y., Yan, P., Li, X.: Manifold regularized sparse NMF for hyperspectral unmixing. IEEE Trans. Geosci. Remote Sens. **51**(5), 2815–2826 (2013)
20. Ma, W.K., Bioucas-Dias, J., Chan, T.H., Gillis, N., Gader, P., Plaza, A., Ambikapathi, A., Chi, C.Y.: A signal processing perspective on hyperspectral unmixing: insights from remote sensing. IEEE Sig. Process. Mag. **31**(1), 67–81 (2014)
21. Nascimento, J., Bioucas-Dias, J.: Vertex component analysis: a fast algorithm to unmix hyperspectral data. IEEE Trans. Geosc. Remote Sens. **43**(8), 898–910 (2005)
22. Paatero, P., Tapper, U.: Positive matrix factorization: a non-negative factor model with optimal utilization of error estimates of data values. Environmetrics **5**(2), 111–126 (1994)
23. Qian, Y., Jia, S., Zhou, J., Robles-Kelly, A.: Hyperspectral unmixing via $l_{\frac{1}{2}}$ sparsity-constrained nonnegative matrix factorization. IEEE Trans. Geosci. Remote Sens. **49**(11), 4282–4297 (2011)
24. Rajabi, R., Ghassemian, H.: Spectral unmixing of hyperspectral imagery using multilayer NMF. IEEE Geosc. Remote Sens. Lett. **12**(1), 38–42 (2015)
25. Vavasis, S.A.: On the complexity of nonnegative matrix factorization. SIAM J. Optim. **20**(3), 1364–1377 (2009)
26. Yuan, Y., Fu, M., Lu, X.: Substance dependence constrained sparse NMF for hyperspectral unmixing. IEEE Trans. Geosci. Remote Sens. **53**(6), 2975–2986 (2015)
27. Zhu, F., Wang, Y., Fan, B., Meng, G., Xiang, S., Pan, C.: Spectral unmixing via data-guided sparsity. IEEE Trans. Image Process. **23**(12), 5412–5427 (2014)

Stroke Width Transform for Linear Structure Detection: Application to River and Road Extraction from High-Resolution Satellite Images

Moslem Ouled Sghaier[1]([✉]), Imen Hammami[2], Samuel Foucher[3], and Richard Lepage[1]

[1] École de Technologie Supérieure, Montréal, QC, Canada
ouled.sgaier.moslem@gmail.com, richard.lepage@etsmtl.ca
[2] Faculté des Sciences de Tunis, Campus Universitaire 2092, El Manar Tunis, Tunis, Tunisia
imen.hammammi@gmail.com
[3] Centre de Recherche Informatique de Montréal, Montreal, QC, Canada
samuel.foucher@crim.ca

Abstract. The evaluation of lines of communication status in normal times or during crises is a very important task for many applications, such as disaster management and road network maintenance. However, due to their large geographic extent, the inspection of the these structures surfaces using traditional techniques such as laser scanning poses a very challenging problem. In this context, satellite images are pertinent because of their ability to cover a large part of the surface of communication lines, while offering a high level of detail, which makes it possible to discriminate objects forming these linear structures. In this paper, a novel approach for extracting linear structures from high-resolution optical and radar satellite images is presented. The proposed technique is based on the *Stroke Width Transform (SWT)*, which allows parallel edges extraction from the input image. This transform has been successfully applied in the literature to extract characters from complex scenes based on their parallel edges. An adaptation of this transform to solve the problem of rivers extraction from *Synthetic Aperture Radar (SAR)* images and roads identification from optical images is described in this paper, and the results obtained show the efficiency of our approach.

Keywords: Stroke width transform · Connected components · Satellite images · Road and river extraction

1 Introduction

Linear structures extraction from satellite images represents one of the most active fields of research in remote sensing. The growing interest in this problematic is mainly due to the diversity of its applications and to the remarkable

© Springer International Publishing AG 2017
F. Karray et al. (Eds.): ICIAR 2017, LNCS 10317, pp. 605–613, 2017.
DOI: 10.1007/978-3-319-59876-5_67

improvement in the resolution of satellites orbiting the Earth. Based on the linear shape of such structures [3,9], their homogeneous texture [8,13] and/or their spatial neighborhood [14], several techniques have been proposed in the literature to solve the problem posed by their extraction. The work of Mena [12] presents a complete literature review of linear structures extraction existing approaches. These techniques provide very promising results when applied to straight-shaped structures, but fail in the identification of locally straight structures presenting significant curvatures.

Among the different techniques dedicated to linear structures identification, approaches based on straight line extraction, such as Hough or Radon transform, remain the most used. They assume that extracting a line from an image amounts to determining the point that has the most votes in the accumulator of the parameter space, with the number of points voted representing the importance of this line in the image space. In [2], an improved Radon transform, called Cluster Radon Transform (CRT), was developed to extract linear features from satellite images. This variant differs from the original transform by its low sensitivity to noise, but is still unable to locate the position of these lines in the image. Approaches which rely on multiscale-based image analysis to extract straight lines have also been applied to identifying linear structures. These techniques use image multiscale decomposition to identify the most appropriate scale characterizing linear objects contained in the image. Moslem and Richard [16] propose a very interesting technique for road extraction from optical images based on texture analysis and multiscale beamlet transform [6,15]. In their approach, local information is first extracted using the *Structural Feature Set (SFS)* texture measurement [11,17], after which global information based on beamlet transform is introduced to distinguish main road axes at coarse scales, and local segments from fine scales, which are aggregated to reconstruct the road network. The results obtained are promising, but the proposed technique cannot be generalized to cover linear structures with high curvature, such as rivers, for example. The application of mathematical morphology is generally restricted to the preprocessing phase, and consists in filtering the noise present in the image through a structuring element. However, in [10], a structuring element in the form of a path is applied to extract linear structures and objects that have local linearity. The results obtained by applying the proposed technique to solve the problem of identifying rivers from radar images are very interesting, but choosing an adequate size for these paths remains problematic.

Contrary to these approaches, our method is mainly based on parallel edges extraction to identify linear structures. It has the advantage of allowing the extraction of these structures at any scale of resolution, while minimizing false alarm rate.

2   Methodology

Within the same image, linear structures can have different textures and/or irregular shapes, but maintain a fixed or almost fixed width between their two sides. Starting from this assumption, we based our reasoning on this feature to extract these structures. In this section, we introduce our linear structure extraction algorithm, as shown in Fig. 1: (1) First, we describe the preprocessing step based on the Canny edge detector; (2) We then define the notion of Stroke Width Transform, and (3) Finally, we explain our fitness function which is capable of distinguishing between linear structures and noise characterizing radar and optical high-resolution images.

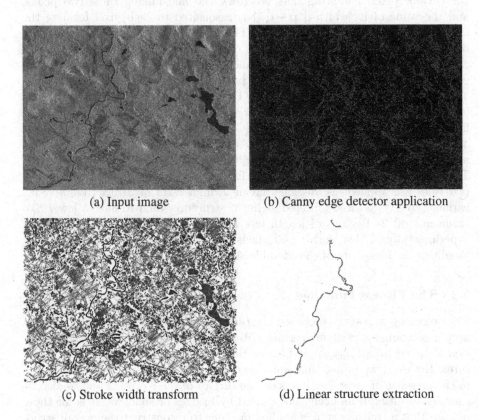

(a) Input image   (b) Canny edge detector application

(c) Stroke width transform   (d) Linear structure extraction

Fig. 1. Results of each step of our approach applied to the input SAR image

2.1   Edge Detection

Applying the Stroke Width Transform on an image requires the use of an edge map as input. Therefore, the Canny edge detector [1] is applied to the input image in order to generate an edge map. For the Canny algorithm, we use a Gaussian filter with $\sigma = \sqrt{2}$, which allows to retain all the significant edges of the image, while suppressing those resulting from noise. This step will provide a list of candidate pixels, which include the desired parallel edges, as well as edges extracted from other objects present in the image.

2.2   The Stroke Width Transform

The application of the SWT on the edge map as described in [7] consists in applying the following steps. For each pixel, p, of the edge map, the gradient direction is computed. Then, we follow this direction in search of another edge pixel, q, having a roughly opposite edge direction, and verifying the following condition: $d_q = -d_p \pm \pi/6$, with d being the gradient direction computed from the current pixel. Following that, we draw the line linking these two pixels, and the value of its width $|| \overrightarrow{p - q} ||$ is associated to each pixel forming the line, including the two ends, p and q. This procedure is repeated for all edges detected from the input image, and the output of the SWT is an image of size equal to the size of the input image, and such that the value of each pixel is equal to the stroke width value at this point. In order to interpret pixel values obtained by the application of the SWT transform in the previous step, it is important to group the pixels in order to identify candidate regions forming linear structures contained in the image. Therefore, the connected component algorithm [5] is applied to cluster pixels that have similar SWT values. Two neighboring pixels are grouped together if their SWT ratio does not exceed 3.0. To eliminate irrelevant regions from the list of connected component candidates, two rules are applied: **1-** Regions having a high ratio of *Stroke Width (SW)* variance to SW mean are filtered as linear structures tend to have a lower SW variance, and **2-** Regions of length less than a certain threshold, which is set experimentally in this work to 50 pixels, are filtered since objects of interest sought in the image are of considerable size.

2.3   The Fitness Function

The preceding step of the proposed algorithm was mainly designed to detect connected components verifying certain criteria, such as the shape and the dimension, denoted by $SW_{current}$, and allows the extraction of almost all linear structures. However, we notice that some junctions were not correctly detected due to the presence of several artifacts on the surface of these objects and to discontinuities resulting from occlusions caused by bridges or shadows. It is for these reasons that a tracking step is applied in order to reconstruct the overall structure of the object of interest. A fitness function is defined to select potential candidates from connected components retained in the previous step denoted by $SW_{candidate}$ and verifying only the first criteria given by:

$$\mathcal{F}_{new} = \mathcal{F} + \min(fitness_{sw}) \tag{1}$$

with $\mathcal{F}$ being the fitness function computed at a connected component retained from the previous step, $\mathcal{F}_{new}$ the new fitness function and $fitness_{sw}$ being the fitness function computed at each connected component verifying only the first criterion, and such that:

$$fitness_{sw} = fitness_{dir} + \alpha fitness_{hom} \tag{2}$$

with $fitness_{dir} = (\overrightarrow{SW}_{candidate} \wedge \overrightarrow{SW}_{current})/180$, $fitness_{hom} = \sigma_{sw_{current+candidate}} - \sigma_{sw_{current}}$ (σ the standard deviation computed from the region occupied by the connected component of the input image) and α a real number lower than 1 if we are looking for roads and greater than 1 if we are looking for rivers.

3 Experimentation

This work was conducted as part of an internship at the Computer Research Institute of Montreal (CRIM), Quebec, Canada, aimed at updating and mapping inland water such as lakes and rivers surfaces across Canada. In order to implement our method, we were mainly based on the Stroke Width Transform code[1] developed using the OpenCV library to create an application based on the Optical and Radar Federated Earth Observation ToolBox (OTB) open-source C++ library [4], which was subsequently incorporated in the Quantum GIS (Qgis) software.

3.1 Results Obtained

Figure 2(a) represents a subimage cropped from an orthorectified high-resolution RADA-RSAT-2 (6 meters of resolution, stored in a single-look complex (SLC) format) of the Grand River in Ontario, Canada, and Fig. 2(b) shows a panchromatic subimage extracted from the satellite image of the city of Port-au-Prince in Haiti using the GeoEye 1 satellite. Figures 2(c) and (d) demonstrate the results obtained by applying the Klemenjak et al. [10] and Moslem et al. [16] methods, respectively. Finally, Figs. 2(e) and 2(f) show our method applied to study areas 1 and 2, respectively.

3.2 Linear Structure Extraction Method Evaluation

To further evaluate the performance of the proposed method, we compared binary images resulting from the application of our algorithm for linear structure extraction to vectors representing river and road surfaces used as a ground truth. The associated confusion matrix is given by Fig. 3. From this confusion matrix, three different measurements are computed:

$$Accuracy = \frac{tp + tn}{tp + tn + fp + fn} \qquad Precision = \frac{tp}{tp + fp}$$

$$Recall = \frac{tp}{tp + fn}$$

As shown in the Table 1, in terms of precision, our algorithm achieves better results than the Klemenjak et al. and Moslem et al. methods, with 0.988 versus 0.996, and 0.93 versus 0.901 for study area 1 and 2, respectively. In terms

---

[1] https://github.com/aperrau/DetectText.

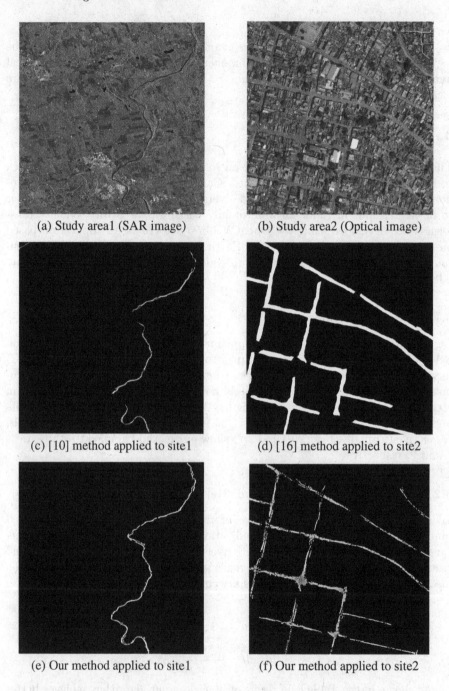

(a) Study area1 (SAR image)   (b) Study area2 (Optical image)

(c) [10] method applied to site1   (d) [16] method applied to site2

(e) Our method applied to site1   (f) Our method applied to site2

Fig. 2. Results of the three methods applied to test images

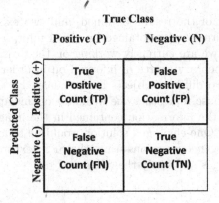

True Class

		Positive (P)	Negative (N)
Predicted Class	Positive (+)	True Positive Count (TP)	False Positive Count (FP)
	Negative (-)	False Negative Count (FN)	True Negative Count (TN)

Fig. 3. Confusion matrix

Table 1. Performance of our linear structure extraction method compared to existing techniques

Method	Study area 1			Study area 2		
	Accuracy	Precision	Recall	Accuracy	Precision	Recall
Klemenjak et al. method [10]	0.839	0.996	0.598	-	-	-
Moslem et al. method [16]	-	-	-	0.83	0.901	0.951
Proposed method	0.922	0.988	0.874	0.8	0.93	0.965

of recall, our method and the Moslem et al. method obtain almost the same result when applied to study area 2, but we achieved a remarkably better result versus the Klemenjak et al. method when the methods were applied to study areas 1, with 0.874 as compared to 0.598. This can be explained by the application of a tracking step that improves the overall performance of our algorithm as compared to the Klemenjak et al. method. In terms of accuracy, Moslem et al. obtain a better result than our approach applied to study area 2. This low performance can be explained by the presence of artifacts on the road surface, which are detected as false alarms due to the high spatial resolution of the used optical image.

4   Conclusion and Future Works

A novel approach for linear structure extraction from high-resolution satellite images using the Stroke Width Transform is introduced in this paper. The approach described is based on parallel edge-identification in a bid to extract potential candidates, and a filtering step is applied by evaluating a fitness function that reflects the quality of the extracted object in order to distinguish between linear structures and false alarms. Several optical and radar images are used to assess the quality of the results obtained, and two existing techniques are implemented to confirm results obtained by visual interpretation. As a continuation of this work, we are currently working on the possibility of integrating texture as a complimentary source of information in order to avoid extracting sides belonging to two different objects in the image. Further, we believe that the edge detection phase represents one weakness of this approach, as it allows the extraction of all the information contained in the image without requiring any prior knowledge. One envisaged solution would be to apply a pretreatment step to extract homogeneous regions using the *SFS* texture measurement, and then extract their edges and apply the *SWT*.

References

1. Canny, J.: A computational approach to edge detection. IEEE Trans. Pattern Anal. Mach. Intell. **8**(6), 679–698 (1986)
2. Chen, Y., Li, Y., Zhang, H., Tong, L., Cao, Y., Xue, Z.: Automatic power line extraction from high resolution remote sensing imagery based on an improved radon transform. Pattern Recogn. **49**, 174–186 (2016)
3. Cheng-Li, J., Ke-Feng, J., Yong-Mei, J., Gang-Yao, K.: Road extraction from high-resolution SAR imagery using hough transform. In: IEEE International In Geoscience and Remote Sensing Symposium (IGARSS), Seoul, Korea (South), 25–29 July, vol. 1, pp. 336–339 (2005)
4. Christophe, E., Inglada, J.: Open source remote sensing: Increasing the usability of cutting-edge algorithms. IEEE Geosci. Remote Sens. Soc. Newslett. **150**, 9–15 (2009)
5. Dillencourt, M.B., Samet, H., Tamminen, M.: A general approach to connected-component labeling for arbitrary image representations. J. ACM **39**(2), 253–280 (1992). http://doi.acm.org/10.1145/128749.128750
6. Donoho, D.L., Huo, X.: Beamlets and multiscale image analysis. In: Barth, T., Chan, T., Haimes, R. (eds.) Multiscale and Multiresolution Methods. LNCSE, vol. 20, pp. 149–196. Springer, Heidelberg (2002). doi:10.1007/978-3-642-56205-1_3
7. Epshtein, B., Ofek, E., Wexler, Y.: Detecting text in natural scenes with stroke width transform. In: IEEE Computer Society Conference on Computer Vision and Pattern Recognition (CVPR), San Francisco, USA, pp. 2963–2970, 13–18 June 2010
8. He, X., Wu, Y., Wu, Y.: Texture feature extraction method combining nonsubsampled contour transformation with gray level co-occurrence matrix. J. Multimedia **8**(6), 675–684 (2013)
9. Herumurti, D., Uchimura, K., Koutaki, G., Uemura, T.: Urban road extraction based on hough transform and region growing. In: 19th Korea-Japan Joint Workshop on Frontiers of Computer Vision, (FCV), Incheon, Korea, 30 January–1 February, pp. 220–224 (2013)
10. Klemenjak, S., Waske, B., Valero, S., Chanussot, J.: Automatic detection of rivers in high-resolution SAR data. IEEE J. Sel. Top. Appl. Earth Obs. Remote Sens. **5**(5), 1364–1372 (2012)
11. Liangpei, Z., Xin, H., Bo, H., Pingxiang, L.: A pixel shape index coupled with spectral information for classification of high spatial resolution remotely sensed imagery. IEEE Trans. Geosci. Remote Sens. **44**(10), 2950–2961 (2006)

12. Mena, J.: State of the art on automatic road extraction for GIS update: a novel classification. Pattern Recogn. Lett. **24**(16), 3037–3058 (2003)
13. Mena, J., Malpica, J.: An automatic method for road extraction in rural and semi-urban areas starting from high resolution satellite imagery. Pattern Recogn. Lett. (Neth.) **26**(9), 1201–1220 (2005)
14. Pellizzeri, T.M., Gamba, P., Lombardo, P., Dell'Acqua, F.: Multitemporal/multiband SAR classification of urban areas using spatial analysis: statistical versus neural kernel-based approach. IEEE Trans. Geosci. Remote Sens. **41**(10), 2338–2353 (2003)
15. Salari, E., Zhu, Y.: A road extraction method using beamlet transform. In: IEEE International Conference on Electro/Information Technology (EIT), pp. 1–4, May 2012
16. Sghaier, M.O., Lepage, R.: Road extraction from very high resolution remote sensing optical images based on texture analysis and beamlet transform. IEEE J. Sel. Top. Appl. Earth Obs. Remote Sens. **9**(5), 1946–1958 (2016)
17. Xin, H., Liangpei, Z., Pingxiang, L.: Classification and extraction of spatial features in urban areas using high-resolution multispectral imagery. IEEE Geosci. Remote Sens. Lett. **4**(2), 260–264 (2007)

Applications

Real Time Fault Detection in Photovoltaic Cells by Cameras on Drones

Alessandro Arenella[2], Antonio Greco[1,2], Alessia Saggese[1,2(✉)],
and Mario Vento[1,2]

[1] Department of Information Engineering,
Electrical Engineering and Applied Mathematics (DIEM),
University of Salerno, Fisciano, Italy
{agreco,asaggese,mvento}@unisa.it
[2] A.I. Tech s.r.l., 83100 Avellino, AV, Italy
arenella@aitech.vision
http://www.aitech.vision

Abstract. Hot spots are among the defects of photovoltaic panels which may cause the most destructive effects. In this paper we propose a method able to automatically detect the hot spots in photovoltaic panels by analyzing the sequence of thermal images acquired by a camera mounted on board of a drone flighting over the plant. The main novelty of the proposed approach lies in the fact that color based information, typically adopted in the literature, are combined with model based one, so as to strongly reduce the number of detected false positive. The experimentation, both in terms of accuracy and processing time, confirms the effectiveness and the efficiency of the proposed approach.

Keywords: Photovoltaic systems · Hot spot detection · Drones

1 Introduction

In the last decade we have assisted to a growing interest towards renewable energy, with particular reference to photovoltaic (PV) plants [12]. The large amount of PV plants to be monitored has led to an increasing interest of the scientific community towards those solutions able to monitor automatically, or at least semi automatically, the performance of the panels of each plant so as to detect potential faults. One of the main defects of the PV panels are the so called *hot spots* [10], corresponding to those areas in PV panels characterized by the higher temperature: indeed, in cases a cell in a panel is affected by this kind of fault, it starts dissipating power in the form of heat instead of producing electrical power [11]. This power dissipation occurring in a so small area results in a local overheating, namely the hot spot, which may cause very destructive effects: cell or glass cracking, sintering, or excessive degradation of the solar cell and thus of the related PV panel.

© Springer International Publishing AG 2017
F. Karray et al. (Eds.): ICIAR 2017, LNCS 10317, pp. 617–625, 2017.
DOI: 10.1007/978-3-319-59876-5_68

In order to automatically detect hot spots, images acquired by thermal cameras have been widely used [2]: the main reason is that this technique is contactless and not invasive, thus it can be performed during the normal life cycle of PV systems without requiring its shut down. There are two types of thermal cameras that are typically adopted, namely radiometric and non radiometric ones (typically simply referred to as thermal cameras). The former enables the measurement of the absolute temperature, so that the intensity level of each pixel in the image corresponds to a temperature [5,9]. Of course, this kind of information may strongly simplify the analysis of the images: indeed, detecting a hot spot could be simply performed by identifying the pixels in the images with the highest temperature (above a given threshold). However, this approach is only very rarely used due to the high cost of the radiometric device. Furthermore, the accuracy in the measurement of the temperature by radiometric cameras is strongly dependent on several environmental factors (i.e. the humidity, the atmospheric temperature and the distance from the PV panels), thus a preliminary and continuous calibration of the camera is required [8]. As evident, the device calibration is a time consuming (and thus expensive) operation, especially for large PV plants, where there is a need to repeat several time this procedure.

On the other side, the usage of a (non radiometric) thermal camera does not require any calibration step. However, thermal images only provides information concerning the difference in temperature of the objects in the scene (in this case, the panels and the hotspots, as well as any other objects taken by the camera, such as trees of buildings). Hot spots pixels, due to their higher temperature, are characterized by a higher luminance than the ones of the panel to which the hot spot belongs to (but not necessarily to the other panels). Thus, the intensity distribution inside a single panel is a very powerful way for detecting possible defects. Most of the approaches proposed in the literature are based on the usage of a single thermal camera: indeed, the human operator manually takes a picture of each panel, moving the camera panel by panel. The acquired images are then automatically processed by using color based information, typically evaluating the histogram of each panel manually isolated over each image by the human operator [13]. However, this is still a time consuming operation.

In the last years, drones have been exploited for automatically acquiring the sequence of images related to the whole plant [4,12]: a drone, provided with a thermal camera, flights over the plant and acquires a video, which is manually analyzed by the human operator so as to discover the hot spots [14]. As evident, the combination between drones and thermal cameras for PV monitoring is faster (and then surely less expensive) than traditional monitoring techniques based on visual inspection on site as well as any other PV hardware devices for parameters measurement [3,7]. The manual inspection of the sequence of images, also in this case, is a very time consuming operation, thus the introduction of systems able to automatically analyze the videos and detect the hot spots becomes mandatory [1,5]. To the best of our knowledge, there are not methods able to automatically process the whole video sequence acquired by flying cameras in real time using an embedded board.

In this paper we propose an accurate and efficient method for detecting hot spots in real time analyzing videos acquired by cameras mounted over flying drones. Differently from the state of the art approaches, our method combines a *color* based analysis with a *model* based one; this choice is justified by the fact that in correspondence of the panels' junctions several high temperature points can be formed, thus implying high luminance regions in that parts of the panel, which may cause for their color intensity nature a false positive. Starting from this consideration, the proposed algorithm detects the panels in the PV module by exploiting geometry and structural information of the panels, then evaluates the color of each panel, independently on the other ones. Finally, the position of the potential hot spot is analyzed with respect to its panel so as to filter out those hot spots close to the junctions. The accuracy of the proposed approach is not paid in terms of execution time. Indeed, another contribution of this paper lies in the fact that the algorithm has been designed for running over System on a chip (SoC) architectures: the video is not acquired and sent to an external server, but instead is processed directly on board of the SoC mounted on the drone, so that only the position of the hot spots (and a couple of images associated to it) is notified to the human operator driving the drone, for an immediate check.

2   The Proposed Method

The method is based on the following three steps, whose output is shown in Fig. 1: (i) during the Preprocessing step, the lines in the images (white lines in Fig. 1b) are extracted and used to align the image and to (ii) find out the panels in the modules (identified by the white rectangles in Fig. 1c). Finally, for each detected panel, the (iii) detection of the hot spots is performed (the red rectangle in Fig. 1d).

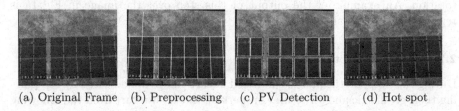

(a) Original Frame   (b) Preprocessing   (c) PV Detection   (d) Hot spot

Fig. 1. The different steps of the proposed approach. (Color figure online)

2.1   Preprocessing

During the preprocessing step, the lines inside each image are detected by using the Hough transform (see Fig. 2a). As we can see from the figure, the mere application of this technique is not sufficient since the following problems arise: several spurious lines are detected, while only horizontal and vertical lines should be considered; for each junction, more than one line is detected; finally, several

(a) Lines   (b) Horizontal   (c) Vertical

Fig. 2. Preprocessing step: (a) application of the Hough transform on the image; (b) horizontal and (c) vertical lines extracted by the proposed approach.

lines are longer than the corresponding panels. In order to deal with these problems, the proposed approach is based on the following steps: (i) depending on the slope of the lines, two sets are formed, namely horizontal and vertical lines. (ii) Then, the average slope $\bar{m}$ of each set is computed and each line $y = m \cdot x + n$ is evaluated; in case $|\bar{m} - m| > T$, the line is filtered out since it is considered as an error of the previous step. (iii) In order to merge all those lines corresponding to the same junction, a hierarchical clustering algorithm based on the distance between the lines is then applied. Finally, (iv) the alignment of the image is performed:

$$A = \begin{bmatrix} \cos\alpha & \sin\alpha \, (1 - \cos\alpha) \cdot x_0 - \sin\alpha \cdot y_0 \\ -\sin\alpha & \cos\alpha \sin\alpha \cdot x_0 + (1 - \cos\alpha) \cdot y_0 \end{bmatrix} \tag{1}$$

The transformation matrix A performs an elementary rotation around the z axis of an angle α, computed as the average rotational angle of the horizontal lines. This rotation is carried out with respect to the center of the image, namely (x_0, y_0). In more details, the rotated image RI is obtained starting from the original image I as follows: $RI(x, y) = I(A_{11}x + A_{12}y + A_{13}, A_{21}x + A_{22}y + A_{23})$. The lines (both horizontal and vertical) are rotated by using the same transformation matrix A, so as to avoid their re-computation after the image rotation. An example of the output of this step over the image in Fig. 1a is reported in Figs. 2b and c.

2.2 Panels Detection

During this step, all the panels contained in each image are detected by recognizing their geometric and structural properties. Indeed we start from the following assumptions: (i) a vertical line can be considered as the border of a panel (and then a *true line*) if the distance with respect to the other vertical lines is approximatively a multiple value of the panel width. (ii) The same assumption holds for horizontal lines. (iii) The whole grid has exactly four borders: two external vertical true lines and two external horizontal true lines. Thus, only the true lines respecting the above constraints are considered and the external borders of the grid are extracted. The above actions are able to only deal with false positive lines, but not with misses lines. In order to face with this problem, the proposed approach exploits structural information of the panels: indeed, missing lines, both horizontal and vertical, are added so as to make uniform the grid of panels in the image, as shown in the two examples reported in Fig. 3.

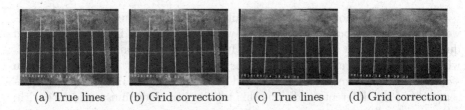

(a) True lines   (b) Grid correction   (c) True lines   (d) Grid correction

Fig. 3. Panels detection step: (a, c) true lines in white; (b, d) red lines are added by the proposed approach so as to correct the grid. (Color figure online)

Once the lines have been corrected, the rectangles, each corresponding to a panel, are identified: the intersections between horizontal and vertical lines are considered and used as the corners of the rectangles. An example is reported in Fig. 1c. In order to be considered a *true panel*, the size of the panel should be within a range identifying the expected size of the panel, which is a priori known.

2.3   Hot Spot Detection

For each panel identified during the previous step, the algorithm verifies the presence of the hotspots, corresponding to those higher temperature areas far from the junctions of the panels. In more details, the Sobel Edge Detector is applied in each panel so as to compute the gradient, in each panel independently on the other ones. Differently from other operators, Sobel has been chosen due to its efficiency. Starting from the gradient, each cell is thresholded and the corresponding mask is computed (see Fig. 4); thus, the connected components are found and their positions are evaluated; components located on the borders are filtered, as shown in Fig. 4b.

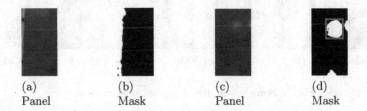

(a)   (b)   (c)   (d)
Panel   Mask   Panel   Mask

Fig. 4. PV panels (a, c) and the corresponding masks (b, d). The connected component in (b) is filtered thanks to the evaluation of the position based information. The red box in (d) is the hot spot detected by the system. (Color figure online)

3   Experimental Dataset

In this section we will discuss the results obtained by the proposed approach, analyzing both the accuracy and the processing time over SoC architectures.

Since there are not any datasets available in the literature for this purpose, we decided to use our own dataset. Five different videos have been acquired during real PV plant inspections by an Italian company, namely Topview s.r.l., which builds drones and provides PV inspection services. The videos have been acquired with a thermal camera mounted over a quadcopter DJI. They are representative of the different conditions this typology of algorithms could face with; indeed, they differ in terms of resolution, frame rate and length (corresponding to different dimensions of the plant). Furthermore, the cameras used for the acquisition are different (thus the optical properties of the camera), as well as the height of the drone' flight. Such properties impact on the average number of panels that can be seen in a single frame (from 8 to 30). An overview of the features of the datasets are reported in Table 1, where details concerning the orientation of the panels (V: vertical; H: horizontal), together with the direction of the movement of the drone (EW: from east to west; NS: from north to south; SN: from south to north), are also detailed. An example for each video is reported in Fig. 5.

Table 1. Description of the dataset used in our experimentations.

Video	# Frame	FPS	Resolution	Movement	# Panels	Orientation
Video 1	400	25	640×480	EW	14	V
Video 2	100	25	336×256	SN	24	H
Video 3	525	25	336×256	NS	8	H
Video 4	85	1	336×256	EW	30	H
Video 5	61	1	336×256	NS	8	H

(a) Video 1   (b) Video 2   (c) Video 3   (d) Video 4   (e) Video 5

Fig. 5. Some examples of the considered dataset.

The performance of the proposed approach has been computed in terms of True Positive (TP), False Positive (FP), Precision (P), Recall (R) and F-Score (F) by using the Pascal Criterion [6]. In more details, given the bounding box T_i associated to the hotspot at the ith frame and the ground truth box, GT_i, the overlap between such boxes is evaluated. T_i can be considered a TP if it satisfies the following condition: $\frac{area(T_i \cap GT_i)}{area(T_i \cup GT_i)} \geq K$, where K has been fixed to 0.4. All the detected hotspots T_i which do not satisfy the previous equation can be considered FP. Given TP and FP, the remaining indices can be evaluated as

Table 2. Performance of the proposed approach (combining color and model based information) compared with a traditional color based approach.

Video	Proposed approach						Color based approach						Improvement
	TP	FP	FN	P	R	F	TP	FP	FN	P	R	F	
Video 1	64	87	49	0.42	0.58	**0.49**	64	931	49	0.10	0.58	0.12	308%
Video 2	179	215	19	0.45	0.90	**0.61**	178	952	20	0.16	0.90	0.27	126%
Video 3	309	295	140	0.51	0.69	**0.59**	309	547	140	0.36	0.69	0.47	25%
Video 4	25	7	7	0.78	0.78	**0.78**	25	140	7	0.15	0.78	0.25	212%
Video 5	17	12	3	0.59	0.85	**0.69**	17	31	3	0.35	0.85	0.50	38%

follows: $P = \frac{TP}{TP+FP}$, $R = \frac{TP}{TP+FN}$, $F = 2 \cdot \frac{P \cdot R}{P+R}$. The obtained results are summarized in Table 2. Since there are not publicly available datasets that consist of videos or images for this specific purpose, we compare the proposed method with the baseline approach, which only evaluates the color based information. From the table, we can note that the main difference between the two algorithms pertains the number of false positives that the proposed approach is able to filter out thanks to the introduction of the model based information. The overall improvement with respect to the baseline approach is significant, ranging from 25% up to 308%, thus confirming the effectiveness of the proposed approach.

In order to also provide a qualitative evaluation of the proposed approach, the following link[1] shows its behavior in action.

In order to verify the effectiveness of the proposed approach even from a computational point of view, we performed an analysis in terms of time required for the computation. The analysis has been conducted over a SoC Intel Joule 570X, equipped with a CPU quad core Intel Atom T5700@1,7 GHz (max 2,4 GHz) and 4 GB RAM LPDDR4. Thanks to its size, this device can be mounted directly on board of the drone, so as to allow a real time processing. The results obtained by the proposed approach, expressed in terms of frame rate, are reported in Table 3. It is worth to point out that five different values have been reported due to the different typologies of videos. We can note, for instance, that the processing

Table 3. Time analysis of the proposed approach.

Video	Preprocessing	PV detection	Hot spot detection	FPS
video 1	95.32%	0.06%	4.62%	8
video 2	96.28%	0.15%	3.57%	11
video 3	97.12%	0.12%	2.76%	18
video 4	96.46%	0.12%	3.42%	10
video 5	96.87%	0.13%	3.00%	15

---

[1] http://mivia.unisa.it.

of the Video 1, acquired at a 4CIF resolution, is slower than the other videos, acquired at a resolution of 1CIF.

Table 3 also reports the percentage of the time required for the elaboration of each phase (namely preprocessing, PV detection and hot spot detection). As expected, the most time consuming operation is the preprocessing step, mainly due to the Hough transform used for the extraction of the lines. However, the obtained frame rate confirms the possibility for the proposed approach to run over a SoC architecture in real time (from 8 to 18 frames per seconds), even using a full 4CIF resolution. This is an important contribution of the proposed approach, since there are not any methods available in the literature thought for running directly in real time over SoC architectures.

4   Conclusions

In this paper we proposed a method for detecting the hot spots in a PV plant by processing the images acquired by a thermal camera mounted on board of a drone. The main novelty is that a model based approach is combined with a traditional color based one so as to increase the overall reliability of the proposed method. The results, obtained over five videos acquired in real PV plants, show that our method is able to run in real time over SoC architecture, obtaining very promising results. The possibility to run over SoC architectures is another important contribution of this paper.

References

1. Aghaei, M., Grimaccia, F., Gonano, C.A., Leva, S.: Innovative automated control system for PV fields inspection and remote control. IEEE Trans. Ind. Electron. **62**(11), 7287–7296 (2015)
2. Buerhop, C., Schlegel, D., Niess, M., Vodermayer, C., Weißmann, R., Brabec, C.: Reliability of IR-imaging of PV-plants under operating conditions. Solar Energy Mater. Solar Cells **107**, 154–164 (2012)
3. Buerhop-Lutz, C., Scheuerpflug, H.: Inspecting PV-plants using aerial, drone-mounted infrared thermography system. In: 3rd Southern African Solar Energy Conference, South Africa, 11–13 May 2015 (2015)
4. Denio, H.: Aerial solar thermography and condition monitoring of photovoltaic systems. In: Photovoltaic Specialists Conference (PVSC), pp. 613–618. IEEE (2012)
5. Dotenco, S., Dalsass, M., Winkler, L., Brabec, C., Maier, A., Gallwitz, F., et al.: Automatic detection and analysis of photovoltaic modules in aerial infrared imagery. In: 2016 IEEE International Conference on WACV, pp. 1–9 (2016)
6. Everingham, M., Van Gool, L., Williams, C.K., Winn, J., Zisserman, A.: The pascal Visual Object Classes (VOC) challenge. Int. J. Comput. VIs. **88**(2), 303–338 (2010)
7. Grimaccia, F., Aghaei, M., Mussetta, M., Leva, S., Quater, P.B.: Planning for PV plant performance monitoring by means of Unmanned Aerial Systems (UAS). Int. J. Energy Environ. Eng. **6**(1), 47–54 (2015)
8. Jensen, A.M., McKee, M., Chen, Y.: Calibrating thermal imagery from an unmanned aerial system-AggieAir. In: 2013 IEEE IGARSS, pp. 542–545 (2013)

9. Kaplani, E.: Detection of degradation effects in field-aged c-Si solar cells through IR thermography and digital image processing. Int. J. Photoenergy **2012**, 11 p. (2012). Article no. 396792. doi:10.1155/2012/396792

10. Manganiello, P., Balato, M., Vitelli, M.: A survey on mismatching and aging of PV modules: the closed loop. IEEE Trans. Ind. Electron. **62**(11), 7276–7286 (2015)

11. Nguyen, D.D., Lehman, B.: Modeling and simulation of solar PV arrays under changing illumination conditions. In: 2006 IEEE Workshops on Computers in Power Electronics, pp. 295–299. IEEE (2006)

12. Quater, P.B., Grimaccia, F., Leva, S., Mussetta, M., Aghaei, M.: Light Unmanned Aerial Vehicles (UAVs) for cooperative inspection of PV plants. IEEE J. Photo-voltaics **4**(4), 1107–1113 (2014)

13. Tsanakas, J., Botsaris, P.: On the detection of hot spots in operating photovoltaic arrays through thermal image analysis and a simulation model. Mater. Eval. **71**(4), 457–465 (2013)

14. Tyutyundzsshiev, N., Martínez Moreno, F., Leloux, J., Narvarte Fernández, L.: Equipment and procedures for ON-SITE testing of PV plants and BIPV. IES (2014)

Cow Behavior Recognition Using Motion History Image Feature

Sung-Jin Ahn, Dong-Min Ko, and Kang-Sun Choi$^{(\boxtimes)}$

Interdisciplinary Program on Creative Engineering,
Koreatech, Cheonan, South Korea
ks.choi@koreatech.ac.kr

Abstract. In this paper, a cow behavior recognition algorithm is proposed to detect the optimal time of insemination by using the support vector machine (SVM) classifier with motion history image (MHI) feature information. In the proposed algorithm, area information indicating the amount of movements is extracted from MHI, instead of motion direction which has been widely used for person action recognition. In the experimental results, it is confirmed that the proposed method detects the cow mounting behavior with the detection rate of 72%.

1 Introduction

In the livestock farms, it is important to monitor the health and anomalies of animals [9]. One of the measures to check animal health is animal movement. In general, healthy animals are highly active, while poorly healthy animals are much less active.

There have been many attempts to understand the movement of people and animals. Gaussian Mixture Model (GMM) has been used to find the foreground objects by removing the background that the model represents [14]. Although GMM is simple and fast, it has the drawback that it is difficult to accurately extract a desired object in a complex environment.

Motion history image (MHI) proposed by Ahad et al. is a technique that can represent not only the location of motion generated in an image but also the process of how an object moves.[1]. MHI has a simple but robust advantage in representing movements and has been widely used to describe the motion information within the video. Also, it is suitable for expressing human posture, gait, gestures, because it is not sensitive to silhouette noises such as holes, shadows and missing parts [7]. There are many ways to extract a feature vector using MHI. Bradski and Davis presented a method for determining the pose of the object based on the gradient in successively layered MHI silhouettes. Hu moments, *translation- and scale-invariant shape descriptors* [8], were applied for current silhouette to recognize pose [3].

Weinland *et al.* introduced a free-viewpoint motion representation based on MHI. Multiple MHIs are obtained using multiple cameras for free-view points, so-called motion history volumes (MHV). MHV is transformed into a cylindrical

© Springer International Publishing AG 2017
F. Karray et al. (Eds.): ICIAR 2017, LNCS 10317, pp. 626–633, 2017.
DOI: 10.1007/978-3-319-59876-5_69

coordinate system, and then a feature descriptor is obtained using the Fourier transform [2]. Kim *et al.* presented depth motion appearance for describing the global 3-D shape of body movement using a modified MHI. An entire sequence of depth maps is encoded to a 4096-dimensional histogram of gradient (HoG) descriptor for motion recognition [2].

Although most of researches using MHI focus on the recognition of human actions [4–6,10], identifying animal behavior patterns is also an important issue as much as understanding human behavior patterns. However, there are several reasons not to directly apply the aforementioned methods for animal behavior recognition. In a single barn, there usually exist several livestock, which leads to complex and cluttered scenes. In the methods for human action recognition, the video is typically captured for the front or side of humans. From such the view direction, it is difficult to observe many objects (animals) at the same time, because animals are frequently occluded. In addition, the scene is static in the methods, that is, the camera is fixed and the object (human) stays at the front of the camera during actions. However, animals in the barn move around continuously, which requires much larger field of views and tracking.

Therefore, top-down viewed videos are usually employed in animal behavior recognition. A multi-level threshold method was presented to monitor each pig's health status by ascertaining water and feeding behaviors in the top view [13]. By extracting object contour information and utilizing it, the method can observe several objects at the same time, but contour information is severely degraded in complex backgrounds.

Utilizing the movement of livestock, we can't only identify its health status but also detect the state of estrus. Cows in estrus often cry, feel uneasy, and slowly move away from person when a person come up to them. They also have a lot of movement including mounting [11]. The detection of the time-limited estrus is very important to alleviate significant losses in livestock farms.

The mounting action of rapid raising the cow's upper body can be easily detected with the camera installed to see from the side [12]. However, such a camera arrangement increases the number of cameras required and causes frequent occlusion as mentioned above. In addition, video can be overexposed and severely degraded by time-varying outdoor illumination.

In order to overcome the problems, we propose an estrus detection method using top-down view of cow behavior pattern by MHI area information. In addition to determine in real time the estrus state of cow, the proposed method is able to identify behavior pattern of several cattle such as mounting, walking, running, tail wagging, foot stamping in real complex environment. As documented in the experimental results, the proposed method achieves visually pleasing performance in real complex environment.

2 Proposed Method

This paper presents a real-time detection method of cow's behavior patterns using MHI motion areas, recent motion areas, position changes, and gradient of motion areas as a feature vector over several frames in a complex background environment. The overall block diagram of the proposed method is shown in Fig. 1, and the detailed method is as follows.

The proposed system intend to determine an optimum artificial insemination time of many dozens of cattle using fisheye camera in real-time from a 24-h video stream.

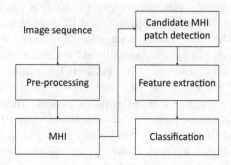

Fig. 1. The block diagram of the proposed method.

2.1 Pre-processing

In the proposed system, a single fisheye camera is utilized to monitor multiple barns to reduce system cost. Since fisheye lens distortion causes inaccurate motion information for the cow, the fisheye lens distortion is corrected and the region of the barns to be monitored is extracted from each video frame F in the pre-processing step. The extracted and undistorted region is denoted by R.

Since the installed camera is fixed, the geometric transformation to correct the lens distortion does not change over time. Therefore, once the coordinate mapping for the transformation is calculated, the coordinate mapping is reused to alleviate huge computation.

2.2 Motion History Image (MHI)

The MHI approach is a view-based temporal template method which is robust in representing motion sequence. Prior to determining MHI, frame difference $D(x, y, t) = R(x, y, t) - R(x, y, t-1)$ is obtained, where (x, y) and t represent pixel position and time index, respectively.

Then, MHI denoted by $H_\Gamma(x, y, t)$ is obtained as follows:

$$H_\Gamma(x, y, t) = \begin{cases} \Gamma, & \text{if } D(x, y, t) > \xi_1, \\ \max(0, H_\Gamma(x, y, t-1) - \delta), & \text{otherwise,} \end{cases} \tag{1}$$

where Γ indicating duration decides the temporal extent of the movement and δ is the decay parameter.

If the object movement occurs, $D(x, y, t)$ becomes greater than a pre-defined threshold ξ_1, and the maximum value of Γ is set to $H_\Gamma(x, y, t)$. Since $H_\Gamma(x, y, t)$ is reduced over time, the motion sequence is represented in H_Γ. That is, The regions where more recently moving pixels have higher values and become gradually lower over time.

Figure 2 shows an example of MHI, where Figs. 2(a) and (b) show R and H_Γ, respectively, and H_Γ is overlaid on R in Fig. 2(c). The red area in H_Γ represents recent movement, while old movement becomes blue.

Let connected pixels with non-zero values of H_Γ be a movement patch. Each patch is labelled.

A patch among several movement patches in Fig. 2(b) represents the mounting action. The size of the patch corresponding to the mounting action tends to be larger than other normal movements. The patch size is an essentially important feature for estrus detection. If the size of a certain patch becomes greater than a pre-defined threshold ξ_2 at the t-th frame and the corresponding patch at the $(t-1)$-th frame is smaller than ξ_2, the patch is regarded as candidate for the mounting action. Whenever each candidate patch O_i is detected, the following features we propose are extracted from the patches corresponding to the candidate patch for a fixed time.

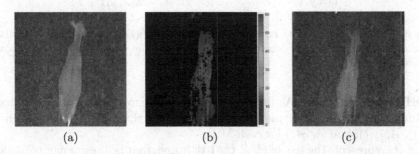

| (a) | (b) | (c) |

Fig. 2. Comparison examples of MHI. (a) An original image. (b) The corresponding MHI. (c) MHI overlaid on the image. (Color figure online)

2.3   Proposed MHI Features

Generally, less moving cows have movements such as tail wagging, foot stamping, walking, running, and mounting. Tail wagging and foot stamping movements take a short time, and thus, yield small-sized MHI patches, while walking, running, and mounting taking a long time induce large MHI patches. Especially, the MHI patches corresponding to the cow mounting action increase rapidly and decrease gradually for 3–7 s.

Based on the observation, once O_i is detected, features are extracted by examining O_i for the next sevral seconds. For this, we compute the Jaccard distances d_j to determine the MHI patch corresponding to the time. d_J is computed as follows:

$$d_J = \frac{|A \bigcup B| - |A \bigcap B|}{|A \bigcup B|}. \tag{2}$$

That is, the larger the overlapping area, the smaller d_j, and the patch with the smallest d_j becomes the corresponding MHI patch (see Fig. 3). Let l be the index of each patch at frame $t+1$. We compute the Jaccard distance (d_J) between its backprojection at time t and all the patches. Then, the patch of minimum Jaccard distances becomes next O_i as follows:

$$O_i(t+1) = argmin_l \left(\frac{|A_l(t+1) \bigcup A_i(t)| - |A_l(t+1) \bigcap A_i(t)|}{|A_l(t+1) \bigcup A_i(t)|} \right). \tag{3}$$

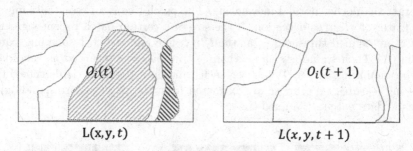

Fig. 3. Example of candidate patch O_i identification. $O_i(t)$ is backprojected, overlapping 2 MHI patches in MHI labeling image $L(x, y, t)$. The patch of minimum Jaccard distance becomes $O_i(t+1)$.

MHI Area Features. As mentioned above, the movement area is the main feature to distinguish cow movements. We extract three area-related features, $A_i(t)$, $A_{i,\xi_3}(t)$, and $A_{i,G}(t)$ for each MHI.

$A_i(t)$ represents the size of O_i at the t-th frame, that is, the amount of motion for the past Γ seconds, while $A_{i,\xi_3}(t)$ represents the amount of motion for the latest ξ_3 seconds ($\xi_3 < \Gamma$) that indicates the area of recent movement.

The ratio of $A_{i,\xi_3}(t)$ and $A_i(t)$ can provide a description of the behavior. If $A_{i,\xi_3}(t)$ is a large portion of $A_i(t)$, the MHI patch represents a fast movement. Therefore, in the cases of mounting and running motion, $A_{i,\xi_3}(t)$ is large, but is small for walking, as compared to $A_i(t)$.

$A_{i,G}(t)$ is the gradient of $A_i(t)$ which describes how the total area of motion changes over time and is obtained as

$$A_{i,G}(t) = \frac{A_i(t) - A_i(t-1)}{2}. \tag{4}$$

MHI Centroid Difference. In order to distinguish mounting from running, we exploit the centroid position information for the candidate patch. For the mounting motion, although the amount of movement is large, the centroid of the patch does not change much. On the other hand, for running motion, both the centroid and amount of movement are large.

To do this, we calculate the centroid position of the patch of interest weighted by the MHI value.

$$x_i(t) = \frac{1}{z_i} \sum_{(x_k, y_k) \in O_i} x_k \cdot H_\Gamma(x_k, y_k, t),$$

$$y_i(t) = \frac{1}{z_i} \sum_{(x_k, y_k) \in O_i} y_k \cdot H_\Gamma(x_k, y_k, t), \tag{5}$$

where $z_i = \sum_{(x_k, y_k) \in O_i} H_\Gamma(x_k, y_k, t)$ is the normalization factor for O_i. Then, the gradient values of $x_i(t)$ and $y_i(t)$ are used as feature, which are obtained as

$$X_{i,G}(t) = \frac{x_i(t) - x_i(t-1)}{2},$$

$$Y_{i,G}(t) = \frac{y_i(t) - y_i(t-1)}{2}. \tag{6}$$

After all, once a candidate patch O_i is detected, the five feature values of $A_i(t)$, $A_{i,\xi_3}(t)$, $A_{i,G}(t)$, $X_{i,G}(t)$, and $Y_{i,G}(t)$ are obtained at every MHI for the next 5 s. Since we generate MHIs at 4 fps, an entire feature vector for O_i becomes a 100-dimensional vector which is fed into SVM classifier to identify the cow behavior pattern (Fig. 4).

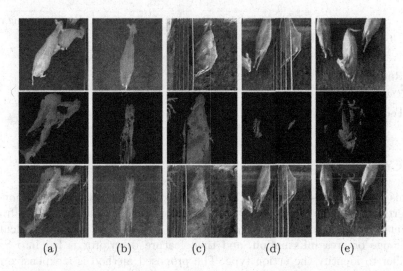

(a)   (b)   (c)   (d)   (e)

Fig. 4. The cow behavior pattern and associated MHI, MHI overlaid on the image: (a) mounting (b) walking (c) running (d) tail wagging (e) foot stamping

3   Experimental Results

For the evaluation, video sequences of frame resolution are used with original frame rate of 20 frames per second. It contains five types of cow actions (mounting, walking, running, tail wagging, foot stamping) performed many times by cows in three different observation time from: 3 s (s1), 5 s (s2), 7 s (s3). This is because the cow mounting action last for 3–7 s.

The pre-defined thresholds ξ_1, ξ_2, ξ_3 the maximum value of Γ, and the decay coefficient δ were set to 20, 2000, 54, 60, and 1, respectively. In order to avoid a high dimensional O_i, we skip every 4 frames.

This experiment used leave-one-out cross validation to evaluate the recognition rate of action classifier in the three different feature extraction times.

Among the three different observation time experiment, s3 (Table 1) revealed the highest recognition rate. That's because each movement has different duration. Generally, dynamic movements, mounting, walking and running, have longer duration than static movements such as tail wagging and foot stamping. Thus, s1 likely misjudged the static movement because it has no enough time to observe the change of movement. As for s3 (Table 1), showed higher recognition rates on static movements and running than other tables. However, too long observation time is prone to be noised by other patches. So, proper observation time is needed to precisely recognition rate. The overall recognition rate of the proposed method in three different observation times is as follows. s1 is 60.83%, s2 is 77.03%, and s3 is 82.83%.

Table 1. Feature matrix of MHI area for s3

	Mounting	Walking	Running	Tail wagging	Foot stamping
Mounting	**80**	15	5	0	0
Walking	35	**60**	0	5	0
Running	0	5	**95**	0	0
Tail wagging	0	0	0	**100**	0
Foot stamping	5	0	0	15	**80**

4   Conclusion

In this paper, we presented MHI-area method for cow action recognition. It converts an video sequence into a motion history image by accumulating frame difference of the sequence. The motion history image patch is used to calculate its change of area information, and then feature descriptor is fed into SVM classifier to identify the action type. The proposed method is inexpensive and can be applied in real complex environment. For future work, we need to develop algorithms for shadows at night, barns swing, and cattle gathering.

References

1. Bobick, A.-F., Davis, J.-W.: The recognition of human movement using temporal templates. IEEE Trans. Pattern Anal. Mach. Intell. **23**(3), 257–267 (2001)
2. Weinland, D., Ronfard, R., Boyer, E.: Free viewpoint action recognition using motion history volumes. Comput. Vis. Image Underst. **104**(2), 249–257 (2006)

3. Bradski, G.-R., Davis, J.: Motion segmentation and pose recognition with motion history gradients. Mach. Vis. Appl. **13**(3), 174–184 (2002)
4. Aggarwal, J.-K., Cai, Q.: Human motion analysis: a review. In: Proceedings of Nonrigid and Articulated Motion Workshop, pp. 90–102. IEEE (1997)
5. Aggarwal, J.-K., Cai, Q.: Human motion analysis: a review. Comput. Vis. Image Underst. **3**(73), 428–440 (1999)
6. Aggarwal, J.-K., Park, S.-H.: Human motion: modeling and recognition of actions and interactions. In: Proceedings of 2nd International Symposium on 3D Data Processing, Visualization and Transmission, 3DPVT 2004, pp. 640–647. IEEE (2004)
7. Liu, J., Zheng, N.: Gait history image: a novel temporal template for gait recognition. In: 2007 IEEE International Conference on Multimedia and Expo, pp. 663–666. IEEE (2007)
8. Hu, M.-K.: Visual pattern recognition by moment invariants. IRE Trans. Inf. Theor. **8**(2), 179–187 (1962)
9. Dawkins, M.-S.: A user's guide to animal welfare science. Trends Ecol. Evol. **21**(2), 77–82 (2006)
10. Ahad, M.A.R., Tan, J.-K., Kim, H.-S., Ishikawa, S.: Human activity recognition: various paradigms. In: International Conference on Control, Automation and Systems, 2008, ICCAS 2008, pp. 1896–1901. IEEE (2008)
11. DuPonte, M.W.: The basics of heat (estrus) detection in cattle. Cooperative Extension Service, College of Tropical Agriculture and Human Resources, University of Hawaii at Mānoa (2007)
12. Chung, Y.-W., Choi, D.-W., Choi, H.-S., Park, D.-H., Chang, H.-H., Kim, S.: Automated detection of cattle mounting using side-view camera. KSII Trans. Internet Inf. Syst. (TIIS) **9**, 3151–3168 (2015). http://www.dbpia.co.kr/Article/NODE 06507183
13. Guo, Y.-Z., Zhu, W.-X., Jiao, P.-P., Ma, C.-H., Yang, J.-J.: Multi-object extraction from topview group-housed pig images based on adaptive partitioning and multilevel thresholding segmentation. Biosyst. Eng. **135**, 54–60 (2015)
14. Zivkovic, Z.: Improved adaptive Gaussian mixture model for background subtraction. In: Proceedings of the 17th International Conference on Pattern Recognition, ICPR 2004, vol. 2, pp. 28–31. IEEE (2004)

Footnote-Based Document Image Classification

Sara Zhalehpour[1]([⊠]), Andrew Piper[2], Chad Wellmon[3], and Mohamed Cheriet[1]

[1] École de Technologie Supérieure, University of Quebec, Montreal, Canada
sara.zhalehpour.1@ens.etsmtl.ca, mohamed.cheriet@etsmtl.ca
[2] McGill University, Montreal, Canada
andrew.piper@mcgill.ca
[3] University of Virginia, Charlottesville, VA, USA
mcw9d@virginia.edu

Abstract. Analyzing historical document images is considered a challenging task due to the complex and unusual structures of these images. It is even more challenging to automatically find the footnotes in them. In fact, detecting footnotes is one of the essential elements for scholars to analyze and answer key questions in the historical documents. In this work, we present a new framework for footnote detection in historical documents. To this aim, we used the most salient feature of the footnotes, which is their smaller font size compared to the rest of the page content. We proposed three types of features to track the font size changes and fed them to two classifiers: SVM and AdaBoost. The framework shows promising results over 80% for both classifiers using our dataset.

Keywords: Visual information retrieval · Footnote detection · Historical documents classification

1 Introduction

In the recent decades, digitizing historical documents to preserve and make them available via electronic media has got considerable attention. Though, the problem of preserving and accessing these documents seems to be mostly solved. The issue of automatic recognition and interpretation of these documents, as well as easy and efficient access to their content, is still a challenge since they are often not equipped with adequate index information.

Visual information retrieval is a research domain concentrated on providing means for retrieving, organizing, indexing and annotating visual information from large data [1]. To perform information retrieval on historical document images, a proper approach is needed for characterization of the document content in a meaningful way. A manuscript can contain various graphical or textual elements which are interesting to retrieve. One of these elements is the footnote, which emerged as an important form of knowledge indexing in Europe in the eighteenth century [2,3]. Using footnotes and in general terms indices is one of the key features of the printed publications in past five centuries. Footnotes are often used to add some additional explanations about a certain content

© Springer International Publishing AG 2017
F. Karray et al. (Eds.): ICIAR 2017, LNCS 10317, pp. 634–642, 2017.
DOI: 10.1007/978-3-319-59876-5_70

of the document. This information could be a definition of a specific topic or citation of another scholar. Thus, finding the footnotes can help understand these documents and also link the different books and scholars during different periods.

There are vast related works in the scope of analyzing historical documents and researchers increasingly address the large-scale collections of historical documents. Automatic identification of visual features like tables and figures in journals and books has been a research topic for several years [4–7]. Page layout analysis is also another area for researchers, where its aim is to detect textual or graphical location information and distinguish among different parts of a page [8–10]. To our best knowledge, this paper is the first research attempts to use image processing and pattern recognition techniques for the purposes of detection footnotes in the historical document pages.

Our main goal in this work is to find the pages containing footnotes. The footnote should give the page a significant visual feature that one might reliably be able to identify. We observed that in over 80% of the document pages the font of the footnote is smaller than the main text. We consider this observation as our hypothesis and reduce the research question to how to find these font size changes. In this work, we proposed three methods for footnote detection from historical documents. These methods use the bounding boxes of characters and projections of lines to track the font size changes and extract features. Then, we used two classifiers to evaluate extracted features.

The organization of the paper is as follows. In Sect. 2, we give the details of our footnote classification method. In Sect. 3, we evaluate our method and discuss the results. Finally, in Sect. 4, we provide conclusions and future works.

2   Methodology

In this section, we first present a new framework for footnote detection for historical document images. We proposed three methods for extracting visual features related to the font size changes behavior on each page of the document images according to some rules. Two of these methods are based on finding the approximate font size for each line and the other is based on finding the dissimilarity between probability distribution of each line in the page. Finally, we used two different classification methods to compare their performance.

2.1   Feature Extraction

As stated before, for extracting features from document images, three methods are proposed in this work. Below, we describe briefly each of these methods.

Bounding Box Based Method (BBox). This method is based on finding the approximate font size of each line. To do that, we tried to find the bounding box around each character in the text lines. We found the smallest rectangle containing the region. To avoid detection small noises or lines as bounding boxes, we use a threshold. Finally, for each line, the average height of bounding boxes are found and normalized to a value between 0 and 1 so the results would be comparable.

After finding bounding boxes, we need a way to compare them and associate their relation to having footnotes or not in the pages. Figure 1 shows an example of font size comparison using these three methods for two cases of having and not having a footnote. We can see from Fig. 1(b) that where there is a footnote there is a large drop in the font size and from Fig. 1(f) that when footnote does not exist, we have oscillations in the same range. Therefore, we looked for some assumption and defined some rules by observing 100 images from both classes. Our basic assumption is that difference of two consecutive lines height must be more than 0.40. We also have some sub-assumptions. The footnote cannot appear in the first three lines. There must be at least four lines in the page so the footnote test could be done. If it detects a small font in the middle lines and again larger fonts come after that line, it cannot be a footnote line. We use the logical [0/1] answers to all the assumptions and their combination to come up with an 18-dimensional feature vector for each image.

Horizontal Projection Based Method (Proj). This method is also based on finding the approximate font size in each line. We calculate the horizontal intensity (horizontal histogram) for each line in the page. Then, this histogram is intersected with a line equal to 0.55 of the difference between maximum and minimum of the histogram. The difference is considered the estimated font size for that line. Final font sizes are normalized between 0 and 1.

Similar to BBox method by studying the behavior of the font size changes, we define some general rules for having footnotes using 100 images from both classes (See Fig. 1(c and h)). The basic assumption is that difference of two consecutive lines height must be more than 0.40. There are also some sub-assumptions. The footnote cannot appear in the first three lines. There must be at least four lines in the page so the footnote test could be done. If a line except the last line selected as the footnote line there must be a difference of 0.25 between the line before and after it. The logical [0/1] answers to all the assumption and their combination are used to create a 24-dimensional feature vector for each page.

Kullback Leibler Divergence Based Method (KLD). This method of feature extraction is based on the assumption that the footnote line is dissimilar from the main text. The horizontal projections obtained for Proj method are normalized to have the probability distribution of the lines. Then, the dissimilarity between the consecutive lines is computed using these distributions. Kullback-Leibler divergence gives us the best results among the dissimilarity methods we tried. It measures the dissimilarity between two probabilistic variables when P is approximated using Q [11]. The Kullback-Leibler divergence is defined as follow:

$$D_{KL}(P,Q) = \sum_i P log \frac{P}{Q} \tag{1}$$

where P is the distribution of line i and Q is the distribution of line i-1. Since KLD is not metric the decision of using $D_{KL}(P,Q)$ over $D_{KL}(Q,P))$ is based on the fact that the line before footnote is the resemblance of the whole page's font and it also empirically showed better results. To handle the length differences of probabilities (histograms) due to the different height of outlines, zeroes are added to the end of the shorter probability distribution.

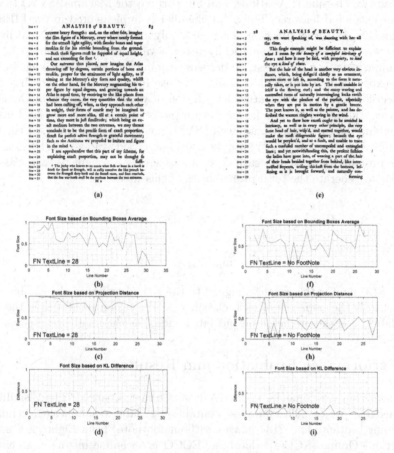

Fig. 1. An example of font size change using three proposed features for pages (a) with a footnote (b) without a footnote. Pairs (b, f), (c, h) and (d, i) are representing the font size change using BBox, Proj and KLD methods, respectively.

We can see from Fig. 1(d and i) that there is a peak when footnote appears and the KLDs are in the same range when there is no footnote. Similar to previous methods some rules are defined with the basic assumption that minimum difference of two consecutive lines must be more than 0.5. There are also some sub-assumptions. There must be at least four lines in the page for having a footnote and footnote cannot be within the first three lines. If there is a peak over

0.35 in the four last lines there is a footnote. The logical [0/1] answers to all the assumptions and their combination are converted to a 25-dimensional feature vector for each page.

2.2   Classification

We used two classifiers: AdaBoost and Support Vector Machine (SVM) to classify our extracted features. The SVM classifier is implemented in the LIBSVM toolbox [12]. As kernel function, a degree 2 polynomial is used and SVM hyperparameters are selected automatically using a validation set equals to 10% of the training set. For AdaBosst, 50 decision stumps are used as weak classifiers.

In experiments using SVM classifier, we use fusing features in both feature and decision levels. In the feature level fusion, features of different methods are simply concatenated to have larger feature vector. For the decision level fusion, we combine the probability outcome of SVM in each method and used the final probability for decision [13]. We use the classic weighted average rule by summing up the probabilities given to each class and assigning a weight to each classifier in order to generate the total probabilities of classifiers:

$$P(\omega_k) = \frac{1}{M} \sum_{i=1}^{M} P(\lambda_i \mid x_i) P(\tilde{\omega_k} \mid x_i, \lambda_i), k = 1, 2. \tag{2}$$

$$\omega^* = \max_k P_{\omega_k}, k = 1, 2. \tag{3}$$

where $P(\lambda_i \mid x_i)$ is the weight assigned to the i-th classifier in the combination and controls the influence of each classifier on the fusion results. ω is the true class label, $\tilde{\omega}$ is the predicted output label and ω^* is the final chosen class.

3   Performance Evaluation and Results

To conduct our experiments, we used a database that is part of "The Visibility of Knowledge"[1] project. This database contains 4322 labeled samples (2138 images containing footnote and 2184 images without footnote) from Eighteen Century Collections Online (ECCO)[2] database. ECCO is an online historical archive of eighteenth century documents in any language that is printed in the United Kingdom between the years 1701 and 1800.

Experiments are done using 10-folds cross-validation setup where all samples were randomly divided into 10 folds. In each testing cycle, one fold is used for testing and the rest of the folds are used for training. The process is repeated 10 times and the average result over all testing cycles is used as the final performance. In any case that we needed a validation set for finding the parameters, we used one fold from the training set.

---

[1] https://txtlab.org/2016/09/the-visibility-of-knowledge/.
[2] http://gdc.gale.com/products/eighteenth-century-collections-online/.

We used SVM and AdaBoost classifiers to conduct our sets of the experiment. Results are given in Tables 1 and 2, which are split into three sections. In the first section of the tables, single feature results are shown using each feature extraction methods of BBox, Proj and KLD. The BBox and Proj methods clearly outperform the KLD method on this database. The second section of Tables 1 and 2 shows the results of feature level fusion, the combination of BBox and Proj features and the combination of features from all three methods. It is obvious that combining the features improves the performance of classifiers since it adds more discrimination power to the feature vector. Finally, the third part of Table 1 shows the results of decision level fusion using two methods classification outputs and three methods. In the case of fusing the output of BBox and Proj classifiers, we give the same weight to both of them. For fusion of the all three methods classifier outputs, we give less weight to the KLD classifier since it has the weaker features and 0.4 as the weight of the other classifiers. It can be well observed that sum-based fusion methods perform better with respect to the single and feature level fusion methods. Also, AdaBoost performs better than SVM in feature level fusion, but applying decision level fusion to SVM boost the results in favor of SVM.

Table 1. Experimental results using SVM over 10-folds cross validation set up. Values are in percent and the bold numbers are indicating the best results.

	F-measure	Recall	Precision
Single method results			
BBox	77.98	82.77	73.92
Proj	78.77	83.8	74.37
KLD	75.3	75.17	74.79
Feature level fusion results			
BBox + Proj	80.71	83.52	78.16
BBox + Proj + KLD	80.6	83.3	78.11
Decision level fusion results			
Fusion(BBox + Proj)	81.5	80.76	**82.36**
Fusion (All)	**82.11**	**85.32**	79.2

One of our observations from the tables is that the value of recall in all cases are higher than precision which means that there are some images with a footnote in the dataset that are hard to detect. We have checked some of these missed images and see that their misclassification is mostly due to their complex layout which makes finding text lines harder. There are images containing two columns, figures, tables, formula and even Greek letters that all reduce the performance of the classifiers. In order to relax the problem of complex structures, we manually reduced about 1000 images of our dataset to 3415 images and tried to only keep the images with a single column, no figure and no tables and repeated all our

Table 2. Experimental results using AdaBoost over 10-folds cross validation set up. Values are in percent and the bold numbers are indicating the best results.

	F-measure	Recall	Precision
Single method results			
BBox	78.88	80.6	77.26
Proj	79.14	80.13	78.25
KLD	69	71.58	66.66
Feature level fusion results			
BBox + Proj	81.22	82.08	80.44
BBox + Proj + KLD	**81.67**	**82.68**	**80.75**

experiments again using SVM and AdaBoost classifiers. We can see that the value of precision along with recall and f-measure are all increased, but this time recall and precision have closer values (see Table 3). AdaBoost in the case of relaxed images set shows weaker result compared to the complete set data. One of the reasons could be the reduction in the data size. Also, AdaBoost can be sensitive to the type of the features it gets as input which are [0/1] here. In general, SVM proves to be the classifier with more robust results in different setups in our case. Comparing relaxed set with the original set, we can see the best performance enhancement happens for the precision as we expected since we eliminated the complex images.

Table 3. Experimental results using SVM over relaxes set with 10-folds cross validation set up. Values are in percent and the bold numbers are indicating the best results.

	F-measure	Recall	Precision
SVM			
Fusion (BBox + Proj)	86.65	83.68	**89.91**
Fusion (All)	**89.38**	**89.67**	89.16
AdaBoost			
Fusion (BBox + Proj)	81.22	82.08	80.44
Fusion (All)	81.67	82.68	80.75

4   Conclusion

In this paper, we presented a method for detecting footnote in historical documents. By considering the fact that footnotes are mostly having smaller font size compared to the main text, we came up with three different visual feature extraction methods. These methods are based on finding the character bounding boxes and the horizontal projection of each line. We used two classifiers,

AdaBoost and SVM and performed fusion in the feature and decision levels. Fusing the output of SVM for all three feature extracting methods gave us the best results where f-measure is over 80% and considering the complexity of our dataset, it is a promising result.

One of the weaknesses of these methods was their dependency to the layout analysis in the first step of feature processing. Hence, investigating more efficient layout methods definitely will increase the performance. It is also good to expand our hypothesis to include other parameters beside the fontsize. Moreover, we are planning to apply other classification methods. One of the ideas that come to mind is using Ada-SVM to enhance the performance so that AdaBoost select the features and SVMs fuse the features to form the final classifier.

Acknowledgments. This publication was made possible by a grant from SSHRC Canada for "The Visibility of Knowledge" project. I would also like to express my gratitude to Ehsan Arabnejad for his detailed and valuable comments.

References

1. Murugappan, A., Ramachandran, B., Dhavachelvan, P.: A survey of keyword spotting techniques for printed document images. Artif. Intell. Rev. **35**, 119–136 (2011)
2. Pasanek, B., Wellmon, C.: The enlightenment index. Eighteenth Century **56**, 359–382 (2015)
3. Grafton, A.: The Footnote: A Curious History. Harvard University Press, Cambridge (1999)
4. Apostolova, E., You, D., Xue, Z., Antani, S., Demner-Fushman, D., Thoma, G.R.: Image retrieval from scientific publications: text and image content processing to separate multipanel figures. J. Am. Soc. Inform. Sci. Technol. **64**, 893–908 (2013)
5. Klampfl, S., Kern, R.: An unsupervised machine learning approach to body text and table of contents extraction from digital scientific articles. In: Aalberg, T., Papatheodorou, C., Dobreva, M., Tsakonas, G., Farrugia, C.J. (eds.) TPDL 2013. LNCS, vol. 8092, pp. 144–155. Springer, Heidelberg (2013). doi:10.1007/978-3-642-40501-3_15
6. Lorang, E., Soh, L.K., Datla, M.V., Kulwicki, S.: Developing an image-based classifier for detecting poetic content in historic newspaper collections. D-Lib Mag. **21**(7) (2015)
7. Khurshid, K., Faure, C., Vincent, N.: Fusion of word spotting and spatial information for figure caption retrieval in historical document images. In: 10th International Conference on Document Analysis and Recognition, ICDAR 2009, pp. 266–270. IEEE (2009)
8. Konya, I., Eickeler, S.: Logical structure recognition for heterogeneous periodical collections. In: Proceedings of the First International Conference on Digital Access to Textual Cultural Heritage, pp. 185–192. ACM (2014)
9. Mehri, M., Héroux, P., Gomez-Krämer, P., Mullot, R.: Texture feature benchmarking and evaluation for historical document image analysis. Int. J. Doc. Anal. Recogn. (IJDAR) **20**, 1–35 (2017)
10. Alaei, F., Alaei, A., Blumenstein, M., Pal, U.: Document image retrieval based on texture features and similarity fusion. In: 2016 International Conference on Image and Vision Computing New Zealand (IVCNZ), pp. 1–6. IEEE (2016)

11. Joyce, J.M.: Kullback-leibler divergence. In: Lovric, M. (ed.) International Encyclopedia of Statistical Science, pp. 720–722. Springer, Heidelberg (2011)
12. Chang, C.C., Lin, C.J.: LIBSVM: a library for support vector machines. ACM Trans. Intell. Syst. Technol. (TIST) 2, 27 (2011)
13. Wang, Y., Guan, L., Venetsanopoulos, A.N.: Kernel cross-modal factor analysis for information fusion with application to bimodal emotion recognition. IEEE Trans. Multimedia 14, 597–607 (2012)

Feature Learning for Footnote-Based Document Image Classification

Sherif Abuelwafa[1]([⊠]), Mohamed Mhiri[1], Rachid Hedjam[1],
Sara Zhalehpour[1], Andrew Piper[2], Chad Wellmon[3], and Mohamed Cheriet[1]

[1] École de Technologie Supérieure, University of Quebec, Montreal, Canada
sherif.abuelwafa.1@ens.etsmtl.ca
[2] McGill University, Montreal, Canada
[3] University of Virginia, Charlottesville, VA, USA

Abstract. Classifying document images is a challenging problem that
is confronted by many obstacles; specifically, the pivotal need of hand-
designed features and the scarcity of labeled data. In this paper, a new
approach for classifying document images, based on the availability of
footnotes in them, is presented. Our proposed approach depends mainly
on a Deep Belief Network (DBN) that consists of two phases, unsupervised pre-training and supervised fine-tuning. The main advantage of
using this approach is its capability to automatically engineer the best
features to be extracted from a raw document image for the sake of
generating an efficient representation of it. This feature learning approach takes advantage of the vast amount of available unlabeled data
and employs it with the limited number of labeled data. The obtained
results show that the proposed approach provides an effective document
image classification framework with a highly reliable performance.

Keywords: Unsupervised feature learning · Hierarchical representation
learning · Document image classification

1 Introduction

Protecting humanity's cultural heritage is highly needed to understand our
past and prepare for the future; therefore, digitizing historical manuscripts and
printed books becomes an essential approach to guarantee a well preserved history and a widely accessible content to thousands of researchers around the
globe. A great example that shows the adaptation of this global movement is the
Eighteenth Century Collections Online (ECCO)[1] that contains around 200,000
volumes (32 million image pages) of online archive related to the eighteenth-
century printed books.

In fact, introducing an efficient data-driven approach that can understand
such huge amount of widely available historical data will play a vital role in

---

[1] http://www.gale.com/primary-sources/eighteenth-century-collections-online.

© Springer International Publishing AG 2017
F. Karray et al. (Eds.): ICIAR 2017, LNCS 10317, pp. 643–650, 2017.
DOI: 10.1007/978-3-319-59876-5_71

unveiling the secrets of such precious content. Generally, understanding a historical document image involves a wide spectrum of subprocesses and tasks that range from layout analysis and document image classification to optical character recognition (OCR). In this work, the main focus will be on the task of classifying document images based on the presence of footnotes in them. Considering the valuable information that is usually contained in a footnote and the strong ties it creates with other documents; footnotes are believed to be reflecting how ideas can be circulated and exchanged between various manuscripts and books throughout centuries and civilizations [1,2]. Therefore, obtaining document images with footnotes has been raised to form the main focus of this research paper.

The performance of the document image classification process dependents highly on the used representation of the document image, where learning how to map the intensity values of the document images' pixels into a relevant decision (i.e., the presence of a footnote in the image or not) is critical. In order to obtain an expressive representation of the document image, the best features have to be captured from it. Such features help in getting better high-level representations of the raw data in a way that explicates the document image's main properties and facilitates the subsequent classification process.

Most of the traditional systems for document image classification depend on carefully hand-designed features (e.g., SIFT [3], SURF [4] and HOG [5]). Those features are being engineered by experts relying on their prior knowledge regarding the used data and the desired application, which is a very complex process. Actually, such hand-designed features are labor-intensive, time consuming and cannot generalize well to new problems. Due to the difficulties related to engineering hand-designed features, we focus in this work on feature learning approaches [7] that can take advantage of the increasing amount of available informative data to automatically learn even better feature representations than the hand-designed ones. Utilizing feature learning in the field of document image classification has been adopted in some recent works, such as the work of Kang et al. [6]. But those works have relied heavily on using only labeled data for training their feature learning algorithms.

In order to train feature learning algorithms either labeled (supervised learning) or unlabeled (unsupervised learning) data can be used. Apparently, acquiring more data leads to better performance regardless of the used learning approach [8], where most of the recent breakthroughs in the results of machine learning approaches are actually due to the availability of a large amount of training data. And since obtaining enough labeled data to perform supervised feature learning is often a very difficult and expensive task due to the required time and labor for labeling, while a vast amount of unlabeled data is available and easily accessible; we are investigating in this research paper the capacity of incorporating an unsupervised feature learning algorithm alongside a supervised one to achieve an optimal approach for document image classification.

In the light of the previously mentioned challenges of hand-designed features and the scarcity of labeled data, the main contribution of this paper is in proposing and evaluating a feature learning approach for document image classification that is capable of the following; generating the best representation of an input raw image through automatically engineer the best features using a trainable feature extractor instead of using hand-crafted features. This is performed while depending mainly on the largely available unlabeled images besides the restrictedly available labeled images.

The rest of the paper is organized as follows. The proposed approach is outlined in Sect. 2. In Sect. 3, the used dataset and the obtained experimental results are reviewed alongside some observations and related discussion. Finally, Sect. 4 demonstrates the conclusion and the future work.

2   The Proposed Approach

Since a footnote is the center of interest in our document classification process; and through studying the document images that contain footnotes, we observed a compelling common feature. We found that document images with footnotes usually contain two different font sizes between their main text and their footnotes text, a property that turned out to be at the core of our hypothesis.

To train a classifier to differentiate efficiently between document images with footnotes and document images with no footnotes, we need to provide it with many positive and negative examples of document images that contain footnotes and do not contain footnotes, respectively. And since we have a very limited amount of labeled data with ground-truth; while, on the other hand, unlabeled data are widely available, it will be effective to utilize an efficient approach that can leverage such wide availability of unlabeled data. As a result of the above factors, our document classification algorithm is based mainly on teaming up an unsupervised pre-training phase with a supervised fine-tuning phase. In fact, this setup has proved to be very efficient in the case of scarcity of labeled data [9]. According to [7,10], exploiting the process of unsupervised pre-training in initializing a later supervised classification process can be certainly helpful for this classification process.

Generally, our proposed document classification approach consists of 4 stages, Fig. 1. This approach depends at its core on a feature learning model that is based on Deep belief network (DBN) architecture [12,13] and composed of two main phases (unsupervised pre-training and a supervised fine-tuning). The following subsections will provide more insights about each stage of them.

2.1   Preprocessing

Considering our observations and hypothesis, each document image is represented by a concatenated image of its two top text-lines and two bottom text-lines, Fig. 2. In order to obtain these text-lines, a projection-based text-line segmentation method is used [11]. Afterward, each concatenated image is negated, normalized then resized to 45×500 for a faster performance.

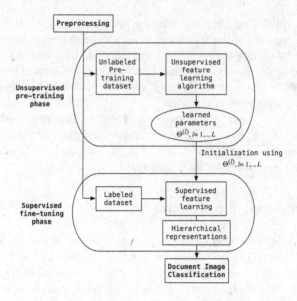

Fig. 1. The proposed approach pipeline.

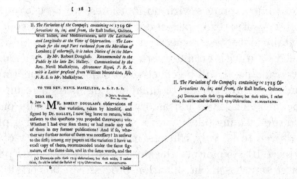

Fig. 2. An example from ECCO dataset for a concatenated image of two top text-lines and two bottom text-lines.

2.2   Unsupervised Pre-training Phase

In this phase, an unlabeled pre-training dataset that contains a large amount of document images is utilized as an input to our unsupervised feature learning algorithm, which is based on DBN architecture. DBN is a generative model in which the dependencies between the nodes in one layer is being statistically encoded in the layer above it by using Restricted Boltzmann machines (RBMs). To train our model, a layer-wise greedy learning algorithm is exploited, where the inputs of a higher layer are the calculated activations of the layer below it. Specifically, an RBM is being trained once per time, where the obtained parameters θ^l are being frozen once the training process is finished; then, another

RBM layer is stacked into that network, and a new training process starts on that level. This process is repeated until the last layer is trained.

For each RBM layer, the following energy function $E(v, h)$ is defined to express the negative log likelihood (cost function) of this layer [13][2]:

$$ -\log P(v, h) = E(v, h) = \frac{1}{2} \sum_i v_i^2 - \sum_{i,j} v_i w_{ij} h_j - \sum_j b_j h_j - \sum_i c_i v_i \quad (1) $$

where c_i is the bias of the visible node v_i, b_j is the bias of the hidden node h_j and w_{ij} is the weight between v_i and h_j. After the training process, a set of parameters $\theta^l = (w_{ij}, b_j, c_i)$ is being learned.

Training a DBN with L layers results in a set of L learned parameters $\theta^l, l = 1, \ldots, L$, which shall contain implicitly some information about the characteristics of used document images.

2.3 Supervised Fine-Tuning Training Phase

A labeled fine-tuning dataset is utilized as an input to a supervised feature learning algorithm. This algorithm can be perceived as a simple Multi-Layer Perceptron (MLP) with the same architecture as the utilized DBN and initialized using the set of learned parameters θ^l obtained at the pre-training phase. In particular, after training our DBN at the previous stage, the parameters of each layer θ^l are used in the initialization process of the same corresponding parameters at our neural network in the current phase. A fine-tune process to the previously learned parameters θ^l is conducted and results in learning high-level hierarchical representations of each document image.

2.4 Classification

A logistic regression classifier is added to our fine-tuning network, as an output layer, in order to classify the used document images into two classes. The fine-tuning dataset is utilized in the classification training and testing processes using the document images' high-level representations obtained from the fine-tuning stage.

3 Experimental Results and Discussion

Many experiments have been conducted in order to assess our proposed approach. Besides using f-measure as an evaluation metric, the following two steps are utilized. First, a cross-validation technique with 10-folds has been used to evaluate the final classification performance. Additionally, in order to investigate the effect of images' layout complexity, we reconducted the experiments using a relaxed version of our dataset. In fact, the obtained results show that our approach is notably effective in classifying images based on the presence of footnotes in them even with images with complex layouts.

---

[2] Considering having real values at the visible nodes (input document image).

3.1   Datasets

The images used in our experiments are part of the ECCO dataset used for "The Visibility of Knowledge"[3] project. We utilize two subgroups of this dataset within our proposed approach; an unlabeled dataset that is utilized at the pre-training phase; in addition, a labeled dataset is exploited at the fine-tuning phase (i.e., this includes the processes of training, validation and testing).

Pre-training Dataset. An unlabeled dataset that contains 6895 document images is utilized to learn features in an unsupervised-manner in the pre-training stage of our proposed approach.

Fine-Tuning Dataset. For fine-tuning and training our proposed approach, a labeled dataset that contains the ground-truth of document images classes is utilized. This dataset includes 4322 labeled samples of ECCO dataset (2138 images contain footnotes and 2184 images without footnotes). In order to study the effect of the images' layout complexity on the results, about 1400 images with complex structures have been removed. This has led to a relaxed version of the dataset with 2894 images. The relaxed dataset only contains images with one column and does not contain figures, tables or formulas. Figure 3 shows some examples of images with complex layouts.

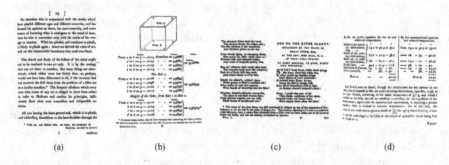

(a)	(b)	(c)	(d)

Fig. 3. Examples of document images with complex layouts: (a) a simple page (b) a page with formulas and figures (c) a page with two columns (d) a page with tables.

3.2   Experimental Setup

In implementing our experiment, we used Python and Theano [14]. Our DBN network composed of 2 layers, each layer contains 1000 hidden units. Specifically, the final architecture can be described as 45×500 - 1000 - 1000 - 2. In this case, 2 represents the number of classes (i.e., 0: an image with no footnote and 1: an image with a footnote). The utilized learning rates for pre-training and fine-tuning are 0.01 and 0.05, respectively. In addition, a batch size with a value 10 is set while considering 20 epochs for pre-training and 400 epochs for fine-tuning.

---

[3] https://txtlab.org/2016/09/the-visibility-of-knowledge/.

Furthermore, we used a 10-fold cross-validation setup to conduct our experiments. In each cycle, 8 folds are assigned to training phase, 1 fold is used at the validation phase to tune the model's hyper-parameters and the final fold is utilized as a test set to calculate the generalization performance of our proposed model when applied to unseen test images.

3.3   Results

As demonstrated in Table 1, our approach has an overall f-measure of 81.37% using the original dataset; a value that has been increased by 4.46% after relaxing the problem. Investigating more, we can observe a big difference between the precision and the recall values using the original dataset; while, on the other hand, the difference between these values is so slight when it comes to the relaxed set. This clearly indicates that the original set contains images with footnotes that are hard to be detected; therefore, relaxing the problem and filtering out the dataset of its complex images has noticeably contributed in improving the value of precision (around 7%). These observations imply the critical role of images with complex structures in affecting the classification overall performance, and raise the need to tackle them.

Table 1. Experimental results using both the fine-tuning original dataset and its relaxed version. Values are in percent.

	Precision	Recall	F-measures
Original set	78.75	84.37	81.37
Relaxed set	85.63	86.16	85.83

4   Conclusion and Future Work

We have proposed a document image classification framework that is significantly suitable for classification problems associated with a limited availability of labeled data. The proposed approach aims to take advantage of the largely available unlabeled data through incorporating a DBN-based unsupervised feature learning procedure. Our cross-validation-based experimental results demonstrated empirically that our approach can attain an efficient generalization performance on classifying document images based on the availability of footnotes in them. Although this framework is capable of acquiring many tangled features, it finds challenges in dealing with document images with complex structures. The upcoming step towards a more efficient classification model that can tackle these challenges is to exploit more pre-training data and investigate the criticality of the unlabeled data in reinforcing the overall classification performance of our approach.

Acknowledgments. This publication was made possible by a grant from SSHRC Canada for "The Visibility of Knowledge" project. The statements made herein are solely the responsibility of the authors.

References

1. Grafton, A.: The Footnote: A curious history. Harvard University Press, Cambridge (1999)
2. Pasanek, B., Wellmon, C.: The enlightenment index. Eighteenth Century **56**(3), 359–382 (2015)
3. Lowe, D.G.: Object recognition from local scale-invariant features. In: The Proceedings of the Seventh IEEE International Conference on Computer Vision, vol. 2. IEEE (1999)
4. Bay, H., Tuytelaars, T., Gool, L.: SURF: speeded up robust features. In: Leonardis, A., Bischof, H., Pinz, A. (eds.) ECCV 2006. LNCS, vol. 3951, pp. 404–417. Springer, Heidelberg (2006). doi:10.1007/11744023_32
5. Dalal, N., Triggs, B.: Histograms of oriented gradients for human detection. In: IEEE Computer Society Conference on Computer Vision and Pattern Recognition, CVPR 2005, vol. 1. IEEE (2005)
6. Kang, L., et al.: Convolutional neural networks for document image classification. In: 2014 22nd International Conference on Pattern Recognition (ICPR). IEEE (2014)
7. Bengio, Y., Courville, A., Vincent, P.: Representation learning: a review and new perspectives. IEEE Trans. Pattern Anal. Mach. Intell. **35**(8), 1798–1828 (2013)
8. Banko, M., Brill, E.: Scaling to very very large corpora for natural language disambiguation. In: Proceedings of the 39th Annual Meeting on Association for Computational Linguistics. Association for Computational Linguistics (2001)
9. Glorot, X., Bordes, A., Bengio, Y.: Deep sparse rectifier neural networks. Aistats. **15**(106), 315–323 (2011)
10. Erhan, D., et al.: Why does unsupervised pre-training help deep learning? J. Mach. Learn. Res. **11**(Feb), 625–660 (2010)
11. dos Santos, R.P., et al.: Text line segmentation based on morphology and histogram projection. In: 10th International Conference on Document Analysis and Recognition, ICDAR 2009. IEEE (2009)
12. Hinton, G.E., Osindero, S., Teh, Y.-W.: A fast learning algorithm for deep belief nets. Neural Comput. **18**(7), 1527–1554 (2006)
13. Lee, H., Ekanadham, C., Ng, A.Y.: Sparse deep belief net model for visual area V2. In: Advances in Neural Information Processing Systems (2008)
14. James, B., Olivier, B., Frédéric, B., Pascal, L., Razvan, P.: Theano: a CPU and GPU math expression compiler. In: Proceedings of the Python for Scientific Computing Conference (SciPy) (2010)

Analysis of Sloshing in Tanks Using Image Processing

Rahul Kamilla[✉] and Vishwanath Nagarajan

Indian Institute of Technology Kharagpur, Kharagpur 721302, India
rahulkam3009@gmail.com

Abstract. Sloshing is referred to as the violent movement of liquid in a partially filled tank that undergoes dynamic motion. There are several examples of such types of motions. In ships, sloshing motion occur in oil tankers, liquefied natural gas carriers and large fuel oil tanks. In case of rockets, it happens in the liquid hydrogen (LOH) and liquid oxygen (LOX) containing external fuel tanks. The sloshing motion is mainly due to large dimensions of tanks with smooth plane surfaces in contact with the liquid. The tank layout fails to damp the sloshing motion of the liquid. The sloshing motion becomes more violent when the parent vehicle's motion contains energy in the vicinity of the natural frequencies for liquid motion inside the tank. Determination of these frequencies is critical to determine the nature of fluid motion inside the tank and thereby predict impact load on the structure holding the liquid. The determination of hydrodynamic pressure on the tank walls due to liquid sloshing motion finds application in the design and construction of liquid tanks in rockets and ships. In some cases, determination of liquid motion inside the tank is also critical as it can get coupled with the parent vehicle's motion dynamics. This paper deals with extraction of data from video recording of liquid sloshing motion inside a rectangular tank. Image processing techniques are used for this purpose. The important fluid dynamics properties which can be determined by image processing are discussed in the paper. The analysis presented is mainly for 2D motions.

Keywords: Sloshing · Image processing · Photogrammetry

1 Introduction

Liquid sloshing is a dynamic phenomenon which needs to be analyzed in time domain. Due to sloshing, very high impact force may come on the structure holding the liquid. To know the hydrodynamic impact force, a time series of variation of several fluid dynamics parameters is required. In this paper, extraction of time series data is based on image processing.

Liquid sloshing occurs in cargo tanks of oil tankers and gas carriers, large fuel oil tanks of ships and liquid hydrogen and liquid oxygen tanks of rockets. In oil tankers the liquid is carried at ambient atmospheric temperature. However, in cryogenic application (liquid gas carrier), the liquid is at low temperature between −55 °C to −200 °C. At low

© Springer International Publishing AG 2017
F. Karray et al. (Eds.): ICIAR 2017, LNCS 10317, pp. 651–659, 2017.
DOI: 10.1007/978-3-319-59876-5_72

temperature, the structural material is already thermally stressed. In this condition extra hydrodynamic load due to liquid sloshing may impact the structure adversely. Therefore, liquid sloshing needs to be investigated in detail for vehicles containing cryogenic liquids. Liquefied Natural Gas (LNG) ships are of two types, spherical tank type and membrane type. Membrane type tanks are prismatic and have clean, wide space. These factors lead to vigorous sloshing in membrane type tanks. This becomes a major problem when the ship undergoes 6 - degrees of freedom motion in a seaway. Determining sloshing loads is important as it can cause structural damage to the tank walls. By sloshing analysis, the hydrodynamic impact loads on the structure can be predicted. Thus, designing suitable membrane and the associated insulation is of vital importance. It also has significant effects on the parent vehicle's motions, as there may be coupling between the sloshing fluid and the vehicle's motion.

2 Problem Formulation

For deriving the natural frequencies of liquid motion in a tank, zero tank excitation case is considered, i.e. there is no external force on the tank. Assuming two-dimensional case of rectangular planar tanks in which the wave profile is same at any y-location (Fig. 1), any parameter can be described as a function of x, z and t.

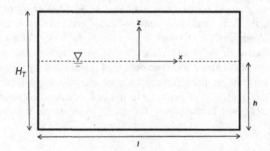

Fig. 1. Tank-fixed co-ordinate system.

The fluid inside the tank is assumed to be inviscid, incompressible and the flow is irrotational. Thus, the fluid should satisfy the following Laplace equation:

$$\nabla^2 \phi(x, z, t) = 0 \tag{1}$$

where ϕ is the velocity potential.

This is a boundary value problem, it should satisfy the following boundary conditions,

$$-\omega^2 \varphi + g\varphi_z = 0 \; on \; z = 0 \tag{2}$$

$$\frac{\partial \varphi}{\partial x} = 0 \ on \ |x| = \frac{l}{2} \tag{3}$$

$$\frac{\partial \varphi}{\partial z} = 0 \ on \ z = -h \tag{4}$$

$$\int_{-\frac{1}{2}l}^{\frac{1}{2}l} \varphi(x,0)dx = 0 \tag{5}$$

Equation (2) is the linearized free surface boundary condition, Eq. (3) is the boundary condition for the sides of the tank, Eq. (4) is the bottom boundary condition and Eq. (5) is for the conservation of volume, where $\varphi(x,0)$ is the free surface elevation. Solving the Laplace equation and applying the boundary conditions, the following solution is obtained,

$$\varphi(x,z) = cos\left(\frac{\pi i}{l}\left(x + \frac{l}{2}\right)\right) \frac{\cosh\left(\frac{\pi i(z+h)}{l}\right)}{\cosh\left(\frac{\pi i h}{l}\right)} \tag{6}$$

for the ith mode.

And the natural frequencies and time periods of the tank came out to be,

$$\omega_i = \sqrt{g\frac{\pi i}{l} tanh\left(\frac{\pi i}{l}h\right)}, T_i = \frac{2\pi}{\omega_i} \tag{7}$$

for the ith mode.

3 Analysis Methodology

For studying the behavior of liquid motion inside the tank, experiments were conducted in a tank of length 300 mm, height of 150 mm and width of 200 mm. The tank was made of perspex glass and the experiment was conducted for three fill levels, 40%, 62% and 76% of the height of the tank. The tank was open at the top.

The tank (filled partially with colored water), was oscillated in a swing by means of random forced oscillation. The tank walls being transparent and the liquid being colored, the movement of liquid in the tank was clearly visible. Time domain variation of wave profile of liquid was captured in videos. For each sample of data, a video of the setup was recorded for a duration of 20 min. The specifications of video are,

- Frame rate: 60 per second
- Width: 1280 pixels
- Height: 720 pixels.

Three samples were recorded for each fill height, i.e. one-hour time series of data was extracted. The swing was levelled on both sides before the start of experiment and the fill height was measured using a scale with least count of 1 mm (Figs. 2 and 3).

Black dots were placed at the four corners of the tank to locate its boundaries. To validate the image processed data, an ideal case was taken, as shown in Fig. 3. The required light intensity was obtained using normal fluorescent tube lights. The tank is in still condition, so is the liquid at 11 cm fill height, as can be seen in the scale.

Fig. 2. Transverse levelling of the swing.

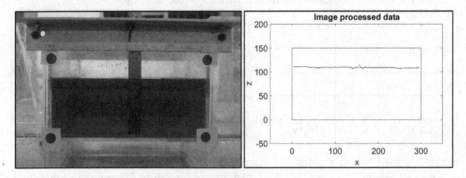

Fig. 3. Validation of image processed data.

The image was analyzed in MATLAB along with thousands of other frames for extracting the wave height data. The image processed data generated for this image is shown in Fig. 3. It can be seen from the figure that the data points gave a mean fill level of 110 mm.

Similarly, when the tank is undergoing motion and the liquid is sloshing, the wave height data is generated from the pixel information at the planes x_1, x_2 and x_3. Using MATLAB, the variation in RGB co-ordinates along the lines x_1, x_2 and x_3 was determined. The precision was kept as 2 pixels i.e. 1 mm. The generated wave profile is plotted in Fig. 4.

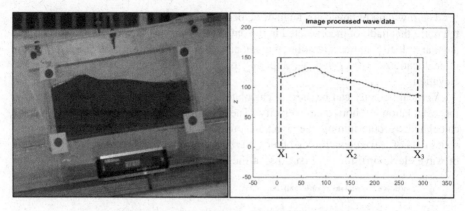

Fig. 4. Wave data points obtained by processing sample frame.

4 Results

The swing's angular displacement, at every frame, was measured using the reference dots. Swing's angular displacement was also measured using a digital inclinometer which logged data at 1.876 Hz. As the time interval between two frames is known, the swing's angular velocity and acceleration was calculated using backward difference scheme and plotted, as shown in Fig. 5.

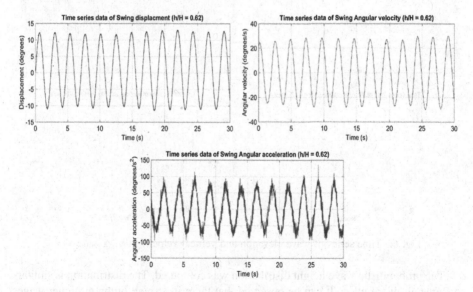

Fig. 5. Time series of swing's angular displacement, angular velocity and angular acceleration.

The wave elevation was measured at the three locations, x_1, x_2 and x_3 as marked in Fig. 4. It was measured with an accuracy of 1 mm, using the pixel information at these vertical planes. Due to the effect of dye in water, there is a significant change in the RGB values at the free surface of the liquid.

The raw data obtained for angular velocity and acceleration was filtered to remove the noise that had got incorporated in the data due to various factors. Similar to swing's angular velocity and acceleration, the velocity of water elevation was calculated at the three locations, x_1, x_2 and x_3, applying backward difference method on the wave elevation.

Vertical velocity can be used to calculate the velocity potential, which will lead to the calculation of horizontal velocity. The sloshing load on the bulkheads can be calculated by determining the dynamic pressure using velocity potential. The time series data of wave elevation and fluids' vertical velocity (which is equal to the velocity of wave elevation), for the first 30 s, at the three planes are show in Fig. 6.

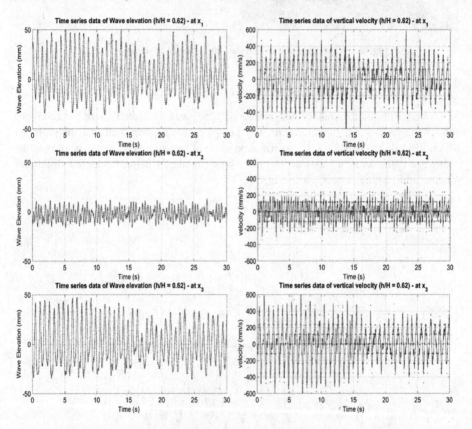

Fig. 6. Time series of wave elevation and vertical velocity at x_1, x_2 and x_3.

The probabilistic wave height distribution was computed. The distribution is similar to Rayleigh distribution. It can be observed that there is a small hump at higher wave heights. This is mainly due to the interference of the incident waves with the waves reflected back from the tank's walls. This interference makes the calculation of dynamic pressure complicated as the flow has significant scattered wave potential along with the incident wave potential. Wave height distribution calculated from the time series data is plotted in Fig. 7.

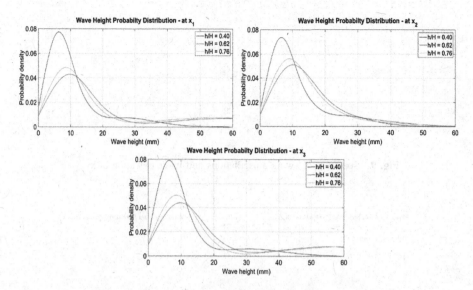

Fig. 7. Probabilistic wave height distribution at x_1, x_2 and x_3.

Information of variation in center of gravity of liquid inside tanks is vital. This is because, it causes a dynamic variation in the global longitudinal/transverse center of gravity of the ship. This affects the ship's response in waves. It may either damp out the motions or enhance them. This temporal variation in the location of CG is shown in Fig. 8.

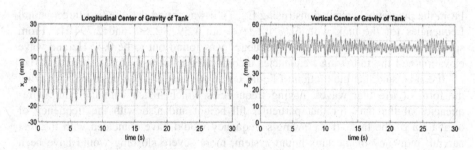

Fig. 8. Time series data of Center of gravity of tank's liquid.

It can be seen that the shift of CG is more significant in the longitudinal direction, indicating the fact that, the pitching motion of the ship will get affected. As there are a number of prismatic tanks in an LNG carrier, the resultant shift in the Longitudinal Centre of Gravity of the ship may be substantial.

Using FFT, spectral analysis was performed on the wave elevation and swing's angular displacement time series data. The spectrum represents the distribution of energy of the system over the range of frequencies. They are shown in Figs. 9 and 10.

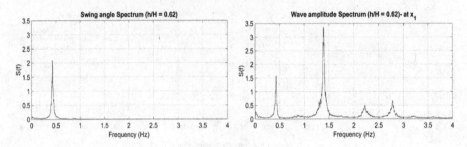

Fig. 9. Spectrum of swing's oscillations and wave elevation at x_1.

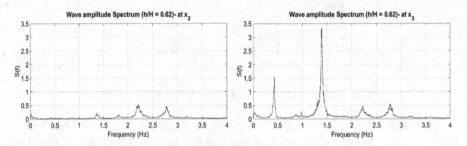

Fig. 10. Spectrum of wave elevation at x_2 and x_3.

5 Inference

From the theoretical calculations (using Eq. 7), it was found that the first three natural frequencies for the tank with 62% fill level are 1.392, 2.238 and 2.794 Hz. From spectral analysis, swing oscillation frequency was found out to be 0.42 Hz and wave elevation had the following frequencies incorporated in it: 0.42, 1.4, 2.2 and 2.8 Hz.

It can be said that the motion of liquid inside the tank can be considered as having the form of irregular waves, having frequencies which match with the natural frequencies of the tank for that particular fill height and also with the frequency of oscillation of the tank. If the swing's frequency would have coincided with the first natural frequency of the tank-liquid system, more severe sloshing would have been observed.

Two ranges of wave heights were dominant, as can be seen in the probabilistic wave height distribution, mainly focusing on the fact that the scattered wave superimposes itself on the incident wave resulting in larger wave heights, and the smaller ranges represents the incidents waves. From the time series data and spectrum, it can be concluded that the wave heights at x_1 and x_3 were much higher than that at x_2, due to scattering of waves at the walls of tank. Further research is in process and is directed towards including the 3D flow characteristics in the sloshing phenomenon using image processing of aerial-view photographs captured from multiple cameras. Mathematical theories can be applied to the wave elevation and its velocity to determine the velocity potential and hence the dynamic pressure on the walls of the tank.

References

1. Abramson, H.: The dynamic behavior of liquids in moving containers. Washington DC (1966)
2. Casasent, D.: Optical Data Processing: Applications. Carnegie-Mellon University, Pittsburgh (1978)
3. Faltinsen, O.M., Timokha, A.N.: Sloshing. Cambridge University Press, Cambridge (2014)
4. Ibrahim, R.A.: Liquid Sloshing Dynamics. Cambridge University Press, Cambridge (2005)
5. Liu, T., Cattafesta, L.N., Radeztsky, R.H., Burner, A.W.: Photogrammetry Applied to Wind Tunnel Testing. High Technology Corporation, Hampton (2000)
6. Vreeburg, J.P.B., Vogels, M.E.S.: Liquid Motion in Partially Filled Containers. National Aerospace Laboratory, Amsterdam (1986)

Light Field Estimation
in the Ultraviolet Spectrum

Julien Couillaud, Djemel Ziou, and Wafa Benzaoui[(✉)]

Departement d'informatique, Universite de Sherbrooke,
2500 Boul de l'Universite, Sherbrooke, QC J1K 2R1, Canada
{Julien.Couillaud,Djemel.Ziou,Wafa.Benzaoui}@USherbrooke.ca

Abstract. Ultraviolet cameras are becoming well used, with their applications in botany, dermatology, and recently in photography. In this paper, we develop a novel method of light field estimation from an ultraviolet image. Our UV light field imaging model is obtained by exploiting the radiometry, the optic, and the acquisition geometry. First, we develop an optic simulation model for an UV camera with a thin lens. That model permits to reconstruct a scene image from the correspondent light field. Then, we define a variational formalism that integrates an image of a scene acquired by an UV camera, the depth map of that scene, and the optic simulation model in order to estimate the light field. The experimental results show that it is possible to estimate the light field in the UV spectrum with accuracy.

Keywords: Ultraviolet light field · Ultraviolet images · Optic simulation model · Variational inference

1 Introduction

The human eye perceives a small portion of the light spectrum between 380 nm and 780 nm, called the visible spectrum. To go beyond the visible spectrum, the ultraviolet (UV) cameras are well placed to provide new information about the visible scene. The UV spectrum is separated into three bands: the ultraviolet A between 380 nm and 315 nm, the ultraviolet B between 315 nm and 280 nm, and the ultraviolet C between 280 nm and 100 nm. The UV imaging may be used in various applications including dermatology [1–3], botany [4,5], criminology [6], biology [7,8], volcanology [9,10] and for artistic purposes. The UV imaging starts with the acquisition of 2D images by using an UV camera and UV light source. The choice of the acquisition system depends on the target application and therefore the target UV band. Recently, light field cameras were introduced to the consumer market as a technology facilitating novel user experiences [12]. However, current cameras have been manufactured for sensing light field in visible spectrum [11,13]. It seems that there are no available cameras, models or methods allowing the acquisition of light field in the UV spectrum. In this paper, we propose a method allowing the light field acquisition in the UV spectrum by

© Springer International Publishing AG 2017
F. Karray et al. (Eds.): ICIAR 2017, LNCS 10317, pp. 660–670, 2017.
DOI: 10.1007/978-3-319-59876-5_73

using an UV camera. Such a light field can be applied in all the domains cited previously in order to reconstruct images where a user can choose the plane of focus position [14] or to allow 3D visualisation of the scene in the UV spectrum [15,16]. This paper does not focus on the design of a special camera for the UV light field acquisition, but the inference of a light field by using a 2D UV image and a depth map captured by a depth camera. The remainder of this paper is organised as follows. Firstly, the general light field representation and its scene depth encoding are explained in Sect. 2. Secondly, the optic simulation model of the UV thin lens camera is introduced in Sect. 3. Thirdly, the light field inference method from the UV image and the depth map is described in Sect. 4. Finally, the results of light field inference and images reconstructed by using the optic simulation model are shown and discussed in Sect. 5.

2 What is a Light Field?

A light field represents all light rays propagating within a scene [17]. For a wavelength λ, and at an instant τ a light ray is parametrised by four coordinate (a, b, c, d) under the assumption that the radiance is constant along the light ray [18]. These four coordinate can be represented by the ray intersection coordinate with two parallel planes spaced by a distance d_0. In Fig. 1, a light ray intersects the plane AB at coordinate (a, b) and crosses the plane CD at coordinate (c, d). This representation of the light field only considers the UV light rays reflected by an object. This consideration imposed some limitation on the UV bands studied in this paper. In fact, the smaller the wavelength the more it is scattered by the medium and the object of the scene, as well as the optical device of a UV camera. Therefore, the UV light in band B and C is mostly scattered and not reflected. Hence, we limit this study to the ultraviolet A band of the UV spectrum which is less scattered and therefore the light amount reaching the camera is higher than the two other bands. The signal-to-noise ratio is also higher. A light ray reflected from a scene point, carries a radiance R emitted by this point. This

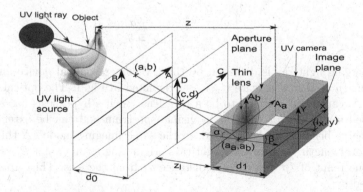

Fig. 1. Projection of UV light rays inside an UV camera.

radiance corresponds to the intensity of a light ray in the UV spectrum which is defined by the plenoptic function $L(a, b, c, d)$ [17]. Dansereau and Bruton [19] showed that all light rays emitted from a scene point P, which is located at a depth z, represent a hyperplane in the light field $4D$ space formed by (a, b, c, d). We call that plane a subspace which is described by:

$$
\begin{cases}
c = \frac{z - z_l}{z - z_l - d_0} a - \frac{z}{d_1} \left(1 - \frac{z - z_l}{z - z_l - d_0} \right) x_p \\
d = \frac{z - z_l}{z - z_l - d_0} b - \frac{z}{d_1} \left(1 - \frac{z - z_l}{z - z_l - d_0} \right) y_p
\end{cases}
\tag{1}
$$

where (x_p, y_p) is the perspective coordinate of the scene point P. This coordinate is determined by the intersection between the image plane Ip and a non-deflected ray, emitted by P and crosses the camera aperture on its centre. Note that d_1 is the distance between the image plane and the aperture plane, and z_l is the distance between the plan aperture and the plane CD.

The Eq. (1) defines lines that identify the subspace. These lines have the same orientation θ which can be determined from the lines slope which can also be written as $\tan(\theta)$. Therefore, the slope angle θ is given by:

$$
\theta = \tan^{-1} \left(\frac{z - z_l}{z - z_l - d_0} \right)
\tag{2}
$$

The Eq. (2) shows that the subspace formed by the light rays emitted by a scene point depends on the depth z of this point. Hence, the light field is built by a set of subspaces which are characterised by the scene depth.

3   Thin Lens Optic Simulation Model

In order to derive the optic simulation model of a UV camera, we analyse the light ray travelling towards the image plane Ip illustrated in Fig. 1. First, the light ray crosses the aperture plane Ap at coordinate (a_c, a_d). By using basic geometry principles, the coordinate (a_c, a_d) is given by:

$$
\begin{cases}
a_c = \frac{z_l + d_0}{d_0} c - \frac{z_l}{d_0} a \\
a_d = \frac{z_l + d_0}{d_0} d - \frac{z_l}{d_0} b
\end{cases}
\tag{3}
$$

The emerging light ray is deflected by the UV camera optical device and projected on Ip at the coordinate (i_x, i_y) as depicted in Fig. 1. The optical device of the UV camera is assumed to be a thin lens. Such a hypothesis has already been used in [20–22] for cameras in visible spectrum and can be extended to UV cameras because they are based on the same imaging model. A thin lens, with a focal length f, focuses the light rays originating from a point P_f at depth z_f, in one point of Ip located at a distance d_1 from the lens. This property is described by:

$$
\frac{1}{f} = \frac{1}{d_1} + \frac{1}{z_f}
\tag{4}
$$

Another characteristic of a thin lens is that a light ray crossing the lens on its center is not deflected. By using basic geometry principles and the mentioned properties, the imaging point at coordinate (i_x, i_y) is determined by:

$$\begin{cases} i_x = \frac{z_l(d_1-f)-fd_1}{fd_0}a - \frac{(d_0+z_l)(d_1-f)-fd_1}{d_0 f}c \\ i_y = \frac{z_l(d_1-f)-fd_1}{fd_0}b - \frac{(d_0+z_l)(d_1-f)-fd_1}{d_0 f}d \end{cases} \tag{5}$$

When a light ray intersects Ip, its radiance is transformed into the irradiance Ir defined as follows:

$$Ir(i_x, i_y) = L(a,b,c,d)\ k\frac{\cos(\alpha)\cos(\beta)^3}{d_1^2}\left(\frac{z_l+d_0}{d_0}\right)^2 dc\,dd \tag{6}$$

$$\alpha = cos^{-1}\left(\frac{z_l}{\sqrt{(a_c-c)^2+(a_d-d)^2+z_l^2}}\right) \tag{7}$$

$$\beta = cos^{-1}\left(\frac{d_1}{\sqrt{(i_x-a_c)^2+(i_y-a_d)^2+d_1^2}}\right) \tag{8}$$

where k is a weight, and α (resp. β) refers to the angle between the incident (resp. the emerging) light ray and Ap normal. The variables dc and dd are the width and the height of an infinitesimal rectangular area around the light ray intersection with the plane CD. Note that the radiance L in the Eq. (6) is already multiplied by a factor η, which models the attenuation caused by the scattering of UV light rays. Thus, for small wavelengths, η is small, and the provided images are dark.

In [13,23], the authors assume that the total energy I generated on a point at coordinate (x, y) on Ip is a linear combination of the irradiance deposited by each ray projected on (x, y). This total energy is given by:

$$I(x,y) = \int_a \int_b \int_c \int_d \alpha_{x,y,a,b,c,d} L(a,b,c,d)\, da\,db\,dc\,dd \tag{9}$$

This latter equation corresponds to the optic simulation model of a camera with a thin lens. In the following sections, I is considered as an UV image formed on Ip. The coefficient $\alpha_{x,y,a,b,c,d}$ is a weight which transforms the radiance into irradiance. The attenuation factor η; and instead of being multiplied by the radiance; can be modeled in the weight α. Thereby, for small wavelengths, η is small, and UV light rays are stopped by the lens, which produce dark images. Note that α also selects light rays projected at (x, y) according to two functions. The first function determines if the light rays are projected on (x, y). It is a unit impulse function $\delta()$ which is one when a light ray intersects Ip on (x, y); otherwise, it is zero. It is satisfied by filtering the light rays in function of their trajectory. This filtering operation can be seen as the numerical simulation micro-lenses [13] embedded in light field cameras. The second function models the camera aperture and informs if a light ray passes through it. We choose to model the aperture as a Gaussian. The Gaussian model has the advantage to approximate the effects of the phenomena caused by the optical device and happened during the image formation, as the diffraction and the dispersion [24]. This Gaussian weighs the rays passing outside of the camera aperture, of radius r, by a value close to zero; otherwise, it generates higher weights. The weight $\alpha_{x,y,a,b,c,d}$ is determined by:

$$\alpha_{x,y,a,b,c,d} = \frac{1}{2\pi(\frac{r}{3})^2}\delta((i_x-x)^2+(i_y-y)^2)\,e^{\left(-\frac{a_c^2+a_d^2}{2(\frac{r}{3})^2}\right)}k\frac{\cos(\alpha)\cos(\beta)^3}{d_1^2}(\frac{z_l+d_0}{d_0})^2 \quad (10)$$

4 Light Field Inference

To reconstruct a light field, we need an image I_b of a scene captured by an UV camera and a depth map of the same scene acquired by a depth camera. These data are then used to solve an inverse problem. This problem is the following: Given a 2D image, we have to determine a light field. In order to solve such a problem, we have to fulfil two requirements. First, by using (9), an accurate reconstruction of I_b must be calculated from a light field estimated from I_b. This requirement is modelled by determining the light field which minimises a reconstruction error described by:

$$F_E(L) = \int_x \int_y \left(I_b(x,y) - \int_a \int_b \int_c \int_d \alpha_{x,y,a,b,c,d}L(a,b,c,d)\,da\,db\,dc\,dd\right)^2 dx\,dy \quad (11)$$

We assume that the radiance of light rays emitted from a scene point are almost equal over consecutive pixels receiving that rays as in [23]. In the light field $4D$ space, it means that the radiance of each light ray represented in a subspace (1) varies smoothly. This allows us to formulate the second requirement as: the light field must minimise the radiance variation between the light rays contained in a subspace. Such a requirement is modelled by the minimisation of the light field gradient in a subspace. These subspaces are identified by the gradient orientation θ defined in (2). The smoothness requirement can be expressed as:

$$F_O(L_a, L_b, L_c, L_d) = \int_a \int_b \int_c \int_d ((L_a(a,b,c,d) + L_b(a,b,c,d))\cos(\theta)$$
$$+ (L_c(a,b,c,d) + L_d(a,b,c,d))\sin(\theta))^2 da\,db\,dc\,dd \quad (12)$$

In order to estimate a light field which fulfils these two requirements, Eqs. (11) and (12) have to be linearly combined. The strength of the second requirement is modulated by a weight γ as in the following equation:

$$min_L\, F_E(L) + \gamma F_O(L_a, L_b, L_c, L_d) \quad (13)$$

In this variational framework, the image I_b is given by the UV camera and the scene depth map is given by a depth camera, e.g. the *Kinect* camera [25]. The UV camera intrinsic parameters are found by using data provided by the manufacturer and the calibration procedure described in [26]. The only unknown variable is the light field L. To solve this inference problem, we choose use the partial derivative equation (PDE) of (13) by using the Gateaux derivates. The PDE of (13) corresponds to the following equation:

$$
\int_x \int_y \Big(\alpha_{x,y,m,n,p,q} I_b(x,y) - \alpha_{x,y,m,n,p,q}
$$
$$
\int_a \int_b \int_c \int_d \alpha_{x,y,a,b,c,d} L(a,b,c,d)\, da\, db\, dc\, dd \Big) dx\, dy
$$
$$
+ \gamma \big((L_{mm}(m,n,p,q) + 2L_{mn}(m,n,p,q) + L_{nn}(m,n,p,q)) \cos(\theta)^2 \qquad (14)
$$
$$
+ 2\sin(\theta)\cos(\theta)\big((L_{mp}(m,n,p,q) + L_{mq}(m,n,p,q) + L_{np}(m,n,p,q)
$$
$$
+ L_{nq}(m,n,p,q)\big) + L_{pp}(m,n,p,q) + 2L_{pq}(m,n,p,q)
$$
$$
+ L_{qq}(m,n,p,q)\big) \sin(\theta)^2 = 0
$$

where L_{mn} is the second derivative of L with regard to m and n. To numerically solve this PDE, one can discretise it. The resulting system of linear equations can be written in matrix format. The matrices representing the discrete image I_b and light field L have a resolution of $X \times Y$ and $A \times B \times C \times D$ respectively, as well as a sampling step of one along their respective dimensions. Let i_b and l be vectors obtained by the vectorisation of those two matrices. We note T, a projection matrix of dimension $XY \times ABCD$, where each row indexes a pixel location in i_b and each column indexes the light ray position in l. Let G be a derivation matrix of dimension $ABCD \times ABCD$ which allows calculating the partial derivatives in (14) by using the explicit finite difference as a numerical scheme. In matrix form, the PDE is written as:

$$
T^t i_b - T^t T l + \gamma G l = 0 \qquad (15)
$$

where T^t is the transpose of T. Assuming $(T^t T - \gamma G)$ is invertible, the light field solution of the PDE is given by the equation:

$$
l = (T^t T - \gamma G)^{-1}(T^t i_b) \qquad (16)
$$

5   Results and Discussion

The light field estimation method and the optic simulation model are tested on several UV images acquired by a UV monochrome camera, *Sony XC-EU50*, and depth maps acquired by a *Kinect* camera. The Table 1 shows the intrinsic parameters of the UV camera and the lens, *Pentax TV lens* 25 mm. The resolution of plane AB resolution is 720×480 and the resolution of plane CD is variable. In the experiments, k is set to one and the distance d_1 is equal to d_0. Moreover, the CD plane is placed on the lens so that z_l is zero. The depth maps acquired by the *Kinect* camera contains gaps where the depth is set to zero. These gaps are filled in with depth values by using an edge confined diffusion algorithm similar to the one explained in [27]. The light field estimation algorithm has been implemented in *C++* and parallised by using *OpenMP* on a 5200 kernels PC cluster of *MOIVRE* laboratory. We tested the light field estimation on various

Table 1. Intrinsic parameters of the camera *Sony XC-EU50* and the *Pentax TV lens* 25 mm.

Parameters	d_0	f	r
	36.84 mm	35 mm	8.75 mm

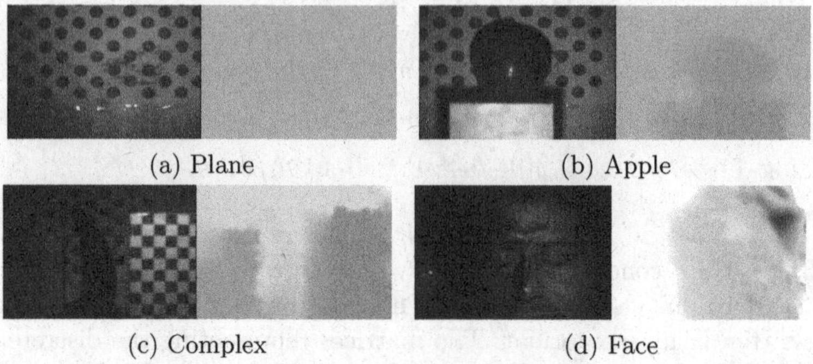

(a) Plane

(b) Apple

(c) Complex

(d) Face

Fig. 2. Greyscale UV images and greyscale depth maps captured with an UV camera and a *Kinect* camera. In the depth map, the darkest to the brightest gray-scale means close depths to further depths with regard to the camera position. In all these scenes, the depth varies between 50 and 123 cm. The scene (a) is composed of a plane object and (b) is an apple placed on a cubic metal object in front of a plane object. The scene (c) contains several plane objects and the UV camera focuses on a banana in the background. The figure (d) is the picture of a person face on which was added sunscreen on a side of the face.

UV images and depth maps. A subset of images is illustrated in Fig. 2. The performance evaluation of the proposed light field estimation method, in the absence of ground truth light fields, comes back to evaluate reconstructed images. Two measures are used to study the influence of the value of the weight γ, in (13), and the resolution of CD plane on the quality of the estimated UV light field. Firstly, the mean absolute error (MAE) is measured between the original UV image I_b and an UV image I reconstructed by applying Eq. (9) on the light field estimated from I_b and the camera parameters in Table 1. The MAE is defined by:

$$MAE = \frac{1}{XY} \sum_{x=1}^{X} \sum_{y=1}^{Y} |I_b(x,y) - I(x,y)| \qquad (17)$$

Secondly, the spatial information of I and I_b are compared. The error of the spatial content reconstruction is measured by the mean cosine of gradient orientation differences Δ between edges of I and I_b. This measure is noted CADE and is given by:

$$CADE = \frac{1}{N} \sum_{n=1}^{N} \cos(\Delta(n)) \qquad (18)$$

Table 2. CADE and MAE measured for several values of γ on images reconstructed from light fields estimated by using images in Fig. 2. The CADE and the MAE are rounded to two digits of precision.

Images	γ	100	50	25	15	10	6	3	2	1.5	1	0.5	0
Plane	CADE	0.86	0.86	0.86	0.86	0.86	0.86	0.86	0.86	0.86	0.86	0.86	0.06
	MAE	0.86	0.86	0.86	0.86	0.86	0.86	0.86	0.86	0.86	0.86	0.86	76.04
Apple	CADE	0.86	0.86	0.86	0.86	0.86	0.86	0.86	0.86	0.86	0.86	0.86	0
	MAE	0.76	0.76	0.76	0.76	0.76	0.76	0.76	0.76	0.76	0.76	0.76	82.08
Complex	CADE	0.81	0.81	0.81	0.81	0.81	0.81	0.81	0.81	0.81	0.81	0.81	0.12
	MAE	0.56	0.56	0.56	0.56	0.56	0.56	0.56	0.56	0.56	0.56	0.56	53.44
Face	CADE	0.85	0.85	0.85	0.85	0.85	0.85	0.85	0.85	0.85	0.85	0.85	−0.03
	MAE	0.62	0.62	0.62	0.62	0.62	0.62	0.62	0.62	0.62	0.62	0.62	37.95

where N is the total number of pixel edges and $\Delta(n)$ is the gradient orientation difference between the edge pixel n of I and I_b.

The Table 2 gives the measures of MAE and CADE for different values of the weight γ and a resolution of plane CD of 7×7.

The MAE has low values around 0.70 and the CADE is close to 0.85 for values of γ between 100 and 0.5 so that γ does not influence the light field reconstruction when it is not zero. For γ equal to zero, the MAE corresponds to a high reconstruction error and the CADE is close to zero which means that I does not resemble I_b. These observations imply that the second requirement in (13) greatly enhances the quality of the estimated light field and γ must be non-null. However, a value of γ greater than zero does not seem to influence the light field quality so that we arbitrarily set it to one in the following experiments.

Table 3 shows the MAE and the CADE measured for five resolutions of plane CD. The CADE gives an error around 0.85. Such a difference shows that there

Table 3. Measured CADE and MAE for several CD plane resolutions ($C \times D$) on image reconstructed from light fields estimated by using images in Fig. 2. The CADE and the MAE are rounded to two digits of precision.

Images	$C \times D$	3×3	5×5	7×7	9×9	11×11
Plane	CADE	0.86	0.86	0.86	0.86	0.86
	MAE	0.90	0.86	0.86	0.86	0.86
Apple	CADE	0.86	0.86	0.86	0.86	0.86
	MAE	0.79	0.76	0.76	0.76	0.76
Complex	CADE	0.81	0.81	0.81	0.81	0.81
	MAE	0.61	0.56	0.56	0.56	0.56
Face	CADE	0.85	0.85	0.85	0.85	0.85
	MAE	0.62	0.62	0.62	0.62	0.62

are errors in the spatial content of the reconstructed image I; however, they are sufficiently small to say that I is close to I_b. The measured MAE shows that each pixel has a small reconstruction error around 0.71. From these two measures, one observes that our method allows a good reconstruction of I_b from the estimated light field. The MAE and the CADE seem invariant for the different CD resolutions. Such an invariance is due to the UV camera depth of field and the configuration of the scene captured.

The estimated light field can be used to reconstruct images with different intrinsic parameters than the ones presented in Table 1. Therefore, the plane of focus and the blur introduced in the image can be changed. Also, variation of these parameters can be used to focus areas of an image at a precise depth. The images in Figs. 3 and 4 are reconstructed from the thin lens optic simulation model of the UV camera in (9) for a 5×5 UV resolution, an aperture radius of 37 mm and various focal lengths.

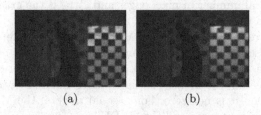

(a)   (b)

Fig. 3. Focus on the foreground (a) and on the background (b) of the scene in Fig. 2c

In Fig. 3, artefacts are visible on the object in foreground which is slightly blurred even when it is in focus. These artefacts are due to bad depth values estimated on this object. However, in Fig. 4, no artefacts are visible because of the accurate depth map estimation from the *Kinect* camera. In these images, a natural blur variation, depending on z_f, is observed. Such a behaviour is realistic because it mimics the focus of a real camera.

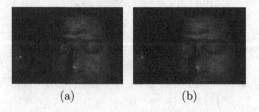

(a)   (b)

Fig. 4. Focus on the face of the person in the foreground (a) and the background (b) of the scene of Fig. 2d

6   Conclusion

In this paper, a light field estimation method by using an UV camera is presented. This method relies on solving an inverse problem based on the light field properties and the thin lens optic simulation model of an UV camera. The quality of a light field is assessed by the comparison of an original image and an image reconstructed from an estimated light field. We make it possible to acquire an UV light field and produce from it UV images with different acquisition configurations. This finding opens the door to the computational photography and strengthens the use of UV in its traditional applications. In future work, we will apply this estimation method in order to create a multi-spectral light field by estimating light fields from an ultraviolet image, an infrared image and a colour image in visible spectrum. Such a light field inference will increase the quantity of information contained in a light field so that it will increase the number of possible applications. In a future study, we will also find a quality assessment method which evaluates the inferred light field quality in a much better way than the CADE and the MAE of a reconstructed image.

References

1. Fulton, J.E.: Utilizing the ultraviolet (UV detect) camera to enhance the appearance of photodamage and other skin conditions. Dermatol. Surg. **23**, 163–169 (1997)
2. Gibbons, F.X., Gerrard, M., Lane, D.J., Mahler, H.I., Kulik, J.A.: Using UV photography to reduce use of tanning booths: a test of cognitive mediation. Health Psychol. **24**, 358 (2005)
3. Stock, M.L., Gerrard, M., Gibbons, F.X., Dykstra, J.L., Mahler, H.I., Walsh, L.A., Kulik, J.A.: Sun protection intervention for highway workers: long-term efficacy of UV photography and skin cancer information on mens protective cognitions and behavior. Ann. Behav. Med. **38**, 225–236 (2009)
4. Midgley, J.J., Stock, W.D.: Natural abundance of $\delta15N$ confirms insectivorous habit of Roridula gorgonias, despite it having no proteolytic enzymes. Ann. Bot. **82**, 387–388 (1998)
5. Moran, J.A., Booth, W.E., Charles, J.K.: Aspects of pitcher morphology and spectral characteristics of six bornean nepenthes pitcher plant species: implications for prey capture. Ann. Bot. **83**, 521–528 (1999)
6. Gray-Ray, P., Hensley, C., Brennan, E.: Violent rape and bite marks: the use of forensic odontology and ultraviolet lighting. Policing: Int. J. Police Strateg. Manag. **20**, 223–234 (1997)
7. Lavigne, D.M., Øritsland, N.A.: Ultraviolet photography: a new application for remote sensing of mammals. Can. J. Zool. **52**, 939–941 (1974)
8. McDaniel, C.S., Wild, J.R.: Detection of organophosphorus pesticide detoxifying bacterial colonies, using UV-photography of parathion-impregnated filters. Arch. Environ. Contam. Toxicol. **17**, 189–194 (1988)
9. Dalton, M.P., Watson, I.M., Nadeau, P.A., Werner, C., Morrow, W., Shannon, J.M.: Assessment of the UV camera sulfur dioxide retrieval for point source plumes. J. Volcanol. Geoth. Res. **188**, 358–366 (2009)

10. Kantzas, E.P., McGonigle, A.J.S., Tamburello, G., Aiuppa, A., Bryant, R.G.: Protocols for UV camera volcanic SO_2 measurements. J. Volcanol. Geoth. Res. **194**, 55–60 (2010)

11. Chia-Kai, L., Tai-Hsu, L., Bing-Yi, W., Chi, L., Chen, H.H.: Programmable aperture photography: multiplexed light field acquisition. ACM Trans. Graph. **27**, 55–65 (2008)

12. Wilkes, T.C., McGonigle, A.J., Pering, T.D., Taggart, A.J., White, B.S., Bryant, R.G., Willmott, J.R.: Ultraviolet imaging with low cost smartphone sensors: development and application of a raspberry Pi-based UV camera. Sensors **16**, 1649 (2016)

13. Ng, R., Levoy, M., Brédif, M., Duval, G., Horowitz, M., Hanrahan, P.: Light field photography with a hand-held plenoptic camera. Comput. Sci. Tech. Rep. **2**, 1–11 (2005)

14. Ng, R.: Fourier slice photography. ACM Trans. Graph. **24**, 735–744 (2005)

15. Jones, A., McDowall, I., Yamada, H., Bolas, M., Debevec, P.: Rendering for an interactive 360 light field display. ACM Trans. Graph. **26**, 40 (2007)

16. Kim, C., Hornung, A., Heinzle, S., Matusik, W., Gross, M.: Multi-perspective stereoscopy from light fields. ACM Trans. Graph. **30**, 190–200 (2011)

17. Adelson, E.H., Bergen, J.R.: The plenoptic function and the elements of early vision. In: Computational Models of Visual Processing, pp. 3–20. MIT Press (1991)

18. Liang, C.K., Shih, Y.C., Chen, H.H.: Light field analysis for modeling image formation. IEEE Trans. Image Process. **20**, 446–460 (2011)

19. Dansereau, D., Bruton, L.: Gradient-based depth estimation from 4D light fields. IEEE Int. Symp. Circ. Syst. **3**, 549–552 (2004)

20. Surya, G., Subbarao, M.: Depth from defocus by changing camera aperture: a spatial domain approach. In: IEEE Computer Society Conference on Computer Vision and Pattern Recognition, pp. 61–67 (1993)

21. Ziou, D., Deschenes, F.: Depth from defocus estimation in spatial domain. Comput. Vis. Image Underst. **81**, 143–165 (2001)

22. Pertuz, S., Garcia, M.A., Puig, D.: Efficient focus sampling through depth-of-field calibration. Int. J. Comput. Vis. **112**, 1–12 (2014)

23. Levin, A., Freeman, W.T., Durand, F.: Understanding camera trade-offs through a bayesian analysis of light field projections. In: Forsyth, D., Torr, P., Zisserman, A. (eds.) ECCV 2008. LNCS, vol. 5305, pp. 88–101. Springer, Heidelberg (2008). doi:10.1007/978-3-540-88693-8_7

24. Hecht, E.: Optics. Addison-Wesley Publishing Company, Boston (1987)

25. Smisek, J., Jancosek, M., Pajdla, T.: 3D with Kinect. In: Fossati, A., Gall, J., Grabner, H., Ren, X., Konolige, K. (eds.) Consumer Depth Cameras for Computer Vision, pp. 3–25. Springer, London (2013)

26. Zhang, Z.: A flexible new technique for camera calibration. IEEE Trans. Pattern Anal. Mach. Intell. **22**, 1330–1334 (2000)

27. Hung, M.F., Miaou, S.G., Chiang, C.Y.: Dual edge-confined inpainting of 3D depth map using color image's edges and depth image's edges. In: Signal and Information Processing Association Annual Summit and Conference, pp. 1–9. Asia-Pacific (2013)

Author Index

Printed in the United States
by Baker & Taylor Publisher Services